T0174138

Infanticide and Parental Care

Ettore Majorana International Life Sciences Series

Edited by A. Zichichi, Director, Ettore Majorana Centre for Scientific Culture.

A series of books from courses and workshops held at the Ettore Majorana Centre.

This book is part of a series. The publisher will accept continuation orders which may be cancelled at any time and which provide for automatic billing and shipping of each title in the series upon publication. Please write for details.

Infanticide and Parental Care

Edited by

Stefano Parmigiani
University of Parma, Italy

and

Frederick S. vom Saal
University of Missouri-Columbia, USA

Proceedings of a workshop held at the
International School of Ethology, Ettore
Majorana Centre for Scientific Culture, Italy,
13–20 June 1990

Routledge
Taylor & Francis Group

LONDON AND NEW YORK

First published 1994 by Harwood Academic Publishers

2 Park Square, Milton Park, Abingdon, Oxfordshire OX14 4RN
52 Vanderbilt Avenue, New York, NY 10017

Routledge is an imprint of the Taylor & Francis Group, an informa business

First issued in paperback 2020

Copyright © 1994 Taylor & Francis

All rights reserved. No part of this book may be reprinted or reproduced or
utilised in any form or by any electronic, mechanical, or other means, now
known or hereafter invented, including photocopying and recording, or in any
information storage or retrieval system, without permission in writing from
the publishers.

Notice:
Product or corporate names may be trademarks or registered trademarks, and
are used only for identification and explanation without intent to infringe.

Library of Congress Cataloging-in-Publication Data

Infanticide and parental care – edited by Stefano Parmigiani and
   Frederick S. vom Saal.
       p.   cm. -- (Ettore Majorana international life sciences series; v. 13)
   Includes bibliographical references and index.
   ISBN 3-7186-5505-5
    1. Infanticide in animals. 2. Parental behavior in animals.
3. Infanticide. 4. Parenting. I. Parmigiani, Stefano. II. Vom Saal, Frederick S.
III. Series.
QL762.5.I53 1994
591.51--dc20   93-41749
                                         CIP

ISBN 13: 978-3-7186-5505-2 (hbk)
ISBN 13: 978-0-367-66978-2 (pbk)

CONTENTS

v

Introduction to the Series

The purpose of this Life Science series is to provide the latest information on the research and development taking place at institutions throughout the world.

The courses, held at the Ettore Majorana Centre for Scientific Culture in Erice, Sicily, are invaluable forums for direct communication between specialists.

It is hoped that the dissemination of information via this series will provide international reference works for researchers, practitioners and students.

Antonio Zichichi
Director of the Ettore Majorana Centre
for Scientific Culture

PREFACE

The right to search for truth implies also a duty; one must not conceal any part of that one has recognized to be true.

Albert Einstein

The chapters in this book were based on papers presented at a workshop "Protection and Abuse of Young in Animals and Man" held at the Ettore Majorana Centre for Scientific Culture in Erice, Sicily. A central theme throughout this book is an evolutionary perspective of infanticide and parental care. Chapters which focus on parental care have been included since we believe that the subject of infanticide cannot be understood without simultaneously discussing strategies for successful parental care. Infanticide here is defined as the killing of infants between the time of birth (or hatching) and either the achievement of independence from the adult in vertebrates (i.e. weaning in mammals) or the end of the larval stage in invertebrates. The use of parental, rather than maternal, care reflects the fact that in many species (monogamous and polygynous), males as well as females contribute both directly and indirectly to the protection and care of infants.

During the last decade the hypothesis that infanticide can, under certain circumstances, be an adaptive behavior has progressed from being a model with little supporting empirical evidence to the point today where empirical support is widespread. Hausfater and Hrdy (1984, p.XI) predicted in the preface to their book *Infanticide: Comparative and Evolutionary Perspectives* that "readers ten years from now may take for granted the occurrence of infanticide in various animal species and may even be unaware of the controversies and occasionally heated debate that have marked the last decade of research on this topic." Indeed, the possibility that all instances of infanticide represent a pathological response to social stress rather than being a behavior which evolved in species with particular social structures is no longer debated. In some instances in which infanticide has been observed, however, there is no apparent increase in the perpetrator's fitness, and, in fact, there may be a decrease in fitness. There is thus also no argument that, in certain circumstances, infanticide and other forms of

infant maltreatment, such as neglect and abuse, represent a pathological response to infants.

Our goal was to provide readers with a multidisciplinary view of infanticide and parental care. The interaction between infants and adult conspecifics is discussed from historical, anthropological, ethological, experimental, psychological and sociological perspectives. The majority of chapters contain information on infanticide and parental care in mammalian species, although discussions of infanticide and parental care in social insects (Turillazzi and Cervo, Chapter 9) and birds (Forbes and Mock, Chapter 10) show the relevance of the comparative approach to the study of these behaviors. Chapters in this book clearly show the remarkable relationships between findings concerning the adaptiveness of infanticide in experimental studies with rodents (Parmigiani *et al.*, Chapter 15; Perrigo and vom Saal, Chapter 16; Elwood and Kennedy, Chapter 17; Gubernick, Chapter 18) and observations in the field of langurs (Sommer, Chapter 7), lions (Pusey and Packer, Chapter 12), dwarf mongooses (Rasa, Chapter 13), and prairie dogs (Hoogland, Chapter 14).

While there were some observations of infanticide in populations of primates prior to 1984, the chapter by Sommer (Chapter 7) on infanticide in langurs reviews the now considerable evidence supporting Hrdy's original hypothesis that infanticide is an adaptive behavior in langurs as well as other primates. The similarities between lions (Pusey and Packer, Chapter 12) and langurs (Sommer, Chapter 7), with regard to the effect of infanticide on reproductive success in males, is striking. An extensive review of female counter-strategies to infanticide by males is also included in the chapter by Pusey and Packer. While infanticide by males was the focus of the sexual selection hypothesis of infanticide, infanticide by females has also been observed in some species, such as dwarf mongooses (Rasa, Chapter 13) and prairie dogs (Hoogland, Chapter 14). Infanticide by females is also common in some domestic house mice and in all wild house mice so far examined (Parmigiani *et al.*, Chapter 15). In social insects infanticide is observed when nest usurpation by a female occurs (Turillazzi and Cervo, Chapter 9). There remain, however, many questions as to the factors which are involved in determining whether females will parent or kill the offspring of other females. In addition, in some species of birds, the phenomenon of killing of young by siblings (siblicide), rather than parents or other adults, is discussed by Forbes and Mock (Chapter 10).

The chapter by Hiraiwa-Hasegawa and Hasegawa (Chapter 6) on infanticide in chimpanzees shows that not all incidents of infanticide are consistent with the hypothesis that infanticide is adaptive (this chapter also pro-

vides a review of infanticide in other primates). Thus, unlike the majority of cases documented by Sommer for langurs, there is, as yet, no satisfactory explanation for the acts of infanticide which have been observed in field studies of chimpanzees. Similarly, instances of wounding and killing of infants in pinnipeds (sea lions and seals) are not consistent with the hypothesis that this is an adaptive behavior (Le Boeuf and Campagna, Chapter 11). Troisi and D'Amato (Chapter 8) report on their observations of a captive Rhesus monkey mother who killed her infant. The pathological behavior of this monkey mother in captivity is viewed by these authors as relevant for an understanding of emotional disorders in women who abuse their infants.

One cannot view infanticide from the perspective that it is an adaptive behavior for males while ignoring the consequences for the fitness of a mother whose offspring are killed. The behavior of pregnant and lactating female mice (house mice and deer mice) is discussed in chapters by Parmigiani *et al.* (Chapter 15), Gubernick (Chapter 18), Svare and Boechler (Chapter 19), and Boechler (Chapter 20). These chapters provide an interesting contrast concerning the relationship between social structure and the proximate and ultimate causes of infanticide in house mice, which are polygynous, and the California mouse (*Peromyscus californicus*) studied by Gubernick, which is monogamous. The chapter by Gubernick begins with a general review of the literature on monogamy in mammals. An interesting contrast (in terms of maternal behavior) is also evident in studies of: 1. the dwarf mongoose by Rasa (Chapter 13) and the prairie dog by Hoogland (Chapter 14), where both authors report that allomothering (nourishment and protection of young by individuals other than the mother) is common, and 2. pinnipeds by Le Boeuf and Campagna (Chapter 11), where allomothering is rare.

In most stocks of house mice (*Mus musculus* or *Mus domesticus*), the majority of males kill unrelated newborn pups. By the early 1980s the basic mechanisms governing the behavior of male house mice toward their own versus alien young had been described, and the hypothesis that infanticide by male house mice is adaptive was supported by a considerable body of evidence. However, during the past decade there has been some confusion concerning the mechanisms which mediate recognition of own versus alien pups in male house mice based on findings from laboratories using different stocks of house mice. The chapters by Parmigiani *et al.* (Chapter 15), Perrigo and vom Saal (Chapter 16), and Elwood and Kennedy (Chapter 17) provide different approaches to the study of the behavior of males toward own versus alien young in different stocks

of mice. A variety of mechanisms have evolved to regulate the behavior of male mice toward young, and it is now understood that all of the mechanisms are present in different genetic stocks of house mice. But, there are differences between stocks in terms of which of these various mechanisms plays the major role in inhibiting males from killing their own offspring.

Studies of behavior of adults toward infants in humans (Hrdy, Chapter 1; vom Saal, Chapter 2; Wilson and Daly, Chapter 3; Ferraris Oliverio, Chapter 4; Daly and Wilson, Chapter 5) provide both interesting parallels and contrasts to laboratory and field studies with other mammals, birds and insects. These chapters reveal that there have been reports of neglect, abuse, abandonment and killing of infants throughout written history. The question which was initially raised by Hrdy concerning the possibility that infanticide can be an adaptive behavior has led to similar questions being asked in relation to human behavior. In particular, the findings of Daly and Wilson (Chapter 5) on the increased likelihood of abuse and infanticide associated with the presence of a stepparent in the home is consistent with information concerning the likelihood of males killing unrelated infants in the many other species covered in other chapters. In another chapter, Wilson and Daly (Chapter 3) discuss the issue of stepparenthood and a variety of other factors which are involved in what they refer to as "discriminative parental solicitude". These two chapters, as well as that by Hrdy (Chapter 1) discuss the behavior of humans toward infants from an evolutionary and comparative perspective.

Hrdy (Chapter 1) discusses the impact of the use of wet nurses on infant survival and overall reproductive success from the perspective that the investment by a parent in a particular infant is influenced by many factors. Thus, what might be the optimal behavior from the perspective of an individual infant is not always the optimal behavior for the parent. However, the finding that in humans, parents neglect, abuse and kill their own infants is often difficult to view within the perspective that these behaviors might actually result in an increase in lifetime reproductive success for the parent and offspring (the fitness of the parent is determined by the number of offspring which survive and reproduce). Instead, behaviors, such as repeatedly beating a child, often render the child less likely to be successful in competing for resources and reproducing, with the result that the parent's fitness is decreased.

It has proven difficult to assess the degree to which children are being victimized today in Western and non-Western societies (Hrdy, Chapter 1), and it is even more difficult to estimate the occurrence of these behav-

iors in prior times. Even though it is typically not possible to quantify the degree to which specific acts occurred in prior times, it is clear that social customs, such as the extensive use of wet nurses in the eighteenth and nineteenth century in Western Europe (Hrdy, Chapter 1), misguided advise of physicians and other experts, such as the promotion of long intervals between nursing and the use of infant formulas (vom Saal, Chapter 2), and various other social and religious customs (Ferraris Oliverio, Chapter 4) have increased the likelihood that infants would be neglected, abused or killed.

A comparison of parenting strategies in Western and non-Western societies shows the degree to which specific social customs can be detrimental to the welfare of infants. While the emphasis of the first five chapters is on information concerning parental behavior and infant welfare in Western societies, comparisons with non-Western societies are also included, particularly by Hrdy (Chapter 1). Unfortunately, a chapter dealing with the treatment of infants in a non-Western society was withdrawn by the author due to concern (shared by many anthropologists) about the possibility that the information might be used to further political and economic agendas, such as relocating a tribal group to allow for economic exploitation of their land. For example, a report that infanticide occurs in a particular tribal group could be used as the basis for relocating the tribe for the supposed purpose of stopping infanticide from occurring. The reality, of course, is that neglect, abandonment and killing of infants are all common occurrences in the societies of the individuals who seek to condemn this practice in other groups. This issue is discussed in more detail by Hrdy (Chapter 1).

Another issue of concern is that there are groups outside of the scientific community opposed to any research involving animals. In addition, even though infanticide is a naturally occurring behavior which is widespread in the animal world, there remain individuals within the scientific community who are opposed to the study of infanticide and thus seek to block publication of articles concerning infanticide; there are even those who are opposed to publications which include observations of infanticide in the field. Concern about these issues, as well as political intrusion into the scientific community in general, led the participants at the workshop to prepare a statement concerning the study of infanticide, which appears below.

We offer our thanks to the director, Professor Antonio Zichichi, director of the Centre, Professor Danilo Mainardi, director of the International School of Ethology, and the staff at the Ettore Majorana Centre for Scientific Culture in Erice for hosting the workshop and provid-

ing travel funds for participants other than from North America. We
also thank the National Science Foundation for providing travel funds for
participants from North America (Grant INT 8915057). Special thanks
are due to Bruce and Mary Alice Svare for assistance with editing the
book and to Michael Boechler for proofreading and editing the chapters. Finally, we are grateful to Ruth Dalke for typing the manuscripts
and tables.

                                                       S. Parmigiani
                                                       F.S. vom Saal

THE SPIRIT OF FREE INQUIRY AND REPORTING OF POLITICALLY-SENSITIVE SCIENTIFIC FINDINGS

Science is the pursuit of knowledge. Scientific inquiry may lead to results
which some individuals or groups may find inconsistent with preconceived
notions or with preexisting social, political or religious agendas. It is also
possible that scientific inquiry will lead to the discovery of information
which has the potential to be used or misused by individuals or groups
seeking to legitimize their actions. If individuals intent on pursuing political goals cannot validate their programs with scientific findings, they will
likely invent the desired information. This is only one of the many reasons
why the censorship of scientific information should be resisted; in the long
run, great harm is done by attempting to suppress the truth.

This workshop at the Centre Ettore Majorana in Erice, Sicily concerns
the social, psychological and biological factors influencing the protection,
abuse and killing of infants. Participants include anthropologists, biologists and psychologists from nine countries. The issue of the abuse and
killing of infants is of immense importance for our understanding of the
biological and cultural bases of behavior. Throughout recorded history in
both Eastern and Western societies, traditional and modern, as well as in
many species of primates and many other mammals and vertebrates, infants have been reported to be abandoned, abused and killed. The finding
that these are widespread behaviors has no bearing on the issue of whether
they should be socially acceptable, justified or condoned. Similarly, the

study and reporting of animal or human behavior by scientists does not imply that we condone or seek to promote the behavior.

We, the undersigned, are deeply concerned with attempts to restrict research and with the possibility of misuse of scientific findings. We are also concerned with the possibility that due to religious, social or political considerations, scientists will be inhibited from seeking answers or publishing findings.

LIST OF CONTRIBUTORS

Michael Boechler
Department of Psychology,
State University of New York
at Albany
1400 Washington Ave.
Albany, New York 12222, USA

Paul F. Brain
Biomedical and Physiological
Research Group,
Biological Sciences,
University College of Swansea
Swansea SA2 8PP, UK

Claudio Campagna
Centro Nacional Patagonico
Puerto Madryn Chubut, Argentina

R. Cervo
Dipartimento di Biologia Animale e
Genetica, Universita di Firenze
via Romana 17, 50125 Firenze, Italy

Martin Daly
Department of Psychology,
McMaster University
Hamilton, Ontario L8S 4K1, Canada

Francesca R. D'Amato
Istituto di Psicobiologia e
Psicofarmacologia del C.N.R.
Rome, Italy

Robert W. Elwood
Division of Environmental and
Evolutionary Biology
School of Biology and Biochemistry,
The Queen's University of Belfast
Belfast BT7 1NN, Northern Ireland

L. Scott Forbes
Department of Biology,
University of Winnipeg,
515 Portage Avenue, Winnipeg,
Manitoba R3B 2E9, Canada

David J. Gubernick
Psychology Department,
University of Wisconsin
Madison, Wisconsin 53706, USA

Toshikazu Hasegawa
Department of Psychology,
The University of Tokyo
Komaba, Meguro-ku, Tokyo 153, Japan

Mariko Hiraiwa-Hasegawa
Institute of Natural Science,
Senshu University
Kawasaki 214, Japan

John L. Hoogland
The University of Maryland,
Appalachian Environmental Laboratory
Frostburg, Maryland 21532, USA

Sarah Blaffer Hrdy
Department of Anthropology,
University of California at Davis
Davis, California 95616, USA

Hazel F. Kennedy
Division of Environmental
and Evolutionary Biology,
School of Biology and Biochemistry,
The Queen's University of Belfast
Belfast BT7 1NN, Northern Ireland

Burney J. Le Boeuf
Department of Biology and
Institute for Marine Sciences
University of California Santa
Cruz, California 95064, USA

Danilo Mainardi
Dipartimento di Biologia e Fisiologia
Generali, Università di Parma
43100 Parma, Italy

Douglas W. Mock
Department of Zoology,
University of Oklahoma
Norman, Oklahoma 73019, USA

Anna Ferraris Oliverio
Cattedra di Psicologia dell
'eta' Evolutiva
University of Rome, Rome, Italy

Craig Packer
Department of Ecology,
Evolution and Behavior
College of Biological Sciences
100 Ecology Building
1987 Upper Buford Circle
St. Paul MN 55108-6097, USA

Paola Palanza
Dipartimento di Biologia e Fisiologia
Generali, Università di Parma
43100 Parma, Italy

Stefano Parmigiani
Dipartimento di Biologia e
siologia Generali
Università di Parma,
43100 Parma, Italy

Glenn Perrigo
Division of Biological Sciences
and The John M. Dalton Research Center
University of Missouri-Columbia
Columbia, Missouri 65211, USA

Anne E. Pusey
Department of Ecology,
Evolution and Behavior
College of Biological Sciences
100 Ecology Building
1987 Upper Buford Circle
St. Paul MN 55108-6097, USA

O. Anne E. Rasa
Abt. Ethologie, Zoologisches
Institut, Universität Bonn
5300 Bonn 1, Germany

Volker Sommer
Institut für Anthropologie der
Georg-August-Universität
Burgerstrasse 50,
D-37073 Gottingen, Germany

Bruce Svare
Department of Psychology,
State University of New York
at Albany, 1400 Washington Ave.
Albany, New York 12222, USA

Alfonso Troisi
Clinica Psichiatrica della II
Università di Roma, Rome, Italy

S. Turillazzi
Dipartimento di Biologia Animale e
Genetica, Università di Firenze
via Romana 17, 50125 Firenze, Italy

Frederick S. Vom Saal
Division of Biological Sciences
Department of Psychology and
The John M. Dalton Research Center
University of Missouri-Columbia,
Columbia, Missouri 65211, USA

Margo Wilson
Department of Psychology,
McMaster University
Hamilton, Ontario L8S 4K1, Canada

PART 1: HISTORICAL, ANTHROPOLOGICAL, SOCIOLOGICAL AND PSYCHOLOGICAL APPROACHES TO THE STUDY OF INFANTICIDE AND PARENTAL CARE IN HUMANS

CHAPTER 1

FITNESS TRADEOFFS IN THE HISTORY AND EVOLUTION OF DELEGATED MOTHERING WITH SPECIAL REFERENCE TO WET-NURSING, ABANDONMENT AND INFANTICIDE

SARAH BLAFFER HRDY

Department of Anthropology, University of California, Davis, California, USA

> Early European accounts of infanticidal savages bristled with ethnocentric moralizing...And yet as we read the tragic accounts...from one society to another, it is not the inhumanity of the unfortunate perpetrators that confronts us, but rather their humanity...
>
> Martin Daly and Margo Wilson, 1988:591.

I INTRODUCTION: MOTHERHOOD AS COMPROMISE

The dilemma confronting working mothers in the western world today has universal dimensions. To imagine that there is anything new in the conflicts faced by modern women is to adopt a mythologized concept of self-sacrificing motherhood. For motherhood has always meant compromise, compromise between subsistence needs of the mother and the time, energy, and resources needed to mate and reproduce (reproductive effort). In addition to conflicts between the mother's own needs and a general commitment to reproduction, iteroparous mothers (breeding over a lifetime) must also partition their reproductive effort among different offspring. Parental investment of time, energy or resources in the production or nurturing of one offspring can diminish a mother's ability to invest in

3

older offspring, or in her ability to produce additional offspring in the future (Trivers, 1972).[1]

As Robert Trivers pointed out (1974), individual infants may attempt to extract greater investment from their parents than the parents have been selected to give. Herein lies the source of the chronic tension between parental commitment to the survival and well being of offspring and parental frustration at the frequency and insistence of infant demands. Although a primate infant should rarely seek to extract more reproductive effort from a mother than is compatible with her survival, it might well seek to extract additional parental investment when it would come at the expense of future siblings, rather than the mother's survival. In this paper I will be dealing with maternal dilemmas and decision making at two levels: 1) at the level of reproductive effort – her own survival versus that of her progeny; and 2) at the level of parental investment – investment in one particular infant versus investment in offspring of another sex, born with different qualities and/or under different circumstances.

In most mammals, and all primates, newborns are dependent on their mother for warmth, protection, locomotion and nutrition. Substitute providers of these functions occasionally crop up in evolutionary history (e.g. van Lawick, 1973 for wild dogs; Hrdy, 1976 and Thierry and Anderson, 1986 for primates) and have even been common during some periods in human history. More often than not however, survival of the infant depends on the mother's survival. From an evolutionary perspective then, it is the mother who is the critical unit of selection and the mother's survival would always have priority except in the case of older or incapacitated mothers with low probabilities of reproducing again, that is, mothers with very low reproductive value. Even here, maternal survival should take priority if survival of the infant depends on her nurturing.

Where parental rank is correlated with the survival and breeding prospects of selected offspring, even the maintenance of parental status may take precedence over the survival of less favored infants. Some of the best documented examples derive from human parents who cloister in convents or actually destroy daughters whose dowry costs jeopardize family socioeconomic status, or destroy daughters who threaten to injure family standing or "honor" (Dickemann, 1979; Manzoni, 1961:134ff; Boone, 1986).

---

[1] Whether individuals are primarily motivated by a desire to perpetuate a biological lineage (i.e. reproductive success) or whether parental investment strategies are geared to perpetuation of the social or economic status of a family or household (containing affines and adopted members) is not always clear. Perhaps this is because over past generations these two outcomes have been so closely interconnected.

In other mammals who produce either sequential young or litters, parents respond to local conditions in evaluating the worth of a particular offspring in terms of probabilistic assessments of future conditions and "as a function of the proportion that this child represents of his total future reproductive prospects" (Dawkins and Carlisle, 1976:132). That is, parents respond with what Daly and Wilson (1980; 1983) refer to as "discriminative parental solicitude" – an amalgam derived from assessments of probable degree of relatedness (clearly more important in the case of males and egg layers than for most female mammals); worth of the offspring in terms of its ability to translate parental investment into subsequent reproduction, and finally consideration of alternate uses to which the parent could devote the resources, such as diverting resources to a stronger child, a child of a preferred sex or sustaining the parent until more favorable opportunities to breed should present themselves. In humans these levels of solicitude are tempered (albeit rarely overridden completely) by cultural ideals, especially ideals about continuation of the household or the lineage, ideals which are in turn shaped through historical time by the changing productive and reproductive value of children (Hrdy, 1990).

Here I focus on fitness tradeoffs made by mothers through human history. In humans, however, as in so many species, maternal subsistence and especially the survival of her offspring are so heavily influenced by other group members that it is impossible to consider the mother in isolation from the web of fitness tradeoffs by other individuals in the social network she is part of (see Hill and Kaplan, 1988 for an exemplary case study exploring how reproductive decisions "mutually constrain one another" and involve complex tradeoffs between alternative behavioral options among Ache hunter-foragers of Paraguay). Relevant individuals may include former and future mates of the mother, biological relatives, affines and unrelated individuals linked to her in either cooperative or competitive arrangements, subordinate individuals she exploits as well as dominant individuals exploiting her. For the purposes of this paper, I will sometimes substitute the term "parental" for maternal when my knowledge of the situation is too limited to separate maternal from paternal interests, or when in fact there is good reason to assume they coincide.

II. FITNESS TRADEOFFS IN DETERMINING THE LEVEL AND TYPE OF SOLICITUDE

For the purposes of this article, I assume that infanticide (were information available) could be documented for virtually all human populations,

although frequencies differ markedly, ranging from near zero to over 40% of live births (section IV-B). The main functional classes of infanticide that have been described for non-human animals (Hrdy, 1979) can all be documented among humans, if only anecdotally. Nevertheless, the patterning of infanticide in humans is considerably different. For example, in other primates unrelated males are the most likely perpetrators of infanticide (Hrdy, 1977, 1979; Leland, Struhsaker and Butynski, 1984). While the close proximity to the mother of males unrelated to her infant (either captors or step-fathers) can represent a threat to human infants (e.g. see Biocca, 1971; Hill and Kaplan, 1988 for Amazonian hunter-foragers; Exodus 1:16 and Matthew 2:16 for ancient Near Eastern pastoralists; Daly and Wilson, 1988 for contemporary North American populations), biological parents are apparently responsible for the largest portion of infanticides, and marriage and inheritance systems, religious beliefs and social norms concerning individual and family honor play central roles in parental decisions to terminate investment in these human infants. Furthermore these parental decisions are informed by a unique awareness of history, the future, and long-term goals for family survival. Hence, although thresholds for parents to invest may be set by evolved motivational processes (Daly and Wilson, 1980; 1988), adjustments in parental investment are consciously calculated to achieve culturally as well as biologically defined goals and are played out in specific demographic and cultural contexts (e.g. see Korbin, 1981; Skinner, 1988 and in press).

By far the most common goal for parents committing infanticide involves the manipulation of family size, composition, or the adjustment of the timing of parental investment by the mother and/or father. In fact, however, infanticide in the sense of Langer's (1974) classic definition ("the willful destruction of newborn babies through exposure, starvation, strangulation, smothering, poisoning or through the use of some lethal weapon") represents only the extreme end of a continuum of behaviors which function to reduce the costs (in terms of time, energy, risk and resources) that offspring impose upon parents. In contrast to rodents and other mammals who may cannibalize supernumerary infants (Day and Galef, 1977; Gandelman and Simon, 1978), thereby recouping nutrients, there are virtually never any benefits to killing one's own offspring apart from perceived benefits in rare cases of child sacrifice (e.g. points to be won with a god for sacrificing a valued son, Genesis 22; see Stager and Wolff, 1984 for ancient Carthage). And quite often, there are costs.

Hence, we would expect that infanticide involving direct destruction of the young occurs only as a last resort when other options for reducing postpartum investment are constrained by legal sanctions which makes aban-

donment riskier than murder, by the lack of supportive kin networks or other potential caregivers, or else by particular sorts of environmental hazards which make abandonment impractical. A different set of constraints may also pertain in the upper reaches of stratified societies where the continued existence of a child may represent a threat to social status, family "honor" or orderly succession. Although it is very common to read in the literature on infanticide that parents were responding to scarcity by eliminating their infants, limited resources by themselves (except in special circumstances, see section IV-A) are no reason to destroy an infant. Scarcity is merely a reason for parents to REDUCE investment in a current infant, perhaps abandon it. There exist a wide array of alternatives to infanticide whose availability varies according to specific historical and ecological contexts.

Compared to infanticide, these alternative means of reducing parental investment in offspring have received less emphasis in the literature. Yet between infanticide and John Bowlby's ideal of a Pleistocene mother linked by an intense emotional bond to her infant, and physically in close contact with this semi-continuously suckling infant (Konner, 1976), there lies a continuum of compromises. Indeed, such retrenchments from the Bowlbian or "primate ideal" are far more common, sociologically and demographically more important, than actual infanticide (exceptions are discussed in section IV-B). I list below seven different ways of dealing with infants, each of which functions to mitigate or terminate parental investment without outright destruction of the infant. This list is by no means exhaustive:

1) Exploitation of the infant as a resource, usually selling the infant, which entails some immediate (usually small) gain to the parents (e.g. see Boswell, 1988:170-171; Fildes, 1986:6). Although prostitution or slavery would be likely fates for offspring sold (Boswell, 1988), some children could conceivably end up with improved prospects of survival or even reproduction.

2) Abandonment of the infant where the parents leave the infant typically out of harm's way so that there is some prospect that the infant will be taken up and cared for by someone else (Trexler, 1973a; Boswell, 1988; see Exodus 2, for the story of Moses). Admittedly, there can be a fuzzy distinction between abandonment and infanticide when real or imagined parental optimism comes up against the realities of infant starvation or hypothermia. Nevertheless, the practice common in Medieval Europe of abandoning infants with identifying tokens indicates – at least for those cases – the existence of a parental mind set where retrieval of the infant one day remained a possibil-

ity. Some parents found solace by fantasizing fabulous destinies of upward social mobility for abandoned progeny (e.g. Romulus, who founds a dynasty; see Boswell, 1988 for other European examples).

3) Fostering out the infant either through arrangements with relatives who can rear children more cheaply (a grandmother in a rural area) or provide them special opportunities (education in an urban area; opportunities to forge extended social networks), or through the more common practice of an "invented" kin tie whereby infants, or more often children past weaning, are sent to a distant household to live and one or both parents pay (in cash, goods or current or future favors) someone (often an older woman) to care for an infant (Goody, 1969; Isiugo-Abanihe, 1985 for West Africa; Pennington, 1989 for South Africa; Radcliffe-Brown, 1922:77 for the Andaman Islands; Shahar, 1990 for Medieval Europe). Although typically infants going to foster mothers are weaned, the distinction between "fostering out" and "wet-nursing" can become blurred when the foster mother does provide milk (Bledsoe and Isiugo-Abanihe, 1989: note 1). Because children fostered out tend to be older, there is also a much higher likelihood that a foster parent (as opposed to a wet nurse) can set them useful tasks, and the use of foster children as household helpers and child minders may explain the fact that among the Herero of Botswanna daughters rather than sons are more often selected (Pennington, 1989).

4) Wet-nursing, when the mother or both parents contract with another woman to suckle their infant. This arrangement encumbers the wet nurse but frees the mother both for status or labor-related pursuits and simultaneously also renders the mother fertile for subsequent pregnancies – an artifact of wet-nursing that may or may not be intended, see next section.

5) Oblation, when one or both parents leaves the children in the custody of a religious institution, usually, but not always, irrevocably (Boswell, 1988:Chapter 8; Fuchs, 1984).

6) Reducing overall reproductive effort, so that parents continue to rear their own children but at a lower level of resource and energy expenditure or a lower level of direct involvement. This may entail delegating care to others, such as locally available kin, particularly older siblings as is characteristic today in much of Bantu East and South Africa (e.g. Weisner, 1987; Draper, 1989).

7) Reducing parental investment in particular children, often weak ones, or else daughters destined to be economic or social liabilities, or in

later born sons in societies with primogeniture, etc. (Cain, 1977; Miller, 1981; DeVries, 1984; Boone, 1986; Das Gupta, 1987; Voland, 1988). Such reduced investment may be motivated either by insufficient resources (e.g. Scheper- Hughes, 1985 for mothers in Brazilian shantytowns), by probable absence or loss of paternal investment or by the advantages which may accrue to families channeling resources towards selected progeny in high density, stratified societies (see Dickemann, 1979; Levine, 1987:293).

In the last five or so thousand years, such "mitigating strategies" accounted for the fates (both survivals and deaths) of far more infants than did infanticide, even though infanticide as a phenomenon has attracted more attention from anthropologists and biologists. Indeed, patterns of continued investment with retrenchments (6 and 7 above) are so common in human societies as to be – at least in their milder forms – completely unremarkable both to the ethnographer in tribal societies or to the sociologist surveying contemporary western populations. These patterns represent the investing end of the continuum that ranges from termination of investment at the one extreme to total self-sacrifice of the parent on behalf of offspring at the other – the fictional account of "Stella Dallas" comes to mind. Although I would expect such maternal self-sacrifice to be more common in fiction than in life, real-life examples can be documented.[2]

Globally, I suspect that retrenchment of parental investment would have pertained to more infant survivals and deaths than all the other strategies, including infanticide, combined. Nevertheless, depending on the time and place one or another of these patterns can become demographically more important. Wet-nursing, which affected large segments of the population in seventeenth and eighteenth century France is a case in point, and is discussed below.

From the perspective of reproductive strategies, abandonment of an infant should be the default divestment strategy for parents terminating investment; infanticide would only be a last resort when this option is curtailed. Hence, when parents destroy their children, it is certainly important to ask what factors diminished parental solicitude toward their offspring (Daly and Wilson, 1984), it is also useful to identify the ecological, social and cultural constraints which prevented parents from abandoning the baby or from attempting to mitigate the cost of rearing an infant through

---

[2] Stephen Lock (1990:397) discusses the case of a mother who risks her life for her unborn child. A twenty weeks pregnant woman informed by her doctor that she has cancer of the cervix must choose between immediate treatment and sacrifice of the pregnancy with an 80 percent chance of complete cure versus delayed treatment until the baby could survive with only a ten percent chance of cure. The mother opted to save the baby.

some other means (Granzberg, 1973). In this paper I focus on the decision-making process of parents, and inquire: what social, economic and environmental factors shape parental cost-benefit analyses? If indeed the various mitigating strategies outlined above are all functionally similar, why do parents opt for one strategy rather than another? What is the role of historical precedents, history and local ecology in these decisions? What factors predispose or forestall pursuit of alternative strategies to mitigate parental effort?[3]

III. WET-NURSING AS A CASE STUDY IN TRADEOFFS

A. Diluting the Costs of Reproductive Effort

It is widely believed that the use of wet nurses in pre-modern Europe was in fact a disguised, nonprosecutable form of infanticide. This interpretation is implicit in terms for wet nurses such as the English "angelmaker" and the German "Engelmacherin". Wet nurses were viewed as "surrogates upon whom parents could depend for a swift demise for unwanted children" (Smith, 1984:64C). "It must have been common knowledge" writes Maria Piers in her book on *Infanticide,* that the wet nurse "was a professional feeder and a professional killer" (1978:52). Critics of wet-nursing, such as the reformer Dr. Alexander Mayer, giving testimony before the Roussel Committee in France at the end of the nineteenth century, just before the Roussel Law of 1874 was passed claimed that the artisans of Paris sent their infants off to wet nurses "with the desire of not seeing them again..." (cited from court records by Sussman, 1982:123).

In fact however, even during the European heyday of wet-nursing at the end of the 18th century when up to ninety percent of infants born in urban centers such as Paris and Lyon were nursed by women other than their biological mother (20,000 of 21,000 infants born in the famous statistic given by Paris lieutenant general of police for Paris, LeNoir 1780:63), wet-nursing is best understood as a strategy to reduce the costs of reproductive effort for individual mothers. For many of these women wet-nursing represented an alternative to worse outcomes (death of infant and maternal destitution) rather than a covert means of destruction. There can be little doubt that wet-nursing varied according to circumstances of parents, and that under some circumstances this form of delegated mothering was associated with high levels of mortality. But these cases involved parents with poor access to resources and dismal alternatives.

---

[3] It is important to note that I do not consider here the infant's "point of view". Nor do I address ethical implications of the behaviors I describe. These are separate issues.

The risks that parents took with the lives of their infants can not be understood without reference to both local customs (e.g. a well established tradition of wet-nursing among elites that had long exposed rural poor in many parts of France to the practice) and prevailing ecological conditions. By the eighteenth century, France was in the throes of a tremendous increase in population (from 20 million in 1720 to 27 million by the end of the century, Hufton, 1974:14). In the countryside, poor harvests and, especially, fractionalization of small landholdings among several sons swelled the numbers of the dispossessed. There was a proliferation of people who found it difficult to provide themselves with the bare necessities. A man by himself, without a working wife, could not expect to earn enough to support more than himself and perhaps, one child. The arrival of additional children could reduce the family to destitution.

Moving to the city could not have been much improvement. Rapid urbanization combined with slow industrialization meant that labor opportunities were few – not only for parents but for children who survived to mature. Incomes were low, rents high. Around sixty percent of these French mothers sending their infants to wet nurses belonged to the large class of artisans. In his detailed analysis of wet-nursing commerce, George Sussman writes that "a majority of the working people of eighteenth century Paris was engaged in a perpetual struggle to avoid insolvency and indigence, a struggle that became more difficult as the cost of living and particularly bread rose faster than wages" (1982:58). Even for this "Bourgeoisie" then, maintenance of that status was precarious. In Sussman's calculation of a typical budget for a family of the artisan class, a husband would earn about 25 livre a month, the wife another fifteen. Of this 40-50 percent would go for food, 15 percent for clothing, 6 percent for light and heat, and another 13 for rent. In addition, each wet-nursed child would cost about 8 livre a month (or 20% of the total budget per child) as long as the parents chose to keep up the payments (see below).

B. Reconstructing the History of Wet-nursing

Such was the ecological context for 18th century France. Let us proceed then by trying to put the phenomenon of wet-nursing in broad evolutionary and comparative perspective, and then examine its antecedents in human history. From a comparative perspective, we can certainly locate examples of communal suckling and "delegated mothering" in other animals. As we further attempt to reconstruct the use of wet nurses in prehistoric times it seems probable that this curious phenomenon initially functioned either to 1) enhance foraging (or labor) opportunities for mothers who would

otherwise be burdened by infants, or 2) to reduce the physiological costs of lactation. In either case, the ultimate outcome would be enhanced reproductive success for the mother, either due to her own or her infant's improved survival prospects and/or her own shorter inter-birth intervals.

In the animal literature, biological relatedness looms large in the evolution of communal suckling (e.g. McCracken, 1984; Lee, 1989). Although there exist primate cases where lactating females adopt unrelated (or distantly related) individuals, under natural conditions adoption of close relatives (siblings, grandchildren) is the more common pattern (Thierry and Anderson, 1986; Goodall, 1986: 101-103; 383-384). As in allomaternal caretaking generally, kinship tends to play an important role (Hrdy, 1976; Sommer, in prep.). Although care of these adopted young may not include suckling, several spectacular cases involve older female monkeys who resume lactating when caring for grand-offspring (see Auerbach, 1981; Auerbach and Avery, 1981 for evidence of induced lactation in women other than the biological mother).

The presence of matrilineal kin might have been problematic however for early humans – just as it is for most contemporary humans. It would be ill-advised to extrapolate to humans from cercopithecine and colobine monkeys which are predominantly matrilineal and female philopatric.[4] By contrast, human societies are most often characterized as patrilineal and patrilocal (Murdock, 1934), that is, humans are most often "male philopatric", a tendency that may be quite ancient (Ghiglieri, 1987), characterizing as it does 4 of five hominoid species. Males stay put and females move between groups in gorillas, common chimps, bonobos and the majority of human societies. If early hominid allomothers only chose to suckle infants born to close kin, they might then have had few opportunities to do so (i.e. allomaternal suckling would have been limited to cases where two sisters were married to the same man or to brothers; or else lactating relatives of the father). Biocca describes for the Amazonian Yanomama a case where the mother had decided not to rear her son. The paternal grandmother intervened and the infant was suckled by the father's sister in addition to her own new infant (1971:299).

It is perhaps not surprising then that in contrast to most examples of communal suckling in animals, co-residence and the potential for reciprocity appear to be more important than kinship in sustaining shared nursing in traditional societies. Where kinswomen are available to nurse, all three factors (relatedness, co-residence and reciprocity) may be at issue (see Tronick et al., 1989 for Ituri Forest pygmies in Central Zaire). Furthermore, it is probably not a coincidence that kin-based fostering out (as op-

---

[4] That is, females remain in their natal group while males migrate away to breed.

posed to commercial day-care and boarding schools, or paid wet-nursing) is most richly developed in West Africa and other parts of the world where matrilineal kinships systems are strongly developed and intact (e.g. see Bledsoe and Isiugo Abanihe, 1989; Draper, 1989).

Assuming that most women in early human societies were living in patrilocal (male philopatric) systems, wet-nursing in its earliest manifestations was probably a reciprocal favor among co-wives or neighbors. Among the Andaman Islanders, for example, lactating women routinely suckled other women's children (Radcliffe-Brown, 1922:76; see also Tronick, Morelli and Winn, 1987, for Efe pygmies in Zaire). In addition to incorporating infants into a network of caretakers, mothers presumably gain from being able to forage more efficiently while another female holds her infant (Hrdy, 1976; Whitten, 1982). Hurtado has recorded significantly lower rates of food acquisition by Ache food gatherers who are lactating (1985); to compensate, women shared food with the inefficient gatherer. Among Solomon Islanders where taboos prevented mothers from nursing their babies in the place where they garden, a mother might leave her infant with a lactating sister-in-law for an hour or so while she went to work in the fields (Akin, 1983 and personal communication). Indeed, among the Arunta, where women nurse one another's children, disputes may arise over which mother is to stay in camp and which is to go and forage (Murdock, 1934:35).

These wet-nursing relationships are best characterized as casual opportunistic cooperation among women – affines, neighbors and blood kin – who are in a position to reciprocate help. Some voluntary wet nurses may look forward to future support from a grown charge who owes his life to her (e.g. Biocca, 1971:214 for the case of a captured woman who nurses the orphaned daughter of a headman). In some societies, wet-nursing arrangements have become more formalized. In Arab culture, such relationships are institutionalized, and Islamic law actually allows for three kinds of kinship: kinship by blood, by marriage and by the happenstance of two individuals having suckled milk from the same woman (see for example Altorki, 1980). The same incest rules pertain to children who suckled from the same woman as for true siblings.

Wet-nursing on a large scale probably did not emerge until there were stratified societies in which one class could command or purchase the services of lower-ranking mothers. But this situation was probably neither that rare nor that recent in human history. Enforced suckling involving females of different dominance statuses occurs in a wide range of animals (e.g. see van Lawick, 1973 for wild dogs; Rood, 1980 for dwarf mongooses; O'Brien, 1988 for cebus monkeys; reviewed in Lee, 1989), and it is likely

that "enforced suckling" among humans predates its appearance in the historical record. In the most famous animal example of enforced suckling, the dominant female in a pack of wild dogs kills all but one of the pups of a subordinate female, designated by van Lawick as "Angel". Angel's single pup is sufficient to sustain the subordinate female's lactation so that at a later date the dominant female's ten-week-old pups can preferentially nurse from her, at the expense of the lone, stunted, survivor who is not competitive for her milk with these older and larger pups (van Lawick, 1973).

Although enforced wet-nursing among humans is no doubt even older, the first written record I can locate of one woman suckling another's infant in a context that was neither kin-based nor rooted in reciprocal coopera- tion derives from a Sumerian lullaby from the late third millennium B.C. As the wife of Shulgi, ruler of Ur sings her son to sleep she promises him first a wife and then a son – complete with wet nurse. "The nursemaid, joy- ous of heart, will sing to him; The nursemaid, joyous of heart, will suckle him" (from Wallis Budge, 1925, cited in Fildes 1986:6). Some centuries later, in the period just prior to the Buddha's lifetime (566? -480 B.C.), references in the Caraka Samhita, an encyclopedic collection of Ayuredic beliefs, made it clear that the use of wet nurses was widespread among the elites and that great care was taken to assure nurses of appropriate caste, color and character. In Hellenistic Egypt, from about 300 B.C., the Greek ruling class used slaves as wet nurses;[5] however, in hard times, possibly ushered in by Roman control of the area, free women as well turned to wet-nursing for income (see Pomeroy, 1984:139 for evidence from Greek papyri). Nursing contracts from the subsequent period of Roman rule in Egypt also survive (Bradley, 1980).

By the second century A.D., wet-nursing in parts of Europe was an organized commercial activity. In Rome it was centered about particular columns in the vegetable market at the Forum Holitorium specifically referred to as "lactaria". From Medieval times onward, wet nurses, paid, indentured, or enslaved, were used by royalty and elites in many European countries. Typically propertied families would hire women to suckle their children under conditions of close supervision, so that only one infant was nursed at a time by a nonpregnant woman with a healthy supply of milk. Although more costly, wet-nursing in this form ensured high rates of infant survival (Sussman 1982 for 18th century France; Klapisch-Zuber 1985 for 15th century Italy). Infant mortality in the French case of in-house wet-

---

[5] Several hundred years later, in the Augustan period, some of these enslaved wet nurses would have been abandoned daughters who had been reared by "foster parents" and subse- quently sold. "Foster parents" in this context are in fact slave dealers (Pomeroy 1984:138).

nursing hovered around 20% – about the same, or only slightly higher than if a French mother of that period suckled her own offspring (see Figure 1, footnote 4).

We know far too little about wet nurses themselves, but there can be little doubt that in most cases their occupation curtailed the opportunities that their own infant had to nurse, and may have led to its death. In Renaissance Italy, nearly 30 percent of the infants sent to foundling homes – where probability of survival would be low – were in fact the offspring of slave women whose milk would subsequently be used or sold to the benefit of her owner (Trexler 1973b:270; Klapisch-Zuber:1985: 141 and 141 n. 33). There are dozens of texts and manuals describing attributes of a good wet nurse. Virtually all advise parents not to select a nurse who is either pregnant or still suckling her own infant – even when they also recommended choosing mothers with very new milk. The wet nurse's own infant, might be farmed out to an even less well paid wet nurse, or else "dry-nursed". Klapisch-Zuber expresses the suspicion that fourteenth and fifteenth century Florentine slave owners may have hastened the death of "certain socially condemned infants" through abandonment in order to obtain a wet nurse (1985:140). In the rich correspondence between Margherita Dattini and her Renaissance merchant husband for whose client she was seeking a suitable wet nurse, Margherita conveys her disappointment that the infant of one prospective candidate had survived after all (Origo 1957:200–201; see also Trexler 1973b for more of the same). In other cases a lactating nurse might simply sustain milk production over years (Jane Austen was the seventh of eight of her siblings to be suckled by the same nurse).

Given that any shift away from breast milk would have introduced new opportunities for infection and lowered survivorship for infants thus deprived of milk their mothers provided to the children of others, wet nurses were directly contributing to the death of their own offspring. Maternal decision- making in these instances must however be examined in social context. The wet nurse's behavior benefited non-relatives more powerful than she (although in some of these instances, fitness tradeoffs will be complicated by biological relatedness between the infant and a male householder who fathered it). The price paid by these mothers for remaining within the system at all (perceived by them with some accuracy as synonymous with survival?) was to redirect their own milk to non-relatives – a forced decision not unlike that made by "Angel", the subordinate wild dog mother.

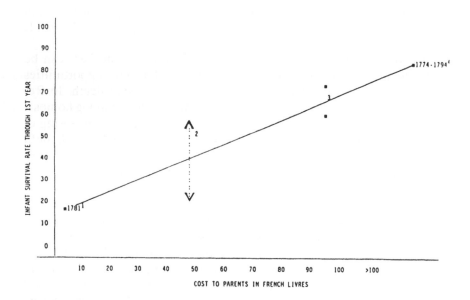

Figure 1 Linear relationship between amount expended by parents to pay wet-nurses and the probability of infant survival.

Notes on how survival rates and costs were calculated:

1. Mortality rates during the first year of life for infants deposited in Parisian foundling hospitals reached 68.5% in 1751, and rose to 85.7% by 1781. Ninety-two percent of these children would die by their 8th birthday (Sussman 1982: 62-64).

2. Roughly 10% of parents who sent children to rural wet-nurses subsequently defaulted on their payments with the result that their infants were eventually deposited in foundling homes (Sussman 1982: 62. Infants abandoned after six months of paid wet-nursing, nevertheless tended to have higher prospects of survival than did those infants abandoned at birth (Delasselle 1975). The cost here is calculated at one half the yearly rate for a rural wet-nurse.

3. A Parisian artisan might earn 20-25 livre per month, his wife one-half that. Seven to eight livre per month went to pay the rural wet-nurse (Sussman 1982: 590. Sussman estimates mortality for these wet-nursed infants at 25-40% (1982: 67). Mortality rates rose somewhat over time as good wet-nurses became increasingly hard to find. By the period 1871-1874 mortality reached 42% according to records kept by the Bureau of Wet-nursing.

4. Infants tended at home by live-in wet-nurses enjoyed roughly the same prospects of survival as infants nursed by their own mothers. Based on information recorded for 11,923 babies born in 19 parishes in suburbs south of Paris and nursed by their own mothers during the period 1774 to 1784, Galliano (1966) estimates that mortality rates were around 18% (Sussman 1982: 67). Data collected by Maurice Garden (1970:125) for Lyon during the period 1785-88 indicates a comparable rate of mortality for mother-nursed infants, only 16% died in their first year of life.

C. Reproductive Consequences of Wet-Nursing

Although it is difficult to precisely document what happened to the wet nurse's own infants, demographic consequences for the women on whose behalf they were hired are better known, largely through family diaries and birth records. Non-nursing mothers gave birth at a much higher rate – as often as annually in some extreme cases instead of every three of four years as one might expect if they nursed their own infants. Maurice Garden's remarkable study of the demography of 18th century Lyon documents nearly annual births in the families of butchers and silk-makers, with mothers routinely producing 12 - 16 children (Garden 1970:95-97). One of the women in his sample of butchers' wives produces 21 children in 24 years (see 1970: Tableau VII "Les enfants de Jacques Gantillon le cadet"). A similar hyper-fertility is documented for upper class British women, but the situation is more complicated for French elites for whom the average number of births per marriage fell from 6.15 in the seventeenth century to 2.79 in the 18th century (Johansson, 1987b) despite continued use of wet nurses.[6] Of fifty women focused on in Judith Lewis' study of childbearing among the British aristocracy between 1760-1860, the Duchess of Leinster was the most fecund, giving birth to her first child at sixteen, only a year after her marriage, and continuing to reproduce until her twenty first and last child was born thirty years later when the Duchess was forty-six (1986: 123-124). For the entire group of fifty, the median span of childbearing from marriage to last birth was eighteen years, resulting in an average of eight children each.[7] Twenty three of these women gave birth to their second child within a year or less of their first. (Contrast this with the four year birth intervals, and average of five children per female thought to have characterized Pleistocene hunter-gatherers during most of human history, Short 1976; Lancaster and Lancaster 1987). There can be little

---

[6] According to Johansson (1987a,b), fixed incomes among continental elites combined with a new enlightenment ethic which called for parents to treat children equally. Anxious to maintain their high social status, these parents reduced fertility, since only by producing few children could they both treat heirs equally and provide them large legacies.

[7] Note that Lewis disagrees with both Stone (1977) and Trumbach (1978) over how late British women continued to use wet nurses. Although Trumbach assumed that most aristocratic women were breastfeeding their own infants by 1780, the data from Lewis' sample causes her to assume that the practice continued much later, and she can come up with no other feasible explanation for these birth intervals. "While there may well have been more women breast-feeding by the 1780s than in earlier generations, it had become by no means a uniform practice..." (Lewis 1986:209). Some women – apparently quite aware of the contraceptive effects of breast-feeding – adjusted their use of wet nurses accordingly – using a wet nurse with the first children and breastfeeding later ones for long periods so as to deliberately avoid additional impregnations (Lewis 1986:212).

doubt that the anemia, poor health, prolapsed uteri and other obstetrical difficulties documented in such lurid detail by Shorter (1982) was in fact the toll taken on women's bodies by nearly annual births during the first decade of marriage, followed by a second decade of production at a slower – but still "unnaturally" high – rate. In the years before the availability to women of birth control, any behavior that circumvented lactational amenorrhea resulted in increased fertility, so that wet-nursing was well-suited to achieve (primarily male?) pronatalist goals.

For those who combined these high levels of fertility with relatively low levels of mortality (i.e. twenty percent or less), wet nurses meant large completed family sizes, such as those recorded for aristocratic British women between 1760 and 1860 (Lewis 1986). There is little reason to doubt Stone's assessment that prior to the use of contraception, the wealthy had larger completed family sizes than the poor (Stone 1977: 64).[8] Even as late as the nineteenth century in parts of Sicily, the landed aristocracy using wet nurses produced significantly more children (an average of 7 per family) and children were born at shorter intervals (2 years) than was the case for families lower in the social hierarchy and among those forego-ing the services of wet nurses (4 children on average, with a birth interval of 4.3 years) (Schneider and Schneider, 1984).

D. Different Fitness Trade-offs In The Social Transformation of Commercial Wet-nursing In Europe

Shorter birth intervals, greater fertility and high infant survivorship were outcomes of wet-nursing for families at the higher end of the social scale whose power and resources permitted them to engage or enforce the highest quality wet nurses and have them perform their services under close supervision in the parents' place of residence. In time, wet nurses themselves may have become a sort of status symbol which members of the sub-elites struggled to retain, even if it meant sending their infants out to wet nurses further and further from the parental abode. Klapisch-Zuber (1986a) documents this transition from a primarily upper class to a middle class child rearing practice for Renaissance Italy. Using data gleaned from domestic diaries or "ricordanze" she shows that between 1302 and 1399 only two of fifteen fathers who put children out to nurse did not come from prominent families. After 1450, however, half the families were of modest rank. From the middle of the fifteenth century onward, nursing by

---

[8] But note that for the reasons given in footnote 7, Stone would probably not concur with my assessment of the role of wet nurses in maintaining this differential – at least not for Britain.

a paid wet nurse or slave was the norm for all but the poorest women. "It is probable" writes Klapisch- Zuber "that the demand for nurses among proper Florentines motivated the popolo minuto – particularly those who ... had professional dealings with the merchants and the families of the highest society – to dispatch their children to the country as soon as they were born so that they could offer the wife's milk to the burghers who found it unthinkable that their own wives be allowed to breast-feed." (1985:138). The practice spread not only by direct emulation, but also through "ricochet" effects as wealthier babies displaced poorer ones and so on down the line.

It seems possible then that this upper class practice, increasingly adopted by sub-elites, served as a model for artisans and marginal members of the bourgeoisie, who for quite different reasons – in order to maintain the wife's labor within the family business – adopted what had begun as an elite pronatalist strategy. Workingwomen began to purchase wet-nursing services from still poorer women (e.g. see Sussman 1982 for France). Garden's data for Lyon documents the use of wet nurses in order to retain a mother's labor. The more involved the wife was in helping her husband in the butcher shop or in silk-making, or other trades – the more likely the family was to use wet nurses (1970:137).

By this point, historical links with traditional patterns of reciprocal wet-nursing as in the Solomon Islands or pygmy cases have become very remote. Enforced wet-nursing such as the exploitation of slaves in Classical and Oriental examples provides a more apt precedent for commercial wet-nursing in pre-modern Europe.

One striking feature of commercial wet-nursing to emerge from the French and Italian case studies is the linear relationship between extent of parental outlay (cost of wet nurse) and infant survivorship (Figure 1). This finding is consistent with the idea that parents were using wet nurses to lower the overall cost of producing infants, but it is important to note that the nature of the costs may have been quite different from group to group. For the elites engaged in this game early on, wet nurses meant shorter birth intervals, and greater reproductive success (since higher fertility was not offset by greater infant mortality). As even the elites confronted greater and greater competition for suitable wet nurses, later born children and non-heirs would have to settle for inferior care. The amount that parents were willing to spend for wet nurses depended on the reproductive value of the nursling. That is, given the system of primogeniture, what was this son's prospect for translating parental investment into subsequent reproduction? Given a system in which dowries were larger for daughters lower in the birth or-

der, what were this daughter's prospects? The firstborn of each sex was likely to beat an advantage over same-sex siblings, and sons favored over daughters.

E. The Role Of Post-partum Sex Taboos

Among the elite, the use of wet nurses clearly circumvented lactational amenorrhea in the mother. Subsequently it seems likely that wet nurses also served to free mothers for status-enhancing social functions. The process by which ambitious women obtained favors for their families by service at court was described in writings from the period (e.g. La Fayette, 1678). An additional function of wet nurses for this group may have included circumvention of postpartum sex taboos, though as yet the sequence and interaction of these factors is not well understood.

Historians tend to regard postpartum sex taboos as a uniquely European phenomenon, but strictures against husbands sleeping with lactating wives are in fact very widespread in traditional societies. These taboos can be documented in cultures as geographically diverse as the Eipo of New Guinea, the !Kung and Herrero of South Africa, the Mende and other tribes of West Africa, the Yanamamo, Nambikwara and other South American tribes as well as the North American Sioux (Schiefenhovel, 1989; Howell, 1979; Pennington, 1989; Isiugo-Abanihe, 1985:72, n. 8; Early and Peters, 1990).

This custom, so common in traditional societies, can be documented in Europe at least by the 2nd century A.D. In his medical writings, Galen ordered all nursing women to abstain completely from sexual relations. The Alexandrian physician Soranus, writing around the same period, maintained that "coitus cools the affection toward nursling by the diversion of sexual pleasure and moreover spoils and diminishes the milk or suppresses it entirely ...by bringing about conception" (cited in Bradley, 1980:322). Similar versions of this postpartum sex taboo persisted through the 18th century in Britain, France and other parts of Europe, although there is some question as to how strictly parents abided by these taboos (Fildes 1986:105). Nevertheless, one reason why the Catholic church so strongly supported the practice of sending babies to wet nurses was that the use of wet nurses (who were also not supposed to be sleeping with their husbands) permitted the women who hired them "to provide for the frailty of her husband by paying the conjugal due" (cited in Fildes 1986:105). (Note that the position of the Catholic church may help explain why wet-nursing was more common in France than in primarily Protestant England).

A common belief associated with postpartum taboos is the notion that the milk of a mother who has intercourse will damage any infant (her own or others') that partakes of it. How these taboos came into existence is simply not well understood but, once practiced, would clearly constrain a couple's sex life. Given that frequent suckling over a 24 hour period would, in any event, contribute to long inter-birth intervals (Konner and Worthman, 1980) the occasionally stated explanation that the taboo was used to prevent closely spaced births seems redundant and impractical. If post-partum sex taboos were really intended to protect babies, it would have made more sense to keep babies with their mothers but permit them to suckle often enough both night and day to suppress ovulation (Konner and Worthman, 1980). It seems an odd premise that babies were sent to wet nurses – often with very detrimental consequences – in order to spare them ingesting the spoilt milk of a sexually active mother.

If protection of the baby was really the point, breaking the taboo seems so much more practical. For this reason, male pronatalist sentiment provides a more convincing explanation for husbands to ship babies out of the house as soon as possible. And indeed, whether for Britain, France or Italy there is little doubt that husbands played key roles in insisting on the use of wet nurses and in lining up the desired service, sometimes seeing that the baby was removed from the house almost before the mother could see it (e.g. the 18th century observations of Madame Rolande, cited in Sussman 1982:80; Klapisch-Zuber 1985:143). Whatever its cause, this early separation of mother and infant would inhibit the bonding of mother with infant. Mothers of the new mother, or mothers-in-law could play key roles in this process, criticizing new mothers who insisted on nursing their own baby (Lewis 1986:61).

Is it possible that the widespread postpartum sex taboo is a cultural outgrowth of male (and lineage) pronatalist interests? The prevalence of these taboos combined with the absence of any obvious rationale for believing that sex spoils milk (although a subsequent pregnancy might, Palloni and Tienda 1986:31) makes this a problem that clearly deserves further research. In any event, whether it was to promote fertility among the elite, or to preserve the woman's labor for family ends among the artisan class, the outcome of wet-nursing was the same: shorter birth intervals for the paying mother, far longer ones for the paid nurse. This higher fertility among elites would have come linked to high survivorship of their offspring, but be linked to high infant mortality among workers. The paid or conscripted wet nurse in this system would suffer both lower fertility and higher mortality for her offspring.

22 S. B. HRDY

F. Differential Treatment Of Wet-Nursed Charges

Among elites, artisans and peasants, parents used wet nurses to reduce the labor-intensive task of nurturing slow-maturing human young. But for each group, the costs and benefits were quite different. Furthermore, the strategies pursued were changing through time in response to social and economic competition, costs of superfluous children (e.g. the increasing cost of providing dowries for elite daughters, see Klapisch-Zuber, 1986a: 215 for Renaissance Italy; Boone, 1986 for Medieval Portugal) and fluctuations in the availability and cost of wet nurses and other resources that mitigated the costs of parenting.

Elites for example were clearly fine-tuning investment in children in line with quite specific social and reproductive needs. One family out of three in Klapisch-Zuber's Florentine sample was more likely to keep sons than daughters at home, to nurse them longer, and to use a higher quality wet nurse. Twenty-three percent of boys were entrusted for relatively long periods to a wet nurse who lived in casa, compared to 12% of girls (1985:138). Conversely, 69% of the daughters born compared to 55% of sons, were sent to wet nurses in the country. Assuming that the nurse in casa cost 18-20 fiorini annually, compared to 9-15 for a nurse in the country (Klapisch-Zuber 1986a:136) parents are clearly paying more for sons. Furthermore, parents were more likely to wean daughters abruptly while paying extra for sons to enjoy a supplementary transition period, which may in part account for the fact that boys spent on average one and one half months longer at wet nurses than daughters did (1985:Table 7:7). Similar preferences can be documented for first versus later-born children.

Given such prejudices, it might appear at first glance surprising that infant boys sent out to wet nurses did not survive any better than girls. In fact, the tendency was slightly in the opposite direction with 18.1 percent of 144 boys and 15.8 percent of 139 girls in Klapisch-Zuber's sample dying. However if we take into account that the primary male heirs were kept at home and not sent out to begin with, these high mortality rates would be impinging upon sons who were already designated "heirs to spare".[9]

Strategic allocation of resources to children vary according to circumstances. Furthermore, within families such strategies might be altered over time in line with specific events. The life of Charles Maurice de Talleyrand-Perigord (1754–1838), the French diplomat and statesman, provides a poignant case in point. Talleyrand's ancient and powerful, but not overly wealthy family, with one son in hand, sent their second son to a less ex-

---

[9] We can not of course rule out the alternative hypothesis that these sons were simply more vulnerable than daughters to harsh conditions.

pensive wet nurse in the suburbs of Paris. Unfortunately, the firstborn son died, and when his parents sent for young Charles Maurice, they learned that sometime between his third or fourth year, the surplus son had fallen from a chest and injured his foot, rendering the newly needed heir crippled for life. Nevertheless, young Talleyrand was brought home to be groomed for his future role. But when his mother gave birth to yet another son, a family council was convened. In the interests of the family, it was decided that the crippled child should forfeit the right of primogeniture. Once again relegated to a secondary position, Talleyrand was educated for a career in the church – a vocation he soon abandoned. The rest, as they say, is history.

IV. CHOOSING BETWEEN ALTERNATIVE MEANS TO REDUCE PARENTAL EFFORT.

If one accepts that wet-nursing, fostering out, abandonment, and infanticide all function to reduce outlays of parental effort (albeit with potentially quite different outcomes for the infant), we are still left with the question of why parents opt for one solution rather than another. Why for example is fostering-out of weaned babies the choice of some 30 to 40% of West African mothers,[10] while among their 18th century French counterparts, parents were electing en masse to hire paid wet nurses? Why do some 20 to 40% of infant births end in infanticide among some Amazonian and Papuan tribes, while abandonment of infants accounted for comparable mortality in pre-modern Europe?

Sociobiologists have been able to identify several predictors for the retrenchment of parental solicitude. They focus on such relationships as the high reproductive value of the mother combined with poor current prospects; low potential for paternal support in environments where such support is critical; or poor prospects of either productive or reproductive returns from investment in the infant (Alexander 1974; Daly and Wilson 1980 and this volume; Hrdy 1987; 1990; Hill and Kaplan 1988).

These predictors probably apply to all types of retrenchment discussed here, and the same "ultimate explanations" may be invoked. For example, Bugos and McCarthy have shown for South American Ayoreo, and Daly

---

[10] One out of three Ghanian women and forty percent of Liberian women between the ages of fifteen and thirty four had a child living in another household. Forty six percent of Sierra Leonian women aged thirty to thirty-four fostered children out. These figures derive from interview data reported in Isiugo-Abanihe, 1985; see also Page, 1989:Table 9.1 for similar data from Cameroon, Lesotho and Ivory Coast, and slightly lower levels of fostering out for Kenya, Nigeria and Sudan.

and Wilson have shown for contemporary North Americans that mothers with high reproductive values, with many additional years of potential reproduction ahead of them, are significantly more likely to commit infanticide than are older mothers nearing the end of their reproductive careers. In a West African setting where infanticide is very rare, this same group of young, high reproductive value mothers, is significantly more likely to send infants to foster homes (Isiugo Abanihe 1985:Table 4 and p. 67). Purely from the perspective of maternal workload, one might expect higher parity birth order mothers, those with a number of children already, to send children away. The fact that it is instead lower parity mothers who reduce their investment in children is consistent with the hypothesis that these young women are keeping open options for future reproductive opportunities. Similarly, lack of male support is a "risk factor" for infanticide for children born in Amazonian and New Guinea societies (see Bugos and McCarthy 1984 for the Ayoreo; Hill and Kaplan 1988 and Hill in prep. for the Ache; Schiefenhovel 1989 for the Eipo) and is also a "risk factor" for fostering in African societies where paternal support is needed. Pennington for example found for patrilineal Herrero pastoralists in Southern Africa that unmarried mothers were nearly twice as likely as those in a stable union to send children to foster mothers (1989).

These parallels illustrate a highly facultative maternal response system that varies in line with life-history stage and socioenvironmental conditions. Examining this response system from a sociobiological vantage may help us understand why a mother would abandon her infant to a foundling home rather than continue to invest heavily by nursing the infant herself.[11] But we must seek explanations at a different level for why a mother chooses to leave her infant at a foundling home versus a relative's rural hut, or why she abandons an infant versus fostering him out? Why bury alive versus abandon, and so forth. Such decisions are made within the framework of imaginable and available options, as well as social and environmental constraints. Hence by identifying constraints which prevent people from selecting alternative tactics for reducing parental effort, we go a long way towards understanding the proximate causes of infanticide. Here sociobiology meets the traditional concerns of cultural ecologists and historians. Or rather, traditional but with an important new twist. Instead of the "group", the focus is upon the decisions made by individu-

---

[11] Although it is a commonly held view, the notion that the facultative withholding of maternal investment proves that maternal responses must be "socially constructed" rather than "biological" in origin (e.g. Badinter, 1980; Scheper-Hughes, 1985) is based on a misunderstanding of what an evolutionary approach means (Hrdy, 1990).

als in accordance with their assessments of maternal survival or lineage prospects.

A. Abandonment Versus Infanticide: Opportunities And Constraints

In his 1988 book *The Kindness of Strangers*, John Boswell documents a widespread traffic in European babies from Antiquity to the Renaissance, as couples without children, and perhaps especially slave-dealers, gathered up those infants that their parents did not choose to rear. High population densities meant that even if no one in the mother's immediate vicinity was known to want her baby there might well be "strangers" willing to rear the foundling. Furthermore, compared to tropical forests with stinging insects and Amazonian jaguars and other predators, there was a reasonable chance that an exposed infant might survive long enough to be found. By the end of the Middle Ages, infant abandonment had become so widespread in the West, that throughout Europe public institutions were formed to cope with this epidemic of foundlings whose supply now clearly exceeded demand.

Between the thirteenth and sixteenth century charitable groups (whose motivations are outside the scope of this paper) set up hospitals and foundling homes. In one particularly well documented case, the commune of Florence, together with the local silk guild, joined to build an asylum called Santa Maria degl'Innocenti. By 1445 the doors were opened to a small flood of "innocents". No doubt numbers of abandoned children rose and fell with economic conditions, but the numbers also increased in response to the opportunity to reduce parental effort without necessarily killing infants that was created by these institutions.

In what was a fairly general pattern for such institutions in continental Europe and Russia, the Innocenti in its early years was a fairly benign environment for infants. Death rates during the first year of life were around 26 percent for 1445, down to 23 percent during the Innocenti's second year (compared to ca. 21% for the population at large). However, by 1448, as the Innocenti became a magnet for abandoned babies from all over the dominion, these rates doubled to 53.6% mortality. By 1451, six years after the institution opened, death rates soared to 57.6% (Trexler 1973b:Table V).

At the outset, in "a hospital of minimum crowding, and with a sufficient supply of *balie* (wet nurses), spared famine and pestilence, the first innocents had as good a chance as any children" (Trexler 1973b:276). And many foundlings left at the Innocenti would definitely have been better off there.

These were the illegitimate children of slaves and servants who, according to Trexler, would have died at triple the rates of legitimate children. The decision by a mother to deposit her baby in the *tour* (rotating barrel) at the foundling home could be construed as in the baby's best interests.

But what of the same decision once mortality rates in foundling homes reached the catastrophic levels which, almost inevitably they eventually did, as more and more parents made the same choice? (See Trexler 1973b for Italy; Dupoux 1958 for France). Dupoux's statistics for Parisian foundling homes at the end of the 18th century indicate that 92% of these children died before their eighth birthday. Death rates of 70-85% were not unusual. Were parents depositing their infants in the *tours* aware of the prognosis for survival of the abandoned baby? Surviving documents from 15th century Florence strongly suggest that parents depositing children at the Innocenti certainly were not only aware of the risks, but were capable of making shrewd assessments concerning the survival chances of a baby kept in Florence versus a baby farmed out by the foundling home to distant wet nurses. "Some parents, in abandoning their child to the Innocenti pleaded that the hospital keep it and not send it to an outside nurse..." (Herlihy and Klapisch-Zuber 1985: 147).

Parents would deposit their infants with various mementos and identifying signs – an indication that they were gambling on a good outcome and that they harbored some hope of one day retrieving their child. Nevertheless, at some point, parents must have become aware of the high death rates suffered by the babies they abandoned.

Volker Hunecke records case studies from 18th and 19th century Milan of a tailor "Filippo A..." who keeps his first son and then deposits the next six (in the space of five and a half years) at the nearest tour. When his first wife dies, Filippo remarries a certain "Cecilia B..." who leaves at the *tour* five infants in five years. A year and a half later, Cecilia tries to retrieve them, but only two had survived long enough for her to be able to. "Francesco G..." and his wife "Amalia S..." similarly produced twelve infants in thirteen years. The first of these died shortly after birth. All the others were left at foundling homes, and only one of them, a girl, survived.

The point here is that these outcomes were not a secret. Parents were in a position to have some sense of the high mortality, and were in a position to communicate what they knew to their neighbors. Granted that it is difficult even for trained social scientists to obtain accurate estimates of infant mortality (and these were after all illiterate people with neither time nor facilities to study the situation), still I don't think we can assume that many parents remained completely ignorant of the prospects for their abandoned children. Yet even if parents were aware of high mortality,

it does not mean that parents opting for the *tour* were merely seeking a legal way to kill an infant. More plausible is the hypothesis that parents were making their calculations based on immediate costs (mother's lost employment; cost of a wet nurse). This information was likely to carry more weight than rumored events behind distant walls. Indeed, Garden's information for silk-makers and butchers in 18th century Lyon suggest that whether or not the mother helped in the family business was a better predictor of whether or not babies were sent to wet nurses than were mortality rates.

Given the long history of child abandonment in the west (or indeed, currently tolerated child mortality rates in industrialized countries like the U.S. today, Gibbs 1990:43) there is a certain irony about the readiness of people from "civilized" (e.g. western-oriented) backgrounds to condemn infanticide among tribal societies. For as I will argue in the next section, infanticide has far more to do with family structure, ecology, and the absence of alternative means of mitigating parental investment than it does with morality.

B. High Versus Low Rates Of Infanticide.

Almost everywhere, infanticide is an uncommon event and tends to be poorly documented. By and large, reviews of infanticide in human societies have used ethnographic accounts to conduct surveys to try to determine whether infanticide is present or absent for a particular culture. Where infanticide occurs, special attention is given to the stated circumstances (Dickeman 1975; Daly and Wilson 1984). In the most extensive such review, Daly and Wilson took a representative sample of sixty cultures described in the Human Relations Area Files; infanticide was reported for 39 of these, and in 35, the circumstances surrounding at least some cases were known. They were able to identify sets of circumstances that pertained to virtually all reported cases; these circumstances were compatible with their sociobiological analysis: the infant killed was probably sired by a man other than the woman's current mate; the infant was defective or considered to be of poor quality or else was one of a pair of twins; there was a problem in the timing of the birth (short interbirth interval); or else for some other reason (mother dead, no male support, poor economic conditions) parental resources were inadequate to rear the child. At the time of these surveys, virtually no data existed on the frequency of infanticide.

Scarcity of information is exacerbated both by the discomfort or grief that perpetrators feel in discussing infanticide (Bugos and McCarthy 1984)

Table 1 Infanticide in nine traditional societies in Africa, Amazonia and New Guinea where available information permits the calculation of rates of infanticide for a sample of liveborn infants.

Culture and location	Subsistence type	No. reported cases infanticide/ no. live births	Proportion infant mortality due to infanticide	Source
1. EFE Ituri Forest, Zaire	specialized hunter-gatherer	≈ 0/530 (≈ 0%)	≈ 0	Bailey 1989; and pers. com. from Bailey and Peacock
2. LESE Ituri Forest, Zaire	horticulture	≈0/777 (≈0%)a	≈0	Bailey 1989; and pers. com. from Bailey and Peacock
3. DATOGA N. Tanzania	pastoralism	0/762 (0%)	0	Borgerhoff Mulder in prep.
4. KIPSIGIS S.W. Kenya	agro-pastoralism	0/2,190 (0%)	0	Borgerhoff Mulder in prep.
5. SAN Kalahari Desert, Botswana	hunter-gatherer	6/500 (1%)	3%	Howell 1979
6. MUCAJI YANAMAMO N. Brazil	horticulture and hunting	17/283 (6%)	44%	Early and Peters 1990
7. ACHE, Paraguay	hunter-gatherer,	26/223 (12%)b,c 11 males (.09) 15 females (.16)	39%c	Hill in prep. and pers. comm.
8. AYOREO S.W. Bolivia and N. Paraguay	horticulture and foraging	54/141 (38%)d 31 males (.41) 16 females (.27)	unknown	Bugos and McCarthy 1984
9. EIPO Highland Central New Guinea	horticulture	20/49 (43%)e 5 males (.21) 15 females (.60)	81%e	Schiefenhovel 1989: Fig 10.8

Footnotes to Table 1

a. It is not possible to state with certainty that infanticide never occurred. It was suspicious for example that in these 777 births, only two sets of twins were reported, and in both cases only one twin survived (one was stillborn, the other twin died shortly after birth). Nevertheless for the purposes of this paper the Efe and Lese qualify as groups with very low rates of infanticide.

b. These 26 Ache cases include children who are killed up to five years of age, and some of these cases involve non-parents.

c. These figures are only for the last decade prior to peaceful contact (1960-70), and represent only a fraction of Hill's total data base. Hill believes that data from this period provide the most accurate estimate of infant mortality since his data suggest an increasing tendency not to report infants who die in cohorts born in the more distant past.

d. This is an inflated rate because Bugos and McCarthy only included in the sample women known to commit infanticide. Apparently most women committed infanticide, but from the article it is impossible to say how many women were left out. Note that sex of infant was unknown in seven cases.

e. These data are for the period 1974-1978. Under mission influence, infanticide rates fell to 10% after 1978.

and the prospect of disapproval by public or religious bodies, or the prospect of legal sanctions. In modern Brazil, a woman may abandon or neglect her infant, thereby indirectly killing it, but if she commits infanticide, she is imprisoned (Scheper-Hughes 1985). This situation is currently complicated by the accusation that anthropologists who attribute violent practices (and infanticide is included as one of these violent practices) to traditional peoples are in fact playing into the hands of forces who wish to manipulate or eliminate these tribes (see Booth 1989 for South American case).[12] For this reason, several of those anthropologists who now for the first time actually have quantitative data on infanticide are reluctant to publish them, and have requested that I delete their data from this paper.

In spite of these difficulties, a limited amount of data have emerged over the last decade, that permit us for the first time to move beyond largely anecdotal ethnographic accounts and examine actual rates. Hence, it is now possible to compare populations with very low rates of infanticide, approaching zero (epitomized here by the various African cases) with those exhibiting high rates of infanticide.[13] Table 1 summarizes information from nine traditional societies in which it has been possible for anthropologists to record the proportion of live births that were killed, yielding minimum rates of infanticide. These rates range from zero or near zero in

---

[12] The notion of violence is apparently crucial to the moral condemnation of infanticide, as lethal levels of child neglect in urban and shantytown areas of the same part of the world (Scheper-Hughes, 1985; Gurson, 1990) are not so condemned.

[13] There is an obvious reporting problem here. Anthropologists tend to go to greater lengths to determine that neonatal deaths are or are not due to infanticide in societies where infanticide is thought to be an important phenomenon or in areas where it is already the subject of debate – as it is for Amazonia (e.g. the famous debate begun by Divale and Harris, 1976 and still ongoing, Chagnon *et. al.* 1979; Early and Peters, 1990). Where infanticide rates are near zero, the matter is not pursued in demographic interviews with the result that investigators then hesitate to publish a rate of infanticide: they assume it is very low, but don't know it for sure.

African hunter-gatherer, horticultural and pastoralist groups and one percent among San hunter gatherers in the Kalahari, to the extraordinarily high rate of 41% of live births among the Eipo.[14] Interviews by ethnographer Wulf Schiefenhovel among the (up to that point) largely uncontacted Eipo tribespeople in the West New Guinean mountains revealed that 31 of the 42 infants killed in the period 1974-1978 were female. These data combined with interview data make it clear that the desire for sons is implicated in this very high rate.[15] Preference for sons is also apparent in the Ache data where sixteen percent of daughters born and nine percent of sons were killed. Other Amazonian groups exhibited much lower rates – five to six percent of live births without such marked son preference. The Ayoreo rate is inflated upwards, not by son preference but by several other factors. In particular, Bugos and McCarthy (1984) only included women in their sample who had committed infanticide, and it is not clear exactly how many (if any) non-infanticidal mothers were excluded.

These high rates of infanticide for New Guinea and Amazonia are consistent with the general ethnographic literature for these areas. That is even though rates were not available, anthropologists were by and large aware that infanticide was going on (e.g. Neel and colleagues estimated that infanticide was occurring at a rate around 15-20 percent among the Yanamamo, Neel 1970). This situation contrasts markedly with the published literature on Africa, where infanticide is rare and largely confined to destruction of defective offspring or twins (Granzberg 1973), although other circumstances are sometimes also cited (see review in Daly and Wilson 1984). For large segments of sub-Saharan Africa, an area where there is an exceptionally high desire for parenthood and a real horror of subfertility (Page 1989), infanticide is unthinkable. Several African ethnographers I asked about infanticide either have no information on its occur-

---

[14] The Eipo and Ache rates may seem unbelievably high to some. For example, in a now famous computer simulated analysis, Schrire and Steiger (1974) demonstrated that if even eight percent of female births are terminated through infanticide the practice will lead to extinction of the population. And indeed, few would like to argue that such high infanticide rates represent a stable situation. Nevertheless, one flaw in such critiques is the assumption that children killed and those kept have equivalent survival rates. Children killed are often those whose survival prospects are in any event compromised.

[15] If the data were comparably quantitative, rates of infanticide per live birth would probably be at least as high for those areas of early China and North India which practiced female-biased infanticide. In particular, in some 19th century North Indian clans, no daughter was ever allowed to live (Cave-Brown, 1857; Parry, 1979; Miller, 1981; Dickemann, 1979). Obviously, such stringent preference for sons would yield infanticide rates on the order of 50%. In spite of the greater public outcry today (Hull, 1990; Rao, 1986) the contemporary practice of female feticide and infanticide in India and China are almost certainly lower than historically they have been.

rence ("the rate is zero") or point out how puzzled their informants would be by the notion (personal communications from R. Bailey for Lese, and M. Borgerhoff Mulder and Lee Cronk for Nilotic Kipsigis and Mukogodo (Ma-speaking) people). The main exceptions to this would be for Africa's nomadic gathering people, who regard infanticide as a mother's right even if it is not one commonly exercised. In the words of demographer Nancy Howell who studied the Khosian-speaking San hunter-gatherers: "infanticide is part of a mother's prerogatives and responsibilities, culturally prescribed for birth defects and for one of each pair of twins" (1979). Of six San cases Howell knew about (out of 500 livebirths) two involved low probability of male support. Anecdotal reports of infanticide in early ethnographies for the Masai, Bemba, Lozi and other African tribal groups (cited in Daly and Wilson 1984) appear to involve similar circumstances. A sense of the cultural difference between San and Bantu in respect to infanticide is conveyed by Nancy Howell's account of a San woman who gives birth to a defective infant. Although traditionally, the delivering mother would have been alone, on this occasion, Bantu women were present. The mother of the defective infant felt it her duty to dispose of the infant, but the Bantu women prevented her (Howell 1979:119-120).

For the present, I accept these findings at face value, and conclude that by and large infanticide is not a salient feature in the lives of many Bantu and other African peoples. The key I think to the general low incidence of infanticide over much of Africa is linked to the same set of factors that lead to the persistent high fertility in contemporary (and presumably also traditional) Africa (Caldwell and Caldwell 1990): 1) children are highly desired for symbolic reasons involving ancestor worship and perpetuation of the lineage; 2) they reportedly cost their parents, particularly their fathers, very little to rear (though this is changing now with more emphasis on the need for education and payment of school fees); and 3) such costs as there are, are borne by the mother and by an assortment of caretakers, the infant's older sibs, real and fictive "grannies", and other patrilineal and especially matrilineal kin. Indeed, the Caldwells' claim in their famous argument (1982;1990) that "wealth flows" from children to parents they claim that children eventually become net assets to parents although to date there are few empirical studies to support this (for an exception see Cain, 1977 for a population in Bangla Desh). Matrilineal social organization combined with female-centered horticultural practices mean that by and large male investment is not critical for child survival and well-being at the same time that the mother's social network makes available to her a range of options for delegating some of the necessary caretaking to other people – either older siblings and other related caretakers (Weisner, 1987;

Draper, 1989) or fostering adults (Page, 1989; Isiugo-Abanihe, 1985). A fifth reason, would be that few of these groups exhibit any strong preference for sons, since daughters are often valued for their labor and the bridewealth they bring,[16] obviating any pressure for sex-biased infanticide.

The African cases contrast markedly with primarily patrilineal and virilocal horticultural/hunting/fishing societies in South America and New Guinea. Male protection and support are essential for the well-being of children, and orphans or children of inappropriate paternity are at high risk of dying before adulthood. Among the Ache horticultural hunters of Venezuela, Hill and Kaplan demonstrate that children whose reported biological fathers die before the children reach 15 years of age are significantly more likely to die (43.3% of 67 such children) compared to children whose fathers remain with the mother (only 19.3% of 171 such children) (Hill and Kaplan 1989:298). Such children are at high risk from being killed by the mother's subsequent mate. Furthermore, because of their poor prospects, infants no longer under the protection of the acknowledged father are at risk of being eliminated by the mother herself (e.g. see Murphy and Murphy 1974: 166 for Mundurucu case;[17] Bugos and McCarthy 1984 for Ayoreo; Hill in press for Ache). In contrast to Africa, men in the South American cases provide the bulk of protein and calories for the village, and children without a male protector are discriminated against (Hill, in press).

As described by Bugos and McCarthy (1984), Ayoreo mothers – caught in transition between war and missionized settlement – confronted es-

---

[16] These generalizations, entrenched as they are in the literature on Africa, should nevertheless be regarded with caution (Turke, 1989). Following the Caldwells, Draper, Page and others I stress the high value placed on children and the prevalence in Africa of fostering. Yet there are signals in the literature that this story may be more complicated. Le Vine and Le Vine (1981) cite a Gusii saying that "another woman's child is like cold mucus" referring to something unattractive which clings. Furthermore, the Le Vines reported that five of eight children under the age of five who died were either illegitimate or being reared by grandmothers. In short although there is general agreement that African children are being reared by a far-flung assemblage of relatives and childless non-relatives, and that this "complex web of dependency weakens the relations between the number of children the woman bears and the number she supports" (Caldwell and Caldwell, 1990), we need more information on precisely which individuals comprise this web, and how much each actually provides to their charges (as proposed by Turke for Ifaluk). At the same time, unpublished information for Yanamamo and other South American tribes (personal communication from Napoleon Chagnon) indicates that there is a great deal of adoption of unwanted children by relatives which may in fact resemble fostering.

[17] In contrast to other Amazonian peoples discussed here, the Mundurucu were probably originally patrilineal and patrilocal, but in the ethnographic present are patrilineal and matrilocal, a situation probably brought about by recent adoption of horticulture.

pecially difficult socioecological conditions which exacerbated and contributed to unstable marriages. One mother, "Asago", with poor prospects of male support from her first three husbands, buries at birth the first six of the ten children she will eventually bear in her lifetime. As extreme as this case is, Asago looses no more children to live-burial than did "Amalia S." to nineteenth century Milanese foundling homes. Hill reports similar (if less extreme) cases for the Ache. Many of these Ache children were either considered defective in some way or stood a high chance of eventually being murdered by a step-father or other male had the mother not eliminated them at birth (Hill and Kaplan 1989; and Hill in press). As in Medieval Europe, compassion for deformed, sick or unwanted children was not a luxury that traditional societies in South America could readily afford, and events which would strike contemporary Europeans, Americans or Africans as astoundingly callous are commonplace (e.g. Biocca, 1971 and Chagnon, personal communication for the Yanamamo; Hill in press for Ache). Among the Ache, children under the age of fifteen who are reared by a woman other than their biological mother suffer higher mortality rates (36.1% of 61 children) than do children reared in intact families (25% of 184 children died) (Hill and Kaplan, 1988:298). For children under two years of age, 100% of the four whose mothers died, also died (contrasted with 33% mortality for children under two whose mothers remained alive). The decision to terminate investment in a fatherless or motherless child, sooner rather than later, can be seen as rational. But why ever smother (e.g. Early and Peters, 1990:77 for Yanamama) or bury an infant alive (a very widespread practice throughout the Amazon, Gregor, 1985:89 for the Mehinacu; Hill, in press for Ache; Milton p.c. 1991 for the Arawete; Wagley, 1977:137 for the Tapirape)? Why this shift in emphasis away from such default strategies as abandonment to either relatives or to "the kindness of strangers"?

The answer I think must be a culturally mediated, and also practical, assessment of what use abandonment could possibly be. If in small, isolated villages, someone was going to take on responsibility for an unwanted child (see Biocca, 1971 for cases involving a grandmother, a sister-in-law and a captured woman) they would have made themselves known. Furthermore, fertility is high and people tend to have as many children as they want. In addition, stringent ecological conditions forestall abandonment. Whether lying on the forest floor or hanging from trees (the classical Greek and early European custom), no infant could survive long within the Amazonian context; any infant left unattended would soon die from the bites of stinging insects (see Hill and Kaplan, 1988; Hurtado et al., 1985) or from predation. Jaguars in this area are a major source of mortality – even for

adult males. An infant left in the forest would not only be doomed but would "condition" jaguars to a small human search image, increasing the predation hazard for wanted children as well. (A dissatisfied child may threaten parents with going off into the forest to be eaten by a jaguar, cited in Johnson 1981:60). If Amazonian infanticide rates seem incredibly high, one must take these ecological conditions into accounts.

Once a tradition of infanticide is developed, customs encouraging psychological distancing between mother and neonate become institutionalized. Hence, even as conditions become altered (e.g.by settlement) infanticide is more likely to remain in the cultural repertoire than if mother-infant bonding were encouraged from the outset. Beliefs which withhold full human identity to newborns until after some specific milestone or ritual (baby takes food; cries; receives a soul; receives a name – the traditional Greek "amphidromia" ceremony comes to mind) or customs which transfer the responsibility for survival to the infant (very different from our own culture where parents hold themselves responsible for infant survival) illustrate ways of looking at the world which facilitate infanticide. Hence even after ecological conditions have changed, or people have migrated into a completely new environment, parents may be culturally preadapted to infanticide, more predisposed to select infanticide as an option than parents in cultures where mother-infant bonding is promoted with little delay or where newborns are regarded as fully "human". Where psychological distancing from the newborn is culturally entrenched, parents are more likely to resort to infanticide as an option rather than inventing new alternatives for mitigating costs of parental effort. Where other alternatives (e.g. giving children away to relatives or unrelated childless adults) also exist, one or the other traditions may become more emphasized. For example, increased contact between native South Americans and outsiders from urban areas has created many more opportunities for indians to give children away. At the same time, infanticide has become less common (Bugos and McCarthy,1984). An obvious conclusion from this analysis is that high rates of infanticide are inversely correlated with alternative opportunities to reduce parental effort.

ACKNOWLEDGEMENTS

I thank M. Borgerhoff Mulder; J. Chisholm; M. Daly; K. Hill; D. Judge; Volker Sommer; and M. Wilson for valuable discussion and criticism. I thank R.C. Bailey, L. Cronk, K. Hill, K. Milton, M. Borgerhoff Mulder, and N. Peacock for permission to cite unpublished data, and Doris Miner for locating references. I am particularly grateful to Fred vom Saal whose

request for an over-view resulted in this chapter, and thank Fred, Stephano Parmigiano and Bruce Svare for organizing the Erice conference. Finally, I offer heartfelt thanks to Lupe de la Concha and Ann Meyer who allowed me to delegate some of my own maternal responsibilities.

REFERENCES

AKIN, K. GILLOGLY (1983) Changes in infant care and feeding practices in East Kwaio, Malaita. Paper presented at Symposium on Infant Care and Feeding in Oceania, Annual Meeting of the Association of Social Anthropology in Oceania, March 9-13, 1983. New Harmony, Indiana.

ALEXANDER, R.D. (1974) The evolution of social behavior. *Annual Review of Ecology and Systematics* **5**, 51-82.

ALTORKI, SORAYA (1980) Milk-kinship in Arab society: An unexplored problem in the ethnography of marriage. *Ethnology* **19(2)**, 233-243.

AUERBACH, KATHLEEN. G. (1981) Extraordinary breast feeding: Relactation/ induced lactation. *Journal of Tropical Pediatrics* **27**, 52-55.

AUERBACH, KATHLEEN. G. and Jimmie Lynne Avery (1981) Induced lactation: A study of adoptive nursing by 240 women. *American Journal of Diseases of Children* **135**, 340-343.

BACHRACH, C.A., Stolley, K.S. and London, K.A. (1992) Relinquishment of premarital births: Evidence from national survey data. *Family Planning Perspectives* **24(1)**, 27-32.

BADINTER, E. (1980) *L'amour en plus*. Flammarion, Paris.

BAILEY, R.C. (1989) The demography of foragers and farmers in the Ituri forest. Paper presented at the 88th Annual Meeting of the American Anthropological Association. Washington, D.C.

BELSKY, J. and L. D. Steinberg (1978) The effects of daycare: A critical review. *Child Development* **49**, 929-49.

BIOCCA, ETORE (1971) *Yanoama: The Narrative of a White Girl Kidnapped by Amazonian Indians*. From a narrative provided by Helena Valero, and translated from the Italian by Dennis Rhodes. E. P. Dutton, New York.

BLEDSOE, C. and U. Isiugo-Abanihe (1989) Strategies of child fosterage among Mende grannies in Sierra Leone. In: *Reproduction and Social Organization in Sub-Saharan Africa*, R. Lesthaeghe (ed.), Berkeley: University of California Press, pp. 442-474.

BOONE, JAMES (1986) Parental investment and elite family structure in preindustrial states: A case study of late Medieval-Early modern Portuguese genealogies. *American Anthropologist* **88(4)**, 859-78.

BOONE, JAMES (1988) Second- and third- generation reproductive success among the Portuguese nobility. Paper presented at the 87th Annual Meetings of the American Anthropological Association. Phoenix, Arizona, November 1988.

BOOTH, WILLIAM (1989) Warfare over Yanomamo Indians. *Science* **243**, 1138-1143.

BOSLER, BARBARA (1990) Underpaid, overworked and from the Philippines. *New York Times*. August 26, 1990.

BOSWELL, JOHN (1988) *The Kindness of Strangers: The abandonment of children in western Europe from late antiquity to the Renaissance*. Pantheon, New York.

BRADLEY, KEITH (1980) Sexual regulations in wet-nursing contracts from Roman Egypt. *Klio* **62**, 321-325.

BRADLEY, KEITH (1986) Wet-nursing at Rome: A study in social relations. In: *The Family in Ancient Rome*. B. Rawson (ed). Cornell University Press, Ithaca.

BUGOS, PAUL, and Lorraine McCarthy (1984) Ayoreo infanticide: A case study. In: *Infanticide: Comparative and Evolutionary Perspectives*. G. Hausfater and S.B. Hrdy (eds.). Aldine de Gruyter, Hawthorne, New York, pp. 503-520.

CAIN, MEAD. T. (1977) The economic activities of children in a village in Bangla Desh. *Population and Development Review*. September issue, pp. 201-227.

CALDWELL, J. (1982) *Theory of Fertility Decline*. Academic Press. New York.

CALDWELL, J. and P. Caldwell (1990) High fertility in sub-Saharan Africa. *Scientific American*, May issue, pp. 118-125.

CAVE-BROWN, J. (1857) *Indian Infanticide: Its Origins, Progress and Suppression*. W.H. Allen, London.

CHAGNON, N. (1973) The culture-ecology of shifting (pioneering) cultivation among the Yanomamo indians. In: *Peoples and Cultures of Native South America*. D.R. Gross (ed.) Doubleday, Garden City, New York.

CHAGNON, N. (1983) *Yanomamo: The Fierce People*. 3rd edition. Holt, Rinehart and Winston, New York.

CHAGNON, N., M. Flinn, and T.F. Melancon (1979) Sex-ratio variation among the Yanomamo Indians. In: *Evolutionary Biology and Human Social Behavior*. N. Chagnon and W. Irons (eds.) pp. 290-320. Duxbury Press, North Scituate, Ma.

CRONK, L. (1989) Low socioeconomic status and female-biased parental investment: The Mukogodo example. *American Anthropologist* **91**(2), 414-428.

DALY, MARTIN and Margo Wilson (1980) Discriminative parental solicitude: A biological perspective. *Journal of Marriage and Family* **42**, 277-288.

DALY, MARTIN and Margo Wilson (1983) *Sex, Evolution and Behavior*. 2nd Edition. Willard Grant Press, Boston.

DALY, MARTIN and Margo Wilson (1984) A sociobiological analysis of human infanticide. In: *Infanticide: Comparative and Evolutionary Perspectives*. G. Hausfater and S.B. Hrdy (eds). Aldine de Gruyter, Hawthorne, New York. pp. 487-502.

DALY, MARTIN and Margo Wilson (1988) *Homicide*. Aldine de Gruyter, Hawthorne, New York.

DAS GUPTA, MONICA (1987) Selective discrimination against female children in rural Punjab, India. *Population and Development Review* **13**(1), 77-100.

DAWKINS, R. and T.R. Carlisle (1976) Parental investment, mate desertion and a fallacy. *Nature* **262**, 131-133.

DAY, C.S.D. and B.G. Galef (1977) Pup cannibalism: One aspect of maternal behavior in golden hamsters. *Journal of Comparative and Physiological Psychology*. **91**(5), 1179-89.

DELASSELLE, CLAUDE (1975) Les enfants abandonnés à Paris au XVIIIe siècle. *Annales: Economies, societés, civilisations*, **30**. Jan.-Feb. 1975, 187-218.

DEVRIES, MARTEN (1984) Temperament and infant mortality among the Masai of East Africa. *American Journal of Psychiatry* **141**(10), 1189-1194.

DICKEMANN, M. (1975) Demographic consequences of infanticide in man. *Annual Review of Ecology and Systematics* **6**, 107-135.

DICKEMANN, M. (1979) Female infanticide and reproductive strategies of stratified societies. In: *Evolutionary Biology and Human Social Behavior*. N. Chagnon and W. Irons (eds). Duxbury, North Scituate, Ma., pp. 321-367.

DICKEMANN, M. (1984). Concepts and classification in the study of human infanticide. In: G. Hausfater and S.B. Hrdy (eds). *Infanticide: Comparative and Evolutionary Perspectives.* Aldine de Gruyter, Hawthorne, New York, pp. 427-437.

DIVALE, W.T. and M. Harris (1976) Population, warfare and the male supremacist complex. *American Anthropologist* **78**, 521-538.

DRAPER, PATRICIA (1989) African marriage systems: Perspectives from evolutionary ecology. *Ethology and Sociobiology* **10**, 145-169.

DUPÛQUIER, J. and Lachiver, M. (1964) Sur les débuts de la contraception en France ou les deux Malthosianismus. *Annales E.S.C.* **6**, 1391-1406.

DUPOUX, ALBERT (1958) Sur les pas de Monsieur Vincent: Trois cents ans d'histoire Parisienne de l'enfance abanonnee. *Revue de l'Assistance Publique*, Paris.

EARLY, J.D. and J.F. Peters (1990) *The Population Dynamics of the Mucajai Yanomama.* New York: Academic Press.

FILDES, VALERIE A. (1986) *Breasts, Bottles and Babies: A History of Infant Feeding.* Edinburgh University Press, Edinburgh.

FUCHS, RACHEL (1984) *Abandoned Children: Foundlings and Child Welfare in Nineteenth Century France.* State University of New York, Albany.

FUCHS, RACHEL (1987) Legislation, poverty and child-abandonment in nineteenth-century Paris. *Journal of Interdisciplinary History*, **XVIII: 1** (Summer 1987), 55-80.

GALLIANO, PAUL (1966) La mortalite infantile (indigene et nourrissons) dans la banlieue sud de Paris a la fin du XVIIIe siecle (1774-1794). *Annales de Demographie Historique.* Paris, Editions Sirey, 1967.

GANDELMAN, RONALD and Neal G. Simon (1978) Spontaneous pup-killing by mice in response to large litters. *Developmental Psychobiology* **11**(3), 235-241.

GARDEN, MAURICE (1970) La démographie lyonnaise: l'analyse des comportements. *Lyon et les Lyonnais au xviii Siecle. Bibliotheque de la Faculte des Lettres de Lyon. D'Edition "Les Belles Lettres"*, Paris, pp. 83-169.

GHIGLIERI, MICHAEL (1987) Sociobiology of the Great Apes and of the hominid ancestors. *Journal of Human Evolution* **16**, 319-357.

GIBBS, NANCY (1990) Shameful bequest to the next generation. *Time Magazine*, October 8, 1990, pp. 42-46.

GOODALL, JANE (1986) *The Chimpanzees of Gombe: Patterns of Behavior.* Harvard University Press, Cambridge.

GOODY, E.N. (1969) Kinship fostering in Gonja: Deprivation or advantage? In: P. Mayer (ed). *Socialization.* International African Institute, London, pp. 137-165.

GOODY, E.N. (1975) Delegation of parental roles in West Africa and West Indies. In: J. Goody (ed.). *Changing Social Structure in Ghana.* International African Institute, London, pp. 137-165.

GRANZBERG, G. (1973) Twin-infanticide – A cross-cultural test of materialist explanation.*Ethos* **1**, 405-412.

GREGOR, THOMAS (1985). *Anxious Pleasures: The Sexual Lives of an Amazonian People.* University of Chicago Press, Chicago.

GURSON, LINDSEY (1990) Remembering a tortured child who lived in the streets of Guatemala City. *New York Times,* October 14, 1990, p. 3.

HERLIHY, D. and Klapisch-Zuber, C. (1985) *The Tuscans and Their Families: A Study of the Florentine Catasto of 1427*, New Haven, Yale University Press.

HILL, K. (In prep) *Life history and Demography of Ache Foragers.* Aldine de Gruyter, Hawthorne, New York.

HILL, K. and H. Kaplan (1988) Tradeoffs in male and female reproductive strategies among the Ache: Parts I and II. In: *Human Reproductive Behavior*. L. Betzig, M. Mulder and P. Turke, (eds.) Cambridge University Press, Cambridge, pp. 277-305.

HOWELL, NANCY (1979) *Demography of the Dobe !Kung*. Academic Press, New York.

HRDY, S. BLAFFER (1976) The care and exploitation of nonhuman primate infants by conspecifics other than the mother. *Advances in the Study of Behavior* VI, 101-158.

HRDY, S. BLAFFER (1977). "The puzzle of langur infant-sharing" In: *The Langurs of Abu: Female and Male Strategies of Reproduction*. Harvard University Press, Cambridge.

HRDY, S. BLAFFER (1979) Infanticide among animals: A review, classification, and examination of the implications for the reproductive strategies of females. *Ethology and Sociobiology* 1, 13-40.

HRDY, S. BLAFFER (1987) Sex-biased investment in primates and other mammals. In: R. Gelles and J. Lancaster (eds.).*Child Abuse and Neglect*. Aldine de Gruyter, Hawthorne, New York, pp. 97-147.

HRDY, S. BLAFFER (1990) Sex bias in nature and in history. *Yearbook of Physical Anthropology* 33, 1-13.

HUFTON, OLWEN (1974) *The Poor in Eighteenth Century France 1750-1789*. Oxford University Press, Oxford.

HULL, T.H. (1990) Recent trends in sex ratios at birth in China. *Population and Development Review* 16(1), 63-83.

HUNECKE, VOLKER (1985) Les enfants trouvés: Contexte européen et cas Milanais (XVIIIe-XIXe Siècles). *Revue d'histoire moderne et contemporaine*. Vol XXXII, January-March 1985, pp. 3-29.

HURTADO, MAGDALENA (1985) *Women's Subsistence Strategies Among Ache Hunter-gatherers of Eastern Paraguay*. Ph.D. thesis, University of Utah.

IRELAND, E. (1989) Why some Waura are killed and not buried, and why some are buried and not killed. Paper presented at the 88th Annual Meeting of the American Anthropological association, Washington, D.C.

IRELAND, E. and T. Gregor (1986) Interring the "thing that created itself". Paper presented at 85th Annual Meeting of the American Anthropological Association, Philadelphia, December 1986.

ISIUGO-ABANIHE, UCHE C. (1985) Child fosterage in West Africa. *Population and Development Review* 11(1), 53-73.

JOHANSSON, S. RYAN (1987a) Centuries of childhood/Centuries of parenting: Philippe Aries and the modernization of privileged infancy. *Family History* 12(4), 343-365.

JOHANSSON, S. RYAN (1987b) Status anxiety and demographic contraction of privileged populations. *Population and Development Review* 13(3), 439-470.

JOHNSON, O. (1981) The socioeconomic context of child abuse and neglect in native South America. In: J. Korbin (ed.). *Child Abuse and Neglect*. University of California Press, Berkeley, pp. 56-70.

KEESING, ROGER (1970) Kwaio Fosterage. *American Anthropologist*. 72, 991-1020.

KLAPISCH-ZUBER, C. (1986a) Blood parents and milk parents: Wet-nursing in Florence, 1300-1530. In: *Women, Family and Ritual in Renaissance Florence*. University of Chicago Press, Chicago, pp. 132-164.

KLAPISCH-ZUBER, C. (1986b) The Griselda complex: Dowry and marriage gifts in the quattrocento. In: *Women, Family and Ritual in Renaissance Florence*. University of Chicago Press, Chicago, pp. 213-246.

KLEINMAN, M.D., Jacobson, Linda, Hormann, E. and Walker, W.A. (1980) Protein values of milk samples from mothers without biologic pregnancies.*The Journal of Pediatrics* **97(4)**, 612-615.

KONNER, M.J. (1976) Maternal care, infant behavior, and development among the !Kung. In: R. Lee and I. DeVore (eds.). *Kalahari Hunter-Gatherers: Studies of the !Kung San and their Neighbors.* Harvard University Press, Boston.

KONNER, M.J. and C. Worthman (1980) Nursing frequency, gonadal function and birth spacing among !Kung hunter-gatherers. *Science* **207**, 788-91.

KORBIN, JILL E. ED. (1981) *Child Abuse and Neglect: Cross-cultural Perspectives.* University of California Press, Berkeley.

LAFAYETTE, MARIE-MADELEINE. (Originally published 1678) *Princesse de Cleves.*

LANCASTER, J. B. and C. S. Lancaster (1987) The watershed: Change in parental-investment and family formation strategies in the course of human evolution. In: J. Lancaster, J. Altmann, A. Rossi and L. Sherrod (eds.).*Parenting Across the Lifespan.* Aldine de Gruyter, Hawthorne, New York.

LANGER, W. (1974) Infanticide: A historical survey. *History of Childhood Quarterly* **1**, 353-365.

LASSELLE, CLAUDE DE (1975) Les enfants abandonnés à Paris au XVIIIe siècle. *Annales E.S.C.* **30(1)**, 187-218.

LAWICK, HUGO VAN (1973) *Solo: The Story of an African Wild Dog Puppy and His Pack,* Collins, London.

LEE, P.C. (1989) Family structure, communal care and female reproductive effort. In: V. Standen and R.A. Foley (eds.). *Comparative Socioecology: The Behavioural Ecology of Humans and Other Mammals.* Blackwell Scientific Publications, London, pp. 323-340.

LELAND, LYSA; T. Struhsaker; T. Butynski (1984) Infanticide by adult males in three primate species of the Kibale forest, Uganda: A test of hypotheses. In: G. Hausfater and S. B. Hrdy (eds). *Infanticide: Comparative and Evolutionary Perspectives.* Aldine de Gruyter, Hawthorne, New York, pp. 151-172.

LENOIR, JEAN-CHARLES-PIERRE. Detail sur quelques etablissemens de la ville de Paris, demande par sa majeste imperiale la reine de Hongrie. *Document in Bibliotheque nationale*, Paris, 68 pp.

LEVINE, N. (1987) Differential child care in three Tibetan communities: Beyond son preference. *Population and Development Review* **13 (2)**, 281-304.

LEVINE, S. and R. LeVine (1981) Child abuse and neglect in sub-Saharan Africa. In: J. Korbin (ed.). *Child Abuse and Neglect: Cross-cultural Perspectives.* University of California Press, Berkeley, California, pp. 35-55.

LEWIS, JUDITH SCHNEID (1986) *In the Family Way: Childbearing in the British Aristocracy 1760-1860.* Rutgers University Press, New Brunswick.

LOCK, STEPHEN (1990) Right and wrong: Book Reviews. *Nature* **345**, 397.

MANZONI, ALESSANDRO (1961, translated from the 1840-42 original) *The Betrothed.* E.P. Dutton, New York.

MCCRACKEN, GARY (1984) Communal nursing in Mexican free-tailed bat maternity colonies. *Science* **223**, 1090-1091.

MILLER, BARBARA. (1981) *The Endangered Sex: Neglect of Female Children in Rural North India.* Cornell University Press, Ithaca.

MONBERG, TORBEN (1970) Determinants of choice in adoption and fosterage on Bellona Island. *Ethnology* **9(2)**, 99-136.

MURDOCK, ROBERT (1934) *Our Primitive Contemporaries.* Macmillian, New York.

MURPHY, YOLANDA and Robert Murphy (1974) *Women of the Forest*. Columbia University Press, New York.

NEEL, J. V. (1970) Lessons from a "primitive" people. *Science* **170**, 815-22.

O'BRIEN, TIMOTHY (1988) Parasitic nursing in the wedge-capped capuchin monkey *(Cebus olivaceus)*. *American Journal of Primatology* **16**, 341-344.

ORIGO, I. R. (1986) *The Merchant of Prato*. Boston, Godine 1957, reprinted 1986.

PAGE, H.J. (1989). Childrearing versus childbearing: Coresidence of sub- Saharan Africa. In: R. Lesthaeghe (ed.).*Reproduction and Social Organization in Sub-Saharan Africa*. University of California Press, Berkeley, pp. 401-441.

PALLONI, ALBERTO and Marta Tienda (1986) The effects of breastfeeding and pace of childbearing on mortality at early ages. *Demography* **23(1)**, 31-52.

PARRY, JONATHAN (1979) *Caste and Kinship in Kangra*. Routledge Kagan Paul, London.

PATTERSON, CYNTHIA (1985) "Not worth the rearing": The causes of infant exposure in ancient Greece. *Transactions of the American Philological Association* **115**, 103-123.

PENNINGTON, R. (1989) Child fostering as a reproductive strategy among Southern African pastoralists. Paper presented at the American Association for Anthropologists. Washington, D.C.

PERLEZ, JANE (1990) In AIDS-stricken Uganda area, the orphans struggle to survive. *New York Times*. June 10, 1990.

PIERS, MARIA (1978) *Infanticide: Past and Present*. W. W. Norton, New York.

POMEROY, S. (1984) *Women in Hellenistic Eygpt*. Schocken Books, New York.

RADCLIFFE-BROWN, A. (1922) *The Andaman Islanders: A Study in Social Anthropology*. Cambridge University Press, Cambridge.

RAMANAMMA, A. and U. Bambawale (1980) The mania for sons: An analysis of social values in South Asia. *Social Science Medicine* **14B**, 107-110.

RAO, R. (1986) Move to stop sex-test abortion. *Nature* **324**, 202.

ROOD, JON. (1980) Mating relationships and breeding suppression in the dwarf mongoose. *Animal Behavior* **28**, 143-150.

SCHEPER-HUGHES, N. (1985) Culture, scarcity and maternal thinking. *Ethos* **13(4)**, 291-317.

SCHIEFENHOVEL, WULF (1989) Reproduction and sex-ratio manipulation through preferential female infanticide among the Eipo, in the highlands of western New Guinea. In: A. Rasa, C. Vogel and E. Voland (eds.). *The Sociobiology of Sexual and Reproductive Strategies*. Chapman and Hall, London, pp. 170-193.

SCHNEIDER, JANE and Peter Schneider (1984) Demographic transitions in a Sicilian rural town. *Journal of Family History*. **Fall**, 245-272.

SCHRIRE, C. and W.L. Steiger (1974) A matter of life and death: An investigation into the practice of female infanticide in Arctic. *Man* **9**, 161-184.

SHAHAR, SHULAMITH (1990) *Childhood in the Middle Ages*. Routledge, London.

SHORT, R. (1976) Lactation–the central control of reproduction. In: *Breast Feeding and the Mother*, **Ciba Foundation Symposium 45**, Holland: Elsevier, pp. 73-86.

SHORTER, EDWARD (1982) *A History of Women's Bodies*. Basic Books, New York.

SIEFF, DANIELA (1990) Explaining biased sex ratios in human populations. *Current Anthropology* **31**, 25-48.

SKINNER, G. WILLIAM. (1988) Reproductive strategies, the domestic cycle, and fertility among Japanese villagers, 1717-1869. Paper presented by Rockefeller Foundation Workshop on Women's Status in relation to fertility and mortality, Bellagio, Italy.

SKINNER, G. WILLIAM (In press) Conjugal power in Tokugawa Japanese families: A matter of life and death. In: B. Miller (ed.) *Sex and Gender Hierarchies*. Cambridge University Press, Cambridge.

SMITH, H. F. (1984) Notes on the history of childhood. *Discovery supplement, Harvard Magazine,* July-August issue, pp. 64G-H.

SOMMER, VOLKER (1989) Infant mistreatment in langur monkeys – sociobiology tackled from the wrong end. In: A. Rasa, C. Vogel, and E. Voland (eds.). *The Sociobiology of Social and Reproductive Strategies.* Chapman and Hall, London, pp. 110-127.

STAGER, LAWRENCE E. and Samuel R. Wolff (1984) Child sacrifice at Carthage: Religious rite or population control? *Biblical Archaeology Review,* January-February 1984.

STONE, LAWRENCE (1977) *The Family, Sex, and Marriage in England 1500-1800.* Harper and Row, New York.

SUSSMAN, GEORGE D. (1982) *Selling Mothers' Milk: The Wet-nursing Business in France 1715-1914.* University of Illinois Press, Urbana.

THIERRY, B. and J.R. Anderson (1986) Adoption in anthropoid primates. *International Journal of Primatology* 7, 191-216.

TREXLER, RICHARD (1973a) Infanticide in Florence: New sources and results. *History of Childhood Quarterly* 1(1), 98-116.

TREXLER, RICHARD (1973b) The foundlings of Florence, 1395-1455. *History of Childhood Quarterly* 1(2), 259-284.

TRIVERS, R.L. (1972) Parental investment and sexual selection. In: B. Campbell (ed.). *Sexual Selection and the Descent of Man.* Aldine, Hawthorne, New York.

TRIVERS, R.L. (1974) Parent-offspring conflict. *American Zoologist* 14, 249-264.

TRONICK, EDWARD Z.; Gilda Morelli; and Steve Winn (1987) Multiple caretaking of Efe (Pygmy) Infants. *American Anthropologist* 89(1), 96-106.

TRUMBACH, RANDOLPH E. (1978) *The Rise of the Egalitarian Family: Aristocratic Kinship and Domestic Relations.* New York: Academic Press.

TURKE, PAUL W. (1989) Evolution and the demand for children.*Population and Development Review* 15(1), 61-90.

VOLAND, EKHART (1988) Differential infant and child mortality in evolutionary perspective: Data from late 17th to 19th century Ostfriesland (German). In: L. Betzig; M. Borgerhoff Mulder; and P. Turke (eds.). *Human Reproductive Behaviour.* Cambridge University Press, Cambridge, pp. 253-261.

WAGLEY, C. (1977) *Welcome of Tears.* Academic Press, New York.

WEISNER, THOMAS (1987) Socialization for parenthood in sibling caretaking societies. In: J.B. Lancaster, J. Altmann, A.S. Rossi and L. R. Sherrod (eds.). *Parenting Across the Lifespan.* Aldine de Gruyter, Hawthorne, New York.

WHITTEN, PATRICIA (1982)*Female Reproductive Strategies Among Vervet Monkeys.* Ph.D. Diss. Harvard University.

WIESCHHOFF, H.A. (1940). Artificial stimulation of lactation in primitive cultures. *Bulletin of the History of Medicine* VIII(10), 1403-1415.

YENGOYAN, ARAM A. (1981) Infanticide and birth order: An empirical analysis of preferential female infanticide among Australian aboriginal populations. *Anthropology UCLA,* vol. 7, pp. 255-273.

CHAPTER 2

THE ROLE OF SOCIAL, RELIGIOUS AND MEDICAL PRACTICES IN THE NEGLECT, ABUSE, ABANDONMENT AND KILLING OF INFANTS

FREDERICK S. VOM SAAL

Division of Biological Sciences, Department of Psychology and John M. Dalton Research Center, University of Missouri-Columbia, Columbia, Missouri 65211, USA

INTRODUCTION

Each society has its own set of norms for the behavior of parents toward infants, and infant rearing practices vary dramatically from one society to another; a behavior which might be labelled as abuse in one society might not be considered to be abuse in another society. There is no optimum strategy for rearing infants (Korbin, 1987), and it is not my intention to be critical of parents who have raised or are currently raising an infant using socially accepted infant-rearing practices. However, parents sometimes engage in behaviors toward infants which appear to be detrimental to the wellbeing of their infants and thus maladaptive. This raises questions as to the factors, both in prior times and today, which have interfered with the development of parent-infant relationships to the detriment, and even death, of infants; the focus of this review will be on the beginning of infancy when an infant is most dependent on parents or other caregivers. I will discuss evidence that many medically accepted infant-rearing practices, policies of governments and corporations, and social and religious customs have increased (both in prior times and today) the likelihood of neglect, abuse, abandonment and killing of infants by parents in Western societies.

The degree to which parents were nurturing and protective of their off-spring in Western societies over the last twenty-five hundred years is con-troversial, and there are arguments among historians as to whether an at-tachment between parents and infants, which most people assume to be "normal" today, was common prior to the twentieth century in Western societies. One view is that the history of the interaction between parents and children in Western societies is a nightmare from which we have only begun to awaken in the twentieth century (Johansson, 1987). In contem-porary Western societies there is a general presumption that most parents want to provide the foundation for a better life, with greater opportunities for success, for all of their children. Lister (1986, p. 1403) has proposed that "a society such as ours, which encourages the birth of new members, has in fact committed itself to their well-being". This optimistic view of parent-infant interaction is supported by evidence that there has been a recent increase in nurturing of infants by fathers over the last few decades in Europe and the United States (Hewlett, 1992a). However, the general assumption that in Western societies today the typical parent is protec-tive, caring, and nurturing of all offspring contrasts sharply with estimates that millions of children are being abandoned, neglected or physically and sexually abused by their parents (Gelles and Cornell, 1983; Lynch, 1985; Gelles, 1987; Creighton, 1988; Green, 1988). Some experts believe that reported child abuse is "just the tip of the iceberg", with the actual inci-dence of abuse being much higher. For example, Green (1988) estimates that there may be 1,500,000 cases of child abuse in the United States annu-ally. Child abuse and neglect are used here to describe situations in which a child suffers identifiable harm that can be attributed to a caregiver (Gelles, 1987; Korbin, 1987).

EFFECT OF WET NURSES AND FORMULAS ON INFANT MORTALITY

While many parents in prior times may have been concerned for the wel-fare of their offspring, the parenting strategy adopted often reduced the chances of the infant surviving and competing in society. For example, throughout recorded history surrogate mothers, referred to as wet nurses, have provided milk for babies when natural mothers were unable or un-willing to nurse their infants. A wet nurse is certainly capable of providing care and nutrition for an infant which is as good as that which the natu-ral mother might have provided. The critical issue is whether it is likely that the care provided by a wet nurse would be commensurate with that

provided by the biological mother (Daly and Wilson; Hrdy; Wilson and Daly, this volume); this is particularly true in cases in which the infant is reared by a wet nurse outside of the home of the natural parent, which, by today's standards would be considered temporary abandonment (the temporary abrogation of parental responsibility for the infant). In fact, it has been estimated that by the time of the industrial revolution, infant mortality associated with the use of wet nurses was extremely high (Hrdy, this volume). Given that this must have been generally known at the time (Cadogan, 1750), it is not unreasonable to consider a practice which had a high probability of leading to the death of an infant as being a form of socially sanctioned infanticide rather than temporary abandonment. An additional issue is that the normal contact between parents and a new-born infant, which has been proposed to contribute to the development of a bond between the parents and infant, would not occur when wet nurses are hired to care for and feed an infant (Bowlby, 1969; Herbert, *et al.*, 1982; Wilson and Daly, this volume).

By the end of the nineteenth century, an increased effort was being made to develop substitutes for breastmilk as an alternative to the use of wet nurses (due to the high mortality of infants reared by wet nurses); this led to the growth of the infant formula industry. For centuries, wealthy families had engaged in the common practice of supplementing breastmilk with mixtures of butter, sugar, spices and other foods (Hardyment, 1983, p. 3). The eighteenth century British physician, William Cadogan (1750) observed that babies receiving these supplements were not as healthy as those only fed breastmilk, which led to the recommendation that infants should only be fed breastmilk; this advice undoubtedly saved the lives of many infants.

During the twentieth century the infant formula industry was highly effective in promoting the use of bottles over breastfeeding in Western societies. Middle-class parents were taught how to properly prepare the formulas, and infants fed formulas were thus not constantly sick, which occurs when unsterile water is used to make the formulas. However, even when commercial formulas are prepared using sterile procedures, problems with infant health have occurred due to a lack of regulation of the nutritional content of the formulas (Palmer, 1988, pp. 209–214). Only recently have attempts been made at regulating the nutritional content of infant formulas in the United States. For example, most parents are probably unaware that soy (commonly used in formulas for infants with lactose intolerance) contains estrogenic chemicals, referred to as phytoestrogens (Whitten and Naftolin, 1991, P. Whitten, personal communication); exposure to estrogenic chemicals during early life can have a profound effect on the course

of fetal and infant development (vom Saal, *et al.*, 1992; Colborn, et al., 1993). Most child-care "experts" in Western societies during the first two-thirds of the twentieth century promoted bottle feeding. It was not until 1978 that the American Pediatrics Association finally officially acknowledged the benefits of breastfeeding.

ROLE OF PHYSICIANS AND OTHER "EXPERTS" IN CHILD ABUSE AND NEGLECT

An eighteenth century physician, William Cadogan (1750), proposed that newborn babies be fed at regular intervals during the day (no more than two feedings per day were recommended), and not at all during the night, rather than "on demand". Cadogan, and subsequent "experts" such as Dr. Frederick Truby King, whose ideas were influential during the early twentieth century, confused the diseases associated with serving infants contaminated supplementary foods with "overfeeding"; the one positive aspect of this erroneous advice was that breastfeeding of infants was promoted as leading to a healthier infant than formulas or other foods. It was thought that an infant's stomach needed to be rested between feedings to clear out microorganisms responsible for causing diarrhea. Second, King could not comprehend why a busy mother would want to waste her time with feedings more frequent than at 4-hour intervals. The benefits of a 3 to 4-hour interval between daytime feedings continued to be accepted by physicians into the latter part of the twentieth century (Hardyment, 1983, pp. 176–181, 214–265).

The expectation that all newborn infants should rapidly adjust to a long-interval feeding schedule during the day and sleeping throughout the night is now known to have been unrealistic. The complexity of the ontogeny of biological rhythms in infants is reflected in considerable individual variability in the establishment of a period of prolonged sleep during the night (Konner, 1991). For example, the development of biological rhythms in infants appears to be influenced by the interaction between the mother and the infant both during pregnancy and after birth (Davis, 1981). Environmental factors also markedly influence the ontogeny of the sleep-wake cycle. In some hospital maternity wards newborn infants have been subjected to continuous bright lights. This may interfere with the normal establishment of the biological rhythms required for eventual adjustment to the demands of diurnal feeding at timed intervals and prolonged nocturnal sleep. For example, there is evidence that for babies kept in nurseries after birth, there was more crying, overall activity, and a delayed establish-

ment of a sleep-wake cycle in which a prolonged period of sleep occurred during the night (Reppert and Rivkees, 1989). Degree of maturity, rather than age from birth, also appears to be an important factor in the onset of biological rhythms, since premature infants develop rhythms later than do infants who were not prematurely delivered (Davis, 1981).

Observations of infant feeding patterns in nonindustrialized societies, as well as a comparative look at feeding schedules in great apes, shows that feeding at 3 to 5 hour intervals is not typical. For example, the interval between feeding bouts in the !Kung who inhabit the Kalahari Desert is approximately 15 minutes during the time that an infant is awake; infants have constant access to the breast (Konner, 1982; Short, 1976, 1984). The use of long intervals between feeding can result in an infant crying prior to the time at which the next feeding is scheduled, since the stomach is empty approximately 1.5 hours after drinking breastmilk (human milk is very dilute). An infant is thus likely to be hungry for hours prior to each meal when fed at intervals as long as 4 hours. Crying can hardly be considered to be an abnormal or unexpected response to a prolonged period without food; in fact, infants exhibit a hunger cry which can be distinguished by mothers from crying for other reasons (Chateau, 1980; Ostwald and Murry, 1985). For some reason William Cadogan (1750) believed that infants only cry when they are sick, never as a result of being hungry. In a situation in which a parent (mother or father) and infant are in almost continuous contact (for example, in some hunter-gatherer societies), infant distress due to an empty stomach is not common (Katz and Konner, 1981; Hurtado and Hill, 1992).

The normal response of a parent when an infant cries should be to alleviate the distress by picking up the infant and, if required, feeding it. In fact, this is the common practice in nonindustrialized societies, such as the Usino in Papua New Guinea (Palmer 1988, p. 115) and the !Kung (Konner, 1991, p. 112). De Meer (1988) also reported that an Aymara mother in Peru would be considered incompetent if her infant showed signs of distress and she did not respond to the infant to stop the distress. In a comparison of middle-class mothers in the United States and Japan, Japanese mothers were reported to find the practice of leaving an infant alone in a cot to cry itself to sleep to be shocking (Elkin and Handel, 1989, p. 39).

With regard to other factors which might lead to crying in infants, in humans and other great apes, infants will hold the breast with their hand or will place the breast in their mouth (without sucking) during times of increased stress, such as the presence of strangers; this is referred to as non-nutritive suckling (Wolff, 1968). Any contact of an infant with the

breast (whether with the hand or mouth) pacifies an infant (Kessen and Leutzendorff, 1963; Palmer, 1988, p. 115). In addition, stimulation of the breast causes the release of prolactin, which stimulates the synthesis of milk (Tucker, 1988; Wakerley, *et al.*, 1988). The inability of many women in Western societies to provide enough milk for their infants using interval feeding (together with wearing clothing which prevents "non-nutritive suckling" breast stimulation) is likely the result of a decrement in prolactin secretion due to inadequate breast stimulation. Also, the contraceptive effect of almost continuous stimulation of the breast is lost and ovulation occurs (Konner, 1982; Short, 1976, 1984). Moreover, it is possible that the physical, verbal and visual interaction between mother and infant during breastfeeding is important for the development of other psychological systems and an attachment between mother and infant, although this is controversial (Svejda, *et al.*, 1982; Kopp, 1989).

The unfortunate and misguided idea that parents should let infants cry and not pick them up or nurture them, which has contributed to problems associated with rearing of infants who could not accommodate to interval feeding, prolonged periods of nocturnal sleep, or were just fussy (Boukydis, 1985), came from physicians and behavioral psychologists. This idea was, until relatively recently, generally accepted by "experts" on child rearing. What is more, some "experts" still believe that infants should be left alone at night rather than sleep with the parents, due to fears that the child would be overprotected or spoiled; the 1985 edition of Dr. Spock's *Baby and Child Care* continued to promote this practice so as not to spoil an infant. At the beginning of his book *Children* (1991, p. 9), Mel Konner identified a number of recent major changes in dogma about child-rearing practices, including the observation that: "having a need for dependency indulged early in life has remarkably little impact, either way, on the likelihood of being overdependent later on."

The belief held by some psychologists that infants would be spoiled if they were attended to when they cry resulted from the inappropriate ap-plication of findings from conditioning experiments with adult pigeons and white rats to human infant-rearing practices. For example, since a white rat can be conditioned to press a lever by receiving a reward as a result of pressing the lever, it was proposed that feeding an infant when it cried would condition the infant to continue crying. In other words, one would spoil the infant. Specifically, John Watson, the founder of behaviorism, believed that the process of forming associations between previously unre-lated events (for example, conditioning a rat to press a lever or a pigeon to peck at a light to obtain food) was relevant to a parent responding to infant crying by attending to the infant and alleviating the distress (Hardyment,

1983, pp. 169-175; O'Donnell, 1985). However, these events (crying by an infant and attending to the infant by a parent) are certainly not unrelated. Behaviorists rejected the idea that there are species-specific differences in the capacity to form associations which exist as a result of evolution. It is now understood that there are profound differences in the capacity for different species to be conditioned to associate events and that the processes involved in conditioning a rat to press a lever do not generalize to all forms of learning (Seligman, 1970; Rozin and Kalat, 1971).

Developmental changes also have to be considered with regard to the hypothesis that attending to a crying infant will result in a spoiled child. For example, it has been proposed that crying by a newborn infant and an older child serve different functions (Murray, 1985; Ostwald and Murry, 1985; Demos, 1986; Kopp, 1989). Furthermore, different types of cry have been identified in infants within the first few weeks after birth (Chateau, 1980; Ostwald and Murry, 1985). Sarah Hrdy (personal communication) identified that in langurs, a very young infant which was loudly crying as a result of being separated from its mother was very rapidly picked up by either the mother or another female; in contrast, crying by an older infant (for example, during weaning) was ignored by adults.

Crying by a newborn infant is an adaptive (evolved) response to distress (due to hunger, fear as a result of separation from the parent, pain or illness), and the adaptive or appropriate response by a parent would be to pick up the infant and alleviate the distress (Lamb and Campos, 1982, p. 179). This does not lead to an increase in crying, as predicted by models based on operant conditioning experiments with rats and pigeons, but instead alleviates the distress of the infant. It is now known that when parents respond to distressed infants, there is a decrease in crying at later stages of development and a general increase in positive interactions between mothers and infants (Chateau, 1980; Bronfenbrenner, 1986).

The physician, Frederick Truby King, and the psychologist, John Watson, proposed that children should be left alone and not overstimulated or overloved. This parenting strategy was promoted as "following the laws of nature" (Hardyment, 1983, pp. 174–178) and consisted of caching infants (hiding them in a safe place) and not touching them for hours; for example, parents in Western societies often leave an infant unattended in a playpen for prolonged periods during the day and completely alone throughout the night. The placing of human infants in swinging seats or playpens (often in front of a television) without much physical contact with other humans during the day and then completely isolating the infant at night should be considered as socially sanctioned neglect of infants.

Caching of infants is the parenting strategy used by the hare (the young remain hidden in a nest as a defense against predators). The hare has extremely concentrated milk and provides its young with the entire daily supply of milk in about one 5-minute feeding bout; there is, as expected, coevolution of a parenting strategy which provides only a few minutes of maternal contact with young each day, an extremely high nutritional content of milk, and the capacity for infants to digest the milk (Blurton Jones, 1972). In contrast, humans have dilute milk relative to the hare in terms of the concentration of nutrients, and as with other great apes, humans obviously evolved to have relatively short intervals between infant feedings (Short, 1976, 1984).

Caching of infants is dramatically different from the parenting strategy observed for great apes (Blurton Jones, 1972) and humans in nonindustrialized countries (Short, 1984; Konner, 1991). One is struck by the fact that mothers and infants are seldom separated in nonindustrialized countries in which the demands of work make this possible. For example, in some parts of Africa, mothers carry infants in slings on their back, and the infant can gain access to the breast virtually on demand (this is made easier in societies in which the breasts are not covered). The baby thus also gets rocking stimulation and physical contact with the mother. In hunter-gatherer societies infants are reported to have a greater amount of physical contact with both their mother and father (as well as other members of the extended family) during the first few years of life relative to agricultural or industrial societies (Katz and Konner, 1981; Hewlett, 1992b; Hurtado and Hill, 1992; MacDonald, 1988, pp. 177–181). The social organization in hunter-gatherer societies is presumed to be similar to that of humans throughout their evolution prior to the advent of agriculture. It may well be that with every cultural "advancement", from hunter-gatherer, to agricultural, urban, and, finally, industrial societies, the nurturing of children has steadily declined.

Most mothers in non-Western societies sleep with their babies rather than caching them throughout the night (Palmer, 1988, p. 115). During the nineteenth century overlaying (or accidental smothering) of infants was believed to be a serious problem (and it was reported to be a common cause of infant death). The accidental smothering of an infant due to a sleeping parent rolling onto the infant is generally considered to have been highly unlikely (Ober, 1986; Palmer, 1988, p. 83). Some infant deaths attributed to overlaying may have been due to sudden infant death syndrome (SIDS). It is likely, however, that overlaying was reported as the cause of death in situations in which the infant was smothered as an overt act of infanticide. For example, reports of suffocation due to overlaying increased by three-

fold relative to population growth between 1850–1880 in London, which most historians attribute to active infanticide (Forbes, 1986; Rose, 1986, pp. 177–181). Social concern about overlaying contributed to the idea that children should not sleep with parents. Keeping the infant isolated from the mother at night (in a separate bed or even another room) obviously was also motivated by other factors, such as the desire by parents to not have an infant interfere with the resumption of sexual relations.

Physicians such as Frederick Truby King saw no reason to expend the energy which demand feeding (during the day or night) required of mothers. Interval feeding during the day and no feeding during the night was intended to limit maternal-infant interactions so that rearing of infants would be more efficient; one objective was to allow the mother more time to care for the father rather than the infant (Hardyment, 1983, pp. 174, 181). Even in situations where a mother was able and willing to feed her infant on demand, in spite of the views of physicians such as King, social mores dictated that this not be done. For example, a problem which has dramatically impacted the choice of women to breastfeed, particularly on a demand schedule, has been the gradual development of proscriptions against public exposure of breasts. Demand breastfeeding is not possible in societies which regard the breasts as a source of sexual stimulation rather than nutrition for infants. An obsession with the breast as a source of erotic stimulation has occurred in the United States (but not all Western cultures) in recent years. This additional cultural factor will complicate any move toward changing the pattern of breastfeeding (at specified intervals) in the United States relative to other societies in which the breast is associated with nurturing infants.

Regardless of whether a woman desires to breastfeed, since the industrial revolution breastfeeding often has not been possible due to constraints associated with work. One of the consequences of the industrial revolution is that women who worked in factories were unable to nurse their babies throughout the work day. Given the demands on women of living in urban environments in the industrial age, one can certainly understand why medically-approved alternatives to demand breastfeeding of infants was appealing. There have thus been numerous factors which have contributed to an interference in Western societies with the evolved pattern of parent-infant interaction, which is likely to be more accurately reflected in contemporary hunter-gatherer than in Western societies (Katz and Konner, 1981).

The striking failure of physicians to recognize the absurdity of some medical practices (such as prolonged intervals between feeding in a species with a relatively dilute and thus rapidly digested milk) has been due to a

lack of understanding that each species has constraints on it's behavior and physiology as a result of evolution. The fact that so many infants could not adjust to interval feeding and a prolonged period of nocturnal sleep without waking up and crying due to hunger or, perhaps, fear of being left alone in the dark did not lead to a re-evaluation of these practices, but instead resulted in development of a medical industry devoted to promoting the strategy of caching infants. Mothers in Western societies whose infants were not able to adjust to these practices were considered to be lacking in mothering skills because they could not get their babies to conform to modern medical practice. Physicians need to appreciate the importance of a basic knowledge of evolution, comparative biology and anthropology to the practice of medicine or they will continue to make the same kinds of mistakes described above. Even a superficial exposure to the literature in comparative biology (Blurton Jones, 1972) and anthropology would have raised doubts about these practices among physicians; in a review of 173 societies in the anthropological record, no societies were found in which infants were placed in a separate room during the night (Konner, 1991, p. 113).

One of the most aversive experiences a person can have is to listen to his or her child crying in distress (Frodi, 1985; Murray, 1985). Sarah Hrdy (personal communication) reported that in her observations of langurs, crying by a very young infant as a result of being separated from its mother appeared to be very distressing to adults, and a crying infant was very rapidly picked up by either the mother or another female. However, Frederick Truby King could not understand why mothers should become distressed by an infant crying for only an hour or so, since he apparently did not find this aversive (Hardyment, 1983, p. 181). Among the Puritans, crying was considered to be an expression of anger by an infant, and parents were encouraged to beat their children to "nip this animosity in the bud" (Hardyment, 1983, p. 8). A socially sanctioned response by parents or other adults to infant distress, as a result of a long interval between feedings or to being left alone in a dark room at night, was thus to hit the infant.

The conflict between social mores dictating that parents not nurture an infant when it cries, and an emotional reaction to infant distress (which is biologically based; Bowlby, 1969; Konner, 1982) has likely resulted in high levels of stress for parents. Periods of high stress are associated with harsh treatment of infants, whereas a more nurturing parenting style is found when adequate resources (associated with low stress) are available (Belsky, et al., 1991). It is not difficult to imagine situations in which parents, particularly those prone to violence or under stress, would hit a crying

infant only to have the infant cry louder, thus leading to an escalation of violence (in other words, child abuse; Frodi, 1985; Browne, 1988). An important issue relates to the factors which predict how parents in Western societies will respond to an infant when it begins crying and they are told by "experts" not to attempt to provide comfort or nutrition. There is no doubt that the likelihood of violence toward infants is increased in this situation (Frodi, 1985; Murray, 1985), particularly if the infant was unwanted (Roberts, 1988). The impact of parental violence is not just on their children, since child abuse tends to be perpetuated within families from one generation to the next, which is referred to as transgenerational child abuse (Ney, 1988; Widom, 1989).

In addition to medical advice and social mores, the social ecology of industrialized societies might contribute to maltreatment of infants. Humans are proposed to have been mildly polygynous throughout their evolutionary history (Alexander, 1979). In terms of parental violence toward infants, one problem which is particularly acute in urban societies is that individuals often do not have the opportunity to observe (as adolescents or young adults) interactions between parents and infants in an extended family or clan group prior to producing their own infants. It is likely that the isolation of urban family units from relatives and the absence of other support groups to assist with caring for an infant are reasons why parents respond with violence to the stress of having to cope with a distressed infant (Bronfenbrenner, 1986). In monkeys, the absence of observation of parenting severely impacts individuals when they have infants and places the infant at risk for neglect or harm (Harlow, et al., 1966; Altmann, 1980). With regard to physicians giving mothers advice about child rearing, the typical medical student and resident who will become a practicing pediatrician has very likely had little or no first-hand experience with parenting and is thus totally dependent on textbooks ("experts") for advice concerning how to treat infants.

While many of the infant-rearing practices described above are no longer considered appropriate by authors of recent pediatrics textbooks and books on child care, any change in attitude among physicians (such as the recent change in physician's attitudes concerning the antimicrobial activity of breastmilk, demand feeding rather than fixed schedules, sleeping with an infant, and the contraceptive effectiveness of breastfeeding) may be countered by the perpetuation of ideas by grandparents when they are available. For example, a grandmother may pressure her children not to adopt infant-rearing practices which are unfamiliar to the grandmother. Unfortunately, this very issue has provided the basis for perpetuating the denigration of women with regard to the rearing of infants by each new

generation of physicians. For example, William Cadogan (1750, p. 2) began his essay on the management of children by stating:

> It is with great pleasure that I see the preservation of children become the care of men of sense. ... In my opinion, this business has been too long fatally left to the management of women, who cannot be supposed to have proper knowledge to fit them for such a task. ... They may presume upon the examples and transmitted customs of their great grand-mothers, who were taught by the physicians of their unenlightened days; when physicians, as appears by late discoveries, were mistaken in many things.

The cycle of "experts" advising each new generation of mothers to ignore their own intuition or the ideas of their parents and follow the practices proposed by "modern experts" is unlikely to be easily broken, although organizations such as the La Leche League will certainly help in this process. Among other things, breaking this cycle will require a dramatic change in the attitude of male physicians toward women.

GOVERNMENT AND CORPORATE POLICIES AND INFANT MORTALITY

Of considerable importance to the health and welfare of children are laws and policies pursued by governments and corporations. A recent documented example of a corporate policy which has clearly lead to dramatic increases in infant disease and death concerns the aggressive marketing (through advertising and the use of free samples) of infant formulas in undeveloped countries. For example, Nestle, a Swiss food company, has for years aggressively promoted the use of infant formulas and bottle feeding in undeveloped countries where it is often not possible for parents to sterilize contaminated water used to prepare the formulas (Palmer, 1988, p. 201). Infant formula manufacturers (the Nestle company is the largest) distribute free samples through hospitals and health officials to women after delivery; just enough formula is provided free to ensure that resumption of lactation is unlikely. The mother thus has no option other than continuing to use the formula, which must then be purchased. Bacteria in contaminated water used to prepare the formulas can lead to the death of infants through dehydration (typically very slowly with tremendous suffering). Newborn infants do not yet have the ability to produce antibodies to fight diseases which result from drinking contaminated water (antibodies which attack microorganisms in the gut are produced by the mother and transported to the infant via her breastmilk; Victoria, *et al.*, 1987).

The governments of many undeveloped countries have encouraged the Western practice of bottle feeding, since it is viewed as a sign of modernization; it also has freed women with infants to participate in the workforce (see Palmer, 1988, p. 119 for an example of this with slaves). A dramatic increase in infant mortality due to gastro-intestinal diseases resulted from this practice; this eventually led to the WHO/UNICEF code of Marketing of Breastmilk Substitutes, which was passed at the World Health Assembly in 1981. There was broad international recognition that it was essential to ban the aggressive marketing practices of companies selling infant formulas in undeveloped countries.

The United States was the only UN member to vote against adoption of the WHO/UNICEF code under pressure from companies producing formulas. Nestle executives also managed to have a discussion of the infant formula issue cancelled at a symposium on bioethics hosted by the National Institutes of Health in 1983 (Marshall, 1983). Officials of the Nestle company agreed to abide by the guidelines contained in the UN resolution after the company suffered substantial losses due to a worldwide boycott of Nestle's products. But, promotion of formula use in undeveloped countries continued and is the subject of yet another UN initiative to stop this practice. The International Association of Infant Food Manufacturers has now promised to stop distributing free formula samples by the end of 1992.

The catastrophic effect of these policies on infant mortality in undeveloped countries (due to lack of training and facilities for the proper preparation of infant formulas) had been documented at the time the decision to aggressively market infant formulas was being made by corporate officials (Campbell, 1984; Habicht, et al., 1986; Victoria, et al., 1987; Palmer, 1988, pp. 198–244). These corporate management policies (leading to the slogan "bottles kill babies"), and the acceptance of these policies by government and health officials, should thus be considered as corporate- and government-sponsored infanticide.

Breastfeeding is highly effective (greater than 95%) as a contraceptive when the infant is allowed continuous access to the breast. When demand suckling occurs, which is typical of the nursing pattern exhibited by great apes, including humans in hunter-gatherer societies, the typical birth interval is between 3.5 to 4 years. This contrasts with the 1.5-year interval between births found in urban societies where there is widespread use of formulas either as the sole source of nutrition or as a supplement to breastmilk; a long interval between feeding may require supplements due to inadequate stimulation of the breast (Konner, 1982; Short, 1976, 1984). Thus, in addition to infant health, of considerable interest to representa-

tives at the World Health Assembly in 1981 was the impact of the shortened birth interval due to the widespread use of infant formulas on population growth. The statistics on the proportion of women in undeveloped countries, particularly in urban areas of Latin America, who breastfeed their babies beyond a few months, if at all, are disheartening (less than 30%; Popkin, *et al.*, 1982). The widespread use of formulas in undeveloped countries has thus led to a loss of the contraceptive effect of lactation, resulting in a short birth interval (Short, 1984), and has also had a devastating effect on infant health and survival.

The abnormally short interval of 1.5 years between babies observed in urban areas denies infants of the normal period of intensive interaction with the mother. An increased birth rate leads to a decrease in the investment of time and resources by parents in each child (Trivers, 1974). Decreased parental investment and stressful conditions are likely to lead to parenting styles that produce psychological characteristics such as high aggressiveness (MacDonald, 1988, pp. 183-235), which can be detrimental to success in an industrialized society. In other words, these practices have led to an increase in the number of individuals who are psychologically unsuited for the demands of an industrialized society. There is a correlation between a short interval between the birth of babies and infant abuse and infanticide. In general, increased family size (a low ratio of parents to children, which includes single parent households) and birth of twins increase the risk of child abuse and infanticide (Lithell, 1981; Nelson and Martin, 1985; Lester, 1986; Leventhal and Midelfort, 1986).

The issue of infant health and nutrition in undeveloped countries is quite complex. While the evidence for the devastating effects of the use of improperly prepared formulas is clear and convincing, evidence is also accumulating that pesticides and numerous industrial waste products contain chemicals that are concentrated in breastmilk (the chemicals are lipophilic) and thus pose a serious threat to infant health in contaminated environments (Jensen and Slorach, 1991; Thomas and Colborn, 1992). Many of these chemicals, such as DDT and a host of other chlorinated compounds, profoundly disturb development in embryos and infants (Colborn and Clement, 1992; Colborn, *et al.*, 1993). It is illegal to produce or use DDT in the United States due to the known danger posed by this compound. However, American and non-American companies are producing and promoting the use of DDT and other dangerous chemicals in the same undeveloped countries targeted by the infant formula industry for aggressive marketing; this is possible since there is a general lack of understanding of the dangers posed by the use of many of these chemicals, and thus there are no restrictions on their use by governments in undevel-

oped countries. There is now compelling evidence that these pollutants are being transported around the globe via the atmosphere, and efforts are being made within the United States and other developed nations to have a world-wide ban on the production and use of the chemicals which are already known to pose a danger to human health (and the health of all other animals; Clement and Colborn, 1992).

SOCIAL AND RELIGIOUS CUSTOMS AND INFANT MORTALITY

There are dramatic differences between conditions faced by infants born today into middle class families in developed countries when compared to infants born into poor families in developed countries, infants born in undeveloped countries, and infants born in prior centuries. Children who survived in Greek and Roman antiquity and Medieval Europe are thought to have been treated very harshly by parents. It may have been that only the oldest male child, who would inherit property and carry on the family name, received any significant investment of time and resources by parents (Flandrin, 1979, p. 205; Boswell, 1988, p. 36). Trivers (1974) proposed that the level and type of parental investment varies as a function of the ecological conditions which are encountered.

Throughout recorded history the issue of whether newborn infants are somewhat less than human has been debated (Moseley, 1986). This was particularly true in cases where there was some clear physical deformity or obvious mental defect (Viljoen, 1959). The Athenians in ancient Greece solved this dilemma by not considering a child as a full human being until a ceremony called Amphidromia was performed (Montag and Montag, 1979). Romans were not legally required to raise their children. In general, unlike the situation after the adoption of Christianity as the state religion in the fourth century A.D., the pre-Christian Roman government did not seek to legislate morality. The father (paterfamilias) was the head of the Roman family, over which he exercised absolute authority; this included the right to kill or abandon his offspring or an infant which he presumed to be illegitimate (Boswell, 1988, p. 57).

Quality of life vs. sanctity of life: the debate over life and death of deformed Infants

There are those who view human life as sacred, regardless of the quality of the life. An issue that is the subject of considerable debate concerns the humanist vs. religious views on the use of heroic measures to save the

lives of terminally ill or severely deformed infants when it is clear that to do so will result in suffering by the infant (in many of these cases lifespan is, at best, only a few months). The issue concerning quality of life is not considered relevant by those with the religious view that one can never take any life, regardless of the circumstances. In contrast, humanists view the quality of one's life as a primary issue. There is thus no real basis for a dialogue. The sanctity of all life, whether of the infant or aged, is an integral part of Judeo-Christian theology, and the argument between the humanist and religious philosophies has been fought on many fronts over many issues. This is one more area where the social costs (referred to as social utilitarianism) and the "will of God" (as divined by theologians) are in direct and unresolvable conflict (Long, 1988). This same issue was debated thousands of years ago in Greece by the great philosophers (Viljoen, 1959), and it will likely be debated a thousand-years hence.

Throughout history the quality of an infant has been used as a basis for decisions concerning infanticide, abandonment and neglect, and, in fact, social and religious customs have promoted the killing or abandonment of deformed infants. For example, in ancient Greece the mother of a deformed child may have been ostracized or even killed, since the production of a deformed infant was taken as an omen that the mother was a witch. The mother would thus have been compelled to kill the infant to hide its deformity from the community (Moseley, 1986). In Medieval Europe during the thirteenth century, the production of a deformed infant was thought to reflect a mother's sins, which must have increased the likelihood of infanticide to avoid the social consequence of producing a deformed child (Bloch, 1988; Boswell, 1988, p. 338). Today, among the Aymara Indians in Kimamo in Southern Peru, very few children with congenital malformations are observed, suggesting that they are killed. This would not be surprising as it appears that killing of healthy infants also occurs as a means of controlling family size (de Meer, 1988).

In recent times the position of the Catholic Church on the use of extraordinary measures to save deformed children was addressed by Pope Pius XII in 1957 in an essay entitled "The Prolongation of Life" (cited in Lister, 1986). The Pope stated that only ordinary means should be used to prolong life, with the burden on others of continued life of deformed infants being considered when determining what course of action to take. Thus, a recent Pope has acknowledged the importance of quality of life in decisions concerning the life or death of an infant. In contrast, others have argued that whether or not the parents might suffer hardship cannot be the determining factor in decisions concerning letting defective infants die, since this represents infanticide and is thus morally wrong.

Currently, it is legal for severely deformed children to be allowed to die passively from starvation or dehydration, although any direct action which leads to death is illegal (Meade and Brissie, 1985; Lister, 1986). The public silence about the involvement of physicians in decisions to allow deformed infants to die was broken by Duff and Campbell (1973); they acknowledged participating in the decision to withdraw treatment in cases in which infants died. The issue of whether this should be allowed has generated considerable debate (Kuhse and Singer, 1987; Long, 1988), and opposition to this practice on ethical grounds again pits religious ideology vs. humanistic philosophy.

Gender-biased infanticide

The rise of agricultural societies increased the value of sons; among other factors, males are bigger and stronger than women. In contrast, in savanna hunter-gatherer societies, women and men play a more equal role in acquiring food and providing for the tribal group. (For a review of sex-biased parental investment in mammals see Hrdy, 1987.) Christian (Patristic) writers of the third to fifth centuries A.D. had an extremely negative view of women, which has also influenced attitudes toward women in Western societies from Medieval times up to today; they proposed that women were the source of lust and were thus evil, an idea which was considered to be radical at the time it was proposed but was incorporated into Canon law during the twelfth century (Brundage, 1987, pp. 62, 173, 184). As discussed above, an extremely condescending attitude toward women has been evident in the writings of physicians (Cadogan, 1750; Hardyment, 1983, p. 181) and still persists today.

The lower value of women than men in agricultural societies, since men were more suited for agricultural work, is generally considered to have increased the likelihood that female infants would be abandoned or killed (Boswell, 1988, pp. 58-60). For example, throughout Greek and Roman antiquity abandonment and infanticide were more common for females than males (Viljoen, 1959; Golden, 1981; Harris, 1982); one's obligation to the state during the Roman empire was to produce male heirs. The markedly higher number of males than females during the early Middle Ages suggests that the practice of female infanticide continued to be prevalent due to the low value and low status of women as well as the custom of dowry. These factors also are implicated in the killing of female infants today in some social groups in India (Saxena, 1975; Miller, 1981; Jeffery, et al., 1984). Similarly, in rural China there have been recent reports of widespread killing of female infants in response to the Chinese

government's policy of penalizing families that have more than one child (Bongaarts and Greenhaigh, 1985).

Abandonment

Abandonment does not involve actively harming one's infant. Abandonment may involve leaving an infant in a place which makes it either likely or unlikely to be found. But in either case the welfare of the infant which is abandoned becomes uncertain, since the parents are no longer responsible for the abandoned child's welfare. Abandonment may also involve directly turning over control of an infant to individuals or organizations, such as the state or the church. What behaviors qualify as abandonment? For example, did sending your child to almost certain death in the children's crusade in 1212 represent abandonment? Also, was the practice of oblation (the offering of one's child to a monastery to be raised in the service of God) during Medieval times a form of abandonment by parents (Boswell, 1988)? The motivation in either case above could have ranged from a desire to abrogate responsibility for the child to stoically performing a religious duty.

It is difficult to know whether abandonment and the Greek "exposure" are synonymous, and for some time there has been considerable controversy concerning whether exposure with the intent and the consequence of children dying was common in ancient Greece and Rome. Exposure is a term which appears to be the best translation from ancient Greek to describe the practice of placing infants in places where they would die of exposure or be killed by animals. Alternatively, exposed infants might be picked up and raised as slaves, prostitutes, beggars, or adopted and reared with the care expected of adoptive parents today.

One problem encountered when attempting to quantify abandonment, abuse and infanticide in ancient times or throughout the Christian era in Europe is that there are very few recorded data available which provide the type of information needed to accurately determine the degree to which these activities occurred (Scrimshaw, 1984). Indeed, even today trying to assess the real incidence of these behaviors is very difficult (Daly and Wilson, 1984; 1988). It is thus not surprising that there has been controversy concerning the likelihood that abandonment or infanticide involving healthy infants was common in ancient Greece and Rome (Harris, 1982). For example, the possibility that exposure of infants was common in ancient Athens is not accepted by all historians, even though it appears that there were fewer children in Athenian families than one might expect based on other contemporary societies (Bolkenstein, 1922; Engels, 1980;

Golden, 1981). In contrast, others have proposed that exposure of infants (both healthy and deformed) was quite common in antiquity, particularly after the fourth century B.C. (Viljoen, 1959; Harris, 1982). Viljoen (1959) cited a passage by Polybius in which the shortage of soldiers throughout Greece in the second century B.C. was attributed to the refusal to rear more than one child born into a family. Aristotle argued for limiting family size through the use of exposure or abortion, but there did appear to be public opinion opposing these practices.

Boswell (1988, p. 139) identified three aspects relating to parent-child interaction in the Old Testament which may have had an influence on attitudes of early Christians toward children: 1. child sacrifice, 2. selling of children, and 3. abandonment. One certainly is struck by the degree to which children were victimized in the Old Testament, with passages even involving the sacrifice of children. It is not suggested that Hebrews engaged in or condoned child sacrifice, but just that there were stories in the Old Testament in which there was violence toward children (some of these stories may have served as examples of the requirement to obey God).

The selling of children, particularly females, was apparently common in pre-Christian Jewish culture; presumably this was mainly done by the poor who needed money and could not afford to raise their children. Viljoen (1959) discusses Solon's law forbidding parents in Athens from selling their children, which suggests that this practice was also not uncommon in ancient Greece. It is generally assumed that the majority of abandoned infants were used as slaves. Many laws enacted in Medieval Europe related to the status of infants of free parents who were abandoned or sold and raised as slaves (Boswell, 1988, p. 205). Even if abandoned infants were adopted and reared by foster parents, they may not have fared as well as children that were reared by their biological parents. There is evidence that in contemporary society (and in prior times) step children have a higher likelihood of abuse and death, and for a variety of reasons have a lower reproductive fitness, than genetically related offspring within families (Flinn, 1988; Daly and Wilson; Hrdy; Wilson and Daly, this volume).

While there is still controversy about this subject, there is considerable evidence that abandonment was common throughout Greek and Roman antiquity, the early Christian era, the Middle Ages, and particularly during the industrial revolution in Western societies (Boswell, 1988). A review of the evidence concerning the fate of abandoned infants who did survive in Western societies in prior centuries leads to the conclusion that life was probably so harsh for many abandoned infants that they might have been

better off had they died. It is likely that since most abandoned infants were used as slaves, they were often maltreated. For example, during Roman antiquity Seneca (the elder) described a man who mutilated exposed children, since this would enhance their value as beggars; supposedly this also occurred in China during the Ming dynasty (Boswell, 1988, pp. 60, 113). In addition to physical abuse, slaves were used as prostitutes or otherwise sexually exploited (Boswell, 1988, p. 113). Until very recently there was very little understanding of the magnitude of the problem of sexual victimization of children (Cohen, 1985; Browne and Finkelhor, 1986). However, this problem needs to be viewed within the context of the Greco-Roman and Judeo-Christian tradition of selling and abandoning children and public awareness that they were sexually abused and used as prostitutes.

Abandonment of illegitimate infants as an alternative to contraception and abortion

The practice of abandoning or killing infants suspected of being illegitimate dates back at least to ancient Greece, where a husband had the legal right to abandon or kill a newborn if there was reason to suspect that the child was not his (Radin, 1925; Viljoen, 1959). Factors relating to paternity predict infanticide, abuse and neglect of infants today (Daly and Wilson, 1988; this volume). Viljoen (1959) stated that in ancient Greece unwed mothers typically exposed (abandoned) their newborn infants, and the interest was in concealing the disgrace of becoming pregnant out of wedlock as opposed to having to conceal the act of abandonment, *per se*; as discussed above, abandonment appears to have been common during Greek and Roman antiquity and throughout the Christian era in Europe. Beginning in the fourth century A.D., the laws which were enacted throughout the Roman Empire reflected Christian doctrine, and as in previous times, the major concern was again with the production of illegitimate children, not with their disposition after birth. The major proscription in Greek, Roman and Christian cultures was thus against premarital sex, there did not appear to be widespread concern with child welfare. Women who became pregnant out of wedlock were harshly treated in Medieval Europe, and this may have resulted in the hiding of pregnancy and then abandonment or killing of the newborn infants to avoid prosecution for committing the sin of fornication; this would have been particularly true for nuns who became pregnant (Brundage, 1987, p. 151; Boswell, 1988, p. 210).

While today there continues to be a stigma associated with becoming pregnant out of wedlock, there is no stigma associated with abandonment (giving one's infant to a state or church agency) of an infant by a single

mother (or even married parents). Abandonment of the child by a parent who is unable or unwilling to provide care for an infant is, and has traditionally been, acceptable. Actual killing of an infant by a parent or abandonment with the intent of the infant dying has not been viewed as acceptable in Christian societies, but even today the killing of a newborn is considered to be quite different from the murder of an older child or an adult.

Leaders of many of the major Christian religions today (most notably the Catholic Church) continue to oppose both the use of contraceptives to avoid pregnancy and elective abortion after detection of an unwanted pregnancy. This represents a continuation of the policy of actively promoting carrying an unwanted fetus to term and then relinquishing control of the infant (i.e., abandonment) while strongly opposing any attempts at controlling reproduction. The obsession by the Catholic clergy with abstinence as opposed to controlling fertility is based on the equating of sex with sin, although this idea was not incorporated into Catholic theology until centuries after the death of Jesus (Brundage, 1987).

The Catholic Church has provided support to institutions (foundling homes) for abandoned infants since early Medieval times (Bloch, 1988; Boswell, 1988, p. 219). A fundamental assumption made by opponents of allowing women to choose whether or not to become pregnant (by using contraceptives) or continue an unwanted pregnancy (by elective abortion) is that any child produced will be cared for by foster parents, the Church or the State. However, the promotion by the Catholic Church of abandonment in prior times and today must be viewed within the context of the evidence reviewed above concerning the likelihood that, throughout recorded history, abandoned infants who survived have had a high risk of being neglected and abused. Pressuring a woman to keep an unwanted child is not always the optimum alternative strategy, since unwanted children who are not abandoned may face an increased likelihood of abuse by their parents (Roberts, 1988).

A major issue in the welfare of infants is whether the infant is a valued commodity. Because abortion is now legal in the United States, infants are a valued commodity (in the United States there are currently 2 million couples trying to adopt babies). Were abortion to become illegal, there would once again be a substantial surplus of infants relative to potential adoptive parents (which is why prior to legal abortions there were foundling homes for abandoned infants). A critical issue today is the fate of the children of women who are drug addicts (these children are often addicted to the drugs that their mothers used while pregnant). Should pregnant women who are addicted to drugs be encouraged to carry an unwanted

fetus to term? Who will provide and care for these children? During the eighteenth and nineteenth centuries children ceased to be a valued commodity, and foundling homes and wet nurses became agents of infanticide (Hrdy, this volume).

One tragic example of a situation today where the consequences of the opposition to the control of pregnancy by the Catholic Church has reached tragic proportions is in Brazil; there are tens of thousands of abandoned, homeless children living like wild animals on the streets of cities. While this was certainly not the consequence envisioned by the Church fathers of the policy opposing birth control and elective abortion, it is the reality which must be faced as the we move into the twenty-first century while undergoing a staggering rate of population growth in undeveloped countries.

CONCLUSION

A review of the treatment of infants in Greek and Roman antiquity and throughout the Christian era in Western societies suggests that abandonment and selling of infants was not uncommon. Most infants who were sold or abandoned and survived had a very harsh existence. Social and religious customs increased the likelihood of abandonment of infants. During Greek and Roman antiquity, but particularly during the Christian era, there was a much greater concern with regulating the sexual activity of women than with the welfare of infants, particularly those which were illegitimate. Christian leaders opposed to controlling reproduction (by using contraceptives or elective abortion) today continue to ignore the consequences for infants who are produced and then abandoned.

Corporate and government policies have also impacted the health and welfare of infants. Government officials have allowed corporations to aggressively market infant formulas in undeveloped countries. The giving away of free samples of infant formulas to mothers in undeveloped countries, where adequate preparation of the formulas with uncontaminated water is not possible, continues today despite decades of evidence that this greatly increases the likelihood of infant disease and death. The corporate and government officials who have continued to promote the use of infant formulas in undeveloped countries rather than decrease profits are responsible for the death of many infants.

In Western societies one finds infants being subjected to a variety of maladaptive medical practices. Some medical practices have directly led to an increase in infant distress. For example, mothers in Western societies

have been faced with physicians telling them to engage in feeding of infants at intervals of 3 to 5 hours, which may result in crying by an infant due to the lack of food in its stomach; infants allowed constant access to the breast feed at much shorter intervals. Mothers were also told of the advantages of using infant formulas either as the sole source of food or as a supplement to breastmilk; supplements may be needed due to inadequate milk production as a result of long intervals between nursing. Particularly in early infancy, the use of formulas or other foods increases the likelihood of introducing disease-causing bacteria into the digestive system at a time when infants rely on breastmilk to provide maternal antibodies with which to combat bacteria in the gut. The result is that the likelihood of having a crying infant has been greatly increased by these practices. Physicians and psychologists then insisted that infants should be left alone in a crib or playpen without being touched while crying in distress. This advice was based on the misguided and now discredited idea that attending to a crying infant would condition the infant to cry even more, thus leading to a spoiled child. The parents who acquiesced to these misguided infant-rearing practices have found themselves in a double bind. A crying infant is highly aversive, and it is very stressful for parents or other bystanders to listen to a crying infant. The natural response of most parents is to alleviate the infant's distress (thereby also alleviating the parent's distress). However, this requires attending to the infant, which parents have been told will lead to a spoiled child. While there are many factors involved in child abuse, there is no doubt that these infant-rearing practices have contributed to the abuse and death of infants by parents or other adults who responded to prolonged crying by an infant with violence.

For centuries parents have been bombarded with child-rearing advise by well-meaning "experts". Parents should not be afraid to reject advice and change strategies when it becomes obvious that either they or their infant are not responding well to the parenting strategy being used. No optimum parenting strategy could possibly exist given the tremendous individual differences among both parents and infants and the wide variety of environments inhabited by people.

ACKNOWLEDGEMENTS

I thank Sarah Blaffer Hrdy, Hugh Humphrey, and Harvey Rosenfeld for valuable discussions and Michael Boechler, David Geary, and Mark Flinn for reviewing a prior version of the chapter. Support during the writing of this chapter was provided by NSF grants DCB 9004806 and INT 8515097.

REFERENCES

ALEXANDER, R.D. (1979) *Darwinism and Human Affairs*. University of Washington Press, Seattle.

ALTMANN, J. (1980) *Baboon Mothers and Infants*. Cambridge, Harvard University Press.

BELSKY, J., Steinberg, L. and Draper, P. (1991) Childhood experience, interpersonal development, and reproductive strategy: An evolutionary theory of socialization. *Child Development*, **62**, 647–670.

BLOCH, H. (1988) Abandonment, infanticide, and filicide. *American Journal of Diseases of Children*, **142**, 1058–1060.

BLURTON JONES, N. (1972) Comparative aspects of mother-child contact. In: N.B. Jones (ed.), *Ethological Studies of Child Behaviour.* Cambridge University Press, Cambridge, pp. 305–328.

BOLKESTEIN, H. (1922) The exposure of children at Athens. *Classical Philology*, **17**, 222–239.

BONGAARTS, J. and Greenhaigh, S. (1985) An alternative to the one-child policy in China. *Population and Development Review*, **11**, 585–617.

BOUKYDIS, C.F.Z. (1985) Perception of infant crying as an interpersonal event. In: *Infant Crying. Theoretical and Research Perspectives*. B.M. Lester and C.F. Zachariah Boukydis (eds.), Plenum Press, New York, pp. 187–215.

BOWLBY, J. (1969) *Attachment and Loss: Vol. 1. Attachment*. Basic Books, New York.

BOSWELL, J. (1988) *The Kindness of Strangers: The Abandonment of Children in Western Europe from Late Antiquity to the Renaissance*. Pantheon, New York.

BRONFEMBREMMER, U. (1986) Ecology of the family as a context for human development. *Developmental Psychology*, **22**, 723–742.

BROWNE, K. (1988) The nature of child abuse and neglect: An overview. In: *Early Prediction and Prevention of Child Abuse*. K. Browne, C. Daves and P. Stratton, (eds.), John Wiley and Sons, New York, pp. 15–30.

BROWNE, A. and Finkelhor, D. (1986) Impact of child sexual abuse: A review of the research. *Psychological Bulletin*, **99**, 66–77.

BRUNDAGE, J.A. (1987) *Law, Sex, and Christian Society in Medieval Europe*. University of Chicago Press, Chicago.

CADOGAN, W. (1750) Essay upon nursing and the management of children from birth to three years of age. In: *Three Treatises on Child Rearing* (1985). New York, Garland Pub.

CAMERON, A. (1932) The exposure of children and Greek ethics. *The Classical Review*, **46**, 105–114.

CAMPBELL, C.E. (1984) Nestle and breast vs. bottle feeding: Mainstream and Marxist perspectives. *International Journal of Health Services*, **14**, 547–565.

CHATEAU, P. de (1980) Parent-neonate interaction and its long-term effects. In: *Early Experience and Early Behavior: Implications for Social Development*. E.C. Simmel (ed.), Academic Press, New York.

CLEMENT, C. and Colborn, T. (1992) Herbicides and fungicides: A perspective on potential human exposure. In:*Chemically Induced Alterations in Sexual and Functional Development: The Wildlife-Human Connection*. T. Colborn and C. Clement (eds.), Princeton Scientific Pub., Princeton, N.J., pp. 347–364.

COHEN, A. (1985) Sexual abuse of children – a review. *South African Medical Journal*, **67**, 730–732.

COLBORN, T. and Clement, C. (1992) *Chemically Induced Alterations in Sexual and Functional Development: The Wildlife-Human Connection*. Princeton Scientific Pub., Princeton, N.J.

COLBORN, T., vom Saal, F. and Soto, A. (1993) Developmental effects of endocrine disrupting chemicals in wildlife and humans. Environmental Health Perspectives, **101**, 378–384.

COLEMAN, E. (1976) Infanticide in the early middle ages. In: Susan Mosher Stuard (ed.), *Women in Medieval Society*, Philadelphia, University of Pennsylvania Press. pp. 47–70.

CREIGHTON, S.J. (1988) The incidence of child abuse and neglect. In: *Early Prediction and Prevention of Child Abuse*. K. Browne, C. Daves and P. Stratton, (eds.), John Wiley and Sons, New York, pp. 31–41.

DALY, M. and Wilson, M. (1984) A sociobiological analysis of human infanticide. In: *Infanticide. Comparative and Evolutionary Perspectives*. G. Hausfater and S.B. Hrdy, (eds.), Aldine, New York, pp. 487–502.

DALY, M. and Wilson, M. (1988) Evolutionary social psychology and family homicide.*Science*, **242**, 519–524.

DAVIS, F. (1981) Ontogeny of circadian rhythms. In: *Handbook of Behavioral Neurobiology*, *Vol. 4*. J. Ashcroff, (ed.), Plenum Press, New York. pp. 257–274.

DE MEER, K. (1988) Mortality in children among the Aymara indians of southern Peru. *Society of Scientific Medicine*, **26**, 253–258.

DEMOS, V. (1986) Crying in early infancy: An illustration of the motivational function of affect. In: T.B. Brazelton and M.W. Yogman (eds.), *Affective Development in Infancy*. Ablex, Norwood, N.J., pp. 39–74.

DUFF, R.S. and Campbell, A.G.M. (1973) Moral and ethical dilemmas in the special-care nursery. *New England Journal of Medicine*, **289**, 890–894.

ELKIN, F. and Handel, G. (1989) *The Child and Society: The Process of Socialization*. Fifth Edition, Random House, New York.

ENGELS, D. (1980). The problem of female infanticide in the Greco-Roman world. *Classical Philogy*, **75**, 112–120.

FLANDRIN, J.L. (1979) *Families in Former Times: Kinship, Household and Sexuality*. Cambridge University Press, Cambridge.

FLINN, M.V. (1988) Step and genetic parent/offspring relationships in a Caribbean village. *Ethology and Sociobiology*, **9**, 335–369.

FORBES, T.R. (1986) Deadly parents: child homicide in eighteenth- and nineteenth-century England. *Journal of the History of Medicine and Allied Sciences*, **41**, 175–199.

FRODI, A. (1985) When empathy fails. Aversive infant crying and child abuse. In: *Infant Crying. Theoretical and Research Perspectives*. B.M. Lester and C.F. Zachariah Boukydis (eds.), Plenum Press, New York, pp. 263–277.

GEARY, D.C. Sociobiology of human intelligence. Submitted.

GELLES, R.J. (1987) What to learn from cross-cultural and historical research on child abuse and neglect: an overview. In: Gelles, R.J. and Lancaster, J.B. *Child Abuse and Neglect: Biosocial Dimensions*. Aldine de Gruyter, New York, pp. 15–30.

GELLES, R.J. and Cornell, C.P. (1983) International perspectives on child abuse. *Child Abuse and Neglect*, **7**, 375–386.

GOLDEN, M. (1981) Demography and the exposure of girls at Athens. *Phoenix*, **35**, 316–331.

GREEN, A.H. (1988) Child maltreatment and its victims. *Psychiatric Clinics of North America*, **11**, 591–609.

68   F.S. VOM SAAL

HABICHT, J.-P., DaVanzo, J. and Butz, W.P. (1986) Does breastfeeding really save lives or are apparent benefits due to biases? *American Journal of Epidemiology*, **123**, 279–290.

HARDYMENT, C. (1984) *Dream Babies: Child Care from Locke to Spock.* Oxford, Oxford University Press.

HARLOW, H.F., Harlow, M.K., Dodsworth, R.O. and Arling, G.L. (1966) Maternal behavior of Rhesus monkeys deprived of mothering and peer association in infancy. *Proceedings of the American Philosophical Society*, **110**, 58–66.

HARRIS, W.V. (1982) The theoretical possibility of extensive infanticide the Graeco-Roman world. *Classical Quarterly*, **32**, 114–116.

HERBERT, M., Sluckin, W. and Sluckin, A. (1982) Mother-to-infant 'bonding.' *Journal of Child Psychology*, **23**, 205–221.

HEWLETT, B.S. (1992a) Introduction. In: *Father-Child Relations: Cultural and Biosocial Contexts.* B.S. Hewlett (ed.), Aldine de Gruyter, New York, pp. xi–xix.

HEWLETT, B.S. (1992b) Husband-wife reciprocity and the father-infant relationship among Aka pygmies. In: *Father-Child Relations: Cultural and Biosocial Contexts.* B.S. Hewlett (ed), Aldine de Gruyter, New York, pp. 153–176.

HRDY, S.B. (1987) Sex-biased parental investment among primates and other mammals: A critical evaluation of the Trivers-Willard Hypothesis. In: R. Gelles and J. Lancaster (eds.), *Child Abuse and Neglect: Biosocial Dimensions.* New York, Aldine, pp 97–147.

HURTADO, A.M. and Hill, K.R. (1992) Paternal effect on offspring survivorship among Ache and Hiwi hunter-gatherers: Implications for modeling pair-bond stability. In: *Father-Child Relations: Cultural and Biosocial Contexts.* B.S. Hewlett (ed.), Aldine de Gruyter, New York, pp. 31–55.

JEFFERY, R., Jeffery, P. and Lyon, A. (1984) Female infanticide and amniocentesis. *Society of Scientific Medicine*, **19**, 1207–1212.

JENSEN, A.A. and Slorach, S.A. (1991) *Chemical Contaminants in Human Milk.* Boston, CRC Press.

JOHANSSON, S.R. (1984) Deferred infanticide: Excess female mortality during childhood. In: *Infanticide. Comparative and Evolutionary Perspectives.* G. Hausfater and S.B. Hrdy, (eds.), Aldine, New York, pp. 463–485.

JOHANSSON, S.R. (1987) Neglect, abuse, and avoidable death: Parental investment and the mortality of infants and children in the European tradition. In: Gelles, R.J. and Lancaster, J.B. *Child Abuse and Neglect: Biosocial Dimensions.* Aldine de Gruyter, New York, pp. 57–93.

KATZ, M.M. and Konner, M.J. (1981) The role of the father: An anthropological perspective. In:*The Role of the Father in Child Development.* M.F. Lamb (ed.), John Wiley and Sons, New York, pp. 155–185.

KESSEN W. and Leutzendorff, A. (1963) The effect of non-nutritive suckling on movement in the human newborn. *Journal of Comparative and Physiological Psychology*, **56**, 69–72.

KONNER, M. (1982) Biological aspects of the mother-infant bond. In: *The Development of Attachment and Affiliative Systems.* R.N. Emde and R.J. Hammond (eds.). Plenum Press, New York, pp. 137–159.

KONNER, M. (1991) *Childhood.* Little, Brown Pub. Co., Boston.

KORBIN, J.E. (1987) Child maltreatment in cross-cultural perspective: Vulnerable children and circumstances. In: Gelles, R.J. and Lancaster, J.B., *Child Abuse and Neglect: Biosocial Dimensions.* Aldine de Gruyter, New York, pp. 31–55.

KOPP, C.B. (1989) Regulation of distress and negative emotions: A developmental view. *Developmental Psychology*, **25**, 343–354.

KUHSE, H. and Singer, P. (1987) II. Death and dying. 3. Debate: Severely handicapped newborns. For sometimes letting — and helping die. *Law, Medicine and Health Care*, **14**, 149–157.

LAMB, M.E. and Campos, J.J. (1982) *Development in Infancy. An Introduction*. Random House, New York.

LESTER, D. (1986) The relation of twin infanticide to status of women, societal aggression, and material well-being. *Journal of Social Psychology*, **126**, 57–59.

LEVENTHAL, B.L. and Midelfort, H. B. (1986) The physical abuse of children. A hurt greater than pain. *Advances in Psychosomatic Medicine*, **16**, 48–83.

LISTER, D. (1986) Ethical issues in infanticide of severely defective infants. *Canadian Medical Association Journal*, **135**, 1401–1404.

LITHELL, U.-B. (1981) Breast-feeding habits and their relation to infant mortality and marital fertility. *Journal of Family History*, **6**, 182–194.

LONG, T.A. (1988) Infanticide for handicapped infants: sometimes it's a metaphysical dispute. *Journal of Medical Ethics*, **14**, 79–81.

LYNCH, M.A. (1985) Child abuse before Kempe: An historical literature review. *Child Abuse and Neglect*, **9**, 7–15.

MACDONALD, K.B. (1988) *Social and Personality Development. An Evolutionary Synthesis*. Plenum Press, New York.

MARSHALL, E. (1983) Nestle letter stops NIH talk. *Science*, **219**, 469.

MEADE, J.L. and Brissie, R.M. (1985) Infanticide by starvation: Calculation of caloric deficit to determine degree of deprivation. *Journal of Forensic Sciences*, **30**, 1263–1268.

MILLER, B.D. (1981) *The Endangered Sex: Neglect of Female Children in Rural North India*. Cornell University Press, Ithaca.

MONTAG, B.A., and Montag, T.W. (1979) Infanticide: A Historical Perspective. *Minnesota Medicine*, **62**, 368–372.

MOSELEY, K.L. (1986) The history of infanticide in western society. *Issues in Law and Medicine*, **1**, 345–361.

MURRAY, A. (1985) Aversiveness is in the mind of the beholder. Perception of Infant crying by adults. In: *Infant Crying. Theoretical and Research Perspectives*. B.M. Lester and C.F. Zachariah Boukydis (eds.), Plenum Press, New York, pp. 217–239.

NELSON, H.B. and Martin, C.A. (1985). Increased child abuse in twins. Child Abuse and Neglect 9, 501–505.

NEY, P.G. (1988) Transgenerational child abuse. *Child Psychiatry and Human Development*, **18**, 151–168.

OBER, W.B. (1986) Infanticide in eighteenth-century England: William Hunter's contribution to the forensic problem. *Pathology Annual*, **21**, 311–319.

O'DONNELL, J.M. (1985) *The origins of behaviorism: American Psychology 1870–1920*. New York University Press, New York.

OSTWALD, P.F. and Murry, T. (1985) The communicative and diagnostic significance of infant sounds. In: *Infant Crying. Theoretical and Research Perspectives*. B.M. Lester and C.F. Zachariah Boukydis (eds.), Plenum Press, New York, pp. 139–158.

PALMER, G. (1988) *The Politics of Breastfeeding*. Pandora, London.

POPKIN, B.M., Bilsborrow, R.E., and Akin, J.S. (1982) Breast-feeding patterns in low-income countries. *Science*, **218**, 1088–1093.

RADIN, M. (1925) The exposure of infants in Roman law and practice. *The Classical Journal*, **20**, 337–343.

REPPERT, S.M. and Rivkees, S.A. (1989) Development of human circadian rhythms: Implications for health and disease. In:*Development of Circadian Rhythmicity and Photoperiodism in Mammals*. S.M. Reppert (ed.), Perinatology Press, pp. 245–259.

ROBERTS, J. (1988) Why are some families more vulnerable to child abuse. In: *Early Prediction and Prevention of Child Abuse*. K. Browne, C. Daves and P. Stratton, (eds.), John Wiley and Sons, New York, pp. 43–56.

ROSE, L. (1986) *The Massacre of the Innocents: Infanticide in Britain 1800–1939*. Routledge and Kegan Paul, London.

ROZIN, P. and Kalat, J.W. (1971) Specific hungers and poison avoidance as adaptive specializations of learning. *Psychological Review*, **78**, 459–486.

SAXENA, R.K. (1975) *Social Reforms: Infanticide and Sati*. Trimurti Publications, New Delhi.

SCRIMSHAW, S.C.M. (1984) Infanticide in human populations: Societal and individual concerns. In: *Infanticide. Comparative and Evolutionary Perspectives*. G. Hausfater and S.B. Hrdy, (eds.), Aldine, New York, pp. 439–462.

SELIGMAN, M.E.P. (1970) On the generality of the laws of learning. *Psychological Review*, **77**, 406–418.

SHORT, R.V. (1976) The evolution of human reproduction. *Proceedings of the Royal Society of London B.*, **195**, 3–24.

SHORT, R.V. (1984) Breast feeding. *Scientific American*, **250**, 35–41.

SPOCK, B. and Rothenberg, M. (1985).*Baby and Child Care*. Simon and Schuster, New York.

SVEJDA, M.J., Pannabecker, B.J. and Emde, R.N. (1982) Parent-to-infant attachment. A critique of the early "bonding" model. In:*The Development of Attachment and Affiliative Systems*. R.N. Emde and R.J. Hammond (eds.). Plenum Press, New York, pp. 137–159.

THOMAS, K. and Colborn, T. (1992) Organochlorine endocrine disruptors in human tissue. In: *Chemically Induced Alterations in Sexual and Functional Development: The Wildlife-Human Connection*. T. Colborn and C. Clement (eds.), Princeton Scientific Pub., Princeton, N.J. pp. 365–394.

TRIVERS, R. (1974) Parent-offspring conflict. *American Zoologist*, **14**, 249–264.

TUCKER, H.A. (1988) Lactation and its hormonal control. In: *The Physiology of Reproduction*. E. Knobil and J. Neill, *et al.* (eds.). Raven Press, New York, pp. 2235–2263.

VILJOEN, G. van N. (1959) Plato and Aristotle on the exposure of infants at Athens. *Acta Classica*, **2**, 58–69.

VICTORA, C.G., Smith, P.G., Vaughan, J.P., Nobre, L.C., Lombardi, C., Teixeira, A.M.B., Fuchs, S.M.C., Moreira, L.B., Gigante, L.P. and Barros, F.C. (1987) Evidence for protection by breast-feeding against infant deaths from infectious diseases in Brazil. *Lancet* , **2**, 319–324.

VOM SAAL, F., Montano, M. and Wang, M. (1992) Sexual differentiation in mammals. In: *Chemically Induced Alterations in Sexual and Functional Development: The Wildlife-Human Connection*. T. Colborn and C. Clement (eds.), Princeton Scientific Pub., Princeton, N.J., pp. 17–83.

WAKERLEY, J.B., Clarke, G. and Summerlee, A.J.S. (1988) Milk ejection and its control. In: *The Physiology of Reproduction*. E. Knobil and J. Neill, *et al.* (eds.). Raven Press, New York, pp. 2283–2321.

WHITTEN, P.L., Lewis, C. and Naftolin, F. Maternal exposure to a phytoestrogen diet induces the premature anovulatory syndrome in rats. Submitted.

WHITTEN, P.L. and Naftolin, F. (1991) Dietary plant estrogens: A biologically active background for estrogen action. In: *The New Biology of Steroid Hormones*. R. Hochberg and F. Naftolin (eds.), Raven Press, New York, pp. 155–167.

WIDOM, C.P. (1989) Does violence beget violence? A critical examination of the literature. *Psychological Bulletin*, **106**, 3–28.

WOLFF, P.H. (1968) Sucking patterns in infant mammals.*Brain, Behavior and Evolution*, **1**, 354–367.

WOLFF, P.H. (1969). The natural history of crying and other vocalizations in early infancy. In: B. Foss (ed.),*Determinants of Infant Behavior* (Vol 4). Methuen, London, pp. 81–110.

CHAPTER 3

THE PSYCHOLOGY OF PARENTING IN EVOLUTIONARY PERSPECTIVE AND THE CASE OF HUMAN FILICIDE

MARGO WILSON and MARTIN DALY

Department of Psychology, McMaster University, Hamilton, Ontario, L8S 4K1, Canada

Child rearing is effortful. Indeed, the parental investment of time, effort, and resources in children, and the incurring of risk on their behalf, have constituted the major lifetime commitments of most people throughout most of human history. But in spite of the effort, children are fervently desired: Even in industrial societies that have undergone the demographic transition from high to low fertility, a substantial majority of adults want children.

Not only is child rearing in general effortful, but in a low-fertility species like *Homo sapiens,* each child-rearing episode consumes a major proportion of the parent's lifetime efforts. In natural-fertility populations living primarily by foraging, a woman who survived until menopause was likely to give birth to only about 4 or 5 children in her lifetime and to nurse each of them for several years (*e.g.* Lozoff *et al* 1977; Howell, 1979; Blurton Jones, 1989). Low fertility and wide birth-spacing must have characterized our species throughout its long pre-agricultural past.

There is variation among societies in the relative involvement of mothers and fathers in childcare and other forms of parental investment, and this variation warrants study in terms of the differential utilities of women's and men's parental efforts (for sexual division of labor and parenting duties among foraging peoples, see West and Konner, 1976; Hill and Kaplan, 1988a,b; Hewlett, 1988; Blurton Jones, 1989). This variation should not obscure such regularities, however, as the fact that human mothers everywhere assume the major responsibility for physical care-

giving and nourishment of small children, and they do so regardless of whether the conjugal division of labor is such that husbands compensate in spheres other than childcare.

Parenting constitutes a major domain of human effort, then, but it is a remarkably underrepresented domain in discussions of human motivation (Daly and Wilson, 1988b). Arguably, psychology has neglected the subject of parental motives for want of a theoretical framework within which to understand variations therein. The requisite theoretical framework is that of evolution by selection.

EVOLUTIONARY PSYCHOLOGY OF DISCRIMINATIVE PARENTAL SOLICITUDE

Few psychologists would quarrel with the claim that the psyche has evolved by selection, but until recently, too few have derived direction or inspiration from what evolutionary biologists know about the process of evolution by selection. Even comparative psychologists interested in parental motivation have focused mainly on demonstrations of the proximate causal roles of hormones and the stimulus properties of the young in the establishment and maintenance of maternal responsiveness in a few species, without addressing why natural selection has designed the motivational and information-processing mechanisms subserving parental behavior to respond as they do.

Parental behavior is not reflexively evoked by infantile conspecifics. It is instead complexly contingent, and the particular ways in which it is contingent have an adaptive logic as a result of a history of selection. An obvious example concerns the fact that evolved parental motivational systems are typically sensitive to cues of individual identity and not simply to species-typical cues of infancy, since one important determinant of whether a youngster is a suitable vessel for parental investment is whether it is the parent's own (Daly and Wilson, this volume). Moreover, a parent might be expected to behave discriminatively even among its own young according to cues that were predictive of fitness returns in the past. Although offspring are a parent's means to genetic posterity, parent-offspring conflict is an endemic feature of sexually reproducing organisms because the allocation of resources and efforts that would maximize a parent's genetic posterity does not necessarily maximize a particular offspring's (Trivers, 1974, 1985). Parental psyches that have been shaped by selection are therefore discriminative psyches, investing parental effort preferentially where it was likely to have yielded the greatest returns during evolutionary history.

Psychology has been concerned with the characterization of behavioral control mechanisms, not with their evolution, and it is certainly possible for psychologists to discover mechanisms of parental psychology by observation and experiment without recourse to evolutionary thinking. However, psychology's task can be greatly facilitated by consideration of the natural selective circumstances that have shaped the mechanisms modulating parental inclinations and behaviors. This is not to say that such evolutionary theorizing can substitute for empirical research, but simply that theory can suggest what adaptations would be useful under one or another set of conditions, and thus help us decide what sort of mechanisms we should be looking for.

Consider, for example, variability in parental commitment to the defense of helpless young. A simple evolutionary theoretical insight is that fitness benefits will accrue to parents whose psyches assess available predictors of offspring's eventual contributions to parental fitness and adjust defense accordingly. A stickleback fish guarding a nest full of eggs will stand his ground against an approaching predator longer, and dart at the predator more bravely, the more eggs he has in the nest (Pressley, 1981). In effect, the greater fitness value of a larger brood elevates the statistical probability of death that the stickleback is prepared to accept. One correlate of brood size, which might be the cue modulating fear versus bravery in this context, is carbon dioxide production by the eggs, and if so, then it is likely that this cue will prove to mitigate fearfulness only in egg-guarding males. One would be unlikely to discover such contextual variation in the controls of fearfulness without the basic Darwinian insight that even personal survival is a subordinate objective to that of genetic posterity. Because fitness is the ultimate arbiter of adaptation, motivational systems have evolved to respond to whatever cues are predictive of the fitness consequences of behavioral options, and realizing this can direct even the most mechanistically reductionist student of psychological processes to better hypotheses.

A SELECTIONIST PERSPECTIVE ON DISCRIMINATIVE PARENTAL SOLICITUDE

Parents "invest" in their young in various ways such as depleting their bodily reserves in lactation, allocating their time to protection and care, and suffering risks to their own lives on behalf of the young. These and other physiological processes and behavioral activities are all instances of "parental investment" (Trivers, 1972, 1974): their common denominator

is that they all entail the parent's contributing to the expected fitness of
the young who receive the investment at some cost to the parent's expected fitness through other avenues. Moreover, the "strategic" commonality among these various forms of investment provides a rationale for ex-
pecting that there will be some commonality of causation as well. Any
offspring whose characteristics make it a good bet to yield fitness for one
sort of parental investment will usually be a good bet for other sorts, and
divestment from lost causes should apply to all manner of parental investments, too. In other words, we may expect parental motivational systems
to contain processes and structures that function as if mediated by a unitary parameter (an "intervening variable"[1]), which is influenced by a variety of parental, offspring and situational cues of offspring-specific expected
fitness, and which influences in its turn the whole gamut of parental activities. This intervening variable might be called offspring-specific parental
love or solicitude.

Our proposal, then, is that the individualized love that parents feel for
their offspring varies between individuals and over time because such love
is an evolved proxy for the expected contribution of each offspring to
parental fitness. If this is so, then offspring-specific parental love may be
expected to vary according to:

(1)   the degree or certainty of parent-offspring relatedness,

(2)   phenotypic attributes of the child predictive of its eventual fitness,

(3)   situational predictors of an offspring's fitness value, and

(4)   the alternative reproductive and investment opportunities of the
      parent.

These considerations bear directly on questions of how parents are likely to
perceive costs and benefits of their various options for allocating reproductive effort over the lifespan of both parent and offspring (*e.g.* Andersson *et
al* ., 1980; Pugesek, 1983; Daly and Wilson, 1981, 1984, 1988a,b,c; Knight
and Temple, 1986; Montgomerie and Weatherhead, 1988; Redondo, 1989;
Thornhill, 1989).

1. Genetic Relatedness

If a parent were to have any grounds for doubt that a putative offspring
were indeed his or her own, then this doubt would warrant an effective

---

[1]Miller (1959), argued that motivational theorists are justified in postulating abstract con-
structs like "thirst", if the task of summarizing the effects of several independent variables
upon several dependent variables is simplified by assuming that the effects are mediated
through a single "intervening variable".

discounting of the parental perception of the benefits (or costs) of any increments (or decrements) in that offspring's expected fitness. The coefficient of relatedness (r) between parent and offspring is 0.5 if the parents are unrelated, but potentially higher in the case of inbreeding and potentially lower in the case that the putative relationship is uncertain.

The problem of misattributed parenthood is sexually asymmetrical in organisms with internal fertilization. This asymmetrical risk of cuckoldry is probably one reason for the rarity of significant paternal investment in mammals; in birds the males may run the same risks but in most species males participate anyway. Recent genetic studies of biparental, "monogamous" bird species in the field are revealing some possibly surprising levels of non-paternity (Westneat et al 1990). Do males then adjust their parental efforts in relation to probabilistic indicators of paternity? At least two studies (Møller, 1988b; Davies, 1990; Hatchwell and Davies, 1990) indicate that they do, the males in both cases reducing their posthatching feeding efforts in relation to earlier lapses in their surveillance of their fertile mates during the egg-laying phase.

In biparental birds, unlike biparental mammals, females also incur some risk of misdirecting parental investment to rivals' offspring in natural situations, as a result of the occasional "dumping" of eggs in conspecific nests (e.g. Gowaty and Karlin, 1984; Brown, 1984; Quinn et al 1987). We are not aware of evidence that the victims of intraspecific brood parasitism reduce their own investment, but interspecific parasitism can inspire abandoning the clutch altogether.

2. Phenotypic Attributes and Reproductive Value of Offspring

The reproductive value (RV) of the offspring refers to its own expected fitness given its age, sex, and any other aspects of its phenotype that are predictive.

(a) Age of offspring

If, as in many nestling birds, the daily mortality risk of immatures is high and largely unrelated to their individual phenotypes, then simple age is a major determinant of RV. This is so because each day survived increases the nestling's likelihood of fledging and eventually breeding: a songbird 10 days after hatching has a very much higher expected future fitness than a new hatchling, even though both are equally nestbound and helpless. In such a case, parental valuation of the young may be expected to increase conspicuously over days. There is a large body of empirical

work, mostly involving assessment of parental willingness to incur risk to self in defense of eggs and nestlings, and the results are for the most part confirmatory (reviews by Montgomerie and Weatherhead, 1988; Redondo, 1989).

(b) Sex of offspring

Fisher (1958) argued that equal investment in sons and daughters is an evolutionarily stable strategy (*sensu* Maynard Smith, 1978) because the total RV of males must equal that of females in a sexual population, with the result that selection would favor investing in whichever sex was receiving less total parental investment in the population as a whole. This theorem has been amply supported, and is the foundation of an expanding body of empirical and theoretical work on sex allocation (*e.g.* Charnov, 1982; Stamps, 1990).

Fisher's analysis indicates that a generalized favoring of one sex or the other cannot be adaptive and species-typical, but it does not rule out the possibility of adaptive investment decisions[2] in relation to offspring sex. An important addendum to Fisher's theory was Trivers and Willard's (1973) demonstration that individual parents might profit by investing preferentially in one or the other sex notwithstanding the populational equilibrium. In polygynous species, for example, males exhibit a higher variance in fitness than do females, so from a parent's point of view extra investment in a son of good quality may yield greater returns (measured in grandchild production) than comparable investment in a daughter of good quality, while investment in a son of less than average quality may be wasted. Depending upon the sex determining mechanisms of the species, parents may have little control over the sex of zygotes at fertilization, but they can and do invest differentially in daughters versus sons after birth, sometimes exhibiting adaptive sex-biased investment of the sort that Trivers and Willard envisioned (*e.g.* Clutton-Brock *et al* 1985; Austad and Sunquist, 1986).

---

[2]We refer to "decisions" without necessarily implying consciousness or rationality. Natural selection creates mechanisms of adaptive contingent response such that a seed might be said to "decide" when to germinate in response to moisture cues, *etc.* Some of the "decisions" we discuss, such as whether to raise a child or to abandon it, surely are affected in complex creatures like *Homo sapiens* by some sort of mental construction of hypothetical alternative scenarios, but even this sort of scenario-building may or may not be accessible to consciousness or to retrospective reconstruction. The extent to which "decision" processes are capable of exploiting particular kinds of information is in each case an empirical issue.

(c) Health status of offspring

Investment in moribund offspring is wasted investment, and we might expect evolved motivational mechanisms to be such as to avoid it. Some birds eject cracked eggs from the nest, and some parturient rodents assess the responsiveness of their newborns in the process of licking up the birth fluids, eating or rejecting the stillborn on the spot. However, the differential costs of wasting a little time on the moribund versus prematurely abandoning a temporarily unresponsive offspring may select for conservative decision rules, producing such apparently maladaptive phenomena as continued transport of dead babies by primate mothers. More generally, parents might be expected to modulate investment in relation to probabilistic cues of offspring survival and reproductive success, rather than confining themselves to all-or-none decisions. Alexander (1979) has argued that parents should have evolved to be sensitive to offspring "need", and it is important to note the technical and nonintuitive meaning that he attaches to the word "need", namely the offspring's capacity to transform parental investment into personal fitness. Hopelessly deformed young or runts, for example, may be desperately "in need" in ordinary parlance, but not in Alexander's special sense in which need is predicated upon the potential to promote one's fitness.

In principle, cues of offspring RV could be accessible to the parent not only in the offspring's own phenotype, but also in characteristics of the mate indicative of the genetic quality of one's young with respect to phenotypic attributes not yet expressed (Montgomerie and Weatherhead, 1988). Thus, for example, one might work harder to care for offspring (and incur elevated costs to one's own survival prospects) when one's mate is perceived as being an especially good one, and care less well for the young of an apparently poor mate, even if offspring cues and own condition were identical in the two cases. Such effects have been shown in at least a couple of avian species (Burley 1986; MEDller 1988a) and may be widespread. An analogous sensitivity to cues of male quality is shown by a female scorpionfly deciding how many eggs to fertilize from sperm stores from her most recent mate before remating (Thornhill, 1981); in this case young are not the recipients of differential parental investment after the zygote stage, but females do analogously allocate their reproductive efforts with consideration of male qualities presumed to be predictive of the genetic quality of offspring.

3. Situational Predictors of Fitness Value of Offspring

Irrespective of the quality of the offspring, circumstances may portend poor survival prospects. If food is scarce and parents would only be risking their own lives by persisting in efforts to rear dependent offspring, the young may be abandoned without any chance of surviving on their own. Food shortage may reduce the age-specific reproductive value of the young by affecting maternal condition or by starvation, or by increasing vulnerability to pathogens or to bad weather. Storms or an abundance of parasites sure to attack the young later or other such local factors might warrant adaptive abandonment or reduced investment even when food is abundant, and characteristics of the construction and locale of the breeding site as well as the type and density of predators will also affect the probability of rearing the young to maturity, and hence the optimal level of investment (Montgomerie and Weatherhead, 1988; Redondo, 1989).

4. Alternative Reproductive Options of the Parents

The decision that present circumstances are not favorable for the rearing of young is an implicitly relative one. Some more promising alternative must, at the least, be a possibility in the future, or else the present decision is futile. A parent's options are affected by its own age, sex, quality, and situation (including variations in demography of the local population), which in turn affect the parent's residual reproductive value by affecting its probability of surviving, of re-mating, and of successfully rearing additional broods (*e.g.* Pugesek, 1983, 1987, 1990; Clutton-Brock *et al* 1985; Wallin, 1987). Moreover, in some species including our own, new young may be started before investment in their siblings ends, so that parents may have to consider the detrimental impacts of their investments in the present offspring on both their older offspring and their potential future offspring (*e.g.* Zaias and Breitwisch, 1989; Weatherhead and McRae, 1990).

Furthermore, in biparental species in which a second parent's contributions are not so valuable as to double the expected offspring production of a single parent, a potential conflict of interest arises since either parent might prefer that the other tend their brood while the deserter pursues additional fitness prospects elsewhere (Maynard Smith, 1977; Westneat, 1988; Lazarus, 1990). Existing asymmetries in residual reproductive value and "bargaining power" may affect the probability that one parent will desert or otherwise exhibit reduced parental investment (*e.g.* Beissinger and Snyder, 1987; Rogers, 1988; Zaias and Breitwisch, 1989; Davies, 1990; Hatchwell and Davies, 1990; Weatherhead, 1990).

MEASURES OF PARENTAL EFFORT

It is not easy to test the impacts of these cues of expected fitness returns on variations in parental valuation of offspring. One complication is that changes in RV can be hard to separate from concomitant change in offspring dependency and needs. A measure of parental investment that lends itself particularly well to the development and testing of evolutionary models of parental discriminative solicitude is the defense of young whose vulnerability holds fairly constant while their RV varies.

In birds, predation upon eggs and unfledged young is often intense, such that there is a dramatic increase in offspring RV over days. As noted above, many studies indicate that parental willingness to incur risk to self in defending the young tracks this increase. Moreover, insofar as nest defense decisions entail an effective trade-off between parental valuation of the young and of oneself, parents should have evolved to be sensitive to their own reproductive prospects as well as the young's, and there is some evidence that they are (Montgomerie and Weatherhead, 1988; Redondo, 1989; Thornhill, 1989). However, the impact of the parent's own RV upon its parental investment decisions is less clear than the impact of the offspring's RV. Efforts to show effects of parental RV have been somewhat bedeviled by the fact that aging animals change in other possibly relevant ways than just declining in RV. Greater parental effort with age can be confused with effects of experience that make the parent more effective without really incurring greater risk to self or otherwise investing more, for example. However, parental experience effects do not seem to explain away increases in parental effort with age in junglefowl (Thornhill, 1989) or California gulls (Pugesek, 1983, 1987).

Unfortunately, although parental defense against a standard threat provides a convenient and logically compelling metric of the parent's present valuation of the young, it is not necessarily a metric by which one can compare the parental commitment of mother versus father. Despite the common causal core of offspring-specific (or brood-specific) parental solicitude, there are relevant considerations that are specific to particular types of parental investment, too. Optimal incubation effort, for example, will be a different function for mother versus father if sexually selected brightness of males means that incubation is a lower-risk activity for the more cryptic females. Which parent plays the greater role may then be reversed in the domain of aggressive defense of the brood if the sexes differ in size or weaponry in ways that affect their respective effectiveness as defenders (*e.g.* Andersson and Wiklund, 1987; Wallin, 1987; Breitwisch, 1988). Such differences may coexist with

sexual similarity in other spheres such as in efficiency of feeding offspring.

It is difficult and often laborious to determine the costs to parents and benefits to the young of feeding, warming, tutelage and other such activities, over the range of variation that actually occurs. If parents deliver food to growing young at increasing rates, does this necessarily reflect increasing valuation of those young because of their growing RV? Increased delivery rates might follow from increased offspring demand even if parents valued the young identically. Even demonstrating that parental care is in any sense costly can require a major research effort, and the task of determining how nearly optimal are the actual investment decisions of parents is vastly greater. It is because of these complications that research on parental investment has been concentrated on the limited but relatively convenient domain of parental defense.

Paradoxically, infanticide can be used as another assay of parental solicitude with some of the same advantages as parental defense. Any factor that may be expected to influence parental investment decisions should also be relevant to the likelihood of lethal divestment, whether as a result of the parent's strategic termination of a particular reproductive episode or as a result of a dangerous shortfall of concern for the offspring's welfare.

INFANTICIDE IN HOMO SAPIENS

Consideration of adaptive parental decision-making provides a framework for the analysis of the circumstances in which human parents decide not to raise a particular offspring. If parental psychology has been shaped by a history of selection, then we may expect to find a correspondence between the circumstances and rationales of infanticide, as revealed in the ethnographic record, and the circumstances in which evolutionary theories predict mitigation of parental solicitude, as discussed above.

People of most cultures recognize that parents will sometimes choose not to raise a child. The factors that they consider relevant to such a decision vary, to some degree, from one society to another, but the great majority of ethnographic accounts of infanticide in nonindustrial societies appear to reflect a strategic allocation of lifetime parental effort: The principal relevant variables are considerations of genetic relatedness, the child's fitness prospects as indicated by offspring quality and by particular local conditions relevant to the child's survival prospects, and the parents' alternatives (Howell, 1979; Mull and Mull, 1988; Bugos and McCarthy, 1984; Daly and Wilson, 1984; Scrimshaw, 1984; Scheper-Hughes, 1985; Hill and Kaplan, 1988b). In other words, people's infanticidal decisions

can be reviewed in relation to the same themes under which we just discussed discriminative parental solicitude in general.

1. Genetic Relatedness

In people, as in all mammals, the issue of relatedness of child and putative parent is a sexually-asymmetrical issue: maternity is a fact and paternity an attribution (Daly and Wilson, 1982). Suspicion or knowledge of nonpaternity may then be a reason for male disinclination to raise a child. Recurring rationales for infanticide in the ethnographic literature include claims of adulterous conception, impregnation by a prior husband, and phenotypic cues of a nontribal sire.

2. Phenotypic Attributes and Reproductive Value of Offspring

(a) Age of offspring

A child's expected contribution to its parent's fitness is directly related to the child's own expected fitness or RV.[3] In the absence of mishap, one's RV increases steadily from birth until at least puberty. This is so primarily because surviving to maturity cannot be taken for granted. With modern medicine, the early increase in RV is muted by declines in infant and juvenile mortality, but where life-historical mortality and fertility are closer to the levels that must have prevailed for most of human history, the prepubertal increase in reproductive value is not trivial.

We would thus expect parental feelings to have evolved such that parents will seem to value offspring increasingly with age, and we might therefore expect to see an age-related decrease in the likelihood of lapses of parental solicitude. As in the avian cases discussed earlier, increased parental solicitude with offspring age may be difficult to detect because the offspring's dependence is waning at the same time, but parental valuation of the young might in principle be assayed by the parent's declining willingness to

---

[3] Fisher's (1958) RV is usually conceived of as the phenotype-specific expected reproductive success: classical Darwinian fitness or "direct" fitness in Browns' (Brown and Brown, 1981) terminology. A more general concept of "inclusive fitness value" (IFV) would incorporate "indirect" fitness effects as well, *i.e.* the organism's expected impacts upon the posterity of its genes through its effects on collateral as well as descendant kin. A child's expected contribution to its parent's inclusive fitness is directly related to the child's own IFV. Reference to the child's RV as indicative of its fitness value from the parental perspective is seldom seriously misleading, however, since, in general, we would expect that factors associated with an increase in RV would be associated with an increase in IFV, too.

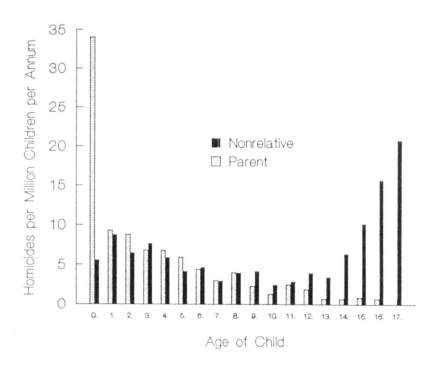

Figure 1 Age-specific rates of homicide victimization among Canadian children 1974–1983, perpetrated by birth parents (N = 352 victims). (Modified from Daly and Wilson 1988a).

tolerate or expose the young to lethal risk. One apparent manifestation of such an age-related change in parental valuation is a monotonic decrease in the risk of filicide (Figure 1), which continues to near zero as the offspring approaches maximal RV in young adulthood. It is especially striking that children become increasingly immune from parental lethal action as they mature, since this maturation entails increasing competitiveness in their interactions with non-relatives, and an increasing overall risk of becoming involved in lethal interpersonal conflict, both as killer and as victim (Wilson and Daly, 1985; Daly and Wilson, 1990).

The steep decline in risk of filicide from the first to the second year of the child's life is expectable upon two considerations: (1) that a major share of the prepubertal increase in the child's RV occurs within the first year postpartum, and (2) that insofar as parental disinclination reflects a "strategic" assessment of the reproductive episode, an evolved assessment

mechanism should be such as to terminate hopeless ventures as early as possible.[4]

(b) Sex of offspring

Female-selective infanticide occurs in many human societies (Daly and Wilson, 1984; Irwin, 1989; Miller, 1988; Schiefenhövel, 1989), whereas unequivocal accounts of male-selective infanticide are hard to find. The rationale for a preference for sons is not immediately obvious, since Fisher's sex ratio theory would seem to imply that preference for either sex would be evolutionarily unstable.

Resolution of this paradox may depend upon the Trivers-Willard hypothesis of adaptive preferences for one offspring sex in certain segments of a population offset by preference for the other elsewhere (Hrdy, 1987). Dickemann (1979) was the first to show that female-selective infanticide is status-graded in stratified societies in which sons are potential polygamists (whether by recognized marriage or by concubinage) and to link this status-grading to hypergyny and dowry competition; thus, human parents are Trivers-Willard strategists. A missing element was any evidence for a switch to preferential treatment of daughters and male-selective infanticide in the lower strata, but of course the behavior of the lower strata is less well documented than that of the wealthy and powerful. Others have since documented the coexistence of a preference for sons in wealthier or otherwise advantaged people and preferential treatment of daughters by the poor, people seeming to flexibly assess and respond to the future marital and reproductive prospects of sons versus daughters (e.g. Cronk, 1991; Mealey and Mackey, 1990; Voland, 1984; Abernethy and Yip, 1990). In extreme cases, such differential treatment amounts to sex-selective filicide by neglect of children of the disvalued sex.

(c) Health status of offspring

It would be an ill-designed organism that delayed or jeopardized future reproduction in order to nurture present offspring whose own reproductive prospects were nil. Thus, although an evolved parental psychology should be responsive to offspring need and dependency, it must be capable of recognizing when further effort will be wasted. Prevalent rationales

---

[4]Organisms that have the means to abort or resorb embryos do so discriminatively in response to cues of the futility of maturing them. Consciously felt inclinations to abort on the part of pregnant women appear to reflect the same strategic logic, responding to the same predictor variables as does infanticide (e.g. David, Dytrych, Matejcek and Schuller, 1988; Devereux, 1955).

for infanticide in the ethnographic literature include conspicuous deformity or illness of the newborn (Dickeman, 1975; Daly and Wilson, 1984).
Where the failing infant is not directly killed or abandoned, the mother's
emotional commitment to the child is likely to be muted with resultant
mortality risk from lesser vigilance and care (*e.g.* Scheper-Hughes, 1985).

3. Situational Predictors of Fitness Value of Offspring

In Daly and Wilson's (1984) review of infanticidal rationales in the ethnographic literature, the largest number of cases fell under this rubric. Famine
is an obvious reason for abandoning a newborn, and the death of either
parent may disincline the survivor to try to go it alone. An unwed mother
finds herself in a situation analogous to the widow's, and a large proportion of neonaticides occur in such circumstances.[5] A baby born to an unwed woman is often a poor fitness prospect by virtue of the mother's limited present ability to raise it, because it compromises the mother's marital
prospects (issue 4, below), and because if she does acquire a husband, he
is likely to discriminate against it (issue 1, above).

A common rationale for infanticide concerns the mother's present inca-
pacity to cope with the demands of child rearing. One circumstance indicative of maternal overburdening, and a frequently cited reason for infanticide, is the birth of twins; the victim might be the second born, the weaker,
or the female. Another circumstance representing a decision about the
allocation of scarce maternal resources is that in which lactational suppression of ovulation has failed, so that a baby is born too soon after the
last one. In this situation, the older child of greater RV is favored over the
newborn.

4. Alternative Reproductive Options of the Parents

The alternatives to a present reproductive venture shrink with time, a fact
of much consequence for women: the older the new mother the lesser her
residual RV. So one would expect a woman to be less and less inclined,
as her reproductive years slip away, to devalue a present offspring in
terms of its compromising effects on her future. It follows that the risk of
maternally perpetrated infanticide might decline as a function of maternal

---

[5] The mother is the parent with the greater allocation of parental effort at least until the infant
is weaned, and is thus the parent who might be expected to be under the greater incentive
to terminate lost causes. It is worth noting that although neonaticides are generally viewed
as acts by the mother, they are likely to be effected by abandonment rather than a directly
lethal act, and the father in such a case abandons the child no less than the mother.

age, and so it does (Bugos and McCarthy, 1984; Daly and Wilson, 1984; 1988a,c).

We would expect that many of the variables that are relevant to changes in maternal solicitude (or lapses therein, as in the case of maternal filicides) should be similarly relevant for fathers. The paternal case, however, is different from the maternal one, in at least three ways, all of which suggest that there will be a sexual asymmetry in the time-course of changes in parental solicitude:

(1) Women's reproductive life spans end before those of men, so the utility of alternative reproductive efforts declines more steeply as a function of own age for women than for men.

(2) Dependent children impose different opportunity costs on mothers and fathers, a nursing infant constraining mother's immediate alternative reproductive prospects much more than father's, for example, and the magnitude of this differential impact upon mother versus father declines with time since birth.

(3) Phenotypic and other evidence of paternity may surface after infancy and is expected to be relevant to paternal, but not maternal solicitude.

These three considerations suggest that a mother's valuation of a child relative to her valuation of herself is likely to rise more steeply with time since the child's birth than is the corresponding quantity for the father. If filicides constitute a sort of reverse assay of parental solicitude, it follows that filicide rate should decline more steeply for mothers than for fathers. Figure 2 presents data confirming this prediction.

PARENTAL SOLICITUDE AND MARITAL CONFLICT

By engaging in sexual reproduction and by the cooperative rearing of offspring, couples forge a powerful commonality of interest at the fundamental level of fitness. This shared interest is analogous to that existing between genetic relatives, but the genetic interests of an exclusively monogamous pair coincide even more closely than those of blood relatives since each's fitness interests are vested in their mutual progeny (Alexander, 1987). However, two considerations act against the evolution of perfect harmony in mated pairs: (1) the possibility of extra-pair reproduction, and (2) the partners' nepotistic interests in the welfare of distinct sets of collateral kin.

The issue of a putative father's disinclination to invest in his wife's children from a prior union or from an adulterous conception was noted

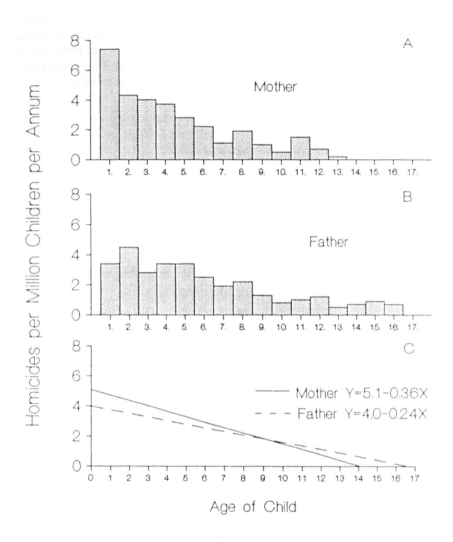

Figure 2 Age-specific rates of homicide victimization among 1–17 year-old Canadian children, 1974–1983: (a) slain by mother (N = 110 victims); (b) slain by father (N = 116); (c) linear regression of homicide rates by mother versus father. Infants < 1 year excluded for linear regression. The slope of the regression of homicide victimization rate on child's age is significantly (t = 2.3, df = 30, P < .02) steeper for mothers than for fathers. (Modified from Daly & Wilson, 1988c).

above as regards infanticides, and is further discussed by Daly and Wilson (this volume). The possibility of varying intensity of paternal solicitude as a function of variation in the putative father's "confidence" (conscious or otherwise) in his paternity is a worthy subject of study. If paternity confidence is sensitive to phenotypic evidence of the child's resemblance to self, for example, one would expect paternal solicitude to vary accordingly. Furthermore, since it is in the mother's interest to promote confidence of paternity, she should be especially motivated to perceive and remark upon paternal similarity, and there is evidence that she is (Daly and Wilson, 1982).

In addition to asymmetries in parent-offspring relatedness, the parental commitment of each parent might be expected to be sensitive to *both* parents' reproductive values. Insofar as a mother, for example, values a child in relation to the expected fitness benefits accruing from her investments, her perception of the father's level of valuation and commitment to the child may be expected to affect her own. The nature of such impacts could vary in direction, since appropriate responses to withdrawal of a partner's investment might be escalated effort on one's own part or giving up altogether or some bet-hedging partial divestment, with the most adaptive course of action depending on many parameters of circumstance, offspring stage and quality, and the parents' alternatives. Lazarus (1990) has considered how selection will shape mated animals' decisions whether to desert and leave the mate to raise the young alone, and has shown that even this binary stay-or-leave issue becomes complex and has non-intuitive equilibria "when strategies are modified in response to the strategy of the partner" (p. 682). The ways in which selection shapes continuous variability in one's level of parental effort relative to one's partner's will be even trickier to analyze, and we need further development of evolutionary models of the parameters and conditions in which there is expected to be variation in each parent's commitment (Winkler, 1987; Montgomerie and Weatherhead, 1988; Knight and Temple, 1988; Weatherhead, 1990).

PARENTAL ATTACHMENT

Parents care profoundly for their children, deriving joy from their welfare and scarcely resenting the prolonged one-way flow of resources. The emotional/motivational/cognitive mechanism that enables such selfless nurture is parental love. Psychologists apparently take parental love so much for granted (in contrast to romantic love, for example) that they have hardly noticed it except in the event of conspicuous lapses therein.

If social psychologists and motivation theorists have largely ignored parental inclinations, however, pediatricians and their applied collaborators have not, and there is a burgeoning literature on parental (mostly maternal) bonding, especially with respect to the effects of postpartum mother-infant contacts. The first wave of empirical studies showed surprisingly powerful and enduring effects of even brief postpartum contact with one's newborn on subsequent maternal responsiveness (Klaus *et al* 1972; Kennell *et al* 1974; Sosa *et al* 1976; de Chateau and Wiberg, 1977a, 1977b; Hales *et al.*, 1977; Carlsson *et al.*, 1978; Kontos, 1978; O'Connor *et al* 1980; Ali and Lowry, 1981; Grossman *et al* 1980). Once mothers had been empowered with regard to hospital birthing practices, the focus of subsequent critical scientific discussion turned to (1) criticisms of the unitary causal model of the maternal bonding process; (2) the question of whether a delimited postpartum period was uniquely important; and (3) the relative importance of postpartum contact as compared to other important variables, including paternal support, whether the pregnancy was strongly desired, and the health status of the baby and the mother. Recent discussants (*e.g.* Herbert *et al* 1982; Egeland, 1982; Lamb and Hwang, 1982; Lamb, 1983; Scheper-Hughes, 1988) often present their analyses as refutations of the "bonding" concept and of the importance of postpartum contact, but the early contact studies have not been invalidated and there is no necessary conflict with demonstrations that situational and other factors also influence maternal feelings (Daly and Wilson, 1987).

The sometimes acrimonious debates about maternal bonding have not been informed by an evolutionary view of parenthood. Thinking about parental psychologies as targets of selection suggests that the development of child-specific parental love is likely to involve at least three distinct processes proceeding over different time courses: an assessment of the quality of the child and the situation, a rapid discriminative attachment to the baby as an individual, and a more gradual deepening of individualized love.

1. Initial "Cues" of Newborn's Fitness Prospects

The first process to be expected is an assessment, in the immediate aftermath of the birth,[6] of the child and of how its qualities and present

---

[6] Information about the health status of the fetus from ultrasound imaging and other modern medical procedures may activate parental assessment mechanisms at a much earlier stage than the birth; *e.g.* Peppers, 1988.

circumstances combine to predict its prospects. If the newborn exhibits deformities or other conspicuous signs of low viability, then the probability of abandoning the infant is raised (see above). There is evidence in the modern West that the birth of a child with major defects commonly evokes an immediate shocked rejection in the parents (*e.g.* Roskies, 1972; Fletcher, 1974; Drotar *et al.*, 1975), a rejection that would undoubtedly lead to quick abandonment in historical settings (*e.g.* Dickeman, 1975; Weir, 1984). Where genetic counselling and termination of pregnancy are available, parents commonly want to abort seriously defective fetuses (*e.g.* Leschot *et al* 1985).

Beliefs that deformed infants are ghosts or demons (or the progeny thereof) are found sporadically the world 'round and such beliefs are invoked to justify infanticide. Analogous superstitious allegations about well-formed, healthy babies are essentially non-existent, indicating that "superstition" is not fairly dismissed as ignorant foolishness but functions instead as an ideological buttress to action with a self-interested rationale (Daly and Wilson, 1988a).

Conspicuously defective newborns are rare, of course, but maternal responsiveness in the immediate postpartum also varies with subtle cues of the infant's quality and health (*e.g.* Klaus and Kennell, 1976; Mann, 1992), including being undersized and premature. Very low weight neonates (<1500g) incur increased risk of abandonment or abuse (*e.g.* Hunter *et al.*, 1978; McCormick, 1985) and when such babies are likely to die, parents may distance themselves emotionally and fail to participate in the infant's hospital care (Newman, 1980). Emotional distancing has also been described among impoverished Brazilian mothers of weak, sickly infants expected to die (Scheper-Hughes, 1985). It should be noted that low birthweight in North America is associated with low socioeconomic status, maternal youth, large family size, and close birth spacing (*e.g.* Zuckerman *et al* 1984), but size and vigor of a newborn may be salient cues modulating the development of parental solicitude. In an observational study of low birth weight twins, the healthier twin was more effective in eliciting maternal responsiveness whereas factors such as duration of postpartum separation and infant smiling did not account for the mother's differential treatment of the twins (Mann, 1992).

Within the first few hours after birth, healthy human infants exhibit a precocious social responsiveness — eye contact and selective attention to maternal speech — that may be a specialized adaptation to advertise quality and elicit maternal commitment during the mother's assessment phase. If circumstances are dicey and the mother is in any way ambivalent, then

the poor responsiveness of a baby might tip the scales toward disinclination to raise the child.

In addition to the quality of the newborn, the present situation of the mother and her residual reproductive value should affect bonding. In one experimental study, mothers' responsiveness to their infants on the day of hospital discharge increased with maternal age (Jones, Green and Krauss, 1980); the age effect was not confounded with mother's marital status, socioeconomic status or amount of postpartum contact. (See also Norr, Roberts and Freese, 1989; Abernethy and Yip, 1990.) Some researchers have argued that socioeconomic status, maternal age, and education, rather than the postpartum bonding experience, account for variations in maternal attachment in the various experimental studies (*e.g.* Campbell and Taylor, 1979; Siegel *et al* 1980; Svejda *et al* 1980), but others have argued that these variables enhance the postpartum contact effects (*e.g.* Anisfeld and Lipper, 1983; Norr *et al* 1989).

From a selectionist perspective on the maternal bonding process, it would not be surprising to discover that an experimental procedure of "extra" postpartum contact had little ameliorative effect on mothers when circumstances such as poverty, lack of paternal support and other indices of maternal overburdening "cue" poor fitness prospects. More work is needed to assess whether situational and other variables (such as the mother's residual reproductive value as measured by her age) interact with and modify the effects of particular postpartum experiences.

Many new mothers experience a transient and brief period of the "blues" within the first several days after giving birth (*e.g.* Cutrona, 1982; Hopkins *et al* 1984). A lesser but considerable number experience a more debilitating postpartum depression associated with concerns about their ability to cope with the baby. Such depression is apparently especially likely when the mother is young, when the mother and father are having marital problems, and when mothers are single or otherwise lack social support (*e.g.* Braverman and Roux, 1978; Cox *et al* 1982; Cutrona, 1982; Hopkins *et al.,* 1984; Kumar and Robson, 1980), as well as when the infant is suffering from poor health (*e.g.* Blumberg, 1980; Grossman *et al* 1980). These circumstances are very similar to the infanticide circumstances described in the ethnographic literature. Women suffering from extreme postpartum depression are sometimes characterized by clinicians as delusional, but the typical content of the "delusions" seems not at all fantastic: concern about own inability to care for the baby, fear of not having enough love for the baby, and guilt aroused by infanticidal thoughts (*e.g.* Herzog and Detre, 1976).

2. Discriminative Bonding to Own Offspring

Parents are highly sensitive to their babies' distinctive features, recognizing them by voice (Formby, 1967; Morsbach, 1980; Rothgänger, 1981) and by smell (Porter *et al* 1983; Russell *et al* 1983) with only minimal exposure. Some have implied that these abilities represent psychological adaptations for discriminative bonding, but of course, people are very good at recognizing individual faces generally (Diamond and Carey 1986; Damasio *et al* 1990); whether there is a specific heightened postpartum infant recognition ability is still to be determined.

Rather than having merely to recognize her own baby, the "task" confronting the new mother is to develop an individualized commitment to it, such that she is emotionally prepared to invest heavily in its welfare without being at the same time vulnerable to parasitism by children generally. Many new mothers report an initial feeling of "indifference" to their babies (perhaps reflecting the initial "assessment" phase as well as the lack of individuation), but very few feel the same way by one week postpartum (*e.g.* Robson and Kumar, 1980). After having had close contact with their infants over the first few days, mothers commonly report developing a feeling that their baby is uniquely wonderful (*e.g.* Kennell *et al* 1975; Klaus and Kennell, 1976).

Perception of paternal resemblance deserves study as a factor contributing to paternal bonding. Paternity is mistakable, and there is a great deal of evidence that men are much concerned with its accurate assessment. There are at least two obvious sources of information contributing to a putative father's confidence that he is indeed the sire: his confidence of the mother's sexual fidelity and his assessment of the phenotypic similarity of the child to himself and his blood relatives. If assessments of phenotypic resemblance were veridical and disinterested, we would expect equal emphases upon maternal and paternal similarities; however, expressed opinions about who newborn babies resemble are biased toward paternal similarities (Daly and Wilson, 1982). Even among the Trobriand Islanders, famous for their apparent ignorance of the male's role in procreation, it is an offense to remark the infant's similarity to uterine kin and polite to assert resemblance to the father (Malinowski, 1929). Researchers have recently become interested in the effects of men's participation in delivery and other postpartum experiences on paternal bonding, but the issue of paternity has not been discussed in this context. An evolutionary psychological view of paternal bonding suggests that perception of paternal resemblance would be correlated with paternal bonding, and evidence of nonpaternity could wash out any ameliorative effects of participating in the delivery. Furthermore, as we noted with respect to the different age

patterns in filicide (Figure 2), phenotypic cues that alter paternity confidence may not become evident for years, and a certain reserve in paternal bonding may therefore be expected. (Of course, a lesser paternal than maternal involvement in the immediate postpartum may have other sources than the sex difference in parental certainty, most notably in the more essential nurturant role of new mammalian mothers as compared to fathers.)

Paternity disputes provide much evidence of men's disinclinations to support children they claim they did not sire (Wilson, 1987). Two major theories in law have been proposed to justify the imposition of a child support obligation on reluctant putative fathers: the theory of delict, whereby the father's liability arises from illicit sexual access to the mother, and the theory of descent, whereby the father's liability is based on his genetic relationship with the child (Sass, 1977). Descent-based laws are generally perceived as more just and are better complied with–facts with interesting implications about people's implicit understandings of parenthood. The imputation that descent is a just basis for allocating obligation implies that the existence and welfare of children are perceived to constitute *benefits* to their genetic parents, even if the latter are uninvolved and thus failing to derive the usual emotional rewards of parenting. He who sires a child and absconds is deemed a cheater because he has failed to pay the cost to which this received benefit obliges him (c.f. Cosmides and Tooby, 1989). The fact that descent-based systems evoke better compliance than delict-based ones further reflects the fact that men are disinclined to invest in children which are not their own and feel themselves to be victims of injustice when obliged to do so.

3. A Gradual Deepening of Parental Love

The third predictable process of parental attachment is a much more gradual one: The strength of parental love may be expected to grow with the child's increasing RV, especially over the first few years when there is the steepest increase in RV. A recent study of the changes in maternal feelings about the well-being of the infant over the first 16 months after birth supports our expectations about a gradual increase in the valuation of child relative to self (Fleming *et al* 1990); the effect was not merely due to changes in maternal condition or situation as the same measures with respect to self and to husband remained relatively stable over the same period. The authors stated that the postpartum growth in the salience and importance of the infant was "reflected by both an increase in total interview time spent talking about the infant (in both positive and negative terms) and an increasingly large proportion of

women reporting such things as feelings of closeness to their infants, being pleased with their infants development, or enjoying child-care activities" (p. 141).

The information that parents garner from their continued monitoring of offspring quality should affect the depth and time-course of their love and commitment, especially over the first few years in which infant mortality risk remains high. Since parental effort is a resource to be invested, not squandered, chronic changes in the infants' responsiveness and robustness, consequent upon the effects of malnutrition, dehydration and pathogens, can be expected to dampen parental love, in spite of the infant's greater "need".

In many societies, newborn babies are not immediately named or officially acknowledged by the community, a practice more or less explicitly linked to their uncertain future. Naming bestows personhood and facilitates the individuation of affection. (Indirect evidence for this claim can be found in observations that naming children after relatives is effective in inspiring namesake investment and inheritance; Smith, 1977; Furstenberg and Talvitie, 1980.) The postpartum delay in recognizing infants' personhood corresponds to a period of high mortality risk, perhaps with the effects of facilitating difficult decisions of disinvestment and lessening the emotional pain should the infant die (Mull and Mull, 1988; Scheper-Hughes, 1985). It is something of a cliché to claim that the valuation of children is a recent Western cultural invention, with tales of child brutalization and parental indifference in history and in other societies introduced as support; those making this argument fail to appreciate that seeming callousness is an understandable response to circumstances that make children poor prospects for survival and reproduction, and that the same mothers who seem indifferent to the plight of one child in one context can be profoundly nurturant to others born at other times in more auspicious circumstances (*e.g.* Pollock, 1983; Bugos and McCarthy, 1984; Vinovskis, 1987).

CONCLUDING REMARKS

Developmental psychologists and pediatricians concerned with variations in parental solicitude have emphasized individual differences and a few experiential factors like postpartum contact and parenting styles experienced in childhood, without coming to grips with the task of producing a general theory of parental motivation (Daly and Wilson, 1987, 1988a,b). We attribute this piecemeal approach to the practitioners' lack of a theoretical

framework that would enable them to make sense of variable parental inclinations. Darwinism provides that framework.

Evolutionary theory is not a substitute for psychological analysis, but
a valuable aid thereto: Understanding how selection operates and what
behavioral control mechanisms have been designed to achieve affords
innumerable hints to their probable organization. Improved maternal
efficacy over the lifespan, to take one example, has routinely been assumed
to reflect the acquisition of skills and/or knowledge, and the "immaturity"
of young mothers. The theory and facts reviewed in this chapter suggest
that such lifespan changes may often reflect adaptive changes in maternal
inclinations as maternal RV and opportunities change, rather than the
mere alleviation of incompetence with experience (Taylor, Wadsworth and
Butler, 1983; McCormick, Shapiro and Starfield, 1981; Abernethy and
Yip, 1990). This alternative view suggests many possible lines of research.

Although evolutionary reasoning is no alternative to psychological accounts of parenting, it does provide grounds for suspecting that certain
prevalent conceptions of parenthood and family relations misrepresent
their essences. The popular focus on families as "systems" and their members as components thereof cannot be correct, for example, insofar as it ignores the fact that family members are agents with only partially congruent
interests and elevates the "system's" objectives above those of its actors.
A quarter century of criticism of naive group selectionism in biology has
clarified why individual organisms are the appropriate level in the hierarchy of life at which to impute integrated agendas, and why the analogizing
of larger groups to self-interested individuals typically fails (e.g. Williams,
1971; Dawkins, 1982). Similarly, the prevalent treatment of parenthood as
a "role" is not adequate to capture its crucial features (Wilson and Daly,
1987; Daly and Wilson, this volume); again, the problem is one of weak
analogy because of a failure to articulate selectively significant aspects of
the phenomenon under consideration.

A crucial contribution of evolutionary reasoning to the study of parent-
offspring relations will come from the diffusion and further development
of the analysis of the individual actors' conflicting agendas, as pioneered
by Trivers (1974). "Socialization" cannot be simply the tutelage of children to function in society, as it is so often portrayed, for although children are the fitness vehicles of parents who have evolved to cherish and
nurture them, their interests are not identical. Evolved ontogenies must
be complex in order to accept information from partly beneficent tutors
while resisting manipulations which would make the young pursue others' interests rather than their own. (Contemporary psychology's geronto-
centric preoccupation with the "socializing" function of parenting reflects

a contemporary circumstance of relative affluence, almost forgetting the fundamental nourishing and protective functions of parental inclinations and repertoires.) Evolved parental psychologies must be equally subtle in the face of self-interested offspring.

Parental motives, emotions, and contingent decision rules have a complex functional interrelatedness that is unlikely to be elucidated without recourse to Darwinism.

ACKNOWLEDGEMENTS

Our homicide research has been supported by the Harry Frank Guggenheim Foundation, the North Atlantic Treaty Organization, the Natural Sciences and Engineering Research Council of Canada, and the Social Sciences and Humanities Research Council of Canada. This chapter was written while the authors were Fellows of the Center for Advanced Study in the Behavioral Sciences with financial support from the John D. and Catherine T. MacArthur Foundation, the National Science Foundation #BNS87-008, the Harry Frank Guggenheim Foundation and the Gordon P. Getty Trust, and while M. Daly was a Fellow of the J.S. Guggenheim Foundation. We wish to thank Stefano Parmigiani, Bruce Svare and Fred vom Saal for organizing this conference.

REFERENCES

ABERNETHY, V. and Yip, R. (1990) Parent characteristics and sex differential infant mortality: the case in Tennessee. *Human Biology*, **62**, 279-290.

ALEXANDER, R.D. (1979) *Darwinism and Human Affairs*. Seattle, Washington, University of Washington Press.

ALEXANDER, R.D. (1987) *The Biology of Moral Systems*. Hawthorne, New York, Aldine de Gruyter.

ALI, Z. and Lowry, M. (1981) Early maternal-child contact: effects on later behaviour. *Developmental Medicine and Child Neurology*, **23**, 337–345.

ANDERSSON, S. and Wiklund, C.G. (1987) Sex role partitioning during offspring protection in the Rough-legged Buzzard *Buteo lagopus*. *Ibis*, **129**, 103–107.

ANDERSSON, M., Wiklund, C.G. and Rundgren, H. (1980) Parental defense of offspring: a model and an example. *Animal Behaviour*, **28**, 536–542.

ANISFELD, E. and Lipper, E. (1983) Early contact, social support, and mother-infant bonding. *Pediatrics*, **72**, 79–83.

AUSTAD, S.N. and Sunquist, M.E. (1986) Sex-ratio manipulation in the common opossum. *Nature*, **324**, 58–60.

BEISSINGER, S.R. and Snyder, N.F.R. (1987) Mate desertion in the snail kite. *Animal Behaviour*, **35**, 477–487.

BLUMBERG, N.J. (1980) Effects of neonatal risk, maternal attitude, and cognitive style on early postpartum adjustment. *Journal of Abnormal Psychology*, **89**, 139–150.

BLURTON Jones, N.G. (1989) The costs of children and the adaptive scheduling of births: towards a sociobiological perspective on demography. In: A.E. Rasa, C. Vogel and E. Voland (eds.), *The Sociobiology of Sexual and Reproductive Strategies*, London, Chapman and Hall.

BRAVERMAN, J. and Roux, J. F. (1978) Screening for the patient at risk for postpartum depression. *Obstetrics and Gynecology*, **52**, 731–736.

BREITWISCH, R. (1988) Sex differences in defense of eggs and nestlings by northern mockingbirds, *Mimus polyglottos*. Animal Behaviour, **36**, 62–72.

BROWN, C.R. (1984) Laying eggs in a neighbor's nest: benefit and cost of colonial nesting in swallows. *Science*, **224**, 518–519.

BROWN, J.L. and Brown, E.R. (1981) Kin selection and individual selection in babblers. In: R.D. Alexander and D.W. Tinkle (eds.), *Natural Selection and Social Behavior*, New York, Chiron Press.

BUGOS, P.E. and McCarthy, L.M. (1984) Ayoreo infanticide: a case study. In: G. Hausfater and S.B. Hrdy (eds.), *Infanticide: Comparative and Evolutionary Perspectives*, New York, Aldine Press.

BURLEY, N. (1986). Sex-ratio manipulation: color-banded populations of zebra finches. *Evolution*, **40**, 1191–1206.

CAMPBELL, S.B.G. and Taylor, P.M. (1979) Bonding and attachment: theoretical issues. *Seminars in Perinatology*, **3**, 3–13.

CARLSSON, S.G., Fagerberg, H., Horneman, G., Hwang, C.-P., Larsson, K., Rodholm, M., Schaller, J., Danielsson, B. and Gundewall, C. (1978) Effects of amount of contact between mother and child on the mother's nursing behavior. *Developmental Psychobiology*, **11**, 143–150.

CHARNOV, E.L. (1982) *The Theory of Sex Allocation*. Princeton, New Jersey, Princeton University Press.

CLUTTON-BROCK, T.H., Albon, S.D. and Guiness, F.E. (1985) Parental investment and sex differences in juvenile mortality in birds and mammals. *Nature*, **313**, 131–33.

COSMIDES, L. and Tooby, J. (1989) Evolutionary psychology and the generation of culture, Part II. Case study: A computational theory of social exchange. *Ethology and Sociobiology*, **10**, 51–97.

COX, J.L., Connor, Y. and Kendell, R. E. (1982) Prospective study of the psychiatric disorders of childbirth. *British Journal of Psychiatry*, **140**, 111–117.

CRONK, L. (1991) Preferential parental investment in daughters over sons. *Human Nature*, **2**: 387–417.

CUTRONA, C.E. (1982) Nonpsychotic postpartum depression: a review of recent research. *Clinical Psychology Review*, **2**, 487–503.

DALY, M. and Wilson, M. (1981) Abuse and neglect of children in evolutionary perspective. In: R.D. Alexander and D.W. Tinkle (eds.), *Natural Selection and Social Behavior*, New York, Chiron Press.

DALY, M. and Wilson, M. (1982) Whom are newborn babies said to resemble? *Ethology and Sociobiology*, **3**, 69–78.

DALY, M. and Wilson, M. (1984) A sociobiological analysis of human infanticide. In: G. Hausfater and S.B. Hrdy (eds.), *Infanticide: Comparative and Evolutionary Perspectives*, New York, Aldine Press.

DALY, M. and Wilson, M. (1987) Evolutionary psychology and family violence. In: C. Crawford, M. Smith and D. Krebs (eds.), *Sociobiology and Psychology*, Hillsdale, NJ, Erlbaum.

DALY, M. and Wilson, M. (1988a) *Homicide*. New York, Aldine de Gruyter. xii + 328 pp.

DALY, M. and Wilson, M. (1988b) The Darwinian psychology of discriminative parental solicitude. *Nebraska Symposium on Motivation*, 35, 91–144.

DALY, M. and Wilson, M. (1988c) Evolutionary social psychology and family homicide. *Science*, 242, 519–524.

DALY, M. and Wilson, M. (1990) Killing the competition. *Human Nature*, 1, 83–109.

DALY, M. and Wilson, M. Stepparenthood and the evolved psychology of discriminative parental solicitude. (this volume).

DAMASIO, A. R., Tranel, D. and Damasio, H. (1990) Face agnosia and the neural substrates of memory. *Annual Review of Neuroscience*, 13, 89–109.

DAVID, H.P., Dytrych, Z., Matejcek, Z. and Schuller, V. (1988) *Born unwanted: developmental effects of denied abortion*. New York, Springer.

DAVIES, N.B. (1990) Dunnocks: cooperation and conflict among males and females in a variable mating system. In: P.B. Stacey and W.D. Koenig (eds.), *Cooperative Breeding in Birds*, Cambridge, Cambridge University Press.

DAWKINS, R. (1982) *The Extended Phenotype*. Oxford, W.H. Freeman.

DE CHATEAU, P. and Wiberg, B. (1977a) Long-term effect on mother-infant behaviour of extra contact during the first hour post-partum. I. first observations at 36 hours. *Acta Paediatrica scandinavica*, 66, 137–143.

DE CHATEAU, P. and Wiberg, B. (1977b) Long-term effect on mother-infant behaviour of extra contact during the first hour post-partum. II. A follow-up at three months. *Acta Paediatrica scandinavica*, 66, 145–151.

DEVEREUX, G. (1955) *A Study of Abortion in Primitive Societies*, revised edition 1976. New York: International Universities Press.

DIAMOND, R. and Carey, S. (1986) Why faces are and are not special: an effect of expertise. *Journal of Experimental Psychology*, 115, 107–117.

DICKEMAN, M. (1975) Demographic consequences of infanticide in man. *Annual Review of Ecology and Systematics*, 6, 107–137.

DICKEMANN, M. (1979) Female infanticide, reproductive strategies, and social stratification: a preliminary model. In: N.A. Chagnon and W. Irons (eds.), *Evolutionary Biology and Human Social Behavior*, North Scituate MA, Duxbury Press.

DROTAR, D., Baskiewicz, A., Irvin, N., Kennell, J. and Klaus, M. (1975) The adaptation of parents to the birth of an infant with a congenital malformation: a hypothetical model. *Pediatrics*, 56, 710–717.

EGELAND, B., Vaughn, B. (1982) Failure of "bond formation" as a cause of abuse, neglect, and maltreatment. *Annual Progress in Child Psychiatry and Child Development 1982*, 188–198.

FISHER, R.A. (1958) *The Genetical Theory of Natural Selection*, second revised edition (orig. 1930). New York, Dover.

FLEMING, A.S., Ruble, D.N. Flett, G.L. and Van Wagner, V. (1990) Adjustment in first-time mothers: changes in mood and mood content during the early postpartum months. *Development Psychology*, 26, 137–143.

FLETCHER, J. (1974) Attitudes toward defective newborns. *Hastings Center Studies*, 2, 21–32.

FORMBY, D. (1967) Maternal recognition of infant's cry. *Developmental Medicine and Child Neurology*, **9**, 293–298.

FURSTENBERG, F.F. and Talvitie, K.G. (1980) Children's names and paternal claims. *Journal of Family Issues*, **1**, 31–57.

GOWATY, P.A. and Karlin, A.A. (1984) Multiple maternity and paternity in single broods of apparently monogamous eastern bluebirds (*Sialia sialis*). *Behavioral Ecology and Sociobiology*, **15**, 91–95.

GROSSMAN, F.K., Eichler, L.S. and Winickoff, S.A. (1980) *Pregnancy, Birth, and Parenthood*. San Francisco, Jossey-Bass.

HALES, D.J., Lozoff, B., Soda, R. and Kennell, J.H. (1977) Defining the limits of the maternal sensitive period. *Developmental Medicine and Child Neurology*, **19**, 454–461.

HATCHWELL, B.J. and Davies, N.B. (1990) Provisioning of nestlings by dunnocks, *Prunella modularis*, in pairs and trios: compensation reactions by males and females. *Behavioral Ecology and Sociobiology*, **27**, 199–209.

HERBERT, M., Sluckin, W. and Sluckin, A. (1982) Mother-to-infant "bonding"? *Journal of Child Psychology and Psychiatry*, **23**, 205–221.

HERZOG, A. and Detre, T. (1976) Psychotic reactions associated with childbirth. *Diseases of the Nervous System*, **37**, 229–235.

HEWLETT, B.S. (1988) Sexual selection and paternal investment among Aka pygmies. In: L. Betzig, M. Borgerhoff Mulder and P. Turke (eds.), *Human Reproductive Behavior*, Cambridge, Cambridge University Press.

HILL, K. and Kaplan, H. (1988a) Tradeoffs in male and female reproductive strategies among the Ache: part 1. In: L. Betzig, M. Borgerhoff Mulder and P. Turke (eds.), *Human Reproductive Behavior*, Cambridge, Cambridge University Press.

HILL, K. and Kaplan, H. (1988b) Tradeoffs in male and female reproductive strategies among the Ache: part 2. In: L. Betzig, M. Borgerhoff Mulder and P. Turke (eds.), *Human Reproductive Behavior*, Cambridge, Cambridge University Press.

HOPKINS, J., Marcus, M. and Campbell, S.B. (1984) Postpartum depression: a critical review. *Psychological Bulletin*, **95**, 498–515.

HOWELL, N. (1979) *Demography of the Dobe !Kung*. New York, Academic Press.

HRDY, S.B. (1987) Sex-biased parental investment among primates and other mammals: a critical evaluation of the Trivers-Willard hypothesis. In: R.J. Gelles and J.B. Lancaster (eds.), *Child Abuse and Neglect: Biosocial Dimensions*, New York, Aldine.

HUNTER, R.S., Kilstrom, N., Loda, E.N. and Kraybill, F. (1978) Antecedents of child abuse and neglect in premature infants: a prospective study in a newborn intensive care unit. *Pediatrics*, **61**, 629–635.

IRWIN, C. (1989) The sociocultural biology of Netsilingmiut female infanticide. In: A.E. Rasa, C. Vogel and E. Voland (eds.), *The Sociobiology of Sexual and Reproductive Strategies*, London, Chapman and Hall.

JONES, F.A., Green, V. and Krauss, D.R. (1980) Maternal responsiveness of primiparous mothers during the postpartum period: age differences. *Pediatrics*, **65**, 579–584.

KENNELL, J.H., Jerauld, R., Wolfe, H., Chesler, D., Kreger, N.C., Alpine, W., Steffa, M. and Klaus, M.H. (1974) Maternal behavior one year after early and extended post-partum contact. *Developmental Medicine and Child Neurology*, **16**, 172–179.

KENNELL, J.H., Trause, M.A. and Klaus, M.H. (1975) Evidence for a sensitive period in the human mother. In: Ciba foundation symposium No. 33. *Parent-infant Interaction*. Amsterdam, Elsevier-Excerpta medica-North Holland.

KLAUS, M.H. and Kennell, J.H. (1976) *Maternal-infant Bonding.* St. Louis, C.V. Mosby.

KLAUS, M.H., Jerauld, R., Kreger, N.C., McAlpine, W., Steffa, M. and Kennell, J.H. (1972) Maternal attachment. Importance of the first post-partum days. *New England Journal of Medicine,* **286**, 460–463.

KNIGHT, R.L. and Temple, S.A. (1986) Nest defense in the American goldfinch. *Animal Behaviour,* **34**, 887–897.

KNIGHT, R.L. and Temple, S.A. (1988) Nest defense behavior in the red-winged blackbird. *Condor,* **90**, 193–200.

KONTOS, D. (1978) A study of the effects of extended mother-infant contact on maternal behavior at one and three months. *Birth and the Family Journal,* **5**, 133–140.

KUMAR, R. and Robson, K. M. (1980) A prospective study of emotional disorders in childbearing women. *British Journal of Psychiatry,* **144**, 35–47.

LAMB, M. (1983) The bonding phenomenon: misinterpretations and their implications. *Journal of Pediatrics,* **102**, 249–250.

LAMB, M.E. and Hwang, C.-P. (1982) Maternal attachment and mother-neonate bonding: a critical review. In: M.E. Lamb and A.L. Brown, (eds.), *Advances in Developmental Psychology* (vol. 2). Hillsdale, New Jersey, Lawrence Erlbaum.

LAZARUS, J. (1990) The logic of mate desertion. *Animal Behaviour,* **39**, 672–684.

LESCHOT, N.J., Verjaal, M. and Treffers, P.E. (1985) A critical analysis of 75 therapeutic abortions. *Early Human Development,* **10**, 287–293.

LOZOFF, B., Brittenham, G.M., Traus, M.A., Kennell, J.H. and Klaus, M.H. (1977) The mother-newborn relationship: limits of adaptability. *Journal of Pediatrics,* **91**, 1–12.

MALINOWSKI, B. (1929) *The Sexual Life of Savages in North-western Melanesia.* London, Routledge.

MANN, J. (1992) Nurturance or negligence: maternal psychology and behavioral preference among preterm twins. In: J. Barkow, L. Cosmides and J. Tooby (eds.), *The Adapted Mind,* New York, Oxford University Press.

MAYNARD SMITH, J. (1977) Parental investment: a prospective analysis. *Animal Behaviour,* **25**, 1–9.

MAYNARD SMITH, J. (1978) *Evolution of Sex.* Cambridge, Cambridge University Press.

MCCORMICK, M.C. (1985) The contribution of low birthweight to infant mortality and childhood morbidity. *New England Journal of Medicine,* **312(2)**., 82–90.

MCCORMICK, M.C., Shapiro, S. and Starfield, B.H. (1981) Injury and its correlates among 1-year-old children. *American Journal of Diseases of Children,* **135**, 159–163.

MEALEY, L. and Mackey, W. (1990) Variation in offspring sex ratio in women of differing social status. *Ethology and Sociobiology,* **11**, 83–95.

MILLER, B.D. (1988) Female infanticide and child neglect in rural north India. In: N. Scheper-Hughes (ed.), *Child Survival: Anthropological Perspectives on the Treatment and Maltreatment of Children,* Boston, D. Reidel.

MILLER, N.E. (1959) Liberalization of basic S-R concepts. In: S. Koch (ed.), *Psychology: A Study of a Science,* Vol. I–2, New York, McGraw-Hill, pp. 196–292.

MØLLER, A.P. (1988a) Female choice selects for male sexual tail ornaments in the monogamous swallow. *Nature,* **332**, 640–642.

MØLLER, A.P. (1988b) Paternity and paternal care in the swallow, *Hirundo rustica. Animal Behaviour,* **36**, 996–1005.

MONTGOMERIE, R.D. and Weatherhead, P.J. (1988) Risks and rewards of nest defense by parent birds. *Quarterly Review of Biology,* **63**, 167–187.

MORSBACH, G. (1980) Maternal recognition of neonates' cries in Japan. *Psychologia: an international journal of psychology in the Orient*, **23**, 63–69.

MULL, D.S. and Mull, J.D. (1988) Infanticide among the Tarahumara of the Mexican Sierra Madre. In: N. Scheper-Hughes (ed.), *Child Survival: Anthropological Perspectives on the Treatment and Maltreatment of Children*, Boston, D. Reidel.

NEWMAN, L.F. (1980) Parents' perceptions of their low birth weight infants. *Pediatrician*, **9**, 182–190.

NORR, K.F., Roberts, J.E. and Freese, U. (1989). Early postpartum rooming-in and maternal attachment behaviors in a group of medically indigent primiparas. *Journal of Nurse-Midwifery*, **34(2)**, 85–91.

O'CONNOR, S., Vietze, P., Sherrod, K., Sandler, H.M. and Altemeier, W.A. (1980) Reduced incidence of parenting inadequacy following rooming-in. *Pediatrics*, **66**, 176–182.

PEPPERS, L.G. (1988) Grief and elective abortion: breaking the emotional bond? *Omega: Journal of Death and Dying*, **18**, 1–12.

POLLOCK, L. (1983) *Forgotten Children: Parent-child Relations from 1500 to 1900*. Cambridge, Cambridge University Press.

PORTER, R.H., Cernoch, J.M. and McLaughlin, F.J. (1983) Maternal recognition of neonates through olfactory cues. *Physiology and Behavior*, **30**, 151–154.

PRESSLEY, P.H. (1981) Parental effort and the evolution of nest-guarding tactics in the threespine stickleback, *Gasterosteus aculeatus* L. *Evolution*, **35**, 282–295.

PUGESEK, B.H. (1983) The relationship between parental age and reproductive effort in the California gull (*Larus californicus*). *Behavioral Ecology and Sociobiology*, **13**, 161–171.

PUGESEK, B.H. (1987) Age-specific survivorship in relation to clutch size and fledging success in California gulls. *Behavioral Ecology and Sociobiology*, **21**, 217–221.

PUGESEK, B.H. (1990) Parental effort in the California gull: tests of parent-offspring conflict theory. *Behavioral Ecology and Sociobiology*, **27**, 211–215.

QUINN, T.W., Quinn, J.S., Cooke, F. and White, B.N. (1987) DNA marker analysis detects multiple maternity and paternity in single broods of the lesser snow goose. *Nature*, **326**, 392–394.

REDONDO, T. (1989) Avian nest defense: theoretical models and evidence. *Behaviour*, **111**, 161–195.

ROBSON, K.M. and Kumar, R. (1980) Delayed onset of maternal affection after childbirth. *British Journal of Psychiatry*, **136**, 347–353.

ROGERS, W. (1988) Parental investment and division of labor in the Midas cichlid (*Cichlasoma citrinelum*). *Ethology*, **79**, 126–142.

ROSKIES, E. (1972) *Abnormality and Normality: The Mothering of Thalidomide Children*. Ithaca, New York, Cornell University Press.

ROTHGÄNGER, H. (1981) Akustisches Wiedererkennen des Säuglingsschreies durch die Mutter. *Zeitschrift für ärztliche Fartbildungen (Jena).*, **75**, 441–446.

RUSSELL, M.J., Mendelson, T. and Peeke, H.V.S. (1983) Mothers' identification of their infants' odors. *Ethology and Sociobiology*, **4**, 29–31.

SASS, S.L. (1977) The defense of multiple access (*exceptio plurium concubentium*) in paternity suits: a comparative analysis. *Tulane Law Review*, **51**, 468–.

SCHEPER-HUGHES, N. (1985) Culture, scarcity, and maternal thinking: maternal detachment and infant survival in a Brazilian shantytown. *Ethos*, **13**, 291–317.

SCHEPER-HUGHES, N., (ed.) (1988) *Child Survival: Anthropological Perspectives on the Treatment and Maltreatment of Children*. Boston, D. Reidel.

SCHIEFENHÖVEL, W. (1989) Reproduction and sex-ratio manipulation through preferential female infanticide among the Eipo, in the Highlands of West New Guinea. In: A.E. Rasa, C. Vogel and E. Voland (eds.), *The Sociobiology of Sexual and Reproductive Strategies,* London, Chapman and Hall.

SCRIMSHAW, S.C.M. (1984) Infanticide in human populations: societal and individual concerns. In: G. Hausfater and S.B. Hrdy (eds.), *Infanticide: Comparative and Evolutionary Perspectives,* New York, Aldine Press.

SIEGEL, E., Bauman, K., Schaefer, E., Saunders, M.M. and Ingram, D.D. (1980) Hospital and home support during infancy: impact on maternal attachment, child abuse and neglect, and health care utilization. *Pediatrics,* 66, 183–190.

SMITH, D.S. (1977) Child-naming patterns and family structure change: Hingham, Massachusetts 1640–1880. *The Newberry Papers in Family and Community History.* Paper 76–5.

SOSA, R., Kennell, J.H., Klaus, M. and Urrutia, J.J. (1976) The effect of early mother-infant contact on breastfeeding, infection and growth. In: CIBA Foundation Symposium No. 45. *Breastfeeding and the Mother,* Amsterdam, Elsevier.

STAMPS, J.A. (1990) When should avian parents differentially provision sons and daughters? *American Naturalist,* 135, 671–685.

SVEJDA, M.J., Campos, J.J. and Emde, R.N. (1980) Mother-infant "bonding": failure to generalize. *Child Development,* 51, 775–779.

TAYLOR, B., Wadsworth, J. and Butler, N.R. (1983) Teenage mothering, admission to hospital, and accidents during the first 5 years. *Archives of Disease in Children,* 58, 6–11.

THORNHILL, R. (1981) Panorpa (Mecoptera: Panorpidae) scorpionflies: systems for understanding resource-defense polygyny and alternative male reproductive efforts. *Annual Review of Ecology and Systematics,* 12, 355–386.

THORNHILL, R. (1989) Nest defense by red jungle fowl (*Gallus gallus spadiceus*) hens: the roles of renesting potential, parental experience and brood reproductive value. *Ethology,* 83, 31–42.

TRIVERS, R.L. (1972) Parental investment and sexual selection. In: B. Campbell (ed.), *Sexual Selection and the Descent of Man 1871–1971,* Chicago, Aldine.

TRIVERS, R.L. (1974) Parent-offspring conflict. *American Zoologist,* 14, 249–264.

TRIVERS, R.L. (1985) *Social evolution.* Menlo Park, California, Benjamin/Cummings.

TRIVERS, R.L. and Willard D.E. (1973) Natural selection of parental ability to vary the sex ratio of offspring. *Science,* 179, 90–92.

VINOVSKIS, M.A. (1987) Historical perspectives on the development of the family and parent-child interactions. In: J.B. Lancaster, J. Altmann, A.S. Rossi and L. Sherrod (eds.), *Parenting Across the Life Span,* New York, Aldine.

VOLAND, E. (1984) Human sex-ratio manipulation: historical data from a German parish. *Journal of Human Evolution,* 13, 99–107.

WALLIN, K. (1987) Defence as parental care in tawny owls (*Strix aluco*). *Behaviour,* 102, 213–230.

WEATHERHEAD, P.J. (1990) Nest defense as shareable paternal care in red-winged blackbirds. *Animal Behaviour,* 39, 1173–1178.

WEATHERHEAD, P.J. and McRae, S.B. (1990) Brood care in American robins: implications for mixed reproductive strategies by females. *Animal Behaviour,* 39, 1179–1188.

WEIR, R.F. (1984) *Selective Nontreatment of Handicapped Newborns: Moral Dilemmas in Neonatal Medicine.* New York, Oxford University Press.

WEST, M.M. and Konner, M.J. (1976) The role of the father: an anthropological perspective. In: M. Lamb (ed), *The Role of Father in Child Development*, New York, Wiley.

WESTNEAT, D.F. (1988) Male parental care and extrapair copulations in the indigo bunting. *Auk*, **105**, 149–160.

WESTNEAT, D.F. (1990) Genetic parentage in the indigo bunting: a study using DNA fingerprinting. *Behavioral Ecology and Sociobiology*, **27**, 67–76.

WESTNEAT, D.F., Sherman, P.W. and Morton, M.L. (1990) The ecology and evolution of extra-pair copulations in birds. In: D.M. Power (ed.), *Current Ornithology, vol. 7*, pp. 331–369.

WIKLUND, C.G. (1990) Offspring protection by merlin *Falco columbarius* females; the importance of brood size and expected offspring survival for defense of young. *Behavioral Ecology and Sociobiology*, **26**, 217–223.

WILLIAMS, G.C., (ed.) (1971) *Group Selection*. Chicago: Aldine-Atherton.

WILSON, M. (1987) Impacts of the uncertainty of paternity on family law. *University of Toronto Faculty of Law Review*, **45**, 216–242.

WILSON, M. and Daly M. (1987) Risk of maltreatment of children living with stepparents. In: R.J. Gelles and J.B. Lancaster (eds.), *Child Abuse and Neglect: Biosocial Dimensions*, Hawthorne New York, Aldine.

WINKLER, D.W. (1987) A general model for parental care. *American Naturalist*, **130**, 526–543.

ZAIAS, J. and Breitwisch, R. (1989) Intra-pair cooperation, fledgling care, and renesting by northern mockingbirds (*Mimus polyglottos*). *Ethology*, **80**, 94–110.

ZUCKERMAN, B.S., Walker, D.K., Frank, D.A., Chase, C. and Hamburg, B. (1984) Adolescent pregnancy: biobehavioral determinants of outcome. *Pediatrics*, **105**, 857–863.

CHAPTER 4

INFANTICIDE IN
WESTERN CULTURES:
A HISTORICAL OVERVIEW

ANNA FERRARIS OLIVERIO

Cattedra di Psicologia dell'Etá Evolotiva,
University of Rome, Italy

PROBLEMS OF DEFINITION

When we take into consideration the problem of human infanticide we are faced with a very complex phenomenon. This complexity is both qualitative (for example, setting the age and the definition of infanticide) and quantitative (for example, the presence of this behavior in different societies and historical ages). A first problem is related to the age of the victim: is infanticide only restricted to neonatal age or to the full first year of life? Shall we speak of infanticide at two, three, five, eight or ten years of life? May we extend the concept of infanticide to include also the prenatal period? There are no unequivocal answers to these questions nor has there been any effort made to find a valid definition which applies to different disciplines. For example, Italian and French laws give a definition of infanticide that is much more limited with respect to that emerging from anthropological studies. Infanticide is only considered to apply to killing the newborn at birth or very shortly thereafter. Furthermore, until a few years ago Italian law gave a rather restricted definition of this crime from a qualitative point of view. If one looks for it through the different Italian Statistical Year Books one discovers that until 1984 the category "infanticide for reasons of honor" is the only one considered. On the contrary, from 1985 to the present, the definition infanticide tout-court is employed, the term infanticide still being related to the neonatal period only (Annuari Statis-

tici Italiani, 1970–89). When children are concerned, the Italian Health Statistics generically refer to homicide and lesions intentionally induced by others (ISTAT, 1983-85). Also French law distinguishes between "infanticide", related to the newborn, and "libéricide" or "filicide" related to grown-up children.

Within the domain of anthropology, on the contrary, the term infanticide is generally employed in a broader sense, the term also being used to indicate the killing of the child at ages which follow or precede birth. The definition of when a life is taken is usually dependent on a cultural definition of when a life begins. In western society, this is often when a fetus can survive outside the mother's uterus; but as this society has the technology to intervene *in utero* to save a life, the failure to intervene could be considered infanticide. "Among the Machigenga, a newborn is not accepted until its mother has nursed it, often a day after birth. Among Andean Indian groups, a child may not be acknowledged as a permanent family member until it has survived its first year. The Peruvian Amahuaca do not consider children fully human until they are 3 years old; among the Japanese in earlier times, naming was delayed until the seventh day after birth" (Scrimshaw, 1984). Also the Christian baptism represents an important stage: it is the moment at which the individual leaves the influence of evil in order to enter the domain of God.

Moreover, different authors tend to include as infanticide not only overt and deliberate actions (such as smothering, drowning, injury) but also abandonment (where survival is possible), "accidents", excessive physical punishment, and neglect: these are forms of passive infanticide that tend to occur some time (months or even years) after birth. "Parents may simply favor some children over others. If resources are scarce, the results can be fatal for the neglected child. In other instances, parents may decide a child is "not for this world" and refuse to seek medical treatment or invest resources in the child. Underinvestment, whether conscious or not, is currently more prevalent than the deliberate killing of a child" (Scrimshaw, 1984, p. 441).

If we accept the definition given by Susan Scrimshaw (1984) - infanticide is a behavior ranging from deliberate to unconscious which is likely to lead to the death of a dependent, young member of the species - we cannot limit infanticide only to that perpetrated at birth nor we can ignore those forms of abuse toward "dependent, young members of the species". Therefore, in this review I will use a broad definition of infanticide, referring to its different forms and to the different ages at which it is perpetrated. Infanticide is known to have been practiced in the past and in different cultures in deliberate and passive forms, consciously and unconsciously.

In this review I will concentrate on the types of infanticide in Western societies.

RITUAL INFANTICIDE

In the ancient Middle Eastern cultures, the Moabites, Phoenicians and Jews practiced child sacrifice for the purpose of beseeching the Gods to bestow specific goods (such as abundant harvests, victory in a battle, protection over a building and so on) or in order to expiate one's sins or to purify a community from a sacrilege committed by a member of the community. Child sacrifice could also be a way for a tyrant to maintain his political power, as in the case of King Herod, or the last attempt to obtain a victory. The same form of sacrifice might also be used for religious or magical reasons.

The most ancient testimony of infanticide refers to the fourteenth century B.C., *i.e.* to the late Bronze Age (it is in a text of Ugarit (Spalinger, 1971), in a Canaanite town in which a sacrifice of a first-born to the God Baal is mentioned). Later on the Bible refers to the child sacrifices of the Canaanites (Jeremiah 19:46) and to Mesha, King of Moab, who sacrificed on the walls his most precious good, that is, his firstborn child. The sacrifice was to express his despair and that of his fellow citizens, and to hope that the Jews would raise the siege (Kings 3:19). The Bible also tells us that Jephte, the judge, sacrificed his own daughter and that Isaac at the very last moment exchanged his son for a lamb. A child sacrifice is represented in an Egyptian bas-relief representing the Canaanite towns between the fourteenth and the twelfth century: in order to horrify the besiegers a priest hurls a six-seven year-old child from the walls (Spalinger, 1971). In Mozia, Spain and Carthage there were the Phoenician tophet sanctuaries (about seventh to eighth century B.C.) which were mainly used for sacrificial purposes (Guzzo, 1986). In Carthage the king might sacrifice his firstborn or his lastborn in an attempt to avert calamity.

Through the sacrifice of their own children (often the most loved one), adults would show their awe of the Gods and their despair. Considering that child mortality was very high, they were giving up their most precious good. But since many human motives are often ambivalent, other reasons may also be envisioned in some of these sacrifices. In many of them, for example, a clear purpose is evident: the king tries to keep his power or to save the community. Through a dreadful deed he intends to produce an empathetic reaction in his subjects or in his enemies. Sometimes this strategy succeeds: in the case of the King of Moab the horrified Jews

raised the siege. A quite different form of child sacrifice was described by Plutarch (45-125 B.C.), who describes a Carthaginian child sacrifice in which children were bought from poor people in order to be sacrificed: ".. with full knowledge and understanding they themselves offered up their own children, and those who had no children would buy little ones from poor people and cut their throats as if they were so many lambs or young birds; meanwhile the mother stood by without a tear or moan; but should she utter a single moan of let fall a single tear, she had to forfeit the money, and her child was sacrificed nevertheless; and the whole area before the statue was filled with a loud noise of flutes and drums so that the cries of wailing should not reach the ears of the people" (Plutarch, 1928, p. 493).

In any case, in all these sacrifices the use of children was instrumental. That is to say, the sacrifice was performed for the benefit of the parents or the community or it served to express the feelings of the adults: fear, rage and despair. From a psychological point of view, the child's sacrifice stresses a process of identification, or a process of role inversion, or both psychological processes. When, as in ancient times, the collective consciousness was stronger than individual consciousness, parents did not consider their child as a person different from them: they loved him, they played with him, but they also considered him as property of which they could dispose. Thus, in specific circumstances the most feeble became the instrument through which adults tried to protect themselves from the fury of the Gods, to cope with their anxieties, their fears and their sense of guilt. In one sense infanticide was a kind of "parental self mutilation".

INFANTICIDE AS A CONSCIOUS REGULATION OF BIRTH IN ANTIQUITY

Until the fourth century B.C. in Greece and in Rome the law and public opinion did not consider infanticide wrong. However, some people, such as the Greek Polybius, condemned the killing of healthy and legitimate children because this reduced the population of Greece: "In our own time the whole of Greece has been subject to a low birth-rate and a general decrease of the population, owing to which cities have become deserted and the land has ceased to yield fruit, although there have neither been continuous wars nor epidemics...as men had fallen into such a state of pretentiousness, avarice and indolence that they did not wish to marry, or if they married to rear the children born to them, or at most as a rule but one or two of them." (Polybius, 1927, p. 30).

However, from the late republican period up to the Christian empire many people believed that the world was already too crowded for its limited resources. Some, like the Roman poet and philosopher Lucretius (98–55 B.C.), supported the so-called theory of cosmic aging (Lucretius, 1975), a theory which was later supported by Christian authors such as Tertullianus (160–220 A.D.). This theory was related to the idea of the end of the world. Also St. Augustine (354–430 A.D.) considered that the world was already packed with people and that human history was close to an end (St. Augustine, 1972). The Christian Fathers refused to admit that there was any social or religious value in population growth. Thus, they fully agreed with the pagans in playing down procreation. The pagans considered that the density of the human population could justify birth control within marriage: this included abortion, exposure of children and infanticide. The Christian Fathers were instead in favor only of virginity (Herlihy, 1989).

Other considerations supported infanticide in the pagan world. In Sparta, Athens and Rome, parents of all social groups generally eliminated children who displayed signs of physical or mental handicaps. As regards this trend Aristotle is very clear: "..the law must forbid the growth of any child presenting any deformity or physical imperfection at birth" (Aristotle, 1960). Seneca (50 B.C. – 40 A.D.) also supported the necessity to eliminate sick infants: "Mad dogs we knock on the head; the fierce and savage ox we slay; sickly sheep we put to the knife to keep them from infecting the flock; unnatural progeny we destroy; we drown even children who at birth are weakly and abnormal. Yet it is not anger, but reason that separates the harmful from the sound" (Seneca, 1963, p. 145). The poor people abandoned children when they were not able to support them. Other people belonging to the middle classes got rid of their children since they did not want the family to degenerate because of offspring receiving a mediocre education, which would have impaired their social standing. Those belonging to the middle classes and to the upper middle class preferred to concentrate their efforts and resources on a smaller number of offspring. In the country it might happen that an already large family gave its surplus children away to another family group (Ariès and Duby, 1987).

The birth of a Roman citizen was not considered a biological event. In Rome, family status was stronger than parenthood and the paterfamilias (the head of the family) was the one who decided if the newborn should enter the family and the society. The paterfamilias also held the right of life and death over children later on. Rather than kill her child, it might happen that the mother gave away her child to somebody who brought him up secretly. This child became a slave or a freedman and in very rare

cases later succeeded in proving his real origins. This was the history of the Emperor Vespasianus (Ariès and Duby, 1987).

Contraception (Hopkins, 1965/66) and abortion were also customary practices in Rome. It is true that some moralists condemned women who aborted. However, less attention was paid to the fetus than by religious groups today. In addition, little attention was paid to the age of the aborted fetus. Infanticide was the safest method of birth control, since it involved no risk of infections or complications for women (Ariès and Duby, 1987).

Because of these different reasons, in ancient Rome and in Greece the number of surviving children was much lower in comparison with the number of newborns. However, the surviving Greek and Roman children, especially those belonging to the upper class, received an education. On the contrary, the barbarian populations living beyond the limes, for example the Germans, kept more children alive in comparison to the Romans or to the ancient Greeks. However, these children, whether slaves or nobles, grew up, as Herlihy said (1989), in "an atmosphere of benevolent indifference". Herlihy also identifies that although the Germans did not try to limit their fertility to the same extent as did the Romans, they practiced abortion from time to time.

In Rome, where culture and education were considered important values, cultural transmission (for example, family status) was very important, therefore, parents did not want to invest their money in children presenting physical or mental handicaps. This also implied that those children who survived were also kept under strict control and faced a severe education. This discipline was based on physical punishment. On this matter St. Augustine wrote: "It would be a terrible option to choose between eternal death and new life as a child! I believe that most people would choose to die" (St. Augustine, 1971, p. 37). The Germanic folks, on the contrary, invested much less in the education of their children.

CONFLICTS BETWEEN THE NEEDS OF PARENTS AND THE NEEDS OF CHILDREN

The population control strategies used in the past were not as clear-cut as we might think today. In fact not all unwanted children were killed at birth: many were exposed and abandoned and some of them somehow survived. The fate of these children was almost invariably harsh. Seneca the Elder wrote about foundlings who were mutilated in order arouse pity when begging. Lattantius wrote that orphans or abandoned children were condemned to slavery or to prostitution (Capul, 1989).

It was in 318 that Roman law began to consider the killing of children as murder. It is also in the same year that emperor Constantine adopted a number of economic regulations in order to discourage the abandonment and the sale of children by their parents. However, in the following years, because of the economic difficulties of the empire, which resulted in the inability of the state to give pecuniary aid to poor parents, Constantine was again obliged to allow fathers to sell their own children (Picca, 1941).

After the Council of Vaison (442), the discovery of an abandoned child had to be announced in Church. However, in Medieval Europe infanticide was not considered as murder to be tried before a secular court, but a less serious crime under the jurisdiction of Church courts and therefore punished with a lighter penalty. In general terms there was no sharp distinction between a non-spontaneous abortion and infanticide, nor between death due to neglect and death caused intentionally. It was difficult to make this distinction since many newborns were smothered to death in the bed where they slept together with the adults (DeMause, 1974).

There are further indications supporting the fact that despite the official prohibition, infanticide and abandonment continued to be practiced in different areas of Europe. For example, the results of some censuses (*e.g.* the one carried out at the beginning of the ninth century in the area of the monastery of Saint-Germain-des Prés (Herlihy, 1989) indicate that there was an imbalance in the sex ratio in favor of the males. It is possible that in Europe a limited period of sexually-selected infanticide existed, as in traditional China or India. In addition to that, in the late Middle Ages some cases of infanticide were also due to dynastic reasons: when a noble or wealthy man had several women or wives, competition sometimes resulted among the half-sibs over the inheritance (Herlihy, 1989).

The official condemnation of deliberate infanticide transformed it into a sin, which very likely resulted in increased abandonment: from a psychological point of view, abandonment is less traumatic, since the adult may always think that somebody will take care of the child. In reality, however, this behavior actually doomed the child to a horrible life. As Maurice Capul (1989) explains in his essay entitled "Abandon et Marginalité", both in Antiquity and under the Ancien Regime, the fate of abandoned infants was nearly always very harsh. Similar conclusions are reached by Picca (1941) in his essay on "Maternity and Infancy" in which he describes the conditions in which children used to live from Antiquity to the past century: children who were abandoned or sold by their parents were used to carry out menial tasks and heavy work or they were used as sexual slaves.

The increased number of abandoned infants led to institutions being established for them. In 787 Dateo of Milan founded the first asylum.

Other countries followed the same pattern. In 1198 Pope Innocent III founded the Holy Ghost hospital in Rome, which had a wheel to introduce the exposed newborns. Gradually, the practice of entrusting abandoned infants to public charity increased, even if a large part of these children did not live beyond the first year of life (Romita, 1965).

Even if evidence may be found of benevolent and tolerant attitudes toward children within the family (Pollock, 1983), the conditions of many children were somewhat precarious. Two facts delayed parental investment in children until child survival seemed insured: the fact that many children died within the first year of life and that death was a constant life experience. In addition, the behavior of the clergy and of many church authorities was often rather ambivalent. Many, in fact, (mostly since the beginning of the sixteenth century) considered children full of "Original Sin", unable to understand, more similar to animals than to human beings. McLaughlin (1976), who studied the period between the ninth and the thirteenth centuries, found two types of child-rearing advice: in one, proponents urged that children should not be beaten, emphasizing the sensitive nature of the child; the other view stressed the importance of discipline and physical correction.

In everyday life, attitudes toward children were strongly ambivalent. Lyman (1976), who studied the period between the second and the eighth century, believed that up until the eighth century parents were ambivalent towards their offspring, viewing them both as a pleasure and an integral part of family life, as well as a bother. He argues that the former was the ideal but that in actual fact the latter more often corresponded to reality. McLaughlin (1976) writes that there was "a conflict between destructive or rejecting and fostering attitudes" on the part of parents toward children for the period spanning the ninth to the thirteenth century. Tucker (1976) concludes from her research into fifteenth and sixteenth century England that children were seen as untrustworthy and as being at the "bottom of the social scale". Tucker states that parents were ambivalent toward their offspring: they were unsure whether to regard them as good or evil, and also whether to include them in adult society.

The practice of wet-nursing may be considered as another ambivalent behavior toward offspring within the family (see chapter by Hrdy). This behavior was already present in ancient Rome but increased rapidly in the middle and upper classes in urban Italy, and in particular in Florence, between the fourteenth and fifteenth centuries. This practice was also quite widespread in France. Ross (1976) argues that mothers were reluctant to breastfeed because they viewed the child as being principally greedy, sucking a vital fluid from a mother's body already weakened by childbirth.

Thus, despite the high mortality rate due to such a practice, upper class French and Italian babies were entrusted to a wet-nurse. Wet-nursing was practiced until the nineteenth century even if many children died because of a lack of hygiene and adequate care. Of course abandoned infants died in much larger numbers. In Lyon, Prost du Royer (1778) wrote "every year 6000 children are born, but 4000 die during nursing". In Paris, the Hôpital des Enfants Trouvés was defined as a machine which gave back only one tenth of the abandoned children.

Wet-nursing behavior shows that in terms of the health and life of the child, a compromise had to be made based on the necessities of the adult world. One reason for wet-nursing of legitimate children of middle and upper class families in Italy and France seems to have been pressure from husbands to resume sexual relations with their wives. It was a widely held belief that semen would curdle a mother's milk and therefore sexual intercourse should not occur while breastfeeding. In the upper classes, with their need for heirs to inherit property, coupled with the lack of alternative foods for babies, the mothers had little choice but to entrust their offspring to a wet-nurse. Moreover, when a mother entrusted her own baby to a wet-nurse she was relieving herself of a burden, more available for social life, house keeping and for the older children, and she could generate other children (Klapisch-Zuber, 1988). However, in many instances, upper class families did not send children away to a wet-nurse and mortality was lower because the nurse came to their home.

Wet-nursing interferes with attachment behavior; in fact, it is during the first period of life that the child attaches him/herself to the mother and the mother becomes attached to the child. At the time of weaning, or perhaps years later, the surviving children reared by a wet-nurse returned to the parental house. In some instances, this fact eventually resulted in psychological maladjustment in adult life (Morelli, 1969). At the same time, the lack of parent-infant attachment increased the possibility of later parental abuse or indifference. This idea had been promoted since ancient times. Aulus Gellius, a second century Latin writer wrote: "When a child is given to another and removed from its mother's sight, the strength of maternal ardor is gradually and little by little extinguished... and it is almost as completely forgotten as if it had been lost by death" (Aulus Gellius, 1966, p. 361).

Swaddling represents a behavior which also shows a conflict between the needs of adults and the needs of children. By reducing the freedom of children, and therefore the risk of accidents, adults had more time for other tasks. However, a swaddled child was touched less often by its mother or caregiver and, as we know today, physical contact is one

of the important features of attachment behavior. There is evidence from the studies of Harlow (1958) on nonhuman primates and those of Bowlby (1969) on children that there is a biological basis for attachment. What is crucial is how far humans can act in opposition to their biological inheritance. Could, for example, socioeconomic and religious factors override genetic influences to the extent that parents not only stop caring for their offspring but also actually decrease the latter's survival value by ill-treatment and abandonment (see Hrdy; Wilson and Daly, this volume)?

Trivers (1974) proposed that because parents and children only share one half of their genes on average, it is possible to predict a certain amount of conflict between the needs of the parent and the needs of the child. Although parental behavior often functions for the good of offspring, it is not always optimal for each infant. There may be circumstances under which parental behavior is not in the best interests of certain offspring and may even lead to the latter's death. Furthermore, organisms may be living in environments to which they are not adapted, and so destructive behavior ensues as, for example, in many caged primates. Therefore, there are a number of external constraints that give priority to the needs of adults and that interfere with the formation of the attachment bond, an effect which may result in increased passive infanticide. Neglect, deprivation and lack of emotional support of children represent different conditions that may result in: a) a disease which may eventually result in its death, b) increased number of accidents, c) lack of resources to cope with the routine difficulties of life.

Of course we must consider that the conditions of children in the past varied extensively with the social status of the family and whether they were legitimate or natural, healthy or sick. In fact, while a child was considered as God's blessing in rich families, on the contrary, it represented a real threat to the survival of the other members in poor families. It was difficult for a single woman to keep a natural child: not only the mother was outlawed but also the child received very severe treatment from society; he or she was often abused and when an adult, would drop-out from society (Stone, 1977).

Children with handicaps or physical disabilities had a double stigma: one written in their body and one, probably more severe, in their soul. They were considered by the clergy as living proof of the sexual sins of their parents (Capul, 1990). Since they were associated with shame for their parents, it is not surprising that in the sixth century, as in the thirteenth and eighteenth centuries, many of them were abandoned or killed at birth. When they survived they were generally the object of mockery and, because they were different, they were often accused of

sorcery (Capul, 1990). Therefore, there were children who were raised in favorable conditions, received care and affection, socialized and went to school, while others lived and died in tragic conditions, were abused during their short life, were persecuted and forced into prostitution or even mutilated or sold.

In this contradictory scenario it is evident that many children of the past received a certain amount of parental care, otherwise the human species in Europe would have faced extinction. Nor we can accept the idea that parents always acted in direct opposition to their biological fitness: the choices of wet-nursing or swaddling, for example, are understandable if we consider them within a number of environmental variables. However, there is no doubt that children belonging to noble, rich or bourgeois families (in particular first-born children) were those who received the most attention. For the other children, life was very difficult: when the living conditions of the adults were very precarious, the risks for their children increased during pregnancy and after childbirth because of different forms of neglect and abuse (Stone, 1977).

Anthropological studies indicate that in very exceptional circumstances, parental care may disappear. Turnbull (1973) studied the Ik tribe in Africa during a famine which was caused by drought. There were signs that before the drought the Ik had been as kind and generous as other human beings, but with increasing starvation, they became ruthless and concerned solely with their own individual survival. Altruism was virtually non-existent; anyone who helped someone else was regarded as a fool. In these circumstances children were neglected by their parents. Turnbull notes that he "never once saw a parent feed a child except when the child was still under three years old" (Turnbull, 1973, p. 114). However, this behavior of the Ik is not an inevitable response to famine.

In Europe the conditions of children underwent radical changes in the sixteenth century when trade increased and people became wealthier. New emphasis was given to child education. Rich families had more financial resources than in the past and were more prone to invest in the cultural formation of their offspring. As in Antiquity, this resulted in increased attention toward children and also to more rigid forms of discipline, ensuring - in harmony with the mentality of that time - the social, cultural and moral formation of the young. Many educators (and parents) believed children needed to be beaten in order to be trained. The puritan educators in particular had the goal of breaking the will of the child (Pollock, 1983).

This increased investment in the child's education did not result - as had already happened in imperial Rome - in a decrease in the rate of infanticide

and abandonment. Tucker (1976) believes that in fifteenth and sixteenth century England "infanticide was woefully common". Shorter (1976) includes abandonment in his list of evidence of the indifference which parents felt towards their children during the seventeenth century. However, he concedes that for some parents poverty was the cause of abandonment and that separation was painful. De Mause claims that infanticide was "an accepted, everyday occurrence" in ancient times and that, by the eighteenth century "there was a high incidence of infanticide in every country in Europe". De Mause further wrote that "once parents began to accept the child as having a soul, the only way they could escape the dangers of their own projection was by abandonment" (De Mause, 1976, p. 28). This not only led to a large number of precocious deaths, but also resulted in a type of infancy different from that of the children belonging to the rich and noble families represented in contemporary paintings. Since the sixteenth century many orphans, foundlings and handicapped children had passed a large part of their life locked up in asylums, hospitals and even in prisons (Capul, 1990). In the absence of sure contraceptive methods, infanticide, abandonment, exposure and neglect continued to be the most widely practiced methods, consciously or unconsciously, of reducing parental investment in offspring.

INFANTICIDE AS PATHOLOGY

At the end of the seventeenth century, some social groups belonging to the European upper middle-class — such as the upper bourgeoisie of Geneva or the English aristocrats — had already adopted birth control methods. Even the theologians admitted that poverty could justify birth control methods (Stone, 1977). Contraceptive practices started to spread in northern Europe around 1750 and may have resulted in a decrease in infanticide (Stone, 1977). Another factor contributing to this behavioral transformation was the spread of a more protective, empathic behavior toward children among some parents; in the seventeenth century St. Vincent de Paul protested against"the slaughter of the innocents" and founded a charitable organization to take care of the problem of abandoned children. In the eighteenth century, J.J. Rousseau, with his pedagogical essays, alerted the public to the problem of infancy and emphasized the mother's role in child rearing; it was at the end of the eighteenth century that the first campaign against wet nurses started (vom Saal, this volume). Rousseau proposed the attainment of a less hierarchical society and family and a different attitude of the expanding national states toward the new genera-

tions, since these were the reservoir for the working force and for the army (Moheau, 1778).

Today, we assume that cases of deliberate infanticide and abandonment are much fewer than in the past, even if we do not know their exact number (Chalou, 1989). In Italy, according to the Year Books of the Central Institute of Statistics (ISTAT), 332 cases of infanticide occurred between 1970 and 1988. However, this figure refers solely to official infanticides found in the criminal records. Out of these 332 cases, only eleven are available in the form of complete records. These refer to findings of infanticide by the Italian Court of Cassation (National Library, Rome, 1990). Six were perpetrated by unmarried mothers; three by prostitutes; two by women who had a partner but lived in conditions of extreme moral and economic precariousness. Out of these eleven case of infanticide, ten were committed by women: five were committed by women acting alone, three were helped by their mothers, one was helped by her brother, one by her partner, who wanted to get rid of the newborn in order to marry his partner's daughter. In one case, the death resulted from the malpractice of a doctor and a midwife. In three cases, the child was abandoned in a garbage bin, in four cases the infant was smothered, in two cases the infant was wounded with a knife, one "fell" into the water closet, and one died of neglect.

If we take into consideration the category "homicide and lesions produced by others", we discover that in the years 1983–1985, 165 cases were heard by the Courts, although nobody can tell exactly how many unrecorded "incidents" there were. On the other hand, we know the reasons of serious child abuse during the first months or years of life. The most common are: 1) social isolation; 2) difficult conditions of life of the parents; 3) severe conflicts within the parental couple; 4) absence of a partner; 5) child handicaps; 6) jealousy of the father who feels excluded from the mother-child dyad; 7) step-parenting (Daly and Wilson, this volume; Wilson and Daly, this volume).

There is no doubt that today the situation is different from the past for many reasons: a) improved economic conditions; b) greater attention focused on infancy; c) birth control practices; and d) high adoption demand. However, a woman may still be tempted to engage in infanticide when she sees in the birth of a child an obstacle to her relationship with a partner, either present or future. While in the past many mothers agreed to wet-nursing since it was a behavior approved by the community, other women probably accepted to send their newborns to a wet-nurse in order not to lose their husbands during the nursing period. They chose to risk losing one of their children in order to maintain family unity. For other reasons

in traditional China, as in medieval Europe, a woman was capable of doing anything in order to eliminate the child of her husband's concubine.

A first point emerging from this review is that there are different forms of infanticide and child abuse and therefore different interpretations related to the different uses and conditions of life. There may be different forms of "unexplainable" infanticide or different forms of abuse or neglect which may be related to psychological motivation linked to the social or family framework. A second point is that in the course of many centuries, the basis for infanticide may have changed.

The adult is driven by many cues in his decisions and in the evaluations of his own actions and responsibilities towards children. These cues are psychological, such as those related to individual sensibility, or biological, such as those involving the protection of a genetically-related helpless child. In addition to these, there are the cues emerging from the presence or absence of environmental resources sufficient to rear a newborn; the scarcity of resources may create a conflict between the interests of the adult and those of the child. Finally, there are cultural cues which have a very important role in the human species. The customs of a society and the concept of infancy and the value given to the single individual are factors exerting an important influence on the adult-child relationship in a given society or in a given historical period.

All these different factors may result in different interactions leading to different behaviors. For example, in our society infanticide and child abuse are strongly disapproved. However, sharing these views does not always mean that the parents will accept their own child, in particular if the child is sick or handicapped. In order to form emotional ties with their infants, many parents, in particular mothers, may need to feel that they are supported by the social microsystem (family, partner, mother) and from the social macrosystem (society, institutions). If this confidence is not present, neither the moral condemnation of society nor individual values are sufficient *per se* to save the newborn from death or the child from abuse in subsequent years.

REFERENCES

ANNUARI Statistici Italiani (1970-1989) "Delitti denunciati all'Autorità giudiziaria dalle forze dell'ordine", Rome.
ARIÈS, P. and Duby, G. (1987) *La vita privata dall'Impero all'anno Mille*. Laterza, Rome.
ARISTOTLE (1960) *Politica*. Lib. VII cap. 14, Garzanti, Milan.
AUGUSTINE, Saint (1972) *De bono coniugali*. Garzanti, Milan.
AUGUSTINE, Saint (1971) *De civitate Dei*. Garzanti, Milan.

AULUS Gellius (1966) *Attic.* pag. 361. Garzanti, Milan.

BOWLBY, J. (1969) *Attachment.* Hogart Press, London.

CAPUL, M. (1989) *Abandon e marginalité.* Privat, Toulouse.

CAPUL, M. (1990) *Infirmité et Hérésie. Les enfants placés sous l'Ancien Régime.* Privat, Toulouse.

CHALOU, S. (1989) *L'Enfance Brisé. Le Pré aux Clercs*, Paris.

DE MAUSE, L. (1976) The Evolution of Childhood. In: *The History of Childhood,* Souvenir Press, London.

GUZZO, M.Q. (1986) La documentazione epigrafica dal tofet di Mozia e il problema del sacrificio Molk. *Studia Phoenicia*, IV, Namur, pp.189–207.

HARLOW, H.F. (1958) The Nature of Love. *American Psychologist*, **13**, 683–685.

HERLIHY, D. (1989) *La famiglia nel Medioevo.* Laterza, Bari.

HOPKINS, K. (1965/66) Contraception in the Roman Empire. *Comparative Studies in Society and History*, vol. **8**, Oct.–July.

ISTAT (1983–85) *Statistiche Sanitarie*, Rome.

KLAPISCH-Zuber, C. (1988) *La famiglia e le donne nel Rinascimento a Firenze.* Laterza, Bari.

LUCRETIUS (1975) *De rerum natura.* Garzanti, Milan.

LYMAN, R. (1976) Barbarism and religion: late Roman and early medieval childhood. In: De Mause, L. (ed.), *The History of Childhood*, Souvenir Press, London.

MCLAUGHLIN, M. (1976) Survivors and surrogates: children and parents from the ninth to the thirteenth centuries. In: De Mause, L. (ed.), *The History of Childhood.* Souvenir Press, London.

MOHEAU, Recherches et considerations sur la population de la France. Paris, 1778. In: Badinter E., *L'Amour en plus.* Flammarion, Paris, 1980.

MORELLI, G. (1969) *Ricordi. Ed.* Vittore Branca, Florence.

NATIONAL Library (1990) Rome.

PICCA, P. (1941) *Maternità e infanzia.* SEI, Rome.

PLUTARCH (1928) *Moralia*, F. Babbitt trans. London, pp. 493.

POLLOCK, L.A. (1983) *Forgotten Children.* Cambridge University Press.

POLYBIUS (1927) *The Histories*, vol. 6, W. Paton trans. London, pp. 30.

PROST du Royer, J. (1778) *Memoire sur la Conservation des enfants.* Paris, pp. 14.

ROMITA, F. (1965) *Evoluzione storica dell'assistenza all'infanzia abbandonata.* Istituto di Storia della Medicina della Università di Roma.

ROSS, J.B. (1976) The middle-class child in Urban Italy, fourteenth to early sixteenth century. In: De Mause, L. (ed.), *The History of Childhood.* Souvenir Press, London.

SCRIMSHAW, S.C.M. (1984) Infanticide in human populations: Societal and individual concerns. In: Hausfater, G. and Hrdy, S.B. (eds.), *Infanticide: Comparative and Evolutionary Perspectives*, Aldine, New York.

SENECA (1963) *Moral Essays*, J.W. Basore trans. Cambridge, Mass., pp. 145.

SHORTER, E. (1976) *The Making of the Modern Family.* W. Collins, London.

SPALINGER, A.J. (1971) A Canaanite Ritual Found in Egyptian Military Reliefs. *Journal of the Society for the Study of Egyptian Antiquities*, **8**, 47–59.

STONE, L. (1977) *The Family, Sex and Marriage in England 1500–1800*, Weindenfeld e Nicolson, London.

TRIVERS R. (1974) Parent-offspring conflict. *American Zoologist*, **14**, 249–264.

TUCKER, M.J. (1976) The child as beginning and end: fifteenth- and sixteenth-century English childhood. In: De Mause, L. (ed.). *The History of Childhood*. Souvenir Press.
TURNBULL, C. (1973) *The Mountain People*. J. Cape, London.

CHAPTER 5

STEPPARENTHOOD AND THE EVOLVED PSYCHOLOGY OF DISCRIMINATIVE PARENTAL SOLICITUDE

MARTIN DALY and MARGO WILSON

Department of Psychology, McMaster University,
Hamilton, Ontario, L8S 4K1, Canada

> Quand la femme se remarie ayant enfants
> Elle leur fait un ennemi pour parent
> — 16th century French proverb (Segalen, 1981)

One need not know anything about 16th century France to find the idea expressed by this proverb familiar: A new stepfather constitutes a threat to children. An even more prevalent and menacing figure in Western lore is the malevolent stepmother. Consider the stories of Cinderella and Snow White, or (less familiar to anglophone readers) la Maratre, who cooked and served to her unwitting husband the child of his former marriage. So pervasive is this image that steprelationships have become a conventional metaphor for the absence of genuine commitment: If I complain that behavioral biology is the "stepchild" of a federal agency funding scientific research, my meaning is immediately understood.

Negative characterizations of stepparents are not peculiar to European traditions. Consult an encyclopedic source such as the *Motif-index of folk literature* (Thompson, 1955), for example, and you will encounter evil stepmothers from around the world. As for stepfathers, Thompson divides his entries, for the sake of convenience, into two categories: folktales about "cruel stepfathers" and those about "lustful stepfathers". Hunter-gatherer

or horticulturalist, tribesman or city-dweller, the stepparent is the villain of every piece.

Negative images of stepparental relationships persist in the contemporary American populace (*e.g.* Bryan *et al.*, 1986; Fine, 1986). The social scientists who have documented these negative images typically refer to them as "myths" and "stereotypes". The implication of these terms (presumably an intended implication; see, *e.g.* Harding, 1968) is that worries about possible risks to stepchildren are unfounded. However, the researchers who have documented the existence of these "myths" and "stereotypes", and who have then decried their pernicious influences, have consistently ignored or obfuscated a crucial question: What, if anything, is their basis in reality?

OWN VERSUS ALIEN

To the Darwinian imagination, the hypothesis that stepparenthood entails genuine risk to children is immediately plausible and deserving of empirical investigation. The reason is as follows. Parental investment ("PI", *sensu* Trivers, 1972) is a precious resource, such that those parental phenotypes which somehow succeed in channeling PI to genetic relatives necessarily enjoy a selective advantage over alternative phenotypes which disperse the benefits of their efforts less discriminatively. This theoretical expectation has been upheld in numerous empirical studies: Animals have evolved diverse discriminatory mechanisms that function to identify own offspring and to direct parental nurture selectively to them (see Daly and Wilson, 1988a). Moreover, comparative studies indicate that parents' recognition of their own young and attendant favoritism are best developed in those species and at those life stages in which there is significant selective pressure in the form of risk that PI will be misdirected (*e.g.* Beecher *et al.*, 1986).

The ubiquity and importance of discriminative parental preferences for own versus alien young were long overlooked by comparative psychologists. This neglect is apparently attributable to an unfortunate happenstance: the fact that the physiological controls of maternal behavior have been investigated primarily with the convenient laboratory rat and mouse. It so happens that these burrow-dwelling rodents are scarcely affected by the own-alien distinction, blithely mothering whatever young they encounter in their nests, including even pups of other species, and because of this peculiarity, mammalian mothering has come to be conceptualized as a motivational state of the mother rather than as an

individualized relationship. Experimenters assess levels of activation of this "maternal state" by measuring responses to "standard stimulus pups", and the physiological and behavioral determinants and manifestations of this maternal state provide seemingly inexhaustible material for experimental investigation (see, *e.g.* various papers in Krasnegor and Bridges, 1990). The reason that this is unfortunate is that a rat-like state of diffused maternal responsiveness is by no means the mammalian norm. In species in which the young are ordinarily nursed in social situations and unrelated same-age young can therefore mingle, mothers form individualized bonds with their own infants in the immediate postpartum; such mothers will attack and even kill aspiring milk thieves despite being fully "maternal" (*e.g.* LeBoeuf and Briggs, 1977 *re* seals; Gubernick, 1981 *re* goats). Burrow-dwelling or hole-nesting species with immobile altricial young, by contrast, have experienced no selection pressure for such early discrimination, with the result that experimenters who transfer pups between nests find the mothers unperturbed. In effect, the nest site and her own odors have become the cues by which a mother rat "recognizes" her young. But in species in which the young become sufficiently mobile to intermingle before they are fully weaned, then even among burrow-dwelling rodents, mothers develop an individualized recognition of their pups and begin to discriminate in their favor at about the time when pup mobility promises an imminent risk that maternal care will be misdirected to alien pups. (It is also worth mentioning here that laboratory rats and mice have undergone intense selection for successful weaning of pups in the presence of an abnormal density of olfactory, auditory and other stimuli of conspecific origin, both in colony rooms and in communal nursing cages.)

The general point is that mammalian motherhood is not typically a generalized state of nurturant inclination toward just any little beggar who happens to present the sign stimuli of a conspecific youngster. Mothers love, nurture, and risk their lives on behalf of those *particular* youngsters who present the mothers with evolutionarily reliable cues that they are *their* youngsters.

NON-HUMAN STEPPARENTING

Non-human animals often find themselves caring "parentally" for young who are not their own progeny. Many such cases represent instances of "brood parasitism" in which the mechanisms of discriminating own from alien young have been circumvented, whether by conspecifics (Rohwer and Freeman, 1989) or by parasitic specialists like the European cuckoo

(Davies and Brooke, 1991). Many others are instances of "cuckoldry" in which the young in question are indeed those of their ostensible mother but have been sired by someone other than their ostensible father (Westneat, Sherman and Morton, 1990); this state of affairs may be no more evident to the cuckolded male than to the human observer.

Paradoxically, although the victims of brood parasitism and cuckoldry regularly make maladaptive investments of parental effort in nonrelatives, it is where these phenomena occur that we find some of the best evidence that selection favors discriminative allocation of PI in favor of own young. In what has been called a "co-evolutionary arms race", species with a long history of victimization by particular brood parasites have often evolved to be more discriminating than unparasitized species, and the brood parasites have often evolved further tactics, such as egg and chick mimicry, and rapid, surreptitious egg deposition, to counter host discrimination (Davies and Brooke, 1991). Similarly, males of species with an evolutionary history of cuckoldry manifest evolved defenses against this recurring threat to fitness, such as mate-guarding specifically confined to the mate's fertile periods and modulation of subsequent paternal effort in relation to the thoroughness of mate surveillance at the time when the young might have been conceived (Møller, 1988; Burke et al.,1989).

Neither brood parasitism nor cuckoldry is closely analogous to stepparenthood, however, because both are cases in which the pseudoparent is in effect "deceived" about parenthood. The misdirected PI in both cases depends upon the alien young's intruding undetected into a situation in which the unrelated adult is prepared to invest in its own young. A closer analogy to stepparenthood would be one in which the investing individual has reliable cues of nonparenthood, but plays parent to a new mate's young from a prior union nonetheless.

Post-zygotic care in many substrate-spawning fishes is provided by the male alone, who may invite multiple females to spawn in his nest. In such cases, eggs already in the nest may be attractive to females, inspiring males to steal eggs or usurp nests as courtship devices. In the process, the usurper assumes the paternal role of guarding and perhaps aerating eggs sired by other males. Sargent (1989) showed that such adopted eggs were less well cared for by male fathead minnows than the usurpers' own and that they suffered higher mortality. This is a nice demonstration of own-alien discrimination, but Sargent's titular claim that "stepfathers discriminate against their adopted eggs" may not be the best way to describe it. To be a "stepfather" is to assume paternal status by virtue of replacing the genetic father as the genetic mother's mate. The fathead minnow "stepfather"

apparently uses unrelated eggs as courtship lures, but he does not mate with those unrelated eggs' mother. A much closer analogue comes from Yanagisawa and Ochi's (1986) study of pair-forming anemonefish, in which the disappearance of one member of a pair leads to the appearance of replacement mates who sometimes help care for their predecessors' fry. In this case, suitable breeding situations are scarce, and Yanagisawa and Ochi argue that stepparental effort constitutes mating effort, a price paid for future reproduction with the surviving genetic parent of the present brood.

The situation is similar in many biparental birds. Rohwer (1986) has reviewed a large number of observations of avian mate replacement and subsequent behavior. He distinguishes three possible responses to the still dependent young of a predecessor: active elimination (infanticide), ignoring them, and the assumption of parental duties. Each of these three responses has been observed in several species, and all three may even be the responses of different individuals within a single species, but there are large species differences in what is typical. Rohwer adopts the adaptationist expectation that typical responses in this situation will be fitness-promoting decisions shaped by past selection, and derives a number of comparative predictions from this view, concerning the expected associations of the three alternatives with such factors as the species' adult sex ratio and its renesting and dispersal practices. The evidence runs mostly in the direction of Rohwer's hypotheses, but he notes many anomalies and a paucity of good evidence on several questions. One idea that seems not to have been systematically tested is that even when they "adopt", stepparents might have different thresholds of tolerable cost than genetic parents in such domains as the defense of the brood against predators.

Underlying Rohwer's analysis is essentially the same argument as Yanagisawa and Ochi's (1986) explanation for stepparental efforts in anemonefish: that where stepparental adoption is prevalent, it represents an investment in the courtship of the genetic parent, elevating the probability and/or the effectiveness of subsequent breeding with the assisted mate. Smuts (1985) invoked much the same explanation for the observation that male baboons indulge infants and juveniles they cannot have sired; certainly, male baboons' behavior is far from universally benign, including considerable intolerance and occasional infanticide, and a male's kindness to a select youngster is part and parcel of his cultivating a "friendship" with the mother. We suggest that human willingness to enter into situations of stepparental obligation is similarly to be explained as a component of courtship of the genetic parent.

HUMAN STEPPARENTING

Substitute parenting in the human animal presents some obvious difficulties as regards the Darwinian prediction of discriminative parental solicitude in relation to cues of relatedness. People regularly undertake the parenting of children under circumstances in which they have reliable cues that those children are not their own. And whereas we may wonder if a nonhuman animal has correctly processed available cues of nonparenthood or has instead been "deceived", there is little question that human beings know that they are not the genetic parents of their adopted or stepchildren; they can tell you so. Must we conclude that human beings lack the discriminative parental solicitude characteristic of most other mammals? And, if so, why?

In this chapter, we shall consider the issues only with respect to stepparenting: the acquisition of pseudoparental obligation as an incidental cost associated with the establishment of a new relationship. Adoption of children unrelated to either "parent" is a different matter, an effort to simulate the genetic nuclear family experience for its own sake rather than as an incidental consequence or attendant cost of the pursuit of other social goals. Such "adoption by stranger" is beyond our present scope (but see Daly and Wilson, 1980; Silk, 1990).

One possible hypothesis to account for stepparental investment might be that human parental solicitude is vulnerable to parasitism by unrelated young because our ancestral circumstances, like those of burrow-dwelling rodents, placed no selective premium on discrimination. Perhaps, during millennia of human evolutionary history, stepparenthood was simply not the sort of recurring adaptive problem that would have inspired the evolution of psychological defenses against it. Nonnutritive saccharine, an evolutionarily unforeseen component of novel environments, tickles our evolved system for the recognition of nutritive sugars. Might stepparenthood constitute a sort of novel social environment: an evolutionarily unforeseen circumstance in which the evolved psychology of parenthood is activated maladaptively?

Such an hypothesis appears to gain plausibility when one turns to the social scientific literature on steprelationships. Cherlin (1978) proposed that stepparenthood is a novel "role" or status whose ground rules have yet to be established, and that difficulties attend steprelationships because of this "incomplete institutionalization" and attendant "role ambiguity". Many writers have embraced and elaborated upon this sort of interpretation (*e.g.* Kompara, 1980; Giles-Sims, 1984; Keshet, 1990), which is in effect a novel social environments argument, albeit a non-Darwinian one. (This "novel

social role" argument has not been articulated in an evolutionarily sophisticated form. An implicit premise of the conceptual framework of its proponents is that social influences and expectations impact upon all roles and relationships in qualitatively similar ways, so that the characterization of the essential distinguishing features of, say, peer *versus* mating *versus* filial *versus* sibling relationships and their respective psychologies is not even part of the analytic agenda. Reconciling this implicit premise that all relationships are essentially alike with elementary principles of social evolution would be difficult if not impossible, but the social scientists who adhere to this premise have not perceived the problems facing their domain-general conception of sociality, let alone confronted them.)

Of course, the fact that the novel social role argument is non-Darwinian does not mean it must be wrong. But even in its own terms, this popular analysis is ahistorical, ethnocentric, and counterfactual. Stepparenthood is *not* a novel circumstance. The mortality levels incurred by tribal hunter-gatherers guarantee that remarriage and stepparenthood have been common for as long as people have formed marital bonds with biparental care; moreover, the ethnographies of recent and contemporary hunter-gatherers abound with anecdotal information on both the prevalence of steprelationships and their predictable conflicts (*e.g.* Shostak, 1981; Hill and Kaplan, 1988). Nor is stepparenthood even newly prevalent in "our society". Historical records indicate that stepparental relationships, consequent upon both widowhood and divorce, have been numerous for centuries in the western world (*e.g.* Dupâquier *et al.*, 1981). Moreover, European historical archives show that having a stepparent was associated with mortality risk in fact and not just in fairy tale (Voland, 1988).

A defender of Cherlin's "incomplete institutionalization" argument might protest that it was offered only as a description and explanation of the most recent American trends. It has become a platitude to claim that steprelationship was recently rare, but that escalating divorce and remarriage are now making it more the norm than the exception. But in fact, Cherlin and his followers have not demonstrated that the allegedly novel and burgeoning phenomenon of stepparenthood was ever very much rarer in American life than it is now. More importantly, their attempt to account for stepfamily conflict in terms of the peculiarities of rapid social change in the contemporary U.S.A. is superfluous: All available evidence suggests that steprelationships are more conflictual than the corresponding genetic relationships in *all* societies, regardless of whether steprelationships are rare or common and regardless of their degree of "institutionalization".

The cross-cultural ubiquity of Cinderella stories reflects certain basic, recurring tensions which have always characterized human society. If a widowed or forsaken parent of dependent children wished to forge a new marital career, then the fate of the children became problematic. A common solution to the dilemma created by stepparents' disinclination to raise the children of their predecessors has been to leave half-orphaned children in the care of post-menopausal female relatives. Alternatively, a widow may retain her children but be blocked from free reentry into the marriage market and be obligatorily remarried to the dead man's brother or other near relative, instead. This practice (the "levirate"), which occurs in a number of patrilineal, patrilocal societies in which the deceased husband and his agnatic kin have paid a "bride-price" for the woman's productive and reproductive services, provides the children with a stepfather who already has a benevolent interest in their welfare, namely their uncle, and thereby at least mitigates the probability or severity of exploitation and mistreatment. In the absence of such practices, children were obliged to tag along as best they could, hoping that their welfare would remain a high priority of the surviving genetic parent.

In tribal societies, the available evidence indicates that the half-orphan who enters the perilous status of stepchild faces a major diminution in the quality and quantity of parental nurture, and a significantly elevated risk of death. An infant's having been fathered by a man other than the mother's present husband is a widely cited rationale for infanticide (Daly and Wilson, 1984), and the hazards extend to older children, too, even if they are not explicitly marked for death. In a study of the foraging Ache in Paraguay, Hill and Kaplan (1988) compared the life trajectories of 67 children raised by mother and stepfather after their natural fathers' deaths to those of 171 children raised by two genetic parents. Twenty-nine (43%) of the stepchildren had died, by a diversity of causes, before reaching the age of 15, as compared to just 19% of those reared by surviving parents.

What about the modern west? Are people in contemporary industrial societies significantly more likely to neglect, assault or otherwise mistreat their stepchildren as compared to their genetic offspring? One might suppose that this rather obvious question would have received considerable attention during the explosion of child abuse research that followed Kempe *et al.*'s (1962) agenda-setting proclamation of "the battered-child syndrome", but the question was curiously overlooked by researchers whose imaginations were not informed by Darwinism. The first published study addressing it did not appear until 1980, when we (Wilson, Daly and Weghorst, 1980) showed that stepchildren constituted an enormously higher proportion of the American Humane Association's "validated" case

reports of child abuse than their numbers in the population-at-large would warrant. Moreover, the over representation of stepchildren was more extreme in assaultive cases than in those that were solely neglectful, and vastly more extreme in the lethal cases. The results of these initial analyses also suggested that steprelationship was not dangerous by virtue of some incidental association with poverty, for there was no such association; steprelationship and poverty instead constituted two independent risk factors for child maltreatment (Wilson *et al.*, 1980; Daly and Wilson, 1981).

Our subsequent research on this question has consistently found even larger differences in the risks to children living with step-plus-genetic-parent *versus* two-genetic-parents than in our initial study. In a local study of child abuse and the household circumstances of children in Hamilton, Ontario, for example, we found that preschoolers living with step-plus-genetic-parent were more than forty times as likely to be victims of severe abuse as those residing with two genetic parents (Daly and Wilson, 1985). The differences were essentially independent of the impacts of such risk factors as low socioeconomic status, large family size, and maternal youth (Daly and Wilson, 1985; Wilson and Daly, 1987). We and others have also demonstrated that abusive stepparents are typically discriminative, sparing their own children within the same household (Lightcap, Kurland and Burgess, 1982; Daly and Wilson, 1985; see also Flinn, 1988); this result refutes the hypothesis (Giles-Sims and Finkelhor, 1984) that excess risk in stepfamilies has nothing to do with steprelationship *per se*, resulting incidentally from an overrepresentation of violent personalities among remarried persons.

The overrepresentation of stepparents in child abuse samples might be explained away as a product of biased detection or reporting, were it not for the fact that stepparents are even more strongly overrepresented in fatal cases, where reporting biases should be minimal. Whereas the sublethal risk differentials between step-plus-genetic-parent and two-genetic-parent homes in our initial American study were only on the order of two- to seven-fold (depending on the child's age), for example, the same data set indicated that the differential in fatal abuse was on the order of 100-fold (Daly and Wilson, 1988c). In Canada, too, the differential risk of being slain by a stepparent versus a genetic parent is even greater than the substantial differential in sublethal abuse noted above (Daly and Wilson, 1988b). Recent English data tell much the same story: One can estimate from Creighton's (1985) child abuse statistics and Wadsworth *et al.*'s (1983) cohort study of children's household circumstances that victimization in step-plus-genetic-parent homes exceeded that

in two-genetic-parent homes by a factor on the order of 30, whereas the fatal baby battering data reported by Scott (1973) apparently indicate that this risk was more than 150 times as great at the hands of a stepfather as compared to a genetic father. In a recent Australian study (Wallace, 1986), the overrepresentation of stepfathers among fatal baby batterers was greater still.

Conflict in steprelationships is not confined to the violent extremes assayed by child abuse and homicide samples. A massive literature on American stepfamilies has developed in recent years. Most of the research has been conducted with volunteer subjects of middle class background, some having sufficient difficulties to have sought help, others apparently thriving. This literature has a single focus: the conflicts and dissatisfactions of stepfamily life, and how people cope with them (see, *e.g.* Anderson and White, 1986; Pasley and Ihinger-Tallman, 1987; Wilson and Daly, 1987; Ihinger-Tallman, 1988; Giles-Sims and Crosbie-Burnett, 1989). Lest we paint too bleak a picture, it is important to stress that people *do* cope; steprelationships clearly can work reasonably well. Some stepparents, albeit a minority, even feel able to profess to "love" their wards (Duberman, 1975). But though steprelationships are not inevitably hostile, the extensive literature is nevertheless unanimous that they are, on average, more distant, more conflictual, and less satisfying than the corresponding genetic parent-child relationships.

It may seem remarkable that steprelationships are ever peaceful, let alone genuinely affectionate. But of course violent hostility is episodic and amicableness is frequent even among nonrelatives. People thrive by the maintenance of networks of social reciprocity and by establishing reputations that will make them attractive exchange partners (Alexander, 1987), with the result that the desire to be generous and humane, and to be *seen* to be generous and humane, is as human as competitiveness and no less functional. Moreover and more specifically, stepparents assume their pseudoparental obligations in the context of a web of reciprocities with the genetic parent, who is likely to recognize more or less explicitly that stepparental tolerance and investment constitute benefits bestowed upon the genetic parent and the child, entitling the stepparent to reciprocal considerations.

There is thus no great conundrum in the fact that people treat their stepchildren for the most part quite tolerantly, nor even in the existence of genuine stepparental investment in the child's welfare at cost to self. The interesting questions are whether the motives and emotions of stepparents *vis à vis* the children ordinarily (or indeed ever) become essentially like those of genetic parents, and, if not, how they differ.

An obvious hypothesis from a Darwinian view of parental motives is that stepparental feelings will indeed differ from those of genetic parents, at least quantitatively and perhaps qualitatively, too. Indulgence toward stepchildren may be a good way to promote domestic solidarity and tranquility, but the circumstances must always have been rare in which a stepchild's welfare was as valuable to the adult's expected fitness as an own child's welfare would be. We wouldn't necessarily expect to see a great deal of abuse of stepchildren, but we would not expect to see stepparents sacrificing as much for them as genetic parents either. Is there a large difference between genetic parents and stepparents in willingness to incur major costs (*e.g.* life-threatening risks) on the children's behalf? We expect that there is, but we know of no relevant study. There is, however, plenty of evidence that stepparents and stepchildren alike view their relationships as less loving and as a less dependable source of material and emotional support than genetic parent-offspring relationships (*e.g.* Duberman, 1975; Perkins and Kahan, 1979; Ferri, 1984; Santrock and Sitterle, 1987; Flinn, 1988).

The dominant framework in the social sciences for discussing steprelationships is "role theory". Parenthood is considered one "role" and stepparenthood another (*e.g.* Cherlin, 1978; Kompara, 1980; Giles-Sims, 1984). The "theory" in "role theory" is surprisingly elusive for something so frequently invoked; Biddle's (1986) review unwittingly suggests that the work of role theorists is devoid of such ordinary signs of theoretical activity as efforts to use the theory to discover something previously unknown. But though the role concept does not really constitute a theory from which expected empirical consequences may be derived, its popularity is not without consequences: It is a metaphor that has directed attention to some issues and away from others. What the role metaphor directs attention to is requisite familiarity with cultural norms or "scripts": You have to know the role in order to act it out. What it directs attention away from is the motivational and emotional aspects of the parental psyche. There is more to the explanation of our choices of social action than mere familiarity with the options. Why do we embrace certain roles and shun others? Parents are profoundly concerned for their children's well-being and future prospects, but human concerns have no part in role theory's explanations of human action (see Biddle, 1986). Stepparents do not, on average, feel the same child-specific love and commitment as genetic parents, and do not reap the same emotional rewards from unreciprocated "parental" investment (Wilson and Daly, 1987). Enormous differentials in the risk of violence are just one particularly dramatic result of this predictable difference in feelings.

ACKNOWLEDGEMENTS

Our homicide research has been supported by the Harry Frank Guggenheim Foundation, the North Atlantic Treaty Organization, the Natural Sciences and Engineering Research Council of Canada, and the Social Sciences and Humanities Research Council of Canada. This chapter was written while the authors were Fellows of the Center for Advanced Study in the Behavioral Sciences with financial support from the John D. and Catherine T. MacArthur Foundation, the National Science Foundation #BNS87-008, the Harry Frank Guggenheim Foundation and the Gordon P. Getty Trust, and while M. Daly was a Fellow of the J.S. Guggenheim Foundation. We wish to thank Stefano Parmigiani, Bruce Svare and Fred vom Saal for organizing this conference.

REFERENCES

ALEXANDER, R.D. (1987) *The Biology of Moral Systems*. Aldine de Gruyter, Hawthorne NY.

ANDERSON, J.Z. and White, G.D. (1986) An empirical investigation of interaction and relationship patterns in functional and dysfunctional nuclear families and stepfamilies. *Family Process*, **25**, 407–422.

BEECHER, M.D., Medvin, M.B., Stoddard, P.K. and Loesche, P. (1986) Acoustic adaptations for parent-offspring recognition in swallows. *Experimental Biology*, **45**, 179–193.

BIDDLE, B.J. (1986) Recent developments in role theory. *Annual Review of Sociology*, **12**, 67–92.

BRYAN, L.R., Coleman, M., Ganong, L. and Bryan, S.H. (1986) Person perception: family structure as a cue for stereotyping. *Journal of Marriage and the Family*, **48**, 169–174.

BURKE, T., Davies, N.B., Bruford, M.W. and Hatchwell, B.J. (1989) Parental care and mating behaviour of polyandrous dunnocks *Prunella modularis* related to paternity by DNA fingerprinting. *Nature*, **338**, 249–251.

CHERLIN, A. (1978) Remarriage as an incomplete institution. *American Journal of Sociology*, **84**, 634–650.

CREIGHTON, S.J. (1985) An epidemiological study of abused children and their families in the United Kingdom between 1977 and 1982. *Child Abuse and Neglect*, **9**, 441–448.

DALY, M. and Wilson, M. (1980) Discriminative parental solicitude: a biological perspective. *Journal of Marriage and the Family*, **42**, 277–288.

DALY, M. and Wilson, M. (1981) Abuse and neglect of children in evolutionary perspective. In: R.D. Alexander and D.W. Tinkle (eds.), *Natural Selection and Social Behavior*, New York, Chiron Press, pp. 405–416.

DALY, M. and Wilson, M. (1984) A sociobiological analysis of human infanticide. In: G. Hausfater and S.B. Hrdy (eds.), *Infanticide: Comparative and Evolutionary Perspectives*, New York, Aldine, pp. 487–502.

DALY, M. and Wilson, M. (1985) Child abuse and other risks of not living with both parents. *Ethology and Sociobiology* 6, 197–210.

DALY, M. and Wilson, M. (1988a) The Darwinian psychology of discriminative parental solicitude. *Nebraska Symposium on Motivation*, **35**, 91–144.

DALY, M. and Wilson, M (1988b) Evolutionary social psychology and family homicide. *Science*, **242**, 519–524.

DALY, M. and Wilson, M. (1988c) *Homicide*. Aldine de Gruyter, Hawthorne NY.

DAVIES, N.B. and Brooke, M. (1991) Coevolution of the cuckoo and its hosts. *Scientific American*, **264**(1), 92–98.

DUBERMAN, L. (1975) *The Reconstituted Family: A Study of Remarried Couples and their Children*. Nelson-Hall, Chicago.

DUPÂQUIER, J., Hélin, E., Laslett, P., Livi-Bacci, M. and Segner, S. (eds.) (1981) *Marriage and Remarriage in Populations of the Past*. Academic Press, London.

FERRI, E. (1984) *Stepchildren: A National Study*. NFER-Nelson, Windsor (UK).

FINE, M.A. (1986) Perceptions of stepparents: variations in stereotypes as a function of current family structure. *Journal of Marriage and the Family*, **48**, 537–543.

FLINN, M.V. (1988) Step- and genetic parent/offspring relationships in a Caribbean village. *Ethology and Sociobiology*, **9**, 335–369.

GILES-SIMS, J. (1984) The stepparent role: expectations, behavior and sanctions. *Journal of Family Issues*, **5**, 116–130.

GILES-SIMS, J. and Crosbie-Burnett, M. (1989) Stepfamily research: implications for policy, clinical interventions, and further research. *Family Relations*, **38**, 19–23.

GUBERNICK, D.J. (1981) Parent and infant attachment in mammals. In: D.J. Gubernick and P.H. Klopfer (eds.), *Parental Care in Mammals*, New York, Plenum, pp. 243–305.

HARDING, J. (1968) Stereotypes. In: D.L. Sills (ed.), *The International Encyclopedia of the Social Sciences,* v. 15, pp. 259–262.

HILL, K. and Kaplan, H. (1988) Tradeoffs in male and female reproductive strategies among the Ache, part 2. In: L. Betzig, M. Borgerhoff Mulder and P. Turke (eds.), *Human Reproductive Behavior*, Cambridge, Cambridge University Press, pp. 291–305.

IHINGER-TALLMAN, M. (1988) Research on stepfamilies. *Annual Review of Sociology*, **14**, 25–48.

KEMPE, C.H., Silverman, F.N., Steele, B.F., Droegemuller, W. and Silver, H.K. (1962) The battered-child syndrome. *Journal of the American Medical Association*, **181**, 105–112.

KESHET, J.K. (1990) Cognitive remodeling of the family: how remarried people view stepfamilies. *American Journal of Orthopsychiatry*, **60**, 196–203.

KOMPARA, D.R. (1980) Difficulties in the socialization process of step-parenting. *Family Relations*, **29**, 69–73.

KRASNEGOR, N.A. and Bridges, R. (eds.) (1990) *Mammalian Parenting: Biochemical, Neurobiological and Behavioral Determinants*. Oxford University Press, New York.

LEBOEUF, B.J. and Briggs, K.T. (1977) The cost of living in a seal harem. *Mammalia*, **41**, 167–195.

LIGHTCAP, J.L., Kurland, J.A. and Burgess, R.L. (1982) Child abuse: a test of some predictions from evolutionary theory. *Ethology and Sociobiology*, **3**, 61–67.

MØLLER, A.P. (1988) Paternity and paternal care in the swallow *Hirundo rustica*. *Animal Behaviour*, **36**, 996–1005.

PASLEY, K. and Ihinger-Tallman, M. (eds.) (1987) *Remarriage and Stepparenting: Current Research and Theory*. Guilford Press, New York.

PERKINS, T.F. and Kahan, J.P. (1979) An empirical comparison of natural-father and step-father family systems. *Family Process*, **18**, 175–183.

ROHWER, F.C. and Freeman, S. (1989) The distribution of conspecific nest parasitism in birds. *Canadian Journal of Zoology*, **67**, 239–253.

ROHWER, S. (1986) Selection for adoption versus infanticide by replacement "mates" in birds. *Current Ornithology*, **3**, 353–395.

SANTROCK, J.W. and Sitterle, K.A. (1987) Parent-child relationships in stepmother families. In: K. Pasley and M. Ihinger-Tallman (eds.), *Remarriage and Stepparenting: Current Research and Theory*, New York, Guilford Press, pp. 273–299.

SARGENT, R.C. (1989) Allopaternal care in the fathead minnow, *Pimephales promelas*: Stepfathers discriminate against their adopted eggs. *Behavioral Ecology and Sociobiology*, **25**, 379–386.

SCOTT, P.D. (1973) Fatal battered baby cases. *Medicine, Science and the Law*, **13**, 197–206.

SEGALEN, M. (1981) Mentalité populaire et remariage en Europe occidentale. In: J. Dupâquier *et al.* (eds.), *Marriage and Remarriage in Populations of the Past*, London, Academic Press, pp. 67–88.

SHOSTAK, M. (1981) *Nisa*. Harvard University Press, Cambridge MA.

SILK, J.B. (1990) Human adoption in evolutionary perspective. *Human Nature*, **1**, 25–52.

SMUTS, B. (1985) *Sex and Friendship in Baboons*. Aldine, Hawthorne NY.

THOMPSON, S. (1955) *Motif-index of folk-literature*. (In 6 volumes.) Indiana University Press, Bloomington, IN.

TRIVERS, R.L. (1972) Parental investment and sexual selection. In: B. Campbell (ed.), *Sexual Selection and the Descent of Man 1871–1971*, Chicago, Aldine, pp. 136–179.

VOLAND, E. (1988) Differential infant and child mortality in evolutionary perspective: Data from late 17th to 19th century Ostfriesland. In: L. Betzig, M. Borgerhoff Mulder and P. Turke (eds.), *Human Reproductive Behavior*, Cambridge, Cambridge University Press, pp. 253–261.

WADSWORTH, J., Burnell, I., Taylor, B. and Butler, N. (1983) Family type and accidents in preschool children. *Journal of Epidemiology and Community Health*, **37**, 100–104.

WALLACE, A. (1986) *Homicide: the Social Reality*. New South Wales Bureau of Crime Statistics and Research, Sydney.

WESTNEAT, D.F., Sherman, P.W. and Morton, M.L. (1990) The ecology and evolution of extra-pair copulations in birds. *Current Ornithology*, **7**, 331–369.

WILSON, M. and Daly, M. (1987) Risk of maltreatment of children living with stepparents. In: R.J. Gelles and J.B. Lancaster (eds.), *Child Abuse and Neglect: Biosocial Dimensions*, New York, Aldine de Gruyter, pp. 215–232.

WILSON, M.I., Daly, M. and Weghorst, S.J. (1980) Household composition and the risk of child abuse and neglect. *Journal of Biosocial Science*, **12**, 333–340.

YANAGISAWA, Y. and Ochi, H. (1986) Step-fathering in the anemonefish *Amphiprion clarkii*: a removal study. *Animal Behaviour*, **34**, 1769–1780.

PART 2: INFANTICIDE IN NONHUMAN PRIMATES

CHAPTER 6

INFANTICIDE IN NONHUMAN PRIMATES: SEXUAL SELECTION AND LOCAL RESOURCE COMPETITION

MARIKO HIRAIWA-HASEGAWA[1] and TOSHIKAZU HASEGAWA[2]

[1]*Institute of Natural Science, Senshu University, Kawasaki, 214, Japan*

[2]*The University of Tokyo, Komaba, Meguro-ku, Tokyo, 153, Japan*

INTRODUCTION

Twenty five years have passed since the first observation of infanticide in nonhuman primates in the wild was reported by Sugiyama (1965). Although his report was largely ignored at that time, the interpretation of the adaptive significance of infanticide has been one of the most controversial topics since the sexual selection hypothesis was presented by Hrdy (1974). Infanticide in nonhuman primates, as well as in other animals and man, has been well known for a long time. Until Hrdy's paper, however, it was usually regarded as abnormal and not worth analyzing. It was partly because the phenomenon was rarely seen and, probably, partly because of our human nature which makes us turn our eyes away from such a cruel incident.

Sugiyama (1965, 1966) himself did not give any ultimate explanation for infanticide. But he did not dismiss the phenomenon as abnormal and made an important remark which was to be developed later. He suggested that infanticide would cause females to resume sexual receptivity and consequently give males the opportunity to mate with females (Sugiyama, 1965). Then, with a clear recognition of the relevance to behavioral ecology, Hrdy (1974) postulated that infanticide was a male reproductive strategy that

137

curtailed the female's investment in the offspring of another male. Infanticide thereby increased the chance of a male siring offspring of his own. In response to Hrdy (1974, 1977), Curtin and Dolhinow (1978) and Boggess (1979, 1984) strongly argued that "infanticide" in langurs was not an established fact but a mere speculation based on fragmentary observations of infants which disappeared after a change in the resident male, and the phenomenon was best interpreted as an abnormal behavior caused by recent habitat disturbance. It was not clear from their argument, however, why only unweaned infants, but not other age-sex class individuals, should disappear, particularly after troop male change in a disturbed habitat. As the actual observations of infanticide from undisturbed populations accumulated, these arguments against Hrdy's hypothesis were eventually refuted.

The other interpretation of infanticide was that of a means of population regulation. This hypothesis stemmed from the fact that the phenomenon was repeatedly demonstrated among laboratory rodents under artificially crowded conditions (*e.g.* Calhoun, 1962), and from the fact that in many human societies the practice of infanticide was interpreted as a means of population regulation (*e.g.* Schrire and Steiger, 1974). The population regulation hypothesis (*e.g.* Rudran, 1973; Ripley, 1980) relies on a group selection model. However, in most of these papers, written mostly in the 1970's, the principle of group selection was taken for granted, and its logic was not questioned. As the idea that selection should act on individuals rather than groups came to be widely accepted (*e.g.* Williams, 1966; Dawkins, 1976), this interpretation of infanticide was also rejected. We do not know of any paper which proposes that group selection is more important than individual selection and then argues that infanticide is a mechanism of population regulation.

Yet another interpretation of infanticide was that it was a means of destroying the results of incestuous matings and thus maintaining genetic diversity (Ripley, 1980; Kawanaka, 1981; Itani, 1982). This hypothesis also relies on the group selection model, which is not generally accepted, and furthermore, detailed analysis of the data revealed that the victims of infanticide were unlikely to result from father-daughter matings (see below).

In 1982, a Wenner-Gren Foundation conference on "Infanticide in Animals and Man" was held at Cornell University, Ithaca, N.Y., which resulted in a comprehensive review of infanticide (Hausfater and Hrdy, 1984). At that time, evidence concerning infanticide in major species of nonhuman primates was already available to researchers. Important contributions since then are the observation of infanticide in langurs in an undisturbed forest habitat by Newton (1986), a detailed analysis of the

long-term records of Jodhpur langurs by Sommer (1987, this volume), evidence of abortion after male takeover in geladas (*Theropithecus gelada*), where infanticide does not occur (Mori and Dunbar, 1985), and within-group infanticide in the chimpanzees of Mahale (summarized here). After a decade of controversy and evaluation of various hypotheses, primatologists appear to be reaching a consensus that, in most cases, infanticide among nonhuman primates is best interpreted as a male reproductive strategy (Struhsaker and Leland, 1987; Hiraiwa-Hasegawa, 1988). The exception is infanticide in chimpanzees, the adaptive significance of which is not yet clear.

OBSERVATIONS OF INFANTICIDE IN NONHUMAN PRIMATES OTHER THAN CHIMPANZEES

The first observation of infanticide in wild nonhuman primates came from a population of hanuman langurs (*Presbytis entellus*) of India (Sugiyama, 1965). Hanuman langurs typically live in groups consisting of only one breeding male, several females and their offspring. Other males form bachelor groups and roam around the one-male groups. Sugiyama observed a group of bachelor males which attacked a one-male group and replaced the harem leader. One of the invading males later took the leader's status, evicted the rest of the males from the breeding unit, and subsequently killed all the suckling infants in the group. Following this event, the females became sexually receptive and mated with the new male.

Contrary to the argument that infanticide in langurs is a maladaptive, abnormal behavior which occurred when there was disturbance of the environment, infanticide in natural conditions has been actually observed so far in twelve species of primates and strongly suspected of occurring in another three species (Table I).

In the redtail monkey (*Cercopithecus ascanius*; Struhsaker, 1977) and the blue monkey (*Cercopithecus mitis*; Butynski, 1982) of Kibale Forest, Uganda, which form one-male breeding units, infanticide by a new, incoming male was observed. In the Campbell's monkey (*Cercopithecus campbelli*) of Ivory Coast, a one-male breeding unit was attacked by poachers, and the leader male left the group after being injured. Then, a new male came into the group and became the leader and subsequently killed two infants born into the group shortly after his takeover (Galat-Luong and Galat, 1979). Infanticide by a new, incoming male after his takeover of the breeding unit was strongly suspected from circumstantial evidence in the purple-faced langur (*Presbytis senex*; Rudran, 1973), in the silvered

Table I　Observations of Infanticide in Nonhuman Primates.

Species	Breeding[a] Unit	Evidence of Infanticide	Situation
Presbytis entellus	S (mat)	Observed	Takeover
Presbytis senex	S (mat)	Suspected	Takeover
Presbytis cristata	S (mat)	Suspected	Takeover
Cercopithecus ascanius	S (mat)	Observed	Takeover
Cercopithecus mitis	S (mat)	Observed	Takeover
Cercopithecus Campbelli	S (mat)	Observed	Takeover
Colobus guereza	S (mat)	Suspected	Takeover
Gorilla gorilla beringei	S (pat)	Observed	Takeover
			Female transfer
Alouatta palliata	M (S)	Observed	Takeover
Alouata seniculus	M (S)	Observed	Takeover
			Status change
Colobus badius	M (pat)	Observed	Status change
Macaca fuscata	M (mat)	Observed	Outsider male
Macaca mulatta	M (mat)	Observed	??
Papio cynocephalus	M (mat)	Observed	Immigrants
Pan troglodytes scweinfurthii	M (pat)	Observed	Between-group male
			Within-group male
			Within-group female

(a)　S (mat): Single-male, matrilineal

　　S (pat): Single-male, patrilineal

　　M (S): Multi-male but matings monopolicd by a single male

　　M (mat): Multi-male matnlineal

　　M (pat): Multi-male, patrilineal

leaf-monkey (*Presbytis cristata*; Wolf and Fleagle, 1977), and in the black and white colobus (*Colobus guereza*; Oates, 1977).

In the mountain gorilla (*Gorilla gorilla beringei*), which also forms a one-male breeding unit, infanticide seems to be rather frequent. When

a breeding unit comes in contact with other breeding units or a lone male, males sometimes kill an infant of a female who belongs to the other group. The female whose infant was killed resumes sexual receptivity soon after her infant's death (Fossey, 1984). However, in this species, infanticide most commonly occurs when the dominant silverback dies, and a female with a suckling infant encounters unfamiliar males (Watts, 1989).

On the other hand, infanticide was also observed in the species that form a multi-male breeding unit. The mantled howler (*Alouatta palliata*; Clarke, 1983) and the red howler (*Alouatta seniculus*; Crocket and Sekulic, 1984), both have a multi-male, multi-female social structure in which only the highest ranking male has access to females during the peak period of sexual receptivity. In these howlers, a male who had recently achieved breeding status was observed to kill infants shortly after his status change. Infanticide by a male after his status change was also observed in the red colobus (*Colobus badius*; Struhsaker and Leland, 1985).

Among Japanese monkeys (*Macaca fuscata*; The Editorial Board of Nihonzaru, 1974; 1976) and baboons (*Papio cynocephalus*; Collins *et al.* , 1984), several cases of infanticide by immigrant males were observed. In the rhesus monkey (*Macaca mulatta*), one case of infanticide by a resident alpha male was observed (Camperio Ciani, 1984).

THE ADAPTIVE SIGNIFICANCE OF INFANTICIDE IN NONHUMAN PRIMATES OTHER THAN CHIMPANZEES

Does the evidence of infanticide in nonhuman primates other than chimpanzees suggest any adaptive significance of this behavior? Close examination of the data reveals the following patterns common to most of the cases:

1. All the cases of infanticide were committed by males.

2. In most cases, infanticide occurred following a change in the breeding status of males. Except in the case of Japanese monkeys, where an outsider male killed an infant but did not immigrate into the group (The Editorial Board of Nihonzaru, 1974; 1976), infanticide was committed either by a male who recently took over a breeding unit (langurs, redtail monkeys, blue monkeys, Campbell's monkeys, gorillas, mantled howlers), by a male who rapidly gained breeding status in his group (red colobus, red howlers), or by a male who recently immigrated into the group (baboons).

3. As is predicted from the above, in most cases it is unlikely that the killer males could have sired the infants which were killed, so that

they most probably did not kill their own offspring. In one striking case in mantled howlers, an infanticidal male who killed infants in the group he took over did not kill the infants of low-ranking females with whom he had mated while he was approaching the group prior to the takeover (Clarke, 1983).

4.   In almost all the cases, the victims were unweaned infants. The continued presence of a suckling infant would have inhibited the mother from rapidly ovulating and mating.

5.   The mother whose infant was killed resumed sexual receptivity within a few days to a few weeks, which is significantly shorter than the cases where the infant survived. Though it has been argued that loss of infants may hardly shorten the time elapsed before the next conception (Boggess, 1979; Harley, 1985), recent reports from a long-term observation of wild populations show that this does occur (Sommer, 1987; this volume). In most cases where the data are available, the killer male subsequently sired his own offspring, although whether a male subsequently mated with the mother after killing her infant is not always reported.

6.   In most cases, the tenure of a male who has achieved breeding status is fairly short; it may range from just a few days to around 5 to 6 years (Table II).

7.   Infanticide is not confined to populations with high density or to animals living in a disturbed habitat, since it also occurs in undisturbed populations. Hanuman langurs distribute themselves widely over the Indian subcontinent and can have both one-male and multi-male breeding units. Newton (1986) showed that, among langurs of India, the occurrence of infanticide was not related to local density but, instead, was related to the ratio of one-male breeding units to the total number of breeding units in the region.

The patterns of infanticide in nonhuman primates summarized above strongly suggest that infanticide evolved as a male reproductive strategy. Infanticide is most frequently observed in species where competition among males for the opportunity to mate is very intense and female physiology enables a male to shorten his waiting time to mate by killing a current suckling infant. Under these circumstances males can improve their reproductive success by killing other males' infants.

From this viewpoint, it is easy to understand why the frequency of infanticide is extremely low or absent among most of the species with multimale breeding units. In such a breeding system few males are actually excluded from the chance of copulation in spite of the male hierarchy, and

Table II Tenure Lengths of the Males of Breeding Status in 5 Species of Nonhuman Primates with Reports of Infanticide.

Species	Sites	Mean Tenure Lengths (Range)	Reference
Presbytis entellus	Jodhpur (India)	26.5 months (a few days–6.2 years)	Sommer (this volume)
Cercopithecus ascanius	Kibale (Uganda)	(11–13 months)	Struhsaker (1977)
Cercopithecus mitis	Kibale (Uganda)	2.2 months (0.4–4.4 months)	Butynski (1982)
Alouatta seniculus	Hato MASA-guaral (Venezuera)	(4.72–7.58 years)	Crocket and Sekulic 1984)
Papio cynocephalus ursinus	Moremi (Botswana)	5.0 months (1–12 months)	Busse and Hamilton (1981)

competition for mating among males is thus milder. In addition, because of this, it would be more difficult for a male to assess which infant is his own and which is not. Furthermore, since any male may have some access to receptive females, if a male kills an infant and thus makes the mother sexually receptive, it is not known whether this would result in an increase in reproductive success of the infanticidal male, as opposed to other, non-infanticidal males.

On the other hand, infanticide does not always accompany male takeover in one-male breeding units. In the hamadryas baboon (*Papio hamadryas*) and in the gelada baboon, infanticide has not been observed after male takeover in the wild. In the hamadryas baboon, however, infanticide repeatedly occurred, as is predicted, when the leader male was artificially removed and a new male was released into the group of females in the zoo or in the field (Kummer *et al.*, 1974; Angst and Thommen, 1977; Gomendio and Colmenares, 1989). Apparently there are some conditions that prohibit infanticide from occurring in natural environments.

A male hamadryas baboon typically forms his own harem by kidnapping young females from other units and waiting for them to mature sexually; according to Kummer *et al.* (1974), over 80 % of mature males had at least one such young female of his own. Therefore, competition among males for the opportunity to reproduce seems to be not as severe as in the case of other species with one-male breeding units. Nevertheless, male takeover

of a breeding unit does occur. In contrast to the artificially induced male changes described above, however, male takeovers of a harem in the wild seem to be a gradual process involving a long period of co-existence of the two disputing males (Kummer et al., 1974; Sigg et al., 1982). This may provide the former leader with the opportunity to protect his own offspring from infanticide.

In the gelada baboon, abortion and unusually quick resumption of post-partum sexual receptivity have been observed after male takeover (Mori and Dunbar, 1985). This brings about the same effect as infanticide for the new male and may be an alternative strategy. Abortion and reduction of the period of postpartum non-receptivity after encountering a novel male have been reported from patas monkeys (*Erithrocebus patas*; Rowell, 1978), anubis baboons (Pereira, 1983), hanuman langurs (Sommer, this volume) and captive hamadryas baboons (Colmenares and Gomendio, 1988).

THE OBSERVATION OF INFANTICIDE IN CHIMPANZEES

Several features of infanticide in chimpanzees (*Pan troglodytes schweinfurthii*) are different from those described above. In chimpanzees, cannibalism of the victim infant is the norm after infanticide. Males as well as females commit infanticide, and the adaptive significance of this behavior is yet unclear.

Since the earliest reports that the chimpanzee, which is our closest relative, killed and ate a conspecific infant (Suzuki, 1971; Bygott, 1972), 13 cases have been actually observed so far (Table III). The observations are from populations of Budongo Forest, Uganda, Gombe National Park, Tanzania, and Mahale Mountains National Park, Tanzania. Infanticide has not been reported from chimpanzee populations in West Africa (Boesch and Boesch, 1989).

Chimpanzees form multi-male social groups in which males typically remain in their natal group, while females transfer between groups (Nishida and Hiraiwa-Hasegawa, 1987). Males form a close bond based on their kinship, patrolling and protecting food resources and females within their home range against the males of different groups. Antagonism between the males of different groups is so severe that extermination of one group by another has been observed in Gombe (Goodall, 1986). Females rarely aggregate, and anestrous females range over a small, preferred area (Wrangham and Smuts, 1980; Hasegawa, 1990). Matings take place promiscuously (Tutin, 1980; Hasegawa and Hiraiwa-Hasegawa, 1990). Chimpanzees rely

Table III Record of Infanticide and Cannibalism in Chimpanzees.

Site	Age of Victim	Sex of Victim	Occasion	Attacker	Cannibalism	Reference
Budongo	Newborn	?	?	AM?	yes	Suzuki (1971)
Gombe	1.5–2 yrs	?	Between	AM	yes	Bygott (1972)
Gombe	3 wks	F	Within	AF,AdF	yes	Goodall (1977)
Gombe	1.5–2 yrs	M	Between	AM?	yes	Goodall (1977)
Gombe	1.5–2 yrs	F	Between	AM	no	Goodall (1977)
Gombe	3 wks	M	Within	AF,AdF	yes	Goodall (1977)
Gombe	3 wks	F	Within	AF,AdF	yes	Goodall (1977)
Mahale	3 yrs	M	Between	?	yes	Nishida et al. (1979)
Mahale	2.5 mos	M	Within	?	yes	Norikoshi (1982)
Mahale	1.5 mos	M	Within	?	yes	Kawanaka (1981)
Mahale	1 mo	M	Within	AM	yes	Takahata (1985)
Mahale	3 mos	M	Within	AM?	yes	Nishida and Kawanaka (1985)
Mahale	6 mos	M	Within	AM	no	Masui (pers. com.)

on fruits for their staple food, but they hunt various vertebrate species, such as monkeys and ungulates (Boesch and Boesch, 1989; Wrangham and van Zinnicq Bergman Riss, 1990). Apparently the chimpanzee is one of the most carnivorous primates other than man.

Infanticide in chimpanzees has been observed in the following 3 contexts: between-group infanticide by males, within-group infanticide by females, and within-group infanticide by males.

1. *Between-group Infanticide by Males*   There are three cases reported from Gombe (Bygott, 1972; Goodall, 1977) and one case from Mahale (Nishida *et al.*, 1979) of male chimpanzees killing an infant of a female in another group. Infanticide occurred at the overlapping area of two groups when a mother-infant pair of one group encountered males of a neighboring group while the males were patrolling their home range. Infants which were killed were apparently sired by males in the mother's group. However, in all these cases, the mother did not subsequently transfer to the killer's group nor mate with the killer.

2. *Within-group Infanticide by Females*   Females have been observed killing an infant of another female in the same group. This was observed only in Gombe, and a particular adult female, named Passion, and her adolescent daughter, Pom, were responsible for all of the killing (Goodall, 1977; 1986). This mother-daughter pair, in collaboration, at least 3 times surrounded another mother with a small infant, took the infant from her, and cannibalized it. After Passion died, Pom ceased to attack and eat other females' infants, probably because Pom alone, without her mother's aid, could not attack another female successfully.

3. *Within-group Infanticide by Males*   Male chimpanzees will kill the infant of a female who recently immigrated into the group. Five cases have been recorded from the M group of Mahale (Kawanaka, 1981; Norikoshi, 1982; Takahata, 1985; Nishida and Kawanaka, 1985; K. Masui pers. comm). All of the victims were the first- or second-born infants of newly-immigrated females, and all of them were male.

Table IV shows the mortality record of infants born to females after their immigration into the M group. Since 1970, 21 females have immigrated into the M group, and 11 of them gave birth in the M group. Among the 12 infants born as the first offspring of the newly-immigrated females, 8 were males, 2 were females, and 1 was of unknown sex. Out of the 8 male infants, 5 were killed as a result of within-group infanticide by males, one was attacked, but the mother successfully guarded the infant, one was killed by inter-group infanticide, and the last one died of an unknown cause. In contrast, the 2 first-born female infants were neither killed nor attacked. Subsequently, 15 more infants, including 7 males, 6 females and 2 unsexed infants, were born to these same females. Among the 7 second-born to fifth-born male infants, only 1 was killed by the M group males, while none of the 6 female infants were killed. In conclusion, male infants born immediately after their mothers' immigration into the M group were more susceptible to within-group infanticide by males than were female infants, suggesting a sex-bias in the incidence of infanticide, although the risk of infanticide declined as the mother's duration of residency in the M group increased.

THE ADAPTIVE SIGNIFICANCE OF INFANTICIDE IN CHIMPANZEES: AN UNSOLVED PUZZLE

In the between-group infanticide by male chimpanzees, males did not gain immediate reproductive advantage by killing infants. In the cases

Table IV Mortality of Infants Born to the Immigrant Females of M group.

Name of Mother	Sex and Mortality within 1 Year of Infants According to Their Birth Order				
	1st	2nd	3rd	4th	5th[a]
ND	?[b] (?)[c]	M(*)	F(s)	F(s)	M(s) M
SA	M(#)	?(d?)	F(s)		
WO	F(s)	M(s)			
WA	M(*)	?(d?)	M (d?)	F(d)	M(s)
WS	M(*?)				
CH	M(*)	F(s)	M(s)		
WD	M(*)	M(s)			
WE	M(**)	F(s)			
TM	M(*)				
PL	F(s)				
OP	M(d?)				

(a) Birth order

(b) Sex of infant: M; male, F; female, ?; unsexed

(c) Mortality: (*); within-group infanticide
(**); between-group infanticde
(*?); suspected within-group infanticide
(#); severely attacked by the group males
(d); died of causes other than infanticide
(d?); died of unknown causes
(s); survived

of gorillas, the mothers whose infants were killed transferred to the killer male afterward, but this was not the case in chimpanzees. However, since the infants which were killed were sired by males of other groups, the behavior cannot be considered to be maladaptive.

Infanticide by females may well be pathological, since it has only been observed only in one mother-daughter pair. It is also conceivable that this type of infanticide may be associated with competition for resources between females (Pusey, 1983). By eliminating the infants of other females of the same group, killer females may secure food and other resources in the environment for themselves and their offspring. However, if this would be the general condition of competition among chimpanzee females, we would expect to have observed more cases of infan-

ticide by females. The description of the behavior of Passion and Pom strongly suggests that the primary purpose of infanticide was to obtain meat (Goodall, 1977; 1986; Hiraiwa-Hasegawa, 1992). It is very difficult, however, for a female to attack another female and take her infant from her, and this may be the reason that infanticide by females is extremely rare.

The cases of within-group infanticide by males are puzzling because males sometimes had copulated with the females whose infants they killed. As was mentioned above, the mothers of the victim infants were newly-immigrated females. It was emphasized that the infanticidal males did copulate with the mother around the time of conception of the victim infant. Therefore, it is possible that a male killed his own infant (Kawanaka, 1981; Takahata, 1985). However, the chance that the killer male actually sired the infant was likely quite low, because females usually copulated at least 100 times with several different adult males during suspected conception cycles (Hiraiwa-Hasegawa, 1987; Hasegawa and Hiraiwa-Hasegawa, 1990).

The newly-immigrated females tended to spend most of their anestrous period in the periphery of the communal range. They copulated with the males of the new group, but they did not meet the adult males of the group on a regular basis (Kawanaka, 1981; Takahata, 1985; Nishida and Kawanaka, 1985); they had little regular contact with the adult males of the new group after conception, during their pregnancy, and for some time after the infant was born. Therefore, it is possible that a female's ranging pattern and the amount of time spent together with males are used by the males as a cue to determine her infant's paternity. When an infant which was the first to be born to a mother after she immigrated into a new group was killed, the mother gradually increased the time associating with the adult males of the new group, and their subsequent infants, which appeared to be sired by the males of the new group, were not killed (Takahata, 1985).

Why should male chimpanzees kill infants of dubious paternity? It may be possible that within-group infanticide by males can be explained as a male reproductive strategy. In the M group there are, on average, 11 to 13 sexually mature males and 10 sexually active females (cycling females, females without suckling infant). Once females give birth, they will not resume estrous cycles for about 3 to 5 years. Therefore, males may kill infants of uncertain paternity (infants of newly-immigrated females) in order to enhance their own chance of reproduction. However, if this is the case, female infants, and not only just male infants, of newly-immigrated females should be killed.

Although the sample size is still very small, male infants of immigrant females appear to be more likely to be killed by resident males than female infants. Nishida and Kawanaka (1985) argued that male infants might be future competitors for adult males (or the adult males' offspring), but female infants may enhance their reproductive success if they were to become their or their relatives' mates; they also stated that it was puzzling why some juvenile males were permitted to transfer to M group with their mothers. However, in chimpanzees, males typically remain in their natal group while females transfer to another group at the time of puberty. Therefore, male infants may serve as their ally in the future but female infants will certainly emigrate and become mates of males of other groups.

LOCAL RESOURCE COMPETITION AND MALE-BIASED INFANTICIDE IN CHIMPANZEES

To explain the male-biased sex ratios observed among galagos, Clark (1978) proposed that local resource competition between a mother and her philopatric daughter was the main cause. Dittus (1977, 1979) reported that juvenile female toque macaques (*Macaca sinica*) received more aggression than their male peers, and that during a drought year, young females were displaced from feeding areas more often than young males. In a colony of bonnet macaques (*Macaca radiata*), Silk *et al.* (1981) observed that immature females received aggression from unrelated, higher-ranking adult females more frequently than did immature males, and the rate of aggression received by immatures was correlated with maternal rank. Moreover, the frequency of injuries sustained was highly correlated with maternal rank for immature females but only weakly correlated for immature males. In conclusion, immature female bonnet macaques receive more aggression and are injured more frequently by unrelated adult females than are immature males of similar maternal rank. In addition, there is a report which showed that female pig-tailed monkeys (*Macaca nemestrina*) carrying female fetuses received more medical treatment, primarily for bite wounds, than females carrying male fetuses (Sackett *et al.* 1975). The same tendency was reported from a captive group of rhesus monkeys (*Macaca mulatta*) in Madingley, Cambridge: females carrying female fetuses received more aggression than females carrying male fetuses (Simpson *et al.*, 1981); however, this finding was not reproduced when the sample size was increased.

Based on these observations, Silk (1983) hypothesized that, in species where females remain in their natal group and thus compete for food re-

source with other females, females reduce the recruitment of immature females into their groups through harassment. Her argument provides an explanation for the skewed sex ratios observed in macaques.

In chimpanzees, the situation is reversed: males remain in their natal group for life and females typically disperse. In this situation, local resource competition, if it ever occurs, would be among males, and the immature male chimpanzees should receive more aggression from unrelated adult males than should immature females.

With respect to competition between different groups, a chimpanzee group which has a larger number of adult males is dominant to a neighboring group that has a smaller number of adult males. Therefore, recruitment of males by birth into the group is critical to the dominance status of the group, and males should achieve benefit by their large number through local resource enhancement (Gowaty and Lennartz, 1985). However, competition among males within the same group is supposed to become more and more intense as the number of adult males in the group increases. There may thus be an optimal size for one group: sixteen is the maximum number recorded so far, and then the group will divide into two groups.

For the cases of inter-group infanticide, the victims were male infants. By killing male infants of neighboring groups, males may reduce the number of future adult males in that group, thus reducing the future strength of the neighboring group. If this is the case, males should not kill female infants born into other groups. However, the sample size is still too small to draw any definite conclusions.

In the cases of within-group infanticide by males in the M group of Mahale, it is possible that local resource competition among males of the same group is the key factor. M group is the biggest chimpanzee group ever recorded; it contains around 12 adult males, and the victims were male infants of newly-immigrated females. The newly-immigrated females were suspected to be unrelated to the adult males of the group.

However, the intensity of competition over resources has not been accurately measured, and the infanticidal individual's gain in fitness in terms of increased access to food is also unclear. Therefore, we have to say that, in the cases of chimpanzee, the adaptive significance of infanticide is still a puzzle. It is possible that infanticide may have different functions in different situations. In order to have a clearer view on this problem, we need more accurate data on competition for mates and for food in different populations of chimpanzees in which infanticide does and does not occur.

REFERENCES

ANGST, W. and Thommen, D. (1977) New data and discussion of infant killing in old world monkeys and apes. *Folia Primatologica*, **27**, 198–229.

BOESCH, C. and Boesch, H. (1989) Hunting behavior of wild chimpanzees in the Tai National Park. *American Journal of Physical Anthropology*, **78**, 547–573.

BOGGESS, J.E. (1979) Troop male membership change and infant killing in langurs (*Presbytis entellus*). *Folia Primatologica*, **32**, 65–107.

BOGGESS, J.E. (1984) Infant killing and male reproductive strategies in langurs (*Presbytis entellus*). In: Hausfater, G. and Hrdy, S.B. (eds), *Infanticide: Comparative and Evolutionary Perspectives*, New York, Aldine, pp. 283–310.

BUTYNSKI, T.M. (1982) Harem-male replacement and infanticide in the blue monkey (*Cercopithecus mitis stuhlmanni*) in the Kibale Forest, Uganda. *American Journal of Primatology*, **3**, 1–22.

BYGOTT, J.D. (1972) Cannibalism among wild chimpanzees. *Nature*, **238**, 410–411.

CALHOUN, J.B. (1962) Population density and social pathology. *Scientific American*, **206**, 139–149.

CAMPERIO Ciani, A. (1984) A case of infanticide in a free-ranging group of rhesus monkeys (*Macaca mulatta*) in the Jackoo Forest, Simla, India. *Primates*, **25**, 372–377.

CLARK, A.B. (1978) Sex ratio and local resource competition in prosimian primates. *Science*, **201**, 163–165.

CLARKE, M.R. (1983) Infant-killing and infant disappearance following male takeovers in a group of free-ranging howler monkeys (*Alouatta palliata*) in Costa Rica. *American Journal of Primatology*, **5**, 241–247.

COLLINS, D.A., Busse, C.D. and Goodall, J. (1984) Infanticide in two populations of savanna baboons. In: Hausfater, G. and Hrdy, S.B. (eds), *Infanticide: Comparative and Evolutionary Perspectives*, New York, Aldine, pp. 193–216.

COlMENARES, F. and Gomendio, M. (1988) Changes in female reproductive condition following male take-overs in a colony of hamadryas and hybrid baboons. *Folia Primatologica*, **50**, 157–174.

CROCKET, C.M. and Sekulic, R. (1984) Infanticide in red howler monkeys (*Alouatta seniculus*). In: Hausfater, G. and Hrdy, S.B. (eds), *Infanticide: Comparative and Evolutionary Perspectives*, New York, Aldine, pp. 173–192.

CURTIN, R.A. and Dolhinow, P. (1978) Primate social behavior in a changing world. *American Scientist*, **66**, 468–475.

DAWKINS, R. (1976) *The Selfish Gene* . Oxford University Press, Oxford.

DITTUS, W.P.J. (1977) The social regulation of population density and age-sex distribution in the toque monkey. *Behaviour* , **63**, 281–322.

DITTUS, W.P.J. (1979) The evolution of behaviours regulating density and age-specific sex ratios in a primate population. *Behaviour* , **69**, 265–302.

FOSSEY, D. (1984) Infanticide in mountain gorillas (*Gorilla gorilla beringei*) with comparative notes on chimpanzees. In: Hausfater, G. and Hrdy, S.B. (eds), *Infanticide: Comparative and Evolutionary Perspectives*, New York, Aldine, pp. 217–236.

GALAT-LUONG, A. and Galat, G. (1979) Consequences comportmentales de perturbations sociales repetees sur une troupe de mones de Lowe *Cercopithecus campbelli lowei* de Cote-d'Ivoire. *Terre et la Vie*, **33**, 54–58.

GOMENDIO, M. and Colmenares, F. (1989) Infant killing and infant adoption following the introduction of new males to an all-female colony of baboons. *Ethology*, **80**, 223–244.

GOODALL, J. (1977) Infant killing and cannibalism in free-living chimpanzees. *Folia Primatologica*, **28**, 259–282.

GOODALL, J. (1986) *The Chimpanzees of Gombe*. Harvard University Press, Cambridge, MA.

GOWATY, P. and Lennartz, M.R. (1985) Sex ratios of nestling and fledgling red-cockaded woodpeckers (*Picoides borealis*). *American Naturalist*, **126**, 347–353.

HARLEY, D. (1985) Birth-spacing in langur monkeys (*Presbytis entellus*). *International Journal of Primatology*, **6**, 227–242.

HASEGAWA, T. (1990) Sex differences in ranging patterns. In: Nishida, T. (ed), *The Chimpanzees of Mahale*, Tokyo, The University of Tokyo Press, pp. 100–114.

HASEGAWA, T. and Hiraiwa-Hasegawa, M. (1990) Sperm competition and mating behavior. In: Nishida, T. (ed), *The Chimpanzees of Mahale*, Tokyo, The University of Tokyo Press, pp. 115–132.

HAUSFATER, G. and Hrdy, S.B. (eds) (1984) *Infanticide: Comparative and Evolutionary Perspectives* . Aldine, New York.

HIRAIWA-HASEGAWA, M. (1987) Infanticide in primates and a possible case of male-biased infanticide in chimpanzees. In: Ito, Y., Brown, J.L. and Kikkawa, J. (eds), *Animal Societies: Theories and Facts*, Tokyo, Japan Scientific Societies Press, pp. 125–139.

HIRAIWA-HASEGAWA, M. (1988). Adaptive significance of infanticide in primates. *Trends in Ecology and Evolution*, **3**, 102–105.

HIRAIWA-HASEGAWA, M. (1992) Cannibalism among nonhuman primates. In: Elgar, M. and Crespi, B. (eds), *Cannibalism: Ecology and Evolution Among Diverse Taxa*, Oxford, Oxford University Press, pp. 323–338.

HRDY, S.B. (1974) Male-male competition and infanticide among langurs (*Presbytis entellus*) of Abu, Rajasthan. *Folia Primatologica*, **22**, 19–58.

HRDY, S.B. (1977) *The Langurs of Abu: Female and Male Strategies of Reproduction* . Harvard University Press, Cambridge, MA.

ITANI, J. (1982) Intraspecific killing among non-human primates. *Journal of Sociology and Biological Structure*, **5**, 361–368.

KAWANAKA, K. (1981) Infanticide and cannibalism in chimpanzees, with special reference to the newly observed case in Mahale Mountains. *African Study Monograph*, **1**, 69–99.

KUMMER, H., Goetz, W. and Angst, W. (1974) Triadic differentiation: an inhibitory process protecting pair bonds in baboons. *Behavior*, **49**, 62–87.

MORI, U. and Dunbar, R.I.M. (1985) Changes in the reproductive condition of female gelada baboons following the takeover of one-male units. *Zeitschrift fur Tierpsychologie*, **67**, 215–224.

NEWTON, P. N. (1986) Infanticide in undisturbed forest population of hanuman langurs, *Presbytis entellus* . *Animal Behavior*, **34**, 785–789.

THE Editorial Committee of Nihonzaru (1974) A case of cannibalism observed among wild Japanese monkeys. *Nihonzaru*, **1**, 123–130.

THE Editorial Committee of Nihonzaru (1976) Unusual aggressive behavior among Japanese monkeys. *Nihonzaru*, **2**, 123–142.

NISHIDA, T. and Hiraiwa-Hasegawa, M. (1987) Chimpanzees and bonobos. In: Smuts, B.B., Cheney, D.L., Seyfarth, R.M., Wrangham, R.W. and Struhsaker, T.T. (eds), *Primate Societies,* Chicago, The University of Chicago Press, pp. 165–178.

NISHIDA, T. and Kawanaka, K. (1985) Within-group cannibalism by adult male chimpanzees. *Primates*, **26**, 274–284.

NISHIDA, T., Uehara, S. and Nyundo, R. (1979) Predatory behavior among wild chimpanzees of the Mahale Mountains. *Primates*, **20**, 1–20.

NORIKOSHI, K. (1982) One observed case of cannibalism among wild chimpanzees of the Mahale Mountains. *Primates*, **23**, 66–74.

OATES, J.F. (1977) The social life of black-and-white colobus monkey, *Colobus guereza* . *Zeitschrift fur Tierpsychologie*, **45**, 1–60.

PEREIRA, M.E. (1983) Abortion following the immigration of an adult male baboon (*Papio cynocephalus*). *American Journal of Primatology*, **4**, 93–98.

PUSEY, A.E.(1983) Mother-offspring relationships in chimpanzees after weaning. *Animal Behavior*, **31**, 363–377.

RIPLEY, S. (1980) Infanticide in langurs and man: adaptive advantage or social pathology? In: Cohen, M.N., Malpass, R.S. and Klein, H.G. (eds), *Biosocial Mechanisms of Population Regulation*, New Haven, Connecticut, Yale University Press, pp. 349–390.

ROWELL, T.E. (1978). How female reproductive cycles affect interaction patterns in a group of patas monkeys. In: Chivers, D.J. and Herbert, J. (eds), *Recent Advances in Primatology*, vol.1, London, Academic Press, pp. 489–490.

RUDRAN, R. (1973) Adult male replacement in one-male troops of purple-faced langurs (*Presbytis senex senex*) and its effect on population structure. *Folia Primatologica*, **19**, 166–192.

SACKET, G.P., Holm, R.A., Davis, A.E. and Fahrenbruch, C.E. (1975) Prematurity and low birth weight in pigtail macaques: incidence, prediction and effects on infant development. In: Kondo, S., Kawai, M., Ehara, A. and Kawamura, S. (eds), *Contemporary Primatology*, Tokyo, Japan Science Press, pp. 189–206.

SCHRIRE, C. and Steiger, W.L. (1974) A matter of life and death: an investigation of the practice of female infanticide in the Arctic. *Man*, **9**, 161–184.

SIGG, H., Stolba, A., Abegglen, J.J. and Dasser, V. (1982) Life history of hamadryas baboons: physical development, infant mortality, reproductive parameters and family relationships. *Primates*, **23**, 473–487.

SILK, J.B. (1983) Local resource competition and facultative adjustment of sex ratios in relation to competitive abilities. *American Naturalist*, **121**, 56–66.

SILK, J.B., Clark-Wheatley, C.B., Rodman, P.S., Samuels, A. (1981) Differential reproductive success and facultative adjustment of sex ratios among captive female bonnet macaques (*Macaca radiata*) . *Animal Behavior*, **29**, 1106–1120.

SIMPSON, M.J.A., Simpson, A.E., Hooley, J. and Zunz, M. (1981) Infant-related influences on birth intervals in rhesus monkeys. *Nature*, **290**, 49–51.

SOMMER, V. (1987) Infanticide among free-ranging langurs (*Presbytis entellus*) at Jodhpur (Rajasthan/India): recent observations and a reconsideration of hypotheses. *Primates*, **28**, 163–197.

STRHUSAKER, T.T. (1977) Infanticide and social organization in the redtail monkey (*Cercopithecus ascanius schmidti*) in the Kibale forest. *Zeitschrift fur Tierpsychologie*, **45**, 75–84.

STRUHSAKER, T.T. and Leland, L. (1985) Infanticide in a patrilineal society of red colobus monkeys. *Zeitschrift fur Tierpsychologie*, **69**, 89–132.

STRUHSAKER, T.T. and Leland, L (1987) Colobines: infanticide by adult males. In: Smuts, B.B, Cheney, D.L., Seyfarth, R.M., Wrangham, R.W. and Struhsaker, T.T. (eds), *Primate Societies*, Chicago, The University of Chicago Press, pp. 83–97.

SUGIYAMA, Y. (1965) On the social change of hanuman langurs (*Presbytis entellus*) in their natural conditions. *Primates*, **6**, 381–417.

SUGIYAMA, Y. (1966) An artificial social change in a hanuman langur troop (*Presbytis entellus*). *Primates*, **7**, 41–72.

SUZUKI, A. (1971) Carnivority and cannibalism among forest-living chimpanzees. *Journal of the Anthropological Society of Nippon*, **79**, 30–48.

TAKAHATA, Y. (1985) Adult male chimpanzees kill and eat a male newborn infant: newly observed intragroup infanticide and cannibalism in Mahale National Park, Tanzania. *Folia Primatologica*, **44**, 161–170.

TUTIN, C.G.E. (1980) Reproductive behavior of wild chimpanzees in the Gombe National Park, Tanzania. *Journal of Reproduction and Fertility Supplement*, **28**, 43–57.

WATTS, D.P. (1989) Infanticide in mountain gorillas: new cases and a reconsideration of the evidence. *Ethology*, **81**, 1–18.

WILLIAMS, G.C. (1966) *Adaptation and Natural Selection* . Princeton University Press, Princeton, NJ.

WOLF, K.E. and Fleagle, J.G. (1977) Adult male replacement in a group of silvered leaf-monkeys (*Presbytis cristata*) at Kuala Selangor, Malaysia. *Primates*, **18**, 949–955.

WRANGHAM, R.W. and Smuts, B.B. (1980). Sex differences in the behavioural ecology of chimpanzees in the Gombe National Park, Tanzania. *Journal of Reproduction and Fertility Supplement*, **28**, 13–31.

WRANGHAM, R.W. and van Zinnicq Bergman Riss, E. (1990) Rates of predation on mammals by Gombe chimpanzees, 1972–1975. *Primates*, **31**, 157–170.

CHAPTER 7

INFANTICIDE AMONG THE LANGURS OF JODHPUR: TESTING THE SEXUAL SELECTION HYPOTHESIS WITH A LONG-TERM RECORD

VOLKER SOMMER

Institut für Anthropologie der Georg-August-Universität,
Bürgerstrasse 50, D-37073 Göttingen, Germany

INTRODUCTION

The Hanuman langur (*Presbytis entellus*) became famous both within and beyond primatological circles (see e.g., textbooks of Wickler and Seibt, 1977; Barash, 1980; Jolly, 1985; Trivers, 1985; Alcock, 1989) because adult male replacements can be accompanied by infanticide. Infant killings have been reported from several study sites in India where langurs live in one-male breeding groups such as Dharwar (Sugiyama, 1965a, b, 1966), Jodhpur (Mohnot, 1971 and others, see below), Mount Abu (Hrdy, 1974), Harihar (Parthasarathy and Rahaman, 1974), Kanha (Newton, 1985) and Jaipur (Mathur and Lobo, 1986). Hrdy (1974, 1977) elaborated the hypothesis that langur infanticide is an adaptation: Since ovulatory cycles are blocked as long as females suckle infants, a new resident male reduces his waiting time to inseminate the mother by killing his predecessor's offspring, and such a male is believed to gain a reproductive advantage over a noninfanticidal counterpart. Moreover, Hrdy (1974, 1979) interpreted the postconception estrus displayed by langur females as a counterstrategy to infanticide in order to confuse paternity : A new resident male copulating with females who were already pregnant during his takeover is believed to tolerate the subsequent infant.

For several other study sites, however, there are no published reports of killings, e.g. in Nepal for Junbesi (Boggess, 1976) and Melemchi (Bishop, 1975), in India for Orcha, Kaukori (Jay, 1965), Singhur (Oppenheimer,

155

1977) and Rajaji (Laws and Laws, 1984), in Sri Lanka for Wilpattu (Muckenhirn, 1972) and Polonnaruwa (Ripley, 1967, 1980). Since langurs have
not been observed to commit infanticide over wide parts of their geograph-
ical range, several scientists concluded that infant killing must be a pathological response to abnormally high population density or human interference (Dolhinow, 1977; Curtin and Dolhinow, 1978; Boggess, 1979, 1980,
1984).

The sexual selection hypothesis, on the other hand, is defended against
such criticism by stressing the following points: (a) Those sites from which
infanticide has not been reported have a predominantly multi-male multi-
female troop structure. Because of the probably low certainty of paternity,
it is not beneficial for males to kill infants. (b) Langur infanticide has been
reported from the completely undisturbed forest habitat of Kanha (Newton, 1985). (c) Infanticide by adult males has meanwhile been reported
from the wild for a total of about 17 nonhuman primate species (for review and references see Struhsaker and Leland, 1987) including various
Colobinae (e.g., *Presbytis entellus, P. senex, P. cristata, Colobus badius*), Cer-
copithecinae (e.g. *Cercopithecus ascanius, C. mitis, Papio cynocephalus*),
Alouattinae (e.g., *Alouatta seniculus)* and Hominidae (*Pan troglodytes, Gorilla gorilla*).

Although a majority of primatologists meanwhile accepts the explanation that infanticide by nonhuman primate males is an adaptation based
on sexual selection, consensus among scientists has not been achieved (see
e.g., Eibl-Eibesfeldt, 1990). Boggess (1984) as a leading critic holds strong
criteria regarding acceptable evidence of infanticide (which is often only
inferred if infants disappeared during male changes; cf. Vogel, 1979) and
data which would support the following predictions of the sexual selection
hypothesis: (a) infanticidal males should not kill own offspring; (b) infanticide should shorten the subsequent interbirth interval of the victim's
mother; (c) the infanticidal male should sire the subsequent infant.

The most detailed data base for free ranging langurs exists for the population living around Jodhpur, which has been studied by various Indian
and German scientists from 1967 onwards. Since 1977, several groups
were subject to almost uninterrupted long-term monitoring of individually known animals. After the initial description of infanticide by an immi-
grating male (Mohnot, 1971), there have been several detailed reports of
fatal and nonfatal attacks by new males on infants (Makwana, 1979; Vogel
and Loch, 1984; Sommer, 1984, 1985, 1987; Sommer and Mohnot, 1985;
Agoramoorthy, 1987; Agoramoorthy and Mohnot, 1988; Agoramoorthy
et al., 1988). The present paper points out some general patterns which
emerged after all data (available by March 1990) for the Jodhpur popula-

tion were pooled. This long-term record is used to test the sexual selection hypothesis.

MATERIAL AND METHODS

Jodhpur is a town in the northwest Indian state of Rajasthan. Located at the eastern edge of the Great Indian desert, the climate is dry with maximum temperatures of up to 50°C during May/June and minimum temperatures around 0°C during December/ January. Ninety percent of the scanty rainfall (annual average 360 mm) is received during the monsoon period (July–September). The town was erected on a hilly sandstone plateau covering approximately 150 km^2, surrounded by flat semi-desert. The natural vegetation is open scrub dominated by xerophytic plants such as *Prosopis juliflora*, *Acacia senegal* and *Euphorbia caducifolia*. There are numerous irrigated parks and fields.

The plateau is inhabited by a geographically isolated population of Hanuman langurs. Within a radius of 100 km, there are no other langur groups. According to censuses, the population increased from about 800 animals in 1967 to about 1300 in 1987. The reproductive units are bisexual one-male troops (harems) with a single adult breeding male. Each troop occupies its own home range of about 0.5 - 1.3 km^2. Troops containing several adult males are rarely found (<10%) and are of short existence. Long-term demographic monitoring of known individuals indicates that (with very few exceptions) females remain in their natal troops throughout their lives. Males emigrate - usually as juveniles - and join all-male bands, whose home ranges can be up to 20 km^2. On average adult males weigh 18.5 kg and adult females 11.7 kg (for details on ecology and demography see Mohnot, 1974; Mohnot *et al.*, 1981; Winkler *et al.*, 1984; Rajpurohit and Sommer, 1991).

Due to fusions and fissions, the number of bisexual troops in the population has varied between 27 and 29, with an average troop size of 38.5 animals (range: 7-93). The number of all-male bands has remained about 13, with an average of 11.8 members (range: 2-47). All-male bands invade home ranges of bisexual troops in an unpredictable pattern, sometimes resulting in replacement of the resident male. Male changes occur during all months of the year. Residencies of single adult males in bisexual troops last from just a few days to at least 74.0 months, with an average of 26.5 months. Jodhpur langurs breed throughout the year, although there is a significant birth peak in March and the fewest births are in November (Sommer and Rajpurohit, 1989). Over the last two decades, the habitat has been greatly influenced by humans, particularly because the popula-

tion of Jodhpur town increased markedly from about 350,000 in 1968 to about 750,000 in 1988. The whole plateau is subject to deforestation, except for a protected reserve of scrub forest with a total area of about 20 km^2. The langurs live in the open scrub forest, but most groups include human habitations (roofs, deserted buildings) and parks and field areas in their home ranges. All permanent water sources (tanks, artificial lakes, wells) are man-made. The diet consists of items from approximately 190 plant species. Most of the troops are fed by local people for religious reasons with wheat preparations, vegetables and fruits. Male bands are less frequently provisioned, because of their unpredictable patterns of movement. Although there are considerable differences between groups, provisioning, as a rule, accounts for one third of the total feeding time (Winkler, 1988; Srivastava, 1989). Additionally, some bands and troops raid crops. As they are considered sacred, however, the langurs are not hunted. Apart from feral dogs, natural predators are absent. The langurs are easy to observe, as they are not shy and spend most of the daytime on the ground.

Infanticidal events which took place at Jodhpur between 1969 - 1987 have been described by different observers. To allow a critical assessment of the sample on which the following conclusions are based, all case histories along with a definition of terms and the sources of information are listed in Appendix I. The present review considers unambiguously documented cases of infanticide together with inferred cases based on circumstantial evidence. However, all appendices, figures and tables allow one to distinguish between different types of evidence :

*** = *Infanticide witnessed* (i.e. actual attack and wounding observed; subsequent death confirmed, because carcass was seen);

** = *Infanticide likely* (i.e. attack and wounding seen, infant disappeared later, but the carcass could not be found);

* = *Infanticide presumed* (i.e. only circumstantial evidence; usually, disappearance shortly after the previous resident was replaced by a male band or a new resident).

+ = *Infanticide attempted* (i.e. infant or a mother with a clinging infant was attacked or wounded, but could either escape or survived).

= *Abortion* (i.e. loss of conceptus in connection with male change).

RESULTS

Langur mothers at Jodhpur display a rather *laissez faire* style of infant guarding if the male who sired their babies is still resident of the troop.

However, they do usually very cautiously avoid new harem residents, probably because these males tend to attack infants fathered by their predecessors. In this respect, it is quite suggestive that in the course of a troop splitting, all lactating and possibly all pregnant females remained with the old resident, whereas only females without infants joined a new male (change D, cases 20-24; for details of case histories, see Appendix II).

In the long-term record for Jodhpur, information is available concerning 17 resident male changes in 12 different troops which occurred between 1969-1985. Details are known about 112 immatures present during male changes or born shortly after. Fifty-seven infants were not attacked ("not harmed"), 14 experienced nonfatal attacks ("attempted infanticide"), 20 infants disappeared in the course of a male change ("presumed infanticide"), 7 infants disappeared after they had been wounded ("likely infanticide"), 12 infants and 2 juveniles died after they were wounded ("witnessed infanticide"). Moreover, 5 abortions have been described in connection with male changes. The actual figure of fetus loss, however, might be higher because it is very difficult to observe abortions under field conditions.

If males attack immatures, they do not restrict their bites to certain foci, but seem to bite their victims on any body part they are able to grasp (Figure 1). Out of 34 described wounds, 23.5% were received in the head region (neck, forehead, face, throat), 44.1% in the groin region (tail, hip, flank), 32.4% at other body regions. Bites into the throat and neck area were the most effective modes of killing (they account for a higher number of "witnessed" cases since the time between attack and death was rather short). Nevertheless, *in toto*, other body parts are more often wounded.

Figure 2 shows that 67.3% of all victims were younger than 6 months and 81.8% were younger than 9 months. The oldest individuals known to be attacked were two males, one wounded when about 25 months old (case 78) and the second killed when about 48 months (case 77). Although these cases were not infanticides in the strict sense, but "juvenilicides", they have been included because they concern the eldest immatures known to be attacked while still living in their *natal* troop. Attacks on weaned individuals cannot any more abbreviate their mother's lactational amenorrhea. In this regard, it is important to note that the two juvenile males were attacked about one year after the harem holder had established himself as troop resident. Hence, it is likely that the function of these two assaults was different from other attacks on infants, which occurred only during takeover periods.

The mean age of all victims including the 2 juveniles was 5.9 months, without them 4.8 months (Table 1). Individuals belonging to the class of

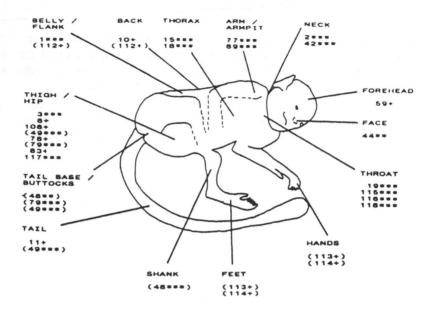

Figure 1 Injuries of infanticide victims at Jodhpur. For case numbers, see Appendix II. Infanticide witnessed ***, likely **, presumed *, or attempted +. Brackets () indicate orientation at several body parts. Left and right orientation is neglected.

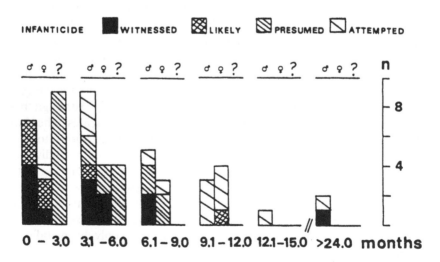

Figure 2 Age-sex distribution of immatures known or believed to be attacked or killed by adult males at Jodhpur (N = 55).

Table 1 Fate of infants present during adult male replacements (sample drawn from Appendix II).

	Males			Females			Sex Unknown			Total			
	N	Age (months) mean	SD	N	Age (months) mean	SD	N	Age (months) mean	SD	N	Age (months) mean	SD	range
Infanticide													
– witnessed	10[a]	8.0	14.3	3	2.3	1.7	–			13	6.7	12.6	0.3 – ~48
– likely	4	1.7	1.8	3	4.4	4.1	–			7	2.8	3.1	0.2 – 9.1
– presumed	4	8.1	1.5	4	5.1	1.8	13+	2.1	1.7	21	4.0	2.8	0.2 – 9.4
– attempted	9[b]	10.4	6.8	5	8.6	3.4	–			14	9.8	5.7	3.0 – ~25
Total	27[c]	7.9	9.7	15	5.5	3.6	13	2.1	1.7	55[d]	5.9	7.4	0.2 – ~48
Survived[e]	18	7.7	3.2	12	7.4	4.5	4	3.5	2.1	34	7.3	3.7	1.0 – >~15.5

[a] If case 77 (male, ~48 months old) is omitted : N = 9, mean 3.5 ± 3.7 months.
[b] If case 78 (male, ~25 months old) is omitted: N = 8, mean 8.6 ± 4.3 months.
[c] If cases 77 and 78 are omitted : N = 25, mean 5.6 ± 4.1 months.
[d] If cases 77 and 78 are omitted : N = 53, mean 4.8 ± 3.7 months.
[e] Excluding those infants sired by the previous resident but born after the male change.

"attempted infanticide" had the highest age average (9.8 months). Sex-differences concerning age could only be recognized if the 2 juvenile males were included in the sample (male victims on average 7.9 months, female victims 5.5 months). However, there was a male bias in the number of victims : Out of 42 immatures with known sex, 27 (64.3%) were male and 15 (35.7%) were female. If tested against the birth ratio (M:F = 1.1:1 ; Rajpurohit and Sommer, 1991) the difference is not statistically significant (goodness-of-fit test, one-tailed, chi-square 2.39, 1 df, p < 0.20). However, if only witnessed cases are taken into consideration (10 males : 3 females), the difference is significant.

A considerable number of infants (44.4% out of a total of 81), which were sired by the previous male and already born before his replacement, were *not* harmed in the course of the male change (Figure 3). The proportion of infants born *after* the resident male replacement, which were not harmed, was even substantially higher: 61.8% out of 34. Although it is possible that some attacks were not observed and the given individuals

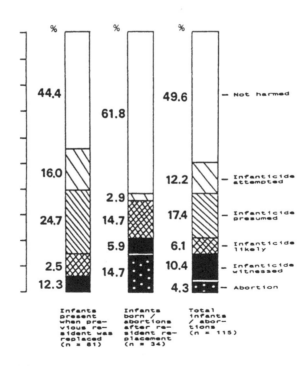

Figure 3 Fate of infants present during resident male changes or born shortly after. "Infant" = individuals ≤ 15.0 months of age; cases 77/78 (Appendix II) omitted.

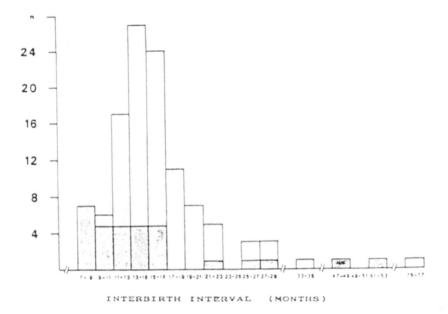

Figure 4 Distribution of interbirth intervals (N = 114). Longitudinal data from long-term monitoring of individually known females of troops Kailana I, Kailana II and Bijolai. Open bars = preceding infant still alive when younger sibling was conceived or died > 10.1 months of age (i.e. mean total IBI [16.7 months] minus mean duration of pregnancy [6.6 months]); shaded bars = intervals after premature loss of preceding infant, i.e. it had already died (< 10.1 months of age) when younger sibling was conceived. Modified after Sommer and Rajpurohit, 1989.

Table 2 Interbirth intervals in months (for sample see Figure 4).

Interbirth interval	N	Mean	SD	Range
Preceding infant still alive when younger sibling was conceived	82	17.2	7.8	10.6 – 76.4
After premature loss of preceding infant before 10.1 months of age	32[a]	15.4	10.4	7.0 – 51.7
Total	114	16.7	8.6	7.0 - 76.4

[a] Without extreme IBI's of 48.5, 51.7 and 22.1 months (cases 52, 51, 46) : N = 29, mean 12.8 ± 5.1 months

were included into the class of "not harmed", there can be little doubt that about half of all infants present during male changes or born shortly after were not attacked.

A crucial factor for a potential reproductive benefit of infanticidal behavior is the abbreviation of the interbirth interval (IBI) of the infant-deprived mothers. Long term monitoring of individually known females revealed a mean IBI of 16.7 months; the shortest IBIs were usually those following the premature death of the previous infant (Figure 4). In total, mean IBIs after premature loss of the preceding infant were 1.8 months (10.5%) shorter than IBIs without infant loss (Table 2). The latter sample, however, includes three long IBIs which are somewhat questionable: (a) Case 52 - An old female, who lost her infant through infanticide, experienced afterwards at least one, but probably two, intermittent ovarian failures of about 5 months each. Ultimately, after a gap of 48.5 months she gave birth to another - her last - offspring. (b) Case 51 - An 8-year old female resumed regular cycling 17 days after her infant was killed. However, despite regular copulations she did not conceive for almost 4 years, since her next offspring was born after 51.7 months. (c) Case 46 - A middle-aged female lost her fetus after a new male took over. Her next infant was born 21.1 months later. Based on her cycle records, however, it is likely that she reconceived after 42 days, but lost her fetus again whilst the abortion went unnoticed. Without this suspected loss, the IBI would have been only 7.9 months. If these three extreme cases are excluded, the average IBI after premature loss of the preceding infant drops down to 12.8 months, i.e. 4.4 months, 25.6% shorter than the average IBI without infant loss. Computation of the median reveals that the IBI was 3.2 months, *i.e.* 20.5% shorter following the loss of an infant than without loss. The shortest IBIs without loss of the preceding infant were 11.2, 11.3 and 11.9 months, respectively. Hence, lactating langur females might conceive 4.6 months after giving birth to an infant.

There is a positive, statistically significant, correlation between the age of an infant during its loss and the subsequent IBI of its mother ($r = 0.652$, $n = 29$, chi-square 4.46, 1 df, $p < 0.05$), i.e. the younger the infant which was lost, the shorter the IBI (Figure 5). However, the correlation coefficient drops to a mere $r = 0.168$ if the three questionable IBIs just described are included.

The likely genetic relationships between infanticidal males, their victims and the subsequent infants born to infant-deprived mothers were inferred from patterns of male-female association and copulation around the period of conception (for details, see Appendix III and Appendix IV). The results are summarized in Figure 6.

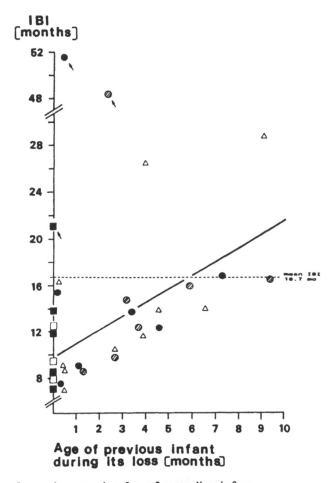

Legend concerning fate of preceding infant:

→   Omitted from calculation of linear correlation

During stable social situation:

Δ   infant death or disappearance

□   abortion.

In connection with male change:

■   abortion;

⊘   infanticide likely / presumed

●   infanticide witnessed

Figure 5 Correlation between age of an infant during its loss and subsequent interbirth interval (IBI) of its mother. Only cases, where infants died or disappeared before 10.1 months of age (cf. Figure 4). Linear correlation r = 0.652, N = 29, chi-square = 4.46, df 1, p < 0.05.

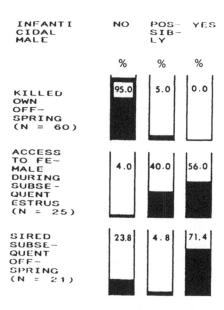

Figure 6 Assumed genetic relationships between infanticidal males, victims and subsequent infants born to infant-deprived females. For sample of case histories see Appendix IV.

(a) Information about the genetic relations between killed infants or aborted fetuses and infanticidal males exists for a total of 60 cases. *During the period of the victim's conception*, the mothers associated or copulated with males other than the attacking male in 95.0% of all cases. In 5.0% of all cases, they associated with multiple males including the attacking male. Hence, these males *might* have killed their own progeny.

(b) Infant deprived females resumed cycling after an average of 11.2 ± 7.2 days (N = 17, range 1-34 days; cf. fourth column in Appendix III). The average period between abortions and onset of estrus was 18.0 ± 10.4 days (N = 5, range 8-34 days; Appendix III). *During the first estrus after infant or fetus loss* the respective females associated or copulated with the attacking male in 56.0% of all 25 known cases. In 40.0% of the cases, they associated or copulated with multiple males including the attacking male, in only one case (4.0%) with other males. Hence, an attacking male had access to the female whose infants he attacked or whom he induced to abort in 90.0% of all cases.

(c)   During the period of conception of the subsequent infant, the proportion of males who were actually able to "earn" the reproductive benefit potentially associated with infanticidal behavior decreased due to further inter-male rivalry in the course of which the attacking males themselves were replaced. Out of a total of 21 cases, for which information is available, 23.8% of the females were likely to have been impregnated by males *other* than the previously attacking male. In another 4.8% of the cases, they associated with multiple males, including the attacking male. However, for a majority of cases (71.4%), it is likely that same male which attacked the previous infant or induced the abortion actually fathered the subsequent offspring.

All infants born within 200 days of a new residency, i.e until one gestation length has passed, are still sired by the predecessor of the current harem holder. Thus, new residents will shorten their waiting time to insemination most effectively if they are able to induce abortions in pregnant females of their newly acquired harem. As a matter of fact, out of 8 cases of abortions (and stillbirths) observed in 3 carefully monitored troops, 5 cases (62.5%) occurred in connection with male changes (cases 16, 46, 50, 58, 99). (Besides the 7 cases described by Agoramoorthy *et al.*, 1988, one additional occurred during a stable residency: female 12 of troop KI had a suspected stillbirth in 1987; Sommer and Srivastava, personal observation).

If abortion does not occur, new males can abbreviate their waiting time to insemination if they kill those infants born after a takeover until one gestation length has passed. Out of 27 infants born within 200 days after the replacement of their sire, 8 (29.6%) were known or believed to be killed by new males (Figure 7). However, the likelihood of a killing was higher if infants were born shortly after the onset of a tenure, and decreased as time passed by: Out of 18 infants born during the first 100 days of a new tenure, 6 (33.3%) were killed, whereas out of 9 born during the second 100 days, 2 (22.2%) were killed, but possibly only 1 (12.5%), since case 5 refers to a "presumed" infanticide.

There are strong indications that infanticidal motivations of some new residents decline with *time*. (a) Male 70 (change N) induced one abortion and killed 4 infants born 28-137 days after he became troop resident (cases 99-103); however, he did not harm an infant born 145 days after onset of his tenure (case 104). (b) Male 119 (change J) induced an abortion and most probably killed an infant born 45 days after his takeover (cases 58, 62) but did not harm infants born after 92 and 109 days (cases 63, 64). There were individual differences concerning the time when these two

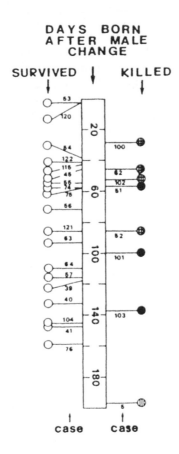

Figure 7 Fate of infants born after a male takeover until one gestation length (200 days) has passed. For case histories, see Appendices. Open circles = not harmed; Circles with vertical lines = infanticide presumed; Circles with cross hatching = infanticide likely; Solid circles = infanticide witnessed.

males stopped killing. However, once the hypothetical "infanticide clock" has gone beyond a critical point, the process of infant killing is irreversibly over. None of the other males resumed killing once they stopped it, and all males started killing infants either during the period of takeover or shortly thereafter, or not at all. (The described attacks on two juvenile males one year after a takeover were obviously special cases; cf. Rajpurohit and Sommer, 1993.)

Nevertheless, the majority of infants born after takeovers survived. In some cases, the "reason" why certain new harem holders spared infants might be that they had sexual contact with the females and might have sired their infants during previous multi-male stages or interim residencies preceding the final outcome of the male-male competition. (a) Male 46 (male change I, cases 53-57) was interim resident in troop KI for successive short periods between October 1982 - January 1983. After he established himself as resident for a tenure of 2.5 years in June 1983, he did not harm any of the 5 infants born within 200 days after his takeover, although he had sired only the *first* of these infants. (b) Male 10 (change K, cases 65-76) neither attacked any of the 9 infants present during the takeover nor the 3 infants born shortly after. He had had contact with troop females during a previous multi-male stage about 173-264 days before they gave birth.

If the previous sexual contact with a female reached back more than 200 days before birth, the given male might have fathered her subsequent offspring, at least theoretically, although the proportion of infants conceived during multi-male stages was only 4.7% in 3 well-studied troops (Sommer and Rajpurohit, 1989). However, the hypothesis has been forwarded that males can be confused about paternity even if they copulate with females which are already pregnant. In order to evaluate the possibility that post-conception estrus (PCE) might function to induce new males to tolerate infants born after their takeover of a troop but sired by their predecessors, the pattern of estrous behavior exhibited during pregnancies has been analyzed (Figure 8). Behavioral estrus displayed during menstrual cycles are confined to a distinct period of about 4 days around the time of ovulation. PCE, however, is extended throughout about day 50 - 140 of pregnancy. Soon after conception there is some minor activity detectable, whereas there is almost no activity during the last 50 days of the gestation period. Most of the estrous solicitations remained unanswered by males (most of the solicitations by cycling females are likewise unsuccessful). In the vast majority of cases, copulations occurred only between about day 50 to about day 90. Females displayed the *same* pattern of PCE behavior to the males who fathered their infants as well as to new males (cf. also Sommer *et al.*, 1992).

DISCUSSION

In The Shadow of Self-Fulfilling Prophecies

The evidence for infanticide among the langurs of Jodhpur is of variable quality. However, taking into account that in 13 "witnessed" cases, the

VOLKER SOMMER

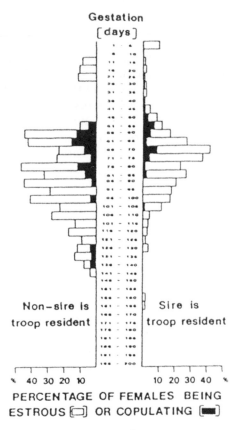

Figure 8 Postconception estrous behavior of langur females at Jodhpur. The gestation period of 200 days is broken down into 40 periods of 5 days each. Data of 31 different pregnancies of 12 individually known females of troop Kailana I between 1981-1987 were synchronized, spanning residencies of 5 different males (observers: V. Sommer, V.K. Dave, C. Borries, A. Srivastava). It is distinguished between "estrous" days (i.e. solicitation of males by head shuddering, lowering of tail and presentation occurred *without* subsequently observed copulation) and "copulation" days. Moreover, it is distinguished if such behavior is displayed towards the sire of the subsequent infant (right bars) or towards a new male ("non-sire") who invaded the troop or installed himself as new resident after the sire was ousted (left bars). For each 5-day period, the total number of observation days was compared to the total number of days females were estrous or observed to copulate. E.g., for behavior displayed towards sires between days 66-70, data were available for 15 different pregnancies; out of a total of 66 days of observation, copulations occurred on 6 days (9.1%) and estrous behaviour on another 22 days (33.3%), amounting to a total of 42.4% of females who showed postconception estrus. However, between e.g. days 131-135, 59 days of observations are available for 13 different pregnancies, but there were no days with postconception estrus. The mean number of days observed on *each* of the 5 days of a given 5-day period is 14.2 ± 3.1 days for behavior displayed towards sires, and 5.1 ± 1.5 days for behavior displayed towards non-sires. There was a minimum of 2 days of observation for *each* of the 200 days available.

whole procedure of killing has been reported, there is good reason to assume that most of the infant disappearances in connection with the 7 cases labelled as "likely" and 21 cases labelled as "presumed" were likewise the result of aggressive behavior of new males. Moreover, I know from my own experience that the act of infanticide can occur within a matter of seconds, making the chance of actually observing it quite unlikely.

Nevertheless, it can hardly be denied that some touches of self-fulfilling prophecies play a prominent role in the controversy about how regular langur infanticide occurs. If a langur researcher witnessed infanticide, even weak evidence for its occurrence at other sites is usually readily accepted. If infanticide was not seen with one's own eyes, one is likely to play down or even deny its existence for other sites (see, e.g., Boggess, 1984, and the reply of Hrdy, 1984). For example, several German students of the Göttingen University did not see any fatal male attacks on infants between 1977-1979, yielding severe criticism of the sexual selection hypothesis by the principal investigator, Christian Vogel (1979). During another adult male replacement in 1980, 6 infants were found wounded and / or disappeared (male change B, cases 6-11). The wounds and disappearances occurred during a temporary absence of the observer. It was emphasized, that "neither infant was endangered by its injury" and "there was no observed indication that infanticide had actually occurred or been attempted" (Vogel and Loch, 1984, p. 249). When I began my own fieldwork during 1981, I was - as a pupil of Vogel - more or less convinced that infanticide does not occur at all. Consequently, it struck me with great surprise when I saw the first infanticide of my study period (case 42 in the Appendices). Since many more "ironclad" cases were reported from 1982 onwards, I put the wounding and disappearances of the 1980 male change in another light and now place them without much doubt into the categories of "presumed" and "attempted" infanticide.

I am stressing the existence of such psychological mechanisms, because "what researchers see is affected by their expectations about the natural world" (Hrdy, 1984). The subjectivity is two-fold. On the one hand, there are people who are not ready to change their mind even if confronted with irrefutable evidence. On the other hand there are the "converts" (like me) which might be even more vulnerable to subjectivity, because a protestant converted to catholicism is likely to be more papal than the pope.

Foci of Attacks

Many mammals direct their killing bite toward the neck or posterior brain case of their prey (Eisenberg and Leyhausen, 1972). This "craniocervical

bite" is also a prominent part of the predatory repertoire in a wide variety of primates (King and Steklis, 1984) and is the predominant mode of
infanticidal killing in several primate genera (King and Steklis, in press).
However, the almost random distribution of injured body parts of infanticide victims at Jodhpur reveals that males obviously bite infants on any
body part they are able to seize (cf. Figure 1). King and Steklis, on the basis
of only sparse data available to them, already noted that in the genus *Presbytis*, infanticidal attacks were not distinctly oriented towards the craniocervical region. They suggested that the pattern of a craniocervical killing
bite inherited from insectivorous primate ancestors was lost during a long
phylogenetic history of folivory in the genus *Presbytis*. If this is true, it
seems likely that infanticide is a rather new phenomenon. This assumption
does not necessarily contradict the assumption that the behavior has an
evolutionary significance, because computer simulations made clear that
in Jodhpur-like populations, the hypothetical trait could have reached an
equilibrium proportion of infanticidal males after a minimum of only 600
generations (Hausfater, 1984).

Male-Bias Among Victims

More immature males (n = 27) than females (n = 15) fell victim to male
attacks. For witnessed cases of killings (10 males, 3 females), the difference
is statistically significant (cf. Table 1). The original reproductive advantage
hypothesis predicts that the reproductive benefit which an infanticidal
male would gain compared to a noninfanticidal counterpart should be
independent of the infant's sex. Although a sample size of 27 cases from
one site is a small one from which to argue for male-biased infanticide
in this species, three factors might be crucial: (a) The troop tenureship
of individual adult males varies between 3 days and at least 74.0 months
(mean 26.5 months; Sommer and Rajpurohit, 1989). The first conception,
on the other hand, occurs with a mean age of 35.9 months (Sommer *et al.*,
1992). In fact, the distribution of 64 resident male tenureships at Jodhpur
revealed that 37.5% lasted longer than 36.0 months. Hence, a substantial
proportion of harem residents could potentially sire offspring with females
who were still infants during the time of takeover. Therefore, a selective
force might exist to spare female infants from infanticide because they can
potentially reproduce with the new male if his residency is long enough.
(b) On the other hand, harem residents might discriminate against future
competitors of their own sons which will be sired with the troop females.
(c) If a residency is long enough, natal males could become as old as 4 to
6 years. Hence, the two attacks on juveniles of about 25 and 48 months of

age (cases 77, 78) are likely to reflect attempts by resident males to expel maturing males that would soon become *direct* competitors. It seems to be much more difficult - if not risky - to expel young males once they have grown to a certain size.

Predictions of The Sexual Selection Hypothesis

The hypothesis, that an infanticidal male, on average, will gain a reproductive advantage compared to a noninfanticidal counterpart provides three basic predictions.

(a) *An infanticidal male should not have sired any infant he attacks or kills.* In 95.0% of the cases where paternity could be assigned, males are assumed not to have killed their own offspring, and in only 5.0% of the cases, males *might* have killed own progeny, because around the time of the victim's conception, the mother associated and copulated with several males, including the infanticidal male (cf. Figure 6). In one of the latter instances, an infant disappeared 6 months after takeover (case 5); its loss is treated as "presumed" infanticide in order to avoid applying a double-standard concerning data that might contradict the sexual selection model. Without this case, the proportion of victims assumed to be genetically non-related raises to 98.3%. Moreover, the sexual selection model gains strong support by the fact that 93.4% out of 55 known or suspected cases of assaults on immatures (cf. Table 1) occurred in connection with adult male replacements. The remaining two instances concern attacks on male juveniles discussed above.

(b) *An infanticidal male should sire the subsequent offspring of the victim's mother.* The chance of an infanticidal male siring the subsequent offspring with the infant-deprived female depends on his ability to monopolize the female after his attack took place. A substantial proportion of attacks occurred before the "final" outcome of the male-male competition, i.e. during temporary interim residencies of single adult males which lasted only some days or weeks, or - less often - during multi-male periods. Consequently, infanticidal males were sometimes already ousted or had to tolerate rivals when the infant-deprived females had resumed estrus. Nevertheless, the male had access to these females in 96.0% of all cases, although they had to "share" access with other males in 40.0% of the cases (cf. Figure 6). However, a female is not likely to conceive during her first estrus after infant loss. As a result of ongoing male-male competition, infantici-

dal males were definitely not present in about one quarter (23.8%) of all cases, when the subsequent infant was sired. In the remaining three quarters, however, they probably fathered the subsequent infant.

(c) *Infanticide should significantly shorten the subsequent interbirth interval (IBI) of the victim's mother.* The upper limit of a demonstrable shortening of the IBI is an infant age of 6.0 months at the time of its loss, because this age corresponds to the average IBI of 16.7 months (cf. Figure 5). In fact, 67.3% of all victims were not older than 6.0 months. On the other hand, the shortest IBI without premature loss of an infant was only 11.2 months. This would correspond to an expected IBI after the loss of an infant of 1.2 months of age. Thus, as has already been pointed out by Winkler *et al.* (1984), even the killing of an infant younger than 3 months does not *necessarily* abbreviate the subsequent IBI of the infant-deprived female. On average, however, the IBI is shortened by 1.8 months after a premature loss of an infant - or even by 4.4 months, if three questionable cases are omitted from the mean (cf. Table 2). Therefore, the infanticidal strategy provides the new resident male with at least a respectable probability of gaining a reproductive advantage by shortening the period of infertility during lactation. As a rule, the younger the victim, the more a male gains from killing infants.

Individuals belonging to the class of "attempted infanticide" had the highest age average (9.8 months). It is not possible to decide if this is due to the fact that older infants (a) were less vulnerable, (b) had a better ability to escape or (c) were less fiercely attacked.

Abortions During Takeovers

Abortions as a result of confrontation with alien males are known to occur, e.g., in rodents (Bruce, 1960; Labov, 1984; Vom Saal, 1984), wild horses (Berger, 1983), lions (Packer and Pusey, 1984), savanna baboons (Pereira, 1983), geladas (Mori and Dunbar, 1985) and hamadryas baboons (Colmenares and Gomendio, 1988). Stress-induced abortions occur in wild langur females trapped for captivity (Ramaswami, 1975). The 5 abortions and stillbirths observed during male changes among the free-ranging langurs of Jodhpur are probably part of the reproductive strategies of both sexes. (a) If an adult male replacement took place, it might be more beneficial for a female to abort instead of investing in a fetus likely to be killed by new resident males, thus reflecting a female counterstrategy to

infanticide. (b) A male who exerts psychological or physical pressure on females who subsequently abort, may considerably reduce his waiting time until a possible new conception by cutting short the gestation period as well as the lactational amenorrhea.

Abortions are difficult to observe in the field. Therefore, it is not known how representative a figure of 5 cases of fetus loss during male changes is. Another 3 incidents occurred during socially stable one-male situations, which are typical for most of a troop's history. Hence, it is likely that the relative rate of fetus loss was in fact much higher during adult male replacements.

In this regard, it is interesting, that the "classical" reproductive-advantage-hypothesis cannot explain why males sometimes attack older, semi-independent infants who no longer hamper a renewed impregnation of their mothers. However, if these mothers are already pregnant again, males may try to induce abortions by attacking their previous infants and, thus, force the pregnant females to protect their offspring in highly aggressive counterattacks - events which have in fact been observed at Jodhpur (Agoramoorthy et al., 1988). It might be an even safer strategy to attack infants instead of the pregnant female herself, because this includes a risk of wounding the potential partner for reproduction.

Infanticidal Versus Noninfanticidal Strategy

Even under the conditions on which the sexual selection hypothesis is based, infanticide can be reproductively disadvantageous if the infanticidal males are themselves replaced after certain tenure lengths (Chapman and Hausfater, 1979). Imagine, e.g., a male kills all unweaned offspring during takeover and sires the subsequent ones. If he will be replaced after one year by another infanticidal male, his infants will be about 3-5 months old and are likely killed. Imagine, on the other hand, a noninfanticidal male who does not kill infants when he takes over. His own progeny can only be sired after the lactational amenorrhea ends. When he is replaced after one year by an infanticidal male, his offspring are still in the womb of their mothers.

The model is based on the assumption that infants born after takeovers are less vulnerable to infanticide. In fact, almost two third of all infants born *after* the male replacement were not harmed compared to a proportion of only 44.4% of those infants which were already *present* when a new male entered the troop (cf. Figure 3).

The overall Jodhpur-record seems to suggest that out of 22 male replacements, 9 (40.9%) were noninfanticidal. However, a close look at

the data reveals that (in modification of my earlier assumptions [Sommer, 1987]) it is very difficult to decide whether there are males which are *per se* noninfanticidal. (a) In several cases labelled as "non-infanticidal", the age-sex distribution of infants is unknown (changes T-W) or it was not observed how the replacement itself took place (change E) or what happened afterwards (change Q). Therefore, nonfatal attacks or even infant-losses might have occurred but went unnoticed. (b) In another case (change F), the infants present during the replacement were between 7.5–13.8 months old and their mothers' lactational amenorrhea had already ended. Hence, there was no "reason" to kill the infants. (c) In some cases, males had previous contact with the troop during multi-male stages or interim residencies, during which they sired (change I) or might have sired (changes E, K) one or more infants. Such males might be more "cautious" and not attack infants born after they succeed in establishing themselves as resident.

The Chapman-Hausfater model is based on the assumption that infanticide is a mixed evolutionary stable strategy involving two different types of males : "killers" and "non-killers". However, infanticidal males at Jodhpur rarely killed *all* alien unweaned offspring. Hence, it would be interesting to compute a model not based on inter-individual "all-or-none" differences concerning the proportion of "infanticidal" and "noninfanticidal" males, but based on the varying proportion of infants killed during different takeovers.

Postconception Estrus - A Female Counterstrategy?

Several primate species (including humans) engage in sexual behavior during the gestation period (see review in Hrdy and Whitten, 1987). Because sexual receptivity during pregnancy cannot enhance the probability of conception, the explanation has been forwarded that postconception estrus (PCE) is an adaptation to confuse paternity and induce new males to assist or tolerate infants born to former mates (Hrdy, 1979). Thus, in species like Hanuman langurs, where immigrant males may commit infanticide, PCE is assumed to inhibit infant killing (Hrdy, 1974). A crucial factor in testing this hypothesis is whether or not the frequency of PCE varies according to the likelihood of a male attacking subsequent infants. Observations of red colobus, another species with male infanticide, suggest that females pregnant at the time of attacks on other infants in the group in fact copulated more frequently and later into their pregnancy than did females either before or after these attacks. An infanticidal male in the study group did not attack their subsequent infants (Struhsaker and Leland, 1985).

However, the broad data base available for the langurs of Jodhpur does not support the hypothesis that PCE evolved to forestall infanticide:

(a) The pattern of PCE was almost identical whether the sire was still resident of the troop or a new male had taken over (cf. Figure 8). Hence, females did not discriminate between "harmless males" (i.e., sires) and potentially infanticidal males (i.e., new residents).

(b) The stereotypical distribution of PCE during as much as 31 pregnancies of 12 different females in the presence of 5 different resident males (and several dozen temporarily intruding males) seems to reflect a rather strict hormonal influence instead of a flexible behavioral response in reaction to certain social conditions.

(c) Eight infants which were born within 200 days after new males established themselves as troop residents became infanticide victims (cf. Figure 7). For 7 cases, it has been reported that their mothers not only solicited the "killer" male but copulated with him (case 5: Mohnot, 1971; cases 51, 52: Sommer, 1987: 186; cases 100-104: Agoramoorthy and Mohnot, 1988: 293; Agoramoorthy, 1987: 107,120). In all these cases, PCE did not forestall infanticide.

(d) It is not true, on the other hand, that mothers of those infants which survived invariably copulated with the new male during their pregnancies. Some infants born after takeovers were not harmed, although the males had killed other infants present during the takeover and despite the fact that the males had *no* observed sexual contact with the females during their pregnancies (cases 45, 63, 64).

(e) Ultimately, infants born during the first 100 days of the residency had a 33.3% risk of being killed; infants born between days 100 - 200 of a new residency had a considerably lower risk of 12.5 - 22.2% (cf. Figure 7). However, despite the increasing risk for mothers which were close to term when a new male took over a troop, the PCE activity was gradually reduced to zero after day 100 of the pregnancy (cf. Figure 8). The prediction of the hypothesis would have been that PCE increases with an increasing risk of killing.

Hence - as has been already pointed out by Sommer (1987) - it is very unlikely that Hanuman langur females developed PCE as a counterstrategy to infanticide. On the other hand, it seems unlikely that sexual receptivity during pregnancy is a mere artifact of endocrinological changes as females incur costs for non-conceptive matings, e.g. by interacting with aggressive harassers (Sommer, 1989) and by limiting the available time for foraging and nonsexual social contacts. As an alternative, it has been proposed

that PCE is an expression of female-female competition for sperm of the harem resident: By "sneaking" sperm away, pregnant females might reduce the probability of conception for other females and thus reduce the future number of resource competitors within the troop (Sommer, 1989). The assumption that sperm is a "limited" resource in the one-male breeding structure observed at Jodhpur is strongly supported by the fact that the probability of conception decreases for a given female when the number of troop mates copulating with the harem holder on the same day increases (Sommer *et al.*, 1992).

The Monkey Grammarian

"Siste Kapitel" ("The Last Chapter"), a novel of the Norwegian Nobel-laureate Knut Hamsun published in 1923, describes a postconception strategy of a young woman. She became pregnant from a man unable to provide secure social conditions to raise a child. Soon after conception, the woman lured an honest and hardworking farmer into a sexual relationship. Because her infant was born only eight months after she had first met him, she appeased her lover by telling him that she was bitten by a snake, causing a premature birth. The farmer tolerated the infant sired by his predecessor. Although another Nobel-laureate, the Mexican poet Octavio Paz, honored the Hanuman langur by attributing the title "El mono gramatico" ("The Monkey Grammarian"; 1974), there is no reason to believe that langur males have read Hamsun's novel and can therefore not any longer be cheated by the trick. Thus, we have to phrase it less poetically: We will need further studies to solve the puzzle of postconception estrus in langur females. The phenomenon of postconception estrus seems not to belong to the other puzzle which I consider solved for most of its parts: Why langur males commit infanticide.

ACKNOWLEDGEMENTS

I am greatly indebted to S. M. Mohnot, Jodhpur, for his support and friendship during all stages of the fieldwork, and to Christian Vogel, Göttingen, for his generous supervision of my langur studies. My own fieldwork was financed by the Deutscher Akademischer Austauschdienst (DAAD) and the Ministry of Education and Social Welfare, Government of India, under the Indo-German Cultural Exchange Programme, and by the Feodor-Lynen Programme of the Alexander von Humboldt-Foundation. At present, support is provided by the German Research Council (DFG, grant no. So 218).

REFERENCES

AGORAMOORTHY, G. (1987) *Reproductive Behaviour in Hanuman Langurs, Presbytis entellus*. Ph.D. thesis, University of Jodhpur, Jodhpur.

AGORAMOORTHY, G. and Mohnot, S. M. (1988) Infanticide and juvenilicide in Hanuman langurs (*Presbytis entellus*) around Jodhpur, India. *Human Evolution* **3**, 279-296.

AGORAMOORTHY, G., Mohnot, S.M., Sommer, V. and Srivastava, A. (1988) Abortions in free ranging Hanuman langurs (*Presbytis entellus*) - a male induced strategy? *Human Evolution* **3**, 297-308.

ALCOCK, J. (1989) *Animal Behavior. An Evolutionary Approach*. 4th ed., Sinauer, Sunderland / Massachusetts.

BARASH, D. P (1980) *Sociobiology and Behavior*. Elsevier, New York.

BERGER, J. (1983) Induced abortion and social factors in wild horses. *Nature* **303**, 59-61.

BISHOP, N. (1975) *Social Behavior of Langur Monkeys (Presbytis entellus) in a High Altitude Environment*. Ph.D. thesis. Univ. of California, Berkeley.

BOGGESS, J. E. (1976) *Social Behavior of the Himalayan Langur (Presbytis entellus) in Eastern Nepal*. Ph.D. thesis, Univ. of California, Berkeley.

BOGGESS, J.E. (1979) Troop male membership changes and infant killing in langurs (*Presbytis entellus*). *Folia Primatologica* **32**, 65-107.

BOGGESS, J.E. (1980) Intermale relations and troop male membership changes in langurs (*Presbytis entellus*) in Nepal. *International Journal of Primatology* **1**, 233-274.

BOGGESS, J.E. (1984) Infant killing and male reproductive strategies in langurs (*Presbytis entellus*). In : Hausfater, G. and Hrdy, S.B. (eds), *Infanticide: Comparative and Evolutionary Perspectives*. Aldine, New York, pp. 283-310.

BRUCE, H. M. (1960) A block to pregnancy in the house mouse caused by the proximity of strange males. *Journal of Reproduction and Fertility* **1**, 96-103.

CHAPMAN, M. and Hausfater, G. (1979) The reproductive consequences of infanticide in langurs : A mathematical model. *Behavioral Ecology and Sociobiology* **5**, 227-240.

CURTIN, R. A. and Dolhinow, P. (1978) Primate social behavior in a changing world. *American Scientist* **66**, 468-475.

COLMENARES, F. and Gomendio, M. (1988) Changes in female reproductive condition following male take-overs in a colony of hamadryas and hybrid baboons. *Folia Primatologica* **50**, 157-174.

DOLHINOW, P. (1977) Normal monkeys? *American Scientist* **65**, 266.

EIBL-EIBESFELDT, I. (1990) Chancen der Freundlichkeit. *Natur* **1**, 70-71.

EISENBERG, J.F. and Leyhausen, P. (1972) The phylogenesis of predatory behavior in mammals. *Zeitschrift für Tierpsychologie* **30**, 59-93.

HAMSUN, K. (1923) *Siste Kapitel*. Kristiana.

HAUSFATER, G. (1984) Infanticide in langurs: strategies, counter-strategies, and parameter values. In: Hausfater, G. and Hrdy, S.B. (eds), *Infanticide: Comparative and Evolutionary Perspectives*. Aldine, New York, pp. 257-282.

HRDY, S. B. (1974) Male-male competition and infanticide among the langurs (*Presbytis entellus*) of Abu, Rajasthan. *Folia Primatologica* **22**, 19-58.

HRDY, S.B. (1977) *The Langurs of Abu*. Harvard Univ. Press, Cambridge, London.

HRDY, S.B. (1979) Infanticide among animals: a review, classification and examination of the implications for the reproductive strategies of females. *Journal of Ethology and Sociobiology* **1**, 13-40.

HRDY, S.B. (1984) Assumptions and evidences regarding the sexual selection hypothesis: a reply to Boggess. In: Hausfater, G. and Hrdy, S.B. (eds), *Infanticide: Comparative and Evolutionary Perspectives*. Aldine, New York, pp. 315-319.

HRDY, S. and Whitten, P. (1987) Patterning of sexual activity. In: Smuts, B.B., Cheney, D.L., Seyfarth, R.M., Wrangham, R.W. and Struhsaker, T.T. (eds), *Primate Societies*. University of Chicago Press, Chicago, pp. 385-399.

JAY, P. (1965) The common langur of north India. In: De Vore, I. (ed), *Primate Behavior - Field Studies on Monkeys and Apes*. Holt, Rinehart, Winston, New York, pp. 197-249.

JOLLY, A. (1985) *The Evolution of Primate Behavior*. Macmillan, New York.

KING, G.E. and Steklis, H.D. (1984) New evidence for the craniocervical killing bite in primates. *Journal of Human Evolution* 13, 469-481.

KING, G.E. and Steklis, H.D. (in press) Craniocervical attack in primate infanticide.

LABOV, J. B. (1984) Infanticidal behavior in male and female rodents: Sectional introduction and directions for future research. In: Hausfater, G. and Hrdy, S.B. (eds), *Infanticide: Comparative and Evolutionary Perspectives*. Aldine, New York, pp. 323-330.

LAWS, J. V. H. and Laws, J. (1984) Social interactions among adult male langurs (*Presbytis entellus*) at Rajaji Wildlife Sanctuary. *International Journal of Primatologie* 5, 31-50

MAKWANA, S. C.(1979) Infanticide and social change in two groups of the Hanuman langur, *Presbytis entellus*, at Jodhpur. *Primates* 20, 293-300.

MATHUR, R. and Lobo, A. (1986) Ecology and behaviour of *Presbytis entellus* with special reference to adult male takeovers and infanticide. *Abstracts of the National Symposium on Conservation and Use of Primates in Biomedical Research*, 19-20 December, 1986, Jaipur, p.18.

MOHNOT, S. M. (1971) Some aspects of social changes and infant-killing in the Hanuman langur, *Presbytis entellus* (Primates: Cercopithecidae), in western India. *Mammalia* 35, 175-198.

MOHNOT, S. M. (1974) *Ecology and Behaviour of the Common Indian Langur, Presbytis entellus*. Ph.D. thesis, University of Jodhpur, Jodhpur.

MOHNOT, S. M., Gadgil, M. and Makwana, S. C. (1981) The dynamics of the Hanuman langur population of Jodhpur, Rajasthan, India. *Primates* 22, 182-191.

MORI, U. and Dunbar, R.I.M. (1985) Changes in the reproductive condition of female gelada baboons following the takeover of one-male units. *Zeitschrift für Tierpsychologie* 67, 215-224.

MUCKENHIRN, N. A. (1972) *Leaf-eaters and Their Predators in Ceylon: Ecological Roles of Gray Langurs, Presbytis entellus, and Leopards*. Ph.D. thesis, Univ. of Maryland.

NEWTON, P. N. (1985) The behavioural ecology of forest hanuman langurs, *Presbytis entellus*. *Animal Behaviour* 34, 785-789.

OPPENHEIMER, J. R. (1977) *Presbytis entellus*, the Hanuman langur. In: Rainier, H. S. H. and Bourne, G. H. (eds), *Primate Conservation*. Academic Press, New York, pp. 469-512.

PACKER, C. and Pusey, A.E. (1984) Infanticide in carnivores. In: Hausfater, G. and Hrdy, S.B. (eds), *Infanticide: Comparative and Evolutionary Perspectives*. Aldine Publishing Company, New York, pp. 31-42.

PARTHASARATHY, M. D. and Rahaman H. (1974) Infant killing and dominance assertion among the Hanuman langur. *Abstracts of the Vth Congress of the International Primatological Society*, August 21-24, 1974, Nagoya, Japan, p. 35.

PAZ, O. (1974) *El mono gramatico*. Seix Barral, Barcelona.

PEREIRA, M. E. (1983) Abortion following the immigration of an adult male baboon (*Papio cynocephalus*). *American Journal of Primatology* 4, 93-98.

RAJPUROHIT, L. S. and Sommer, V. (1991) Sex differences in mortality among langurs (*Presbytis entellus*) of Jodhpur, Rajasthan. *Folia Primatologica* 56, 17-21.

RAJPUROHIT, L.S. and Sommer, V. (1993) Emigration of juvenile male langurs (*Presbytis entellus*) from natal troops: a result of intrasexual competition. In: Pereira, M. and Fairbanks, L. (eds), *Juvenile Primates*. Oxford Univ. Press, New York, pp. 86–103.

RAMASWAMI, L. S. (1975) Some aspects of the reproductive biology of the langur monkeys, *Presbytis entellus* Dufresne. *Proceedings of the Indian National Science Academy* B41, 1-30.

RIPLEY, S. (1967) Intertroop encounters among Ceylon gray langurs (*Presbytis entellus*) In: Altmann, J. (ed), *Social Communication among Primates*. Univ. of Chicago Press, Chicago, pp. 237-254.

RIPLEY, S. (1980) Infanticide in langurs and man: Adaptive advantage or social pathology? In: Cohan, M.N., Malpass, R.S. and Klein, H.G. (eds), *Biosocial Mechanisms of Population Regulation*, Yale Univ. Press, New Haven and London, pp. 349-390.

SOMMER, V. (1984) Kindestötungen bei indischen Langurenaffen (*Presbytis entellus*) - eine männliche Reproduktionsstrategie? *Anthropologischer Anzeiger* 42, 177-183.

SOMMER, V. (1985) *Weibliche und männliche Reproduktionsstrategien der Hanuman-Languren (Presbytis entellus) von Jodhpur, Rajasthan/Indien*. Ph.D. thesis, Georg-August Univ., Göttingen.

SOMMER, V. (1987) Infanticide among free-ranging langurs (*Presbytis entellus*) at Jodhpur (Rajasthan/India): Recent observations and a reconsideration of hypotheses. *Primates* 28, 163-197.

SOMMER, V. (1988) Male competition and coalitions in langurs (*Presbytis entellus*) at Jodhpur, Rajasthan, India. *Human Evolution* 3, 261-278.

SOMMER, V. (1989) Sexual harassment in langur monkeys (*Presbytis entellus*): competition for ova, sperm, and nurture? *Ethology* 80, 205-217.

SOMMER, V. and Mohnot, S.M. (1985) New observations on infanticides among hanuman langurs (*Presbytis entellus*) near Jodhpur (Rajasthan / India). *Behavioral Ecology and Sociobiology* 16, 245-248.

SOMMER, V. and Rajpurohit, L.S. (1989) Male reproductive success in harem troops of Hanuman langurs (*Presbytis entellus*). *International Journal of Primatology* 10, 293-317.

SOMMER, V., Srivastava, A. and Borries, C. (1992) Cycles, sexuality, and conception in free-ranging langurs (*Presbytis entellus*). *American Journal of Primatology* 28, 1–27.

SRIVASTAVA, A. (1989) *Feeding Ecology and Behaviour of Hanuman Langur, Presbytis entellus*. Ph.D. thesis, University of Jodhpur, Jodhpur.

STRUHSAKER, T.T. and Leland, L. (1985) Infanticide in a patrilineal society of red colobus monkeys. *Zeitschrift für Tierpsychologie* 69, 89-132.

STRUHSAKER, T.T. and Leland, L. (1987) Colobines: Infanticide by adult males. In: Smuts, B.B., Cheney, D.L., Seyfarth, R.M., Wrangham, R.W. and Struhsaker, T.T. (eds), *Primate Societies*, Univ. of Chicago Press, Chicago, pp. 83-97.

SUGIYAMA, Y. (1965a) Behavioral development and social structure in two troops of Hanuman langurs (*Presbytis entellus*). *Primates* 6, 213-247.

SUGIYAMA, Y. (1965b) On the social change of Hanuman langurs (*Presbytis entellus*) in their natural conditions. *Primates* 6, 381-417.

SUGIYAMA, Y. (1966) An artificial social change in hanuman langur troop (*Presbytis entellus*). *Primates*, **7**, 41-72.

TRIVERS, R. (1985) *Social Evolution*. Benjamin/Cummings, Menlo Park, California.

VOGEL, C. (1979) Der Hanuman Langur (*Presbytis entellus*), ein Parade-Exempel für die theoretischen Konzepte der Soziobiologie. *Verhandlungen der Deutschen Zoologischen Gesellschaft* **72**, 73-89.

VOGEL, C. and Loch, H. (1984) Reproductive parameters, adult- male replacements, and infanticide among free-ranging langurs (*Presbytis entellus*) at Jodhpur (Rajasthan), India. In: Hausfater, G. and Hrdy, S.B. (eds), *Infanticide: Comparative and Evolutionary Perspectives*. Aldine, New York, pp. 237-255.

VOM SAAL, F. S. (1984) Proximate and ultimate causes of infanticide and parental behavior in male house mice. In: Hausfater, G. and Hrdy, S.B. (eds), *Infanticide: Comparative and Evolutionary Perspectives*. Aldine, New York, pp. 401-425.

WICKLER, W. and Seibt, U. (1977) *Das Prinzip Eigennutz - Ursachen und Konsequenzen sozialen Verhaltens*. Hoffmann und Campe, Hamburg.

WINKLER, P. (1981) *ZurÖko-Ethologie freilebender Hanuman- Languren (Presbytis entellus entellus Dufresne, 1797) in Jodhpur (Rajasthan), Indien*. Ph.D. thesis, Georg-August-Universität, Göttingen.

WINKLER, P. (1988) Feeding behavior of a food-enhanced troop of Hanuman langurs (*Presbytis entellus*) in Jodhpur, India. In: Fa, J.E. and Southwick, C.H. (eds), *The Ecology and Behavior of Food-Enhanced Primate Groups*. Alan R. Liss, New York, pp. 3-24.

WINKLER, P., Loch, H. and Vogel, C. (1984) Life history of Hanuman langurs (*Presbytis entellus*): reproductive parameters, infant mortality, and troop development. *Folia Primatologica* **43**, 1-23.

Appendix I. Definition of terms used In the following tables and appendices

Troop: Bisexual troops are labelled by local topographic names. The key for different abbreviations sometimes used in other publications after 1983 is as follows : Bijolai = B = B21 [before 1983 : B26]; Kailana I = KI = B19; Kailana II = KII = B20; Kaga A = Kaga South = B11; Kaga B = Kaga North = B12.

Period of male change: Replacements of resident males can be gradual processes (sometimes involving 45+ males from several bands and lasting several months with several males as "interim" residents) or rapid changes (sometimes involving a single male competitor who defeats the previous resident within a day; Sommer, 1988).

New resident male: In this paper, the term "new resident male" defines males who are in sole control of a troop for a minimum of one month.

Since 1977, adult males were labelled by figures (*M1, M2, M3* ....), which refer to the entire population and are not confined to a particular troop or band (with the exception of those young males whose natal troops and mothers are known individually). Further exceptions occur with respect to some early publications: *YAM1* for troop Bijolai (Mohnot 1971), *MA1, MA3, MB2, MB3* for troops Kaga A, Kaga B (Makwana 1979); *ExR* = Ex-resident.

Case / evaluation of evidence: Each relevant set of data is labelled by a "case" number. To minimize the problematic issues concerning the variable quality of the evidence for infanticide, emphasis is given on the respective nature of the data base :

> *** = *Infanticide witnessed* (i.e. actual attack and wounding observed; subsequent death confirmed, because carcass was seen);

> ** = *Infanticide likely* (i.e. attack and wounding seen, infant disappeared later, but the carcass could not be found);

> * = *Infanticide presumed* (i.e. only circumstantial evidence; usually, disappearance shortly after the previous resident was replaced by a male band or a new resident).

> + = *Infanticide attempted* (i.e. infant or a mother with a clinging infant was attacked or wounded, but could either escape or survived).

> # = *Abortion* (i.e. loss of conceptus in connection with male change).

Denotation of infant:

    – *M* = male.

    – *F* = female.

    – *inf. I* = "black coated infant", from birth to about 5 months of age until fur color change is completed.

    – *inf. II* = "white coated infant", after completion of fur color change to completion of weaning, i.e. about 5 – 15 months of age.

All members of the three main study troops were known individually (troops Bijolai and Kailana since 1977; troop Kailana I since 1978; troop Kailana II since 1983). The adult females of these troops were named by figures. Their respective infants were denominated by the number of the mother and a second one representing the rank within the birth order during the period of observation. E.g., *6.1* stands for the first known infant born to F6. Consequently, *6.3.1.* designates the first infant of the third child (in this case, daughter) of F6, hence a grandchild of that female. However, *6.1* does not necessarily imply that F6 was a primiparous female at that time. Individuals which delivered their first infants during the study period were : F6.1 in troop Bijolai; F11, F12, F13, F2.3, F3.2, F4.4, F6.3 in troop Kailana I; probably also F4 in troop Kailana II.

Sire : The *sire* of an infant is defined by calculating which male resided in a particular troop 200 days before birth (mean gestation length 200.3 days, SD 3.4 days, n = 31, range 189–203 days; Sommer *et al.*, 1992). *Parenthesis* indicate lack of observations around time of conception ("presumed" sire).

The possibility of males, other than the current harem holder, inseminating troop females seems to be almost negligible. Most troops are geographically isolated and dwell in open habitats with good visibility. Harem holders can easily discover male bands in the vicinity and do not tolerate their approach. Hence, during stable residencies, it is very difficult for extra-troop males to copulate with troop females. These conditions are particularly true for troops Bijolai, Kailana I and Kailana II, from which the bulk of data is drawn. Moreover, since 1977, copulations during the actual conception estrus were monitored in these troops in many cases, further supporting presumptions of paternity. It is extremely rare that females leave their troop during a part of the daytime and contact a neighboring resident (a preliminary estimate by this author runs to less than two

percent of all days of observations). Moreover, during an 18 months study of 3 different male bands, it was never observed that an extratroop male successfully invaded a harem troop's home range and copulated.

Between 1977–1988, the social situation in troops Bijolai, Kailana I and Kailana II was characterized by stable harem holder residencies during 95.4% of the time, whereas periods of social instability (i.e. multi-male stages, interim residencies < 1 month) confined to only 4.6% of the time (total n = 9121 days, Sommer and Rajpurohit, 1989).

For troops at Kaga A, Kaga B and Mandore, which were not subjected to long term monitoring of individually known animals, the conditions were slightly different, because the core area of these troops is not so well separated and the chance of extra-troop males copulating might be higher (also this is not clear).

In sum, under the particular conditions at Jodhpur, it is extremely difficult for extratroop males to "sneak" copulations. Therefore, this author is convinced that assumptions of paternity made by field observers have a high degree of certainty.

Date of birth (abortion): If *births* occurred during break days of observation, the estimated day / month / year is underlined. *Abortions* include only eye-witnessed cases (except case 65, which was inferred on the basis of cycle monitoring).

Born after male change: Includes all births which occurred or – in case of abortions – were due within 6.6 months after the male changed, i.e., all infants which were still sired by the previous resident.

Fate:

> *Attacked* = Aggressive chasing seen, but individual (or its mother with clinging infant) could escape; the attack was invariably executed by the new resident male, if not otherwise indicated;
>
> *Wounded* = Attack *and* wounding by the new resident male;
>
> *Found wounded* = Attack not seen, only wounded individual found;
>
> *Died* = Carcass found;
>
> *Survived unharmed* = Long-term monitoring supports the presump-

tion that the new resident did never attack (although such attacks might have occurred during temporary absence of observers);

Survived = Individual survived despite previous attacks or wounding.

Age: If the month of birth has been estimated, age figures are treated as approximations ( ).

Observer: Refers to the fate of infant; dates of birth and sires might have been recorded by other fieldworkers.

Mo = Surendra Mal Mohnot

Ma = Suresh Chandra Makwana

Lo = Hartmut Loch

Wi = Paul Winkler

So = Volker Sommer

Ag = Goyindasamy Agoramoorthy

Ra = Lal Singh Rajpurohit

Sr = Arun Srivastava

Source: Indicates the most detailed report, because several cases have been repeatedly published.

(*a*) Mohnot (1971), 187–194

(*b*) Mohnot (1974), 251,255

(*c*) Makwana (1979), 256–258

(*d*) Winkler (1981), 203f

(*e*) Vogel and Loch (1984), 248–250

(*f*) Sommer (1985), 460–483

(*g*) Sommer (1987), 167–180

(*h*) Agoramoorthy (1987), 55–59,115–123

(*j*) Agoramoorthy and Mohnot (1988), 282–291

(*k*) Agoramoorthy, Mohnot, Sommer and Srivastava (1988), 304f

(*l*) Srivastava (1989), 26

(*m*) unpublished data, provided by the observer(s)

Appendix II: Fate of infants present during resident male changes or born shortly after (for definitions and abbreviations see Appendix I)

Troop	Period of male change [key-number]	New resident male	Case/evaluation of evidence	Denotation of infant (pregnant female)	(Presumed) sire	Date of birth (abortion)	Born after male change	Fate	Age in months during death(I) disappearance(II) attack(III) or onset of male change(IV)	Observer (Source)
Bijolai	Jul-Aug69 [A]	YAM1 (infanticidal)	1***	M inf.I	(ExR)	May69		Wounded and died 24Jul69	~3.3 (I)	Mo (a)
			2***	M inf.I	(ExR)	May69		Wounded and died 28Jul69	~3.4 (I)	Mo (a)
			3*	F inf.I	(ExR)	May69		Disapp.29Jul69;blood on mother	~4.0 (II)	Mo (a)
			4***	F inf.I	(ExR)	Jun69		Wounded and died 03Aug69	~3.1 (I)	Mo (a)
			5*	? inf.I	YAM?	22Jan70	x	Disappeared 28Jan70	0.2 (II)	Mo (a)
Bijolai	Sep-Nov80 [B]	M22 (infanticidal?)	6*	M5.3	M1	24Dec79		Disappeared 14Sep80	8.7 (II)	Lo (e,m)
			7*	F9.3	M1	13Mar80		Disappeared 15Sep80	6.1 (II)	Lo (e,m)
			8+	F1.3	M1	11Feb80		Found wounded 27Oct80 Disappeared 1981	8.1 (III)	Lo (e,m)
			9*	M6.3	M1	17May80		Disappeared 14Nov80	5.9 (II)	Lo (e,m)
			10+	M7.3	M1	04May 80		Found wounded 22Nov80, survived	6.6 (III)	Lo (e,m)
			11+	F3.3	M1	16Dec79		Found wounded 13Dec80; Disappeared 1981	11.9 (III)	Lo (e,m)
			12	F2.4	M1	06Jul80		Survived unharmed	2.4 (IV)	Lo (e,m)
Bijolai	Jan83 [C]	M38 (infant icidal)	13*	M3.5	(M22)	05May82		Disappeared 17Jan83	8.4 (II)	Ag (h,j)
			14+	M6.1.2	M22	25Oct82		Attacked 4 times Jan83, survived	3.3 (IV)	Ag (j)
			15***	M5.4	M22	06Dec82		Wounded 23Jan83; wounded, died 03Feb83	1.9 (I)	Ag (j)
			16#	(F2)	M22	(24Jan83)		(Fetus ~137 days old); F2 attacked Died 30Jan83; 'wound infection'(?)	aborted (IV)	Ag (k)
			17	F7.4	M22	15Oct81			~15.5 (I)	Ag (h)
			18***	F6.5	M22	29Oct82		Wounded and died 09Feb83	3.4 (I)	Ag (j)
			19***	M1.5	M22	25Sep82		Wounded 10Feb83; died 11Feb83	4.6 (I)	Ag (j)
Kailana	Oct-Dec77 (troop	M2 (old)	20	M6.1	(M2)	15May77		All 6 lactating and 2 (i.e. possibly all) pregnant females remained with old resident M2;	unborn (IV)	Wi (d)
		M3	21	M4.2	M2	15Sep77			unborn (IV)	Wi (d)
			22	M1.1	M2	15Oct77			~0.8 (IV)	Wi (d)

Appendix II: continued

Troop	Period of male change [key-number]	New resident male	Case/evaluation of evidence	Denotation of infant (pregnant female)	(Presumed) sire	Date of birth (abortion)	Born after male change(tion)	Fate	Age in months during death(I) disappearance(II) attack(III) or onset of male change(IV)	Observer (Source)
	splitting) [D]	(new faction)	23	M2.1	M2	15Nov77		14 other females joined new faction with resident M3	~4.8 (IV)	Wi (d)
			24	M7.1	M2	15Dec77			unborn (IV)	Wi (d)
Kailana I	between Aug-Dec78 (exact date unknown) [E]	M4 (non-infant-icidal?)	25	F3.1	M8.1M2	15Jun78	x?	Survived	~2-6 (IV)	Lo (c,m)
			26		M2	01Dec78	x?	Survived	~1/unborn? (IV)	Lo (c,m)
			27-34	8 M inf.I	(M2?/M4?)	12Mar-15Jun79		Survived unharmed, but perhaps already sired by new resident if male change occurred late Aug77	all unborn? (IV)	Lo (c,m)
Kailana I	Oct-Nov81 [F]	M20 (non-infant-icidal)	35	M8.2	M4	22Sep80		Survived unharmed	13.8 (IV)	So (f,g)
			36	M11.2	M4	15Dec80		Survived unharmed	11.0 (IV)	So (f,g)
			37	F1.3	M3	20Dec80		Survived unharmed	10.8 (IV)	So (f,g)
			38	M7.4	M4	01Mar81		Survived unharmed	~7.5 (IV)	So (f,g)
			39	M7.5	(M4?)	13Mar82	x	Unharmed; died 22Mar82 (starvation)	0.3 (I)	So (f,g)
			40	M1.4	(M4?)	27Mar82	x	Survived (but see case 44)	unborn (IV)	So (f,g)
			41	F4.5	(M4?)	11Apr82	x	Survived (but see case 43,48)	unborn (IV)	So (f,g)
Kailana I	Jun82 [G]	M43 (infant-icidal)	42***	M11.3	M20	05Jun82		Wounded 09Jul82; died 10Jul82	1.2 (I)	So (f,g)
			43	43+	F4.5	See case 41		Attacked (?) 12Jul82; survived (but see case 48)	3.0 (III)	So (f,g)
			44**	M1.4	See case 40			Wounded 13Jul82; disapp.19Jul82	3.7 (II)	So (f,g)
			45	F12.1	M20	30Jul82	x	Survived unharmed (but see case 47)	unborn (IV)	So (f,g)
Kailana I	Sep 82-Jan83 [H]	M11 (infant-icidal)	46#	(F7)	M44	(21Oct82)	(x)	(Fetus 35 days old)		So (k)
			47*	F12.1	see case 60			Disapp. 14Nov82 (multi-male-stage)	3.2 (II)	So (f,g)
			48**	F4.5	see case 56			Attacked 06,11,22,25Nov82,12Jan83, found wounded; disapp. 13Jan83	9.1 (II)	So, Da (f,g)
			49***	M6.4	M20	11Oct82		Wounded 25Nov82; died 26Nov82	1.5 (I)	So (f,g)

Appendix II: continued

Troop	Period of male change [key-number]	New resident male	Infants sired by previous males: Case/evaluation of evidence	Denotation of infant (pregnant female)	(Presumed) sire	Date of birth (abortion)	Born after male change	Fate	Age in months during death(I) disappearance(II) attack(III) or onset of male change(IV)	Observer (Source)
			50#	(F4)	M43	(07Dec82) (x)		(Presumed abortion; fetus 108 days old); infant F4.5 attacked (case 48)		So (d)
			51***	F11.4	M43	10Mar83	x	Wounded 21Mar83; died 22Mar83	0.4 (I)	Da (f,g)
			52**	F1.5	M57/M51 or M11	08Apr83 17-25Jun83	x	Attacked 26Apr83; disapp.	2.6 (II)	Da (f,g)
Kailana I	Apr-Jun83 [I]	M46 (non-infanticidal)	53	F4.4.1	M46	02Jul83	x	Survived unharmed; disapp.23Oct84	unborn (IV)	Ra (m)
			54	F6.3.1	M11	07Aug83	x	Survived unharmed	unborn (IV)	Ra (m)
			55	M4.6	M11	23Aug83	x	Survived unharmed	unborn (IV)	Ra (m)
			56	F2.3.1	M11	09Sep83	x	Survived unharmed	unborn (IV)	Ra (m)
			57	M12.2	M11	23Oct83	x	Survived unharmed	unborn (IV)	Ra (m)
Kailana I	Dec85 [J]	M119 (infanticidal)	58#	(F7)	M46	(29Dec85)	(x)	(Fetus ~175 days old)	aborted (IV)	Sr, Ra (k,l)
			59+	M6.3.2	M46	20Jan85		Attacks Dec-Jan, wounded, survived	12.0 (III)	Ra, Sr (k,l)
			60+	F4.4.2	M46	08Feb85		Some attacks Dec-Jan, survived	10.5 (III)	Ra, Sr (k,l)
			61+	F3.2.1	M46	16Mar85		Attacked (?) 23Dec85, mother wounded; survived	9.3 (IV)	Ra, Sr (k,l)
		62**	M2.3.2		M46	05Feb86	x	Attacked 05Feb86, disapp. 11Feb86	0.2 (II)	Ra, Sr (k,l)
		63	M4.4.3		M46	24Mar86	x	Survived unharmed	unborn (IV)	Sr (k,l)
		64	F6.3.3		M46	10Apr86	x	Survived unharmed	unborn (IV)	Ra, Sr (k,l)
Kailana II [K]	Feb82 infant-	M10 (non-	65-71	2F inf.II 5M inf.II	(M3 or or M63)	May-Oct81		No infant attacked;	~4-~9 (IV)	So (f,g)
								1 M Inf.II died 09Apr82 (electrocution)	2.0 (IV)	So (f,g)
		72	M inf.I	(M63)	Dec81			2 M inf.II disappeared, joined probably male band	1.0 (IV)	So (f,g)
		icidal?)	73	F inf.I	(M63)	Jan82			unborn (IV)	So (f,g)
			74	F inf.I	(M63)	Apr82	x		unborn (IV)	So (f,g)
			75	M inf.I	(M63)	Apr82	x		unborn (IV)	So (f,g)
			76	F inf.I	(M63)	Jul82	x			So (f,g)

Appendix II: continued

			Infants sired by previous males							
Troop	Period of male change [key-number]	New resident male	Case/evaluation of evidence	Denotation of infant (pregnant female)	(Presumed) sire (abortion)	Date of birth change	Born after male	Fate	Age in months during death(I) disappearance(II) attack(III) or onset of male change(IV)	Observer (Source)
Kailana II	Jul-Aug83 [L]	M52 (infanticidal)	77***	M2.1	(M3)	Feb79		Wounded 01Feb83; died 09Feb83	~48.0 (I)	Ag (j)
			78+	M juv.I	(M63)	Feb81		Wounded 29Mar83; joined male band 30Mar83; disappeared 02Apr83	~25.0 (III)	Ag (j)
			79***	M6.1	M10	15Jan83		Wounded 23Aug83; died 27Aug83	7.4 (I)	Ag,Ra (j)
			80-87	4M inf.II	M10	Sep82-Mar83		All survived unharmed	~5-11 (IV)	Ag,Ra (h)
			88	M10.1(?)	M10	15Apr83		(but see case 93)	3.9 (IV)	Ag (h)
Kailana II	Sept-Oct83 [M]	M72 (infanticidal)	89***	M3.1	M10	15Feb83		Wounded by M71 (a male band member) 12Oct83; died 13Oct83	~8.0 (I)	Ag (j)
			90	1F juv.	?	?		Died 15Oct83 (reason unknown)	? (I)	Ag (h)
			91+	M9.1	M10	01Sep82		Wounded 20Nov83; (observed in male band 21-26Nov83; disappeared 27Nov83	14.0 (III)	Ag (j)
			92*	F inf.I	M10	Apr-Jun83		Disappeared 10Dec83	~6-8 (II)	Ag (h)
			93*	M10.1(?)	see case 88			Disappeared Jan84	9.4 (II)	Ag,Ra (h)
			94-98	3M inf.II	M10	Sep82-Mar83		All survived unharmed	7-13(IV)	g,Ra (m)
				2F inf.II	M10	Mar83				
Kailana II	Jan84 [N]	M70 (infanticidal)	99#	(F7)	M72	(04Feb84)	(x)	(Fetus 55 days old)	aborted (IV)	Ag (k)
			100** icidal	F5.1	M42	23Feb84	x	Found wounded 02Apr84; disappeared 03Apr84	1.4 (III)	Ag,Ra (j)
			101***	M12.2	M10/M72	04May84	x	Wounded and died 11May84	0.3 (I)	Ag, Ra (j)
			102**	M4.2	M52	18Mar84	x	Found wounded, disapp. 10Jun84	2.7 (III)	Ag, Ra (j)
			103**	M6.2	M72	11Jun84	x	Found wounded17Jun84;died 18Jun84	0.2 (III)	Ra (j)
			104	M9.2	M72	19Jun84	x	Survived; not attacked to at least 08Mar87	unborn (IV)	Ra (m)

Appendix II: continued

Troop	Period of male change [key-number]	New resident male	Infants sired by previous males					Fate	Age in months during death(I) disappearance(II) attack(II) or onset of male change(IV)	Observer (Source)
			Case/evaluation of evidence	Denotation of infant (pregnant female)	(Presumed) sire (abortion)	Date of birth change	Born after male			
Kaga South	Jan71 [O]	New Male (infanticidal?)	105-107*	3 inf.I	(ExR)	?		Disappeared within 5-12 days	< 6.0 (II)	Mo (b)
Kaga A (South)	Apr-May77 [P]	MA3 (infant-icidal)	108+	M inf.II	(MA1)	May76		Found wounded, attacked 4 times, disapp. 08May77(with male band 10May77)	~12.0 (III)	Ma (c)
			109-111*	3 inf.I	(MA1)	Apr-May77		Disappeared 08May77; one mother found slightly wounded	~1.0-2.0(II)	Ma (c)
			112+	M inf.I	(MA1)	Feb77		Found wounded 06Jun77, fate unknown	~4.0 (III)	Ma (c)
			113+	M inf.II	(MA1)	Jun76		Both found wounded 11Jun77(invasion of male band), fate unknown	~12.0 (III)	Ma (c)
			114+	M inf.I	(MA1)	Jan77			~5.0 (III)	Ma (c)
			115	(?)nf.I	(MA1)	May-Jul77	x	Survived unharmed	unborn (IV)	Ma (c)
Kaga B (North)	Apr-May77 [Q]	MB3 (non-infant-icidal)	116-119	4 inf.I (?)	(MB2)	Dec76-May77		All survived unharmed to at least 06May77	~1.0-6.0 (IV)	Ma (c)
			120-124	5 inf.I	(MB2)	May-Jul77	x	Survived unharmed	unborn (IV)	Ma (c)
Mandore (B7)	May70 [R]	New male (infanticidal?)	125-130*	6 inf.I	(ExR)	?		Disappeared within 5-12 days	< 6.0 (II)	Mo (b)
Old Residency	Nov69 [S]	New male (infanticidal?)	131* (unknown number)	? inf.I	(ExR)	?		Disappeared within 5-12 days	< 6.0 (II)	Mo (b)

Appendix II: Continued

Troop	Period of male change [key-number]	New resident male	Infants sired by previous males							Observer (Source)
			Case/evaluation of evidence	Denotation of infant (pregnant female)	(Presumed) sire	Date of birth (abortion)	Born after male change	Fate	Age in months during death(I) disappearance(II) attack(III) or onset of male change(IV)	
Mandore East	Jul68 [T]	New male (non-infanticidal?)	132 (number)	? inf. I (unknown	(ExR)	?		'No infant-killing observed'		Mo (b)
Mandore Fort	Aug68 [U]	New Male (non-infanticidal?)	133 number)	? inf. I (unknown	(ExR)	?		'No infant-killing observed'		Mo (b)
Chandpole Temple	Nov69 [V]	New male (non-infanticidal?)	134 number)	? inf. I (unknown	(ExR)	?		'No infant-killing observed'		Mo (b)
Kadamkandi	Nov-Dec67 [W]	New male (non-infanticidal?)	135 number)	? inf. I (unknown	(ExR)	?		'No infant-killing observed'		Mo (b)

Appendix III: Assumed genetic relationship between infanticidal males with (a) infanticide victims and (b) subsequent offspring born to infant-deprived females. For details of case histories see Appendix II.

	Infanticide victim/abortion		Subsequent estrus		Subsequent conception		Remarks
Case evaluation of evidence	Sire: male was probably present (I), observed to be present (II), solicited (III), or copulated (IV) around time of conception	(Presumed) attacking/ infanticidal male	Period between infant loss and onset of estrus	Males being present (II), solicited (III) or copulating (IV)	Sire: male was probably present (I), observed to be present (II), solicited (III), or copulated (IV) around time of conception	Interbirth interval	
INFANTICIDE WITNESSED, LIKELY OR PRESUMED							
1***	Ex-resident (I)	YAM1	4 days	YAM1, invaders (IV)		>~13 months	
2***	Ex-resident (I)	YAM1	8 days	YAM1(II),invaders (IV)		>~13 months	
3*	Ex-resident (I)	(YAM1)	10 days	YAM1(II),invaders (IV)		>~14 months	
4***	Ex-resident (I)	YAM1	1 day	YAM1, invaders (IV)		>~13 months	
5*	YAM1(IV), invaders (II)	(YAM1)	?			?	
6*	M1 (II)	(M22/invaders)	?	M22 (II)		?	Birth interval 35.4 months, but one birth possibly missed
7*	M1 (II)	(M22/invaders)	?			–	Very old female; became probably menopausal; no more infants until death 13Feb83
9*	M1 (II)	(M22/invaders)	?		M22 (II)	~16.0 months	
13*	M 22 (II)	(M38/invaders)	–			–	Mother died same day when infant lost
15***	M22 (II)	M38		M38 (IV)	M38 (II)	13.8 months	Mother disapp. 8 days after infant loss
18***	M22 (II)	M38	6 days	M38 (IV)	M38 (II)	12.4 months	
19***	M22 (II)	M38	8 days	M43 (IV)	M43 (IV)	9.2 months	
42***	M20 (IV)	M43	12 days	M43 (IV)	M57, M51, M11 (IV)	12.3 months	
44**	M4? (I)	M43	(16 days)	M42, M38, M48,M37,M47 M49 (III), M46 (IV), M11(II)	M11 (II)	14.8 months	Mother resumed cycling before infant loss
47*	M20 (IV)	(M11/invaders)	(13 days)	M11 (II)	M11 (II)	16.4 months	
48**	M4? (I)	M11	(10 days)	M37,M48 (IV) M11 (II)		–	Mother resumed cycling before infant loss
49**	M20 (IV)	M11	34 days	M11 (III)			Infanticidal male copulated 72 days after infant loss; mother menopausal afterwards
51***	M43 (IV)	M11	17 days	?	M119(IV)	51.7 months	Old female, stopped cycling intermittently
52**	M57,M51,M11(IV)	M11	> 53 days	M119 (III)	M119 (IV)	48.5 months	Female disapp. during month 5.1 of pregnancy; interval thus theoretical
62**	M46 (IV)	M119	10 days		M119 (IV)	(9.2 months)	

Appendix III: continued.

	Infanticide victim/abortion			Subsequent estrus		Subsequent conception		Remarks
Case evaluation of evidence	Sire: male was probably present (I), observed to be present (II), solicited (III), or copulated (IV) around time of conception	(Presumed) attacking/infanticidal male		Period between infant loss and onset of estrus	Males being present (II), solicited (III) or copulating (IV)	Sire: male was probably present (I), observed to be present (II), solicited (III), or copulated (IV) around time of conception	Interbirth interval	
INFANTICIDE WITNESSED, LIKELY OR PRESUMED								
77*	M3 (I)	M10		/		/		No lactational amenorrhea of mother
79***	M10 (II)	M52		?		M72/invaders (II)	16.9 months	During conception 7 males including M52 present; M72 interim resident
89***	M10 (II)	M71		?			?	No further details
92*	M10 (II)	(M72)		?			?	
93*	M10 (II)	(M72)		?			?	
100**	M42 (II)	(M70)		10 days		M70 (II)	16.6 months	
101***	M10/M72 (II)	M70		9 days	M70 (IV)	M70 (II)	8.6 months	M70 was not present when victim was sired
102**	M52 (II)	(M70)		7 days	M70 (IV)	M70 (II)	7.6 months	
103**	N72 (II)	(M70)		15 days	M70 (IV)	M70 (II)	9.8 months	
							15.5 months	
105*	Ex-resident (I)	(New male)		?			?	No further details
106*	Ex-resident (I)	(New male)		?			?	No further details
107*	Ex-resident (I)	(New male)		?			?	No further details
109*	MA1 (I)	(MA3)		? (see remarks)			?	"Females" copulated with MA3, MB3 and invaders
110*	MA1 (I)	(MA3)		? (see remarks)			?	"Females" copulated with MA3, MB3 and invaders
111*	MA1 (I)	(MA3)		? (see remarks)			?	"Females" copulated with MA3, MB3 and invaders
125*	Ex-resident (I)	(New male)		?			?	No details
126*	Ex-resident (I)	(New male)		?			?	No details
127*	Ex-resident (I)	(New male)		?			?	No details
128*	Ex-resident (I)	(New male)		?			?	No details
129*	Ex-resident (I)	(New male)		?			?	No details
130*	Ex-resident (I)	(New male)		?			?	No details

Appendix III: continued.

	Infanticide victim/abortion			Subsequent estrus		Subsequent conception		Remarks
Case evaluation of evidence	Sire: male was probably present (I), observed to be present (II), solicited (III), or copulated (IV) around time of conception	(Presumed) attacking/ infanticidal male		Period between abortion and onset of estrus	Males being present (II), solicited (III) or copulating (IV)	Sire: male was probably present (I), observed to be present (II), solicited (III), or copulated (IV) around time of conception	Interval after abortion	

ABORTIONS IN CONNECTION WITH MALE CHANGES

16#	M22 (II)	M38	34 days	M38 (III)	M38 (II)	12.0 months	Female attacked by M38
46#	M44 (IV)	(M11/remarks)	17 days	M46 (III) M11(II)	M46 (II)	21.1 months	M11 infanticidal prior to and after abortion
50#	M20 (III)	M11	21 days	M46 (IV)	M11 (II)	8.5 months	Female's semi-weaned infant attacked
58#	M46 (II)	(M119/remarks)	10 days	M119 (III)	M119 (III)	13.9 months	M119 attacked other female and infants
99#	M72 (II)	(M70/remarks)	8 days	M70 (IV)	M70 (IV)	7.1 months	M70 infanticidal prior to and after abortion

INFANTICIDE ATTEMPTED

8+	M1 (II)	(M22/invaders)
10+	M1 (II)	(M22/invaders)
11+	M1 (II)	(M22)
14+	M22 (II)	M38
43+	M4? (I)	M43
59+	M46 (II)	M119
60+	M46 (II)	M119
61+	M46 (II)	M119
78+	M63 (II)	M10
91+	M10 (II)	M72
108+	MA1 (I)	MA3
112+	MA1 (I)	MA3
113+	MA1 (I)	(MA3/invaders)
114+	MA1 (I)	(MA3/invaders)

Appendix IV: Summary of the assumed genetic relations between infanticidal males, victims and subsequent infants born to infant-deprived females. For analyses of case histories see Appendix III.

Strong cases = observations during estrus/period of conception;
weak cases = no observations; assumed, that established resident was present.

During period of victim's conception	Mother of attacked, killed or aborted infant associated/copulated:						Subtotal
	with male other than attacking/infanticical male		with attacking/infanticidal male		with multiple males including attacking/infanticidal male		
	Strong cases	Weak cases	Strong cases	Weak cases	Strong cases	Weak cases	
(a) Later, infanticide was witnessed (***), likely (**) or presumed (*)	(6*,7*,9*, 13*,15***, 18***,19***, 42***,47*, 49***,51***, 62**,79***, 89***,92*, 93*,100**, 102**,103***)	(1***,2***, 3*,4***,44**, 48**,77***, 105*,106*, 107*,109*, 110*,111*, 125*,165*, 127*,128*, 129*,130*)			(5*,52**)	(101***)	
	N = 19	N = 19			N = 2	N = 1	
(b) Later, abortion (#) occurred	(16#, 46#, 50, 58, 99)						
	N = 5						
(c) Later, infanticide was attempted (+)	(8+, 10+, 11+, 14+, 59+, 60+, 61+, 78+, 91+)	(43+, 108+, 112+, 113+, 114+)					
	N = 9	N = 5					N = 60

Appendix IV: continued.

Strong cases = observations during estrus/period of conception;
weak cases = no observations; assumed, that established resident was present.

	Mother of attacked, killed or aborted infant associated/copulated:						
	with male other than attacking/infanticical male		with attacking/infanticidal male		with multiple males including attacking/infanticidal male		Subtotal
	Strong cases	Weak cases	Strong cases	Weak cases	Strong cases	Weak cases	
During first estrus after infant fetus loss							
(a) Prior, infanticide was witnessed (***), likely (**) or presumed (*)			(18***, 19***, 42***, 44**, 48**, 51***, 62**, 100**, 101**, 102**, 103**) N = 11	(6*) N = 1	(1***, 2***, 3*, 4***, 47*, 49***) N = 6	(109*, 110*, 111*) N = 3	N = 25
(b) Prior, abortion (#) had occurred	(50#) N = 1		(16#, 58# 99#) N = 3		(46#) N = 1	N = 3	
During period of conception of subsequent infant							
(a) Prior, infanticide was witnessed (***), likely (**) or presumed (*)	(44**, 51*** 52**, 93*) N = 4		(9*, 18***, 19***, 42***, 48**, 62**, 100**, 101***, 102**, 103***) N = 10	(47*) N = 1	(79***) N = 1		N = 21
(b) Prior, abortion (#) had occurred	(46#) N = 1		(16#, 50#, 58# 99*) N = 4				

CHAPTER 8

MECHANISMS OF PRIMATE INFANT ABUSE: THE MATERNAL ANXIETY HYPOTHESIS

ALFONSO TROISI[1] and FRANCESCA R. D'AMATO[2]

[1]*Clinica Psichiatrica della II Università di Roma, Roma, Italy*

[2]*Istituto di Psicobiologia e Psicofarmacologia del C.N.R., Roma, Italy*

INTRODUCTION

One might expect that maternal behavior, being a crucial component of reproductive effort leading to the maximization of inclusive fitness, would be displayed by every primate mother in such a way as to guarantee that every infant will receive appropriate care. In reality, both laboratory and field studies have reported that primate mothers can sometimes exhibit behaviors that are harmful to their infants and that may result in the infant's death. Among these behaviors, physical maltreatment of the infant is doubtless one of the most dramatic manifestations.

The primate literature specifically devoted to infant abuse is quite limited: most of the studies have been conducted in highly artificial settings and the few reports of cases occurred in natural or seminatural environments rarely include quantitative behavioral data. As a consequence, we still have a limited knowledge of the specific processes and mechanisms that underlie primate infant abuse.

The aim of this paper is to present a new hypothesis concerning the pathogenesis of primate infant abuse. Based on the results from recent quantitative studies of infant abuse in macaque social groups, we will argue that pathological maternal anxiety may be the emotional mechanism mediating primate infant abuse.

We would emphasize that, in the primate literature, the term infant abuse encompasses several different forms of maltreatment of offspring which involve different perpetrators (the mother, adult females, adult males, juveniles) and various behaviors (physical abuse, neglect, abandonment, kidnapping, infanticide). We believe that these phenomena are very different in terms of both proximate and ultimate causation and that, therefore, efforts to account for such diverse behaviors in terms of a single explanatory paradigm are likely doomed to failure. Only physical abuse perpetrated by the mother will be examined in this chapter. Thus, the conclusions should not be extended to all the other forms of primate infant abuse.

ETIOLOGY AND PATHOGENESIS OF PRIMATE INFANT ABUSE

We have only a partial knowledge of the causal factors (etiology) that induce some primate mothers to abuse their infants. Among the various risk factors that have been implicated in the etiology of primate infant abuse, mother's abnormal early experience, especially maternal loss early in life, seems to play a crucial role in predisposing a female to display later abusive mothering.

The general notion that early life experiences can adversely influence later behavior has received considerable support in the nonhuman primate literature (Mineka and Suomi, 1978). In particular, since the pioneering work of Harlow, many studies have demonstrated that a social separation during a macaque's infancy can result in later deficiencies in maternal behavior (Ruppenthal et al., 1976). Distinguishing between the various pathogenic components of social separation, Suomi (1978) has argued that the absence of the mother early in life is the major factor inducing later abusive mothering. In fact, other kinds of early social experience (i.e. with peers and/or adults other than the mother) apparently do not serve to prevent infant abuse by females raised in the absence of their own mothers. More information about this effect comes from field studies of wild Japanese macaques showing that orphaned primiparous mothers, while displaying no apparent abnormality in their general social behavior, typically maltreat their own infants, who are likely to die (Hasegawa and Hiraiwa, 1980; Hiraiwa, 1981). This in spite of the fact that these females can benefit from the continuing and multiple social experiences that come to all monkeys raised in a natural group.

A major variable when considering potential long-term consequences is not the severity of early separation experience per se, but rather the individ-

ual's response to, and probably perception of, the experience. Rhesus females who encountered frequent social separations during childhood and reacted to them with depression are at subsequent high risk for maternal dysfunction: more than 80% of them neglect or abuse their first offspring. In contrast, low-risk females who encountered the same number of social separations, but who did not react to any of these separations with depression, subsequently turn out to be normally nurturant mothers (Suomi, 1984).

Even less is known regarding the mechanisms (pathogenesis) mediating maternal abusive behavior. What factors in the adult mother mediate the effects of her early deprivation and make it a cause of her present behavior? Different hypotheses have been advanced to explain the pathogenesis of primate infant abuse. The learning defect hypothesis asserts that abusive mothering would be caused by a failure of the early environment to provide experiences critical for basic-perceptual motor development of infant-rearing skills. A different view is that the deficiencies involve not so much specific responses as more general motivational abnormalities, including indifference and/or hyperaggressiveness toward the infant (Harlow and Harlow, 1965). Finally, we have proposed that an emotional disorder, namely abnormal maternal anxiety, is the mechanism causing some primate mothers to physically abuse their infants (Troisi and D'Amato, 1984). Evidence supporting this pathogenetic hypothesis is reviewed below.

THE BEHAVIOR OF ABUSIVE MOTHERS

Classical laboratory studies in which maternal abusive behavior was experimentally induced by atypical rearing procedures have reported that physical abuse of the infant is framed in a broader set of negative maternal responses ranging from avoidance through physical rejection to punishment. Typically, however, the manipulations employed in these studies have been undoubtedly extreme (*e.g.* total or partial social isolation), and the resulting deficiencies pervasive (*e.g.* hyperaggression and general social incompetence). Studies of inadequate mothering occurring under more ecologically valid conditions have drawn a different picture of the quality of mothering in abusive monkey mothers.

Primatologists who reported the natural (*i.e.* nonexperimentally induced) occurrence of infant abuse in confined groups of monkeys pointed out the fact that abusive mothers cannot be usually distinguished from normal animals (Caine and Reite 1983; Schapiro and Mitchell 1983). This finding suggests that, when not engaged in active maltreatment, these

mothers are not so rejecting or indifferent toward their infants as to draw the observers' attention. In this regard, they greatly differ from those monkey mothers who abuse their infants because of severe experimental manipulations suffered during childhood. Still more informative is the finding that, among free-ranging Japanese macaques, some orphaned primiparous mothers typically maltreat their own infants by alternating physical abuse with extremely protective behavior:

> this infant was never observed to ride on its mother's back, nor to move over 5 m from the mother, nor to play with other infants. This was because HG [an orphaned primiparous abusive mother] always held the infant tightly in her arms (Hiraiwa, 1981, p. 320).

In our studies of infant abuse in a colony of confined Japanese macaques, we measured the maternal behavior of abusive mothers (Troisi and D' Amato, 1984; Troisi *et al.*, 1989). The abusive mothers in our studies scored higher on maternal warmth, protectiveness, possessiveness, and scored lower on maternal rejection than did control mothers (Figure 1). Not only these abusive mothers alternated violent abuse with attentive maternal care but, unlike normal mothers, they also had extremely possessive relationships with their infants. Interestingly, some of the abusive rhesus mothers in Harlow's studies, after they stopped abusing their infants, became extremely protective of them (*e.g.* Seay *et al.*, 1964).

The finding that abusive mothers alternate violent abuse with attentive maternal care is not consistent with the learning defect explanation of primate infant abuse. Clearly, these mothers possess the complete repertoire of maternal abilities and display it appropriately for the most part of the time spent in interacting with their infants. The pathogenetic hypothesis of motivational abnormalities, such as maternal indifference and hostility toward the infant, is not compatible with the observation that physical abuse stands out against a background of positive maternal responses, including an extremely low frequency of rejection. Rather, the extremely possessive style of mothering typical of the abusive mothers suggests that these females suffer from an emotional disorder consisting of abnormal levels of maternal anxiety. Consistent with this interpretation is not only the common assumption that, in monkeys, a possessive style of mothering reflects high levels of maternal anxiety (*e.g.* Mitchell and Stevens, 1969; Hooley, 1983) but also the recent finding that maternal possessiveness correlates with an independent behavioral measure of anxiety (Troisi *et al.*, 1991).

In a systematic study of the morphology of monkey maternal abusive behavior, we compared abusive behavior patterns with patterns allocated to other categories of the behavioral repertoire of macaques (Troisi and

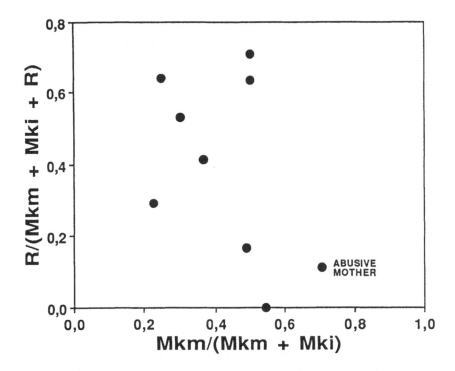

Figure 1 Assessment of rejection/possessiveness in an abusive mother-infant pair and in 8 control pairs based on data from the 11th, 12th, and 13th weeks of infants' lives. Ordinate: proportion of all attempted ventroventral contacts, including those initiated by the mother, that were prevented by the mother (maternal rejection). Abscissa: proportion of ventro-ventral contacts initiated by the mother (maternal possessiveness). Pairs whose points lie high on and close to the ordinate have a relationship involving a rejecting mother, those lying along and close to the abscissa have a relationship involving a possessive mother. The abusive mother proved to be the most possessive in the sample. (Adapted from Troisi *et al.* 1989).

D'Amato, 1983). Two major findings emerged from this morphological analysis. The first finding was that most abusive behavior patterns seem not to be aggressive acts. In fact, neither adult monkeys engaging in agonistic encounters, nor normal mothers punishing their infants, display behavior patterns such as, for example, dragging or sitting on the infant. Among all the abusive behavior patterns, only biting and striking belong to the macaque aggressive repertoire. However, the relative frequency of occurrence of these two patterns during abuse seems to vary greatly from case to case, and, even in the most severe cases of abuse, biting and striking can

be completely absent (Troisi and D'Amato, 1984). The second finding was that motor patterns identical to the abusive behavior patterns are usually performed by monkeys manipulating inanimate objects. Sometimes, adult monkeys also have been observed to direct these patterns to infants in contexts other than abuse: for example, during social play (Dolhinow, 1971), agonistic buffering (Deag and Crook, 1971) or play-mothering (Spencer-Booth, 1968). Obviously, these patterns are potentially harmful to the infant and impress the observer with their roughness. Yet, in those contexts, if the infant experiences pain and thus squeals, the actor promptly corrects his/her behavior or is compelled to do so by other group members (*e.g.* the infant's mother). This is not the case for abuse where frequently the infant's distress signals increase maternal maltreatment (Troisi *et al.*, 1982).

These findings concerning the morphology of maternal abusive behavior have implications for the understanding of the pathogenesis of primate infant abuse. The fact that most of the abusive behavior patterns do not belong to the aggressive repertoire questions the view that abuse is merely a form of dysfunctional maternal aggression toward the infant. Confirming this, several authors have reported that, while maltreating their infants, abusive mothers generally display no signs of hostility as measured by the usual indices of threat, vocalization, and piloerection (Ruppenthal *et al.*, 1976; Dienske *et al.*, 1980; Troisi *et al.*, 1982). The finding that most abusive patterns consist in treating the infant as an inanimate object does not accord with the learning defect explanation of infant abuse either. In the cases of macaque infant abuse that we observed, the abusive mothers alternated patterns of the kind inanimate object manipulation with fully adequate maternal responses (Troisi and D'Amato, 1984; Troisi *et al.*, 1989). Therefore, the lack of infant-rearing skills is not the pathogenetic factor causing abusive mothers to display these inappropriate behavior patterns. Our interpretation is that a latent condition of abnormal emotionality temporarily disorganizes those complex motor patterns that allow normal mothers to cope with anxiety-triggering situations involving their infants. Under these circumstances, abusive mothers would therefore react in a clumsy and excited way by manipulating their infants as inanimate objects.

ELICITATION OF ABUSE

In their studies of infant abuse by isolation-reared rhesus females, Harlow and his associates repeatedly reported the infant's attempts to establish physical contact with the mother as the major condition triggering abuse (*e.g.* Harlow *et al.*, 1966). This observation may convey the idea that

abusive monkey mothers are consistently indifferent and hostile toward their infants. However, even in the laboratory studies, there are some observations regarding the situations that precipitate abuse that suggest a different view of the abusive mothers' emotional attitude toward their infants. Consider, for example, the following reports by Seay *et al.* (1964, p. 347):

> Attempts to remove and hand feed the baby were abandoned since such efforts provoked violent attacks [by the mother] directed against her infant.

> On initial approaches by other infants, MM59 violently attacked her own baby, pulling it to the floor and biting it.

A mother who is indifferent or hostile toward her infant should not react violently to situations, as those above, that drive the infant away from her. However, such a reaction becomes understandable if we hypothesize that the mother is intolerant of any separation from her infant.

We observed monkey infant abuse in a large social group, an environment much more complex and varied than that typical of the laboratory studies (Troisi *et al.*, 1982). Most of the episodes of abuse were preceded by one of the following situations: the infant's behavior of emancipation from the mother, difficulties in mother-infant communication, or emotional disturbance (*i.e.* screaming, convulsive jerk) on the part of the infant. Frequently, when the infant succeeded in leaving her, the mother ran after and abused it. Invariably, when the infant was engaged in a play interaction with a peer, the mother reacted violently by threatening the newcomer and abusing her infant. Also the infant's failure to respond to the mother's retrieval signals precipitated abuse. To the emotional disturbance on the part of the infant, the mother sometimes replied adequately by performing comfort behaviors; however, in most cases, infant's distress signals elicited abuse. What these different situations have in common is the capacity to elicit anxiety in a monkey mother.

PHARMACOLOGICAL TREATMENT OF ABUSE

To test the hypothesis that maternal anxiety is implicated in the pathogenesis of abuse, we treated an abusive monkey mother with diazepam, an anxiolytic drug (Troisi and D'Amato, 1991). Diazepam treatment during the first 8 weeks postpartum was associated with dramatic changes, including elimination of abuse, in the maternal behavior of this macaque mother. More in detail, diazepam treatment resulted in: 1) an elimination of abuse;

2) a decrease in behaviors reflective of maternal anxiety; and 3) an increase in aggressive maternal responses toward the infant that are characteristic of normal mothers (Figure 2). The results of this study give further support to the view that maternal anxiety, an important manifestation of which is possessive mothering, is the pathogenetic factor causing some primate mothers to abuse their infants. In addition, the finding that the abusive mother showed higher levels of maternal rejection when she was receiving the highest dose of diazepam is consistent with our interpretation, derived from the morphological analysis of abusive behavior (Troisi and D'Amato, 1983), that infant abuse is a distinct behavior pattern due to an abnormal emotional condition rather than an exaggeration of normal maternal aggression.

THE ORIGIN OF PATHOLOGICAL MATERNAL ANXIETY

The data reviewed above suggest that maternal anxiety is implicated in the pathogenesis of maternal abuse of offspring in monkeys. Therefore, the question arises as to the origin of the abnormal maternal anxiety of abusive monkey mothers. Based on the data showing the causal link between early separation and later abusive behavior, we think that these females suffer from the emotional disorder that Bowlby (1973) has called anxious attachment. An individual who has experienced the unpleasant experience of separation from an attachment object is not likely to forget it and may well develop a fear (separation anxiety) that at some time in the future separation will again occur. This results in anxious attachment, a syndrome existing when the individual has no confidence that his or her attachment object will remain available and accessible and hence tends to maintain close proximity in order to ensure continued access. In effect, the abusive monkey mothers of our own studies, and also those observed by other investigators, seemed to be in constant anxiety lest they lose their infants, as indicated by their possessive style of mothering and their violent reactions to minor separations from their infants.

An important task of future research will be to identify the critical variables that determine whether early separation will create a predisposition to maternal abusive behavior later in life. Some recent work by Suomi has begun to delineate which factors are critical in creating this predisposition: in rhesus females, there is a relationship between the tendency to react with anxiety and depression to stressful experiences early in life and later inadequate mothering. These individual differences in vulnerability and reactivity to stressors seem to carry a strong genetic component (Suomi, 1984).

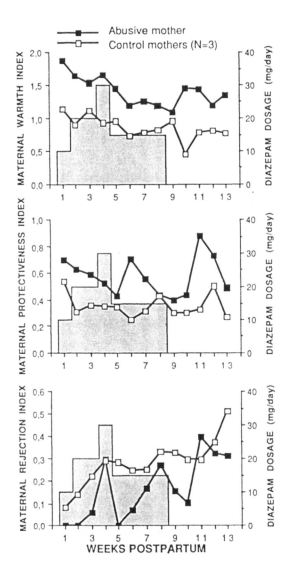

Figure 2 Time course of maternal warmth (maternal role in making ventroventral contact plus proportion of ventroventral time spent embracing the infant plus proportion of ventroventral time spent grooming the infant), maternal protectiveness (maternal initiation of ventroventral contact plus maternal restraining plus maternal embracing), and maternal rejection (proportion of all attempted ventroventral contacts, including those initiated by the mother, that were prevented by the mother) for an abusive mother (during and after diazepam treatment) and for control mothers during the first 13 weeks postpartum. (Adapted from Troisi and D'Amato, 1991).

IMPLICATIONS FOR AN EVOLUTIONARY VIEW OF INFANT ABUSE

The focus of this paper is on the mechanisms of primate infant abuse, that is one aspect of its proximate causation. However, the maternal anxiety hypothesis has relevant implications for an evolutionary understanding of abuse as well. Recently, evolutionary biologists have provided abundant evidence that, under particular circumstances such as scarcity of resources or low reproductive value of offspring, maternal investment in an individual offspring may be reduced or terminated if the mother thereby enhances overall reproductive prospects (Alexander, 1974; Bernds and Barash, 1979). Based on this new conceptual framework, different behaviors that have in common maternal maltreatment of offspring, such as abandonment, neglect, physical abuse, and infanticide, have been lumped and analyzed from a sociobiological perspective (*e.g.* Lenington, 1981; Lightcap *et al.*, 1982; Daly and Wilson, 1987). Underlying the lack of distinction between these behaviors is probably the idea that the adaptive mechanism mediating the discriminative allocation of parental effort is to be sought at the psychological rather behavioral level (Daly and Wilson, 1987). Independently from the actual behavior, the psychological state of a mother confronting unfavorable circumstances should be the opposite of that promoting solicitous maternal care and should therefore consist of indifference, intolerance, or overt hostility toward that particular child. The data reviewed in the present study indicate that, in nonhuman primates, physical abuse can be combined with solicitous maternal care. In these cases, the psychological state of the mother seems to be anxiety, not indifference or hostility. Consistent with this difference in the psychological correlates of physical abuse and other forms of defective maternal behavior are the consequences of these various behaviors in terms of maternal investment. Whereas abandonment, neglect, and infanticide all involve an effective reduction or termination of maternal investment, physical abuse can be in fact associated with an increase in maternal care.

These considerations suggest that the sociobiological construct of discriminative parental solicitude (Daly and Wilson, 1980) is not equally applicable to different behaviors such as abandonment, neglect, infanticide, and physical abuse. An evolutionary understanding of these behaviors requires differential evaluation of both their psychological correlates and behavioral consequences.

REFERENCES

ALEXANDER, R.D. (1974) The evolution of social behavior. *Annual Review of Ecology and Systematics*, **5**, 325–383.

BERNDS, W.P. and Barash, D.P. (1979) Early termination of parental investment in mammals, including humans. In: Chagnon, N.A. and Irons, W. (eds), *Evolutionary Biology and Human Social Behavior*, North Scituate, Massachusetts, Duxbury. pp. 487–506.

BOWLBY, J. (1973) *Separation: Anxiety and Anger*. New York: Basic Books.

CAINE, N. and Reite, M. (1983) Infant abuse in captive pig-tailed macaques: Relevance to human child abuse. In: Reite, M. and Caine, N. (eds), *Child Abuse: The Nonhuman Primate Data*, New York: Alan R. Liss. pp. 19–27.

DALY, M. and Wilson, M. (1980) Discriminative parental solicitude: a biological perspective. *Journal of Marriage and the Family*, **42**, 277–288.

DALY, M. and Wilson, M. (1987) Children as homicide victims. In: Gelles, R.J. and Lancaster, J.B. (eds), *Child Abuse and Neglect: Biosocial Dimensions*, New York, Aldine de Gruyter. pp. 201–214.

DEAG, J.M. and Crook, J.H. (1971) Social behaviour and agonistic buffering in the wild Barbary macaque, *Macaca sylvana, L. Folia Primatologica*, **15**, 183–200.

DIENSKE, H., van Vreeswijk, W. and Koning, H. (1980) Adequate mothering by partially isolated rhesus monkeys after observation of maternal care. *Journal of Abnormal Psychology*, **89**, 489–492.

DOLHINOW, P. (1971) At play in the fields. *Natural History*, **80**, 66–71.

HARLOW, H.F. and Harlow, M.K. (1965) The affectional systems. In: Schrier, A.M., Harlow, H.F. and Stollnitz, F. (eds.), *Behavior of Nonhuman Primates*, vol. 2. New York: Academic Press.

HARLOW, H.F., Harlow, M.K., Dodsworth, R.O. and Arling, G.L. (1966) Maternal behavior of rhesus monkeys deprived of mothering and peer associations in infancy. *Proceedings of the American Philosophical Society*, **110**, 58–66.

HASEGAWA, T. and Hiraiwa, M. (1980) Social interactions of orphans observed in a free ranging troop of Japanese monkeys. *Folia Primatologica*, **33**, 129–158.

HIRAIWA, M. (1981) Maternal and alloparental care in a troop of free-ranging Japanese monkeys. *Primates*, **22**, 309–329.

HOOLEY, J. M. (1983) Primiparous and multiparous mothers and their infants. In Hinde, R. A. (ed.), *Primate Social Relationships*. Oxford: Blackwell, pp. 142–145.

LENINGTON, S. (1981) Child abuse: The limits of sociobiology. *Ethology and Sociobiology*, **2**, 17–29.

LIGHTCAP, J.L., Kurland, J.A., Burgess, R.L. (1982) Child abuse: A test of some predictions from evolutionary theory. *Ethology and Sociobiology*, **3**, 61–67.

MINEKA, S. and Suomi, S.J. (1978) Social separation in monkeys. *Psychological Bulletin*, **85**, 1376–1400.

MITCHELL, G. and Stevens, C.W. (1969) Primiparous and multiparous monkey mothers in a mildly stressful social situation: first three months. *Developmental Psychobiology*, **1**, 280–286.

RUPPENTHAL, G.C., Arling, G.L., Harlow, H.F., Sackett, G.P. and Suomi, S.J. (1976) A 10-year perspective of motherless-mother monkey behavior. *Journal of Abnormal Psychology*, **85**, 341–349.

SCHAPIRO, S.J. and Mitchell, G. (1983) Infant-directed abuse in a seminatural environment: precipitating factors. In: Reite, M. and Caine, N. (eds), *Child Abuse: The Nonhuman Primate Data*, New York: Alan R. Liss, pp. 29–48.

SEAY, B.M., Alexander, B.K. and Harlow, H.F. (1964) Maternal behavior of socially deprived rhesus monkeys. *Journal of Abnormal and Social Psychology*, **69**, 345–354.

SPENCER-BOOTH, Y. (1968) The behaviour of group companions towards rhesus monkey infants. *Animal Behaviour*, **16**, 541–557.

SUOMI, S.J. (1978) Maternal behavior by socially incompetent monkeys: neglect and abuse of offspring. *Journal of Pediatric Psychology*, **3**, 28–34.

SUOMI, S.J. (1984) Individual differences in separation anxiety and depression in rhesus monkeys: biological correlates. *Clinical Neuropharmacology*, **7** (Suppl. 1), 754–755.

TROISI, A., Aureli, F., Piovesan, P. and D'Amato, F.R. (1989) Severity of early separation and later abusive mothering in monkeys: what is the pathogenic threshold? *Journal of Child Psychology and Psychiatry*, **30**, 277–284.

TROISI, A. and D'Amato, F.R. (1983) Is monkey maternal abuse of offspring aggressive behavior? *Aggressive Behavior*, **9**, 167–173.

TROISI, A. and D'Amato, F.R. (1984) Ambivalence in monkey mothering: infant abuse combined with maternal possessiveness. *Journal of Nervous and Mental Disease*, **172**, 105–108.

TROISI, A. and D'Amato, F.R. (1991) Anxiety in the pathogenesis of primate infant abuse: a pharmacological study. *Psychopharmacology*, **103**, 571–572.

TROISI, A., D'Amato, F.R., Fuccillo, R. and Scucchi, S. (1982) Infant abuse by a wild-born group-living Japanese macaque mother. *Journal of Abnormal Psychology*, **91**, 451–456.

TROISI, A., Schino, G., D'Antoni, M., Pandolfi, N., Aureli, F. and D'Amato, F.R. (1991) Scratching as a behavioral index of anxiety in macaque mothers. *Behavioral and Neural Biology*, 56: 307–313.

PART 3: FIELD STUDIES OF INFANTICIDE AND PARENTAL CARE IN INSECTS, BIRDS AND MAMMALS

CHAPTER 9

OOPHAGY AND INFANTICIDE IN COLONIES OF SOCIAL WASPS

S. TURILLAZZI and R. CERVO

Dipartimento di Biologia Animale e Genetica,
Università di Firenze, via Romana 17,
50125, Firenze, Italy

INTRODUCTION

It has long been known that social insects often kill and cannibalize immature individuals in their colonies. Although the first descriptions of such phenomena appeared in the eighteenth century, apart from a short review in Wilson's book (Wilson, 1971), the subject does not appear to have been reviewed in depth. With the increasing importance of infanticide (here defined as the killing of posthatched conspecifics) in animal studies, the occurrence of this behavior needs to be reconsidered in order to furnish a term of comparison with other invertebrate (see Polis, 1984) and vertebrate taxa (see Hrdy and Hausfater, 1984).

As our field of interest concerns social wasps (Figure 1), we shall limit our survey on infanticide to their societies, but similar phenomena have been reported in other social insect colonies where they are a fairly common occurrence (for a review of the main literature see Wilson, 1971, 1975).

Intracolonial Brood Cannibalism

One characteristic of social vespids is that they rear their immature brood in cells of paper (or sometimes mud) nests. Eggs, larvae and pupae are completely defenseless, relying entirely on the adults for food and protection. In vespine wasps, colonies are founded by a single queen at the beginning of the favorable season, but in lower polistine wasps such

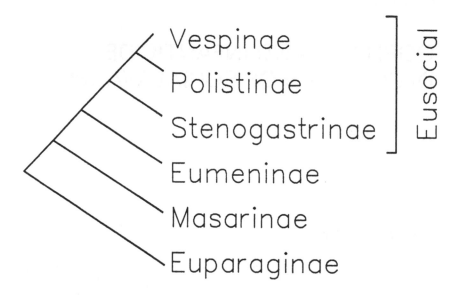

Figure 1 Cladogram of the family Vespidae (after Carpenter, 1982) with indication of social subfamilies.

as *Polistes, Ropalidia, Mischocyttarus etc.* two or more females may found a nest in association. Some tropical species found their nests in swarms (West Eberhard, 1982). In all social wasps, the colonial cycle is divided into distinct phases, more evident in the temperate species but nevertheless still present in tropical wasps. The workers are always the first to emerge, followed by the males and non- workers. Figure 2a shows how the number of immature stages varies over the colonial cycle in the temperate *Polistes gallicus* (Dani, 1990), and Figure 2b shows changes in the adult population of the same wasp (Dani, 1990). Data for tropical polistine species can be found in West (1969) and Jeanne (1972).

The adults largely depend on salivary secretions emitted by the larvae. Maschwitz (1966) claims that the larvae serve as temporary food "reservoirs" when foraging is poor. Cannibalism of the wasps' own brood was first reported as long ago as the eighteenth century by such authors as Reamur (1748), Rossi (1794), Von Siebold (1871), Janet (1895), Rau (1929), Freisling (1942), Duncan (1939) (all quoted by Pardi, 1951). More recently the same phenomenon has been observed in other vespine,

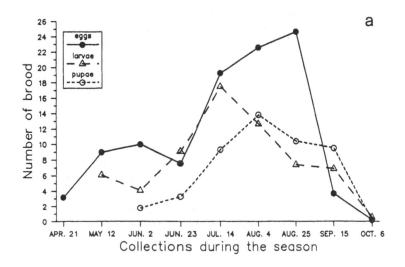

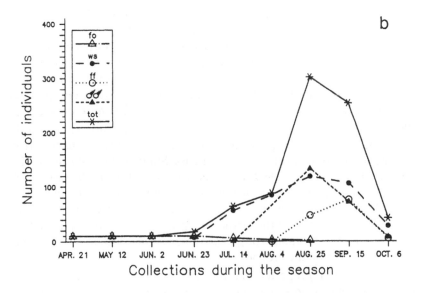

Figure 2 (a) Variations in number of immature brood and b) adult population in colonies of *Polistes gallicus* during the seasonal cycle (after Dani, 1990).

polistine and stenogastrine species (*e.g.* Ishay *et al.* 1968, in *Vespa orientalis*; West, 1969, in *Polistes fuscatus* and *P. erithrocephalus* (= *P. canadensis*); Spradbery, 1971, and Archer, 1981a, 1981b, in *Vespula vulgaris*; Jeanne, 1972, in *Mischocyttarus drewseni*; Litte, 1979, in *M. flavitarsis*; Sekijima *et al.* 1980, in *Parabolybia indica*; Gadagkar and Joshi, 1981, in *Ropalidia cyatyformis*). Hansell (1981), and Turillazzi and Francescato (1989) reported oophagy by males in mature colonies of *Parischnogaster mellyi*, and Turillazzi and Pardi (1982) oophagy in *P. nigricans serrei* (Stenogastrinae).

Larvae destruction is most evident in the vanishing phase of the colony. Deleurance (1948, 1952, 1955) refers to this phase as the "couvain abortif" (abortive brood, when the immature individuals are abnormal and the adults remove them from their cells) as opposed to the early "couvain normal" consisting exclusively of healthy individuals. However, brood suppression can occur at any stage of the colonial cycle. Quantification of the phenomenon has been provided for several species. Archer (1981a, 1981b) observed that larval mortality in *Vespula vulgaris* is a rare event and that it never occurs until the first sealed cells containing the queens appear (in vespine wasps the reproductives are reared in large cells while the workers are reared in smaller ones). Using various parameters, brood mortality was estimated to affect approximately 20% of the larvae and pupae in the small cells and 42% of the larvae and pupae in large cells.

On one nest of *Polistes dominulus* (= *P. gallicus*), Pardi (1951) calculated mortality rate to be 37% in the post emergence stage, as many as 58% of the larvae either died or were suppressed before the abortive nest phase (Figure 3). Sometimes the extracted larvae were already dead, but at other times they were very much alive. Mostly young larvae meet this fate (Figure 4). Pupae are seldom killed, and then invariably decapitated. Similar data can be found in Jeanne (1972) on *Mischocyttarus drewseni*.

Pardi (1951) quotes several theories to explain the phenomenon, mostly concerning scarce food supplies due to external or internal factors (*e.g.* a change in the forager/colonial needs ratio, Maidl 1934, Freisling 1942). In *Polistes fuscatus* West (1969) observed that soliciting adults often mob foragers returning to the nest when the ratio between non-foraging adults and workers on the colony rises. Consequently the larvae do not receive food and the females pull the larvae and pupae out of their cells, either to share with their nestmates or give to other larvae. Otherwise they may simply drop them on the ground.

Ishay *et al.* (1967) reported that they never observed workers of *Vespa corientalis* eating larvae at the end of the season, only ejecting dead larvae from the nest. They presumed that the larvae had died because the nest temperature had dropped below 19°C.

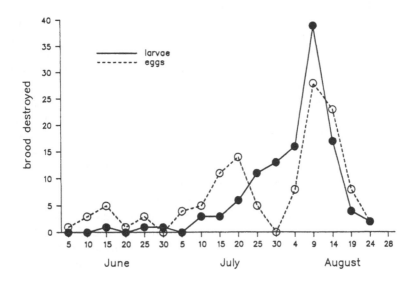

Figure 3 Egg and larva destruction in a *Polistes dominulus* nest in the post emergence period (data from Pardi, 1951).

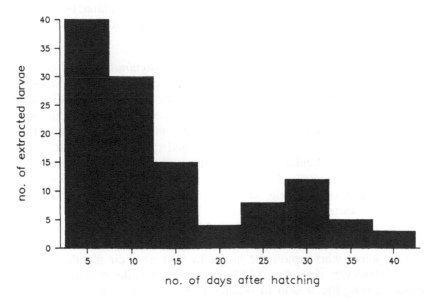

Figure 4 Age of larvae extracted from cells of a *P. dominulus* nest (data from Pardi, 1951).

Deleurance (1955) demonstrated that failure of larval development in *Polistes* did not depend on any of the more obvious factors such as temperature or lack of food supplies. In one experiment he produced similar premature death in the larvae by severing the salivary ducts of the workers, and proposed that the saliva of the nurses contained some substance essential to larval growth which falls in supply at the end of the season. In the field, West (1969) did not observe the same phases in *P. fuscatus* as Deleurance described in *P. dominulus* (= *P. gallicus*) kept in captivity. Similar events only occurred in colonies transferred to the laboratory. West also objected that the reproductive condition of the queen was not taken into account and was of the opinion that the numerous attempts to explain the phenomenon were all inconclusive. West (1969), reported that in the nests she had observed the phenomenon only occurred after the foundress queen had stopped laying eggs. In the tropical wasp *P. erithrocephalus* (= *P. canadensis*), West also found that there was a correlation between phenomena associated with cessation of nest growth and the end of egg laying or disappearance of the queen (West 1969). In one nest under observation, 101 eggs and 47 out of 106 larvae disappeared three weeks after the last cell had been built. This suggests that the presence of a reproductively active queen is essential for normal nest expansion and brood care. In another colony, eggs disappeared after the queen was temporarily removed, suggesting that eggs are neglected in the queen's absence, or else that egg care is a function of the queen herself or is at least stimulated by her presence. Miyano (1986) also reported heavy egg destruction in orphan nests of *P. chinensis antennalis*.

However, Pardi (1948) observed that in *P. dominulus* (= *P. gallicus*) the disappearance of the queen does not implicate colonial decline since she can be substituted by another female. Montagner (1966) dismissed that the cause behind the abortive nest phase in *Vespula* is that older foundresses lay a different type of egg in small cells at the end of the summer, and suggests that quantitative and qualitative trophic factors are responsible. Under-feeding of larvae at the end of the season is caused by three factors: 1) dominance struggles between workers which thus neglect brood rearing duties; 2) the low number of workers (the queen has started laying eggs in the larger regal cells) and their advanced age (they are no longer suitable for nursing larvae, a task carried out by the younger workers); 3) heavy interference from males and young queens who exploit the larvae by sucking their glandular secretions. However, according to Montagner even under-nourished larvae pupate during the season to become small workers. In a series of experiments using radioactive tracers, it was found that certain substances

secreted by the workers were not given to the larvae on a nest in the abortive phase, but were administered to those in non-abortive combs. Montagner concluded that qualitative trophic factors are determinative in larval development and are reinforced by quantitative ones. In the tropical wasp *Mischocyttarus drewseni*, Jeanne (1972) showed that the ratio between non-workers plus males and foragers increases simultaneously with brood eating towards the end of the colonial cycle, and hypothesized that reduced food supplies for the workers leads to brood eating and subsequently to nest abandonment. He also noticed that the queen often laid eggs well into the brood eating phase, and deduced that colonial cycle development could not depend on the queen's reproductive cycle alone. More probably the queen and the workers, together with certain nest properties, all interacted with one another (see Pardi, 1948). Jeanne suggested that as the colony matures, the adult emergence rate increases and the queen fails to dominate all the new females. Fewer emerging females thus become workers with a consequent rapid increase in the number of non-workers and males and a fall in the worker replacement rate, which causes a rise in the non-workers plus males/workers ratio. When this ratio reaches a certain limit, the workers can no longer deliver sufficient food to the nest; larvae and pupae are then aborted to feed the hungry adults. If the ratio remains high long enough, the whole brood is aborted, the adults disperse and the colony declines.

Thus, the larvae serve as a buffer system for the colony during brief periods of food shortage when the adults kill them for food. Jeanne's ideas are in accordance with Roubaud (1916) and Zikan (1951) and with the result of an experiment by Turner (1912) on *Polistes pallipes* and supports the conclusion that brood abortion begins when the adults have insufficient proteinacious food. While Litte (1979) observed brood eating at the end of the season in colonies of *Mischocyttarus flavitarsis*, she also reported that some nests had been abandoned by the adults even though they still contained live larvae.

In *Ropalidia cyathiformis*, Gadagkar and Joshi (1981) observed that only one adult was responsible for eating the first larva or pupa in a colony, but this action was sufficient to trigger cannibalism in the others. The authors could not however offer an explanation as to the cause of infanticide in the first place.

Matsuura (1985), reported that under favorable conditions very few mortality factors affect broods of five Japanese species of the genus *Vespa*, except for cannibalism by adult workers. When adult sexuals in the reproductive period nurse the larvae, even prospering colonies suffer catas-

trophic brood mortality rates, probably because they devote too little attention to the brood. Cannibalism by adults and starvation are the two main mortality factors during this period.

Archer (1981b) observed that in some colonies of *Vespula vulgaris* only male broods sealed in large cells were destroyed. Such selective destruction could reflect the control the workers exert over the queen's male offspring (which are laid in large cells); at the same time they avoid destroying their own male offspring (eggs layed in small cells). Such an expensive method of control would be necessary if the workers could not distinguish the sex of the brood at the larval stage. This agrees with the theory put forward by Trivers and Hare (1976) which foresees a conflict between the queen and workers in haplo-diploid social insects over sex allocation of offspring.

So far only the more obvious phenomenon of brood cannibalism in a colony, especially in the declining phase, has been considered. Clearly the phenomenon necessitates re-analysis as in some cases the causes behind oophagy and infanticide are not completely understood and decisive experiments still wait to be performed. Other types of immature brood destruction may occur in the early colonial stages. Although less conspicuous, this phenomenon is part of the struggle for a supremacy in the colony and is relevant to sociobiology.

In some temperate and tropical species of polistine wasps the nest is founded by more than one fertilized, egg-laying female. This phenomenon was first described by Heldmann (1936). Pardi (1942) demonstrated the linear dominance hierarchy in the associated foundresses of new *P. dominulus* (= *P. gallicus*) nests founded in the spring. At first, all the associated females lay eggs but later only the leader female maintains her egg-laying role and the subordinates, dominated by the alpha female, take over as workers and stop laying eggs. Moreover, as several authors have observed (*e.g.* Pardi, 1942, 1948; Gervet, 1964; West, 1969), most of the eggs laid by the subordinate females never hatch as they are destroyed by the leader and other high ranking females (Figure 5). Once she has devoured the eggs of her subordinates, the leader female may well lay an egg of her own in the empty cell (Pardi, 1942; West, 1969). Differential oophagy has also been observed in *Mischocyttarus* (Jeanne, 1972) and *Belonogaster* (Marino Piccioli and Pardi, 1970).

Oophagy is quite common during the initial period of associative nest foundation. Pardi (1942) reported that 38.3% and 48.4% of eggs were destroyed respectively in two associations each with three females. While Heldmann (1936), Pardi (1942) and Gervet (1964) affirm that *P. dominulus* foundresses not only eat the eggs of the other females, but also their own.

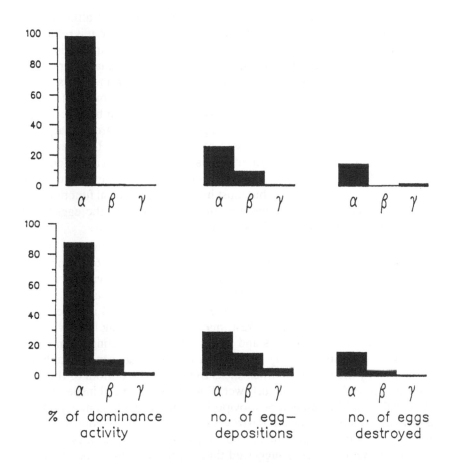

Figure 5   Dominance rank and oophagy in *P. dominulus* (data from Pardi, 1942).

West (1969) considers that one of the characteristics of leader females in *P. fuscatus* colonies is that they never eat their own eggs. West noted that only egg-laying females ate eggs, and that oophagy was never observed in sterile females.

Differential oophagy has been studied in depth by Gervet (1964) in *P. dominulus* (=*P. gallicus*) and was found to be directly related to the dominance hierarchy. The dominant female lays the most eggs but at the

same time also exhibits the highest level of oophagy. According to West (1969), the one leads to the other.

Considering how oophagy varies with time in *P. dominulus*, although the actual number of eggs laid at the beginning of the cycle is low, it is relatively high with respect to cell availability. Thus Pardi (1942) maintains that high oophagy rates reflect the competition between the females for empty cells where they can lay their eggs. As nest construction increases, more cells become available and more eggs are laid. A large number of eggs is still destroyed, but a lower fraction of total eggs laid than before. Later, when the subordinate females stop laying eggs, oophagy drops sharply as competition between the foundresses comes to a halt. However Pardi (1942) noticed that a similar form of oophagy also occurs in nests founded by a single female and concluded that oophagy was not simply caused by competition between females, but must have some biological function too, since the foundresses fed themselves and their larvae on the eggs when adverse meteorological conditions hampered foraging flights.

Pardi observed that in polygynic colonies of *P. dominulus* most of the eggs disappeared within 24 hours of deposition (Figure 6a) (compare this distribution with the age of eggs destroyed in more advanced phases of colonial development (Figure 6b)). West (1969) calculated that the average age of eggs eaten in *P. fuscatus* colonies was 11.4 minutes. An oophagic female, according to West, may avoid devouring her own eggs by recognizing newly laid eggs and limiting herself to eating them when she has not recently laid any herself. Newly laid eggs can probably be recognized by the secretion which sticks the egg to the bottom of the cell, and which in this case is still fairly wet. Post-oviposition cell inspection by the females and heightened post-oviposition aggressiveness may prevent associates from detecting and destroying the eggs. On the contrary in an impressive series of experiments (but unfortunately not always clear or controlled), Gervet (1964) suggested that oophagic *P. dominulus* females learn to distinguish their own eggs from those of other females by cues they detect just after laying, which reinforce an "internal predisposition" ("preconnaissance") towards these eggs. Even the egg laying territory ("territoire de ponte") seems to influence oophagic behavior in egg laying females.

More recently, Downing and Jeanne (1983) hypothesized that the secretion from Dufour's gland released with the egg at oviposition, could mark the eggs of the leader female and inhibit oophagy by subordinates. This gland grows to a maximum and reaches it peak of activity in June, at the beginning of the colonial cycle. It is larger in leader females than in subordinates and lone foundresses.

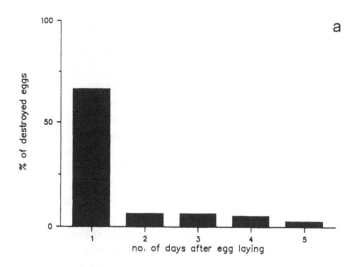

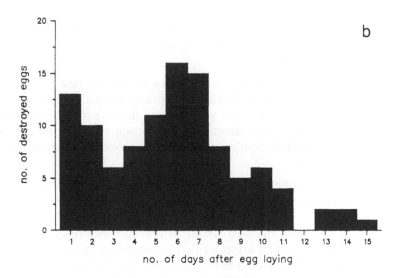

Figure 6   a) Age of the eggs (N = 108) destroyed in an associative foundation of *P. dominulus* and (b) during the entire colony cycle (data from Pardi, 1942).

Intraspecific Nest Usurpation and Brood Cannibalism

The immature brood is also destroyed when an alien conspecific foundress finds herself on a nest other than her own. To verify this occurrence, we carried out nest-exchange experiments on solitary foundresses of *P. gallicus* (Cervo and Turillazzi, 1989) (Figure 7). The technique consists of switching the positions of two nests in the pre-emergence stage belonging to two different foundresses whilst they are out foraging. On their return, the wasps found an alien nest instead of their own. The behavior of the wasps indicated that they did recognize some anomalities in the nest upon landing, but even so about 83% (85 out of 103) of the foundresses accepted the new one. An enormous number of eggs were always destroyed during the initial period on the new comb, and even more were destroyed when the experiment was run with older colonies. However, relatively few larvae and pupae were destroyed.

Klahn and Gamboa (1983) performed similar experiments on solitary foundresses of *P. fuscatus*, in nest-exchanges between sister and non-sister foundresses. The brood was destroyed in 30 out of 34 non-sister exchanges, as opposed to 6 out of 24 sister exchanges. Non-sisters ate significantly more eggs and ate or ejected more smaller larvae than sisters, at all stages of colonial development. It is still not clear whether a female accepts her sister's nest because she cannot distinguish it from her own, or whether she does recognize it as such and tolerates the sister brood. Klahn and Gamboa (1983) maintain that in natural conditions it is important for *Polistes* females to recognize an alien brood from a non-related one in at least two different contexts:

(1) when females gather together in spring to found a new nest. It has often been observed that associative foundresses all come from the same nest they shared the year before, and are therefore sisters. In this case, the subordinate sisters tolerate the immature brood belonging to their dominant sister, thus probably increasing their inclusive fitness.

(2) during conspecific nest usurpation. Intraspecific usurpation is a common form of social parasitism in wasps (see Klahn, 1988 for references), where one foundress is ousted by another female of the same species who uses the host colony to rear her own brood. Massive brood destruction in intraspecific usurpation has been reported by Litte (1979) in *Mischocyttarus flavitarsus*, by Makino in *P. biglumis* (1985) and *P. riparius* (1989), by Klahn in *P. fuscatus* (1988) and by Lorenzi and Cervo (1990) in *P. biglumis bimaculatus*. Although conspecific nest usurpation is widespread among the Vespinae, no data

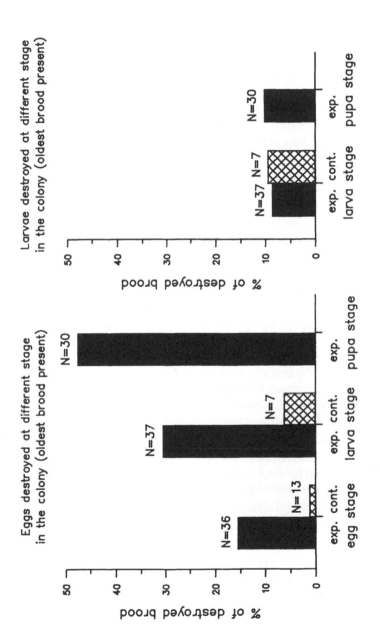

Figure 7 Brood destruction on exchanged nests in *P. gallicus* (after Cervo and Turillazzi, 1989).

are available on the phenomenon of immature brood destruction in the usurped colonies.

Usurper females which destroy the immature brood already on the nest show very similar behavioral patterns to those observed in the nest exchange experiments (particularly non-sister exchanges) (Figure 8).

Most of the brood in a *Polistes* nest the day before usurpation disappeared within only a few days of the newcomer's arrival. Makino (1985, 1989) and Klahn (1988) both observed massive brood destruction within two to five days of usurpation. Klahn (1988) has no doubts that the usurpers, and not the legitimate owners of the nests, are responsible for brood destruction. He observed that defeated females never returned to their colonies, and that in undisturbed nests, immature broods in monogynic colonies were rarely destroyed before the workers emerged, except in severe cases of stress from food shortage. The same author claims there are two "related trends" in brood destruction, which are typical of the nest exchange experiments described above:

(1)   Younger broods are more easily destroyed than older ones. In *P. fuscatus* colonies, for example, which are usurped in an advanced stage of development, initial massive destruction of eggs is followed by destruction of the younger larvae. Older larvae are less frequently killed and the pupa hardly touched at all. However, Klahn points out that in these colonies the usurper devours eggs of all ages, unlike polygynic associations where mostly new laid eggs are destroyed. The same tendencies which Klahn observed in *P. fuscatus* are evident from an analysis of data collected by Makino on brood destruction in *P. riparius* colonies five days after usurpation.

(2)   The second tendency correlated with age concerns the higher destruction rates of broods in more mature colonies. If the nest only contains eggs or eggs and larvae, the usurper leaves most of them intact, as observed in the nest-exchange experiments, but if the nest is in a more advanced stage of development she will destroy a large number of the eggs and larvae.

Klahn and Gamboa (1983) consider the possible adaptive implications of differential brood destruction for the usurper. Their first observation was that the most mature brood on the colony at the time of usurpation (just before the workers emerged) was left practically unharmed. These individuals are destined to become workers and the usurper can exploit them to rear her own brood. By destroying the eggs (and small larvae) on the nest, the usurper also avoids rearing any non-related reproducers which usually emerge later in the colonial cycle. In addition, she can also

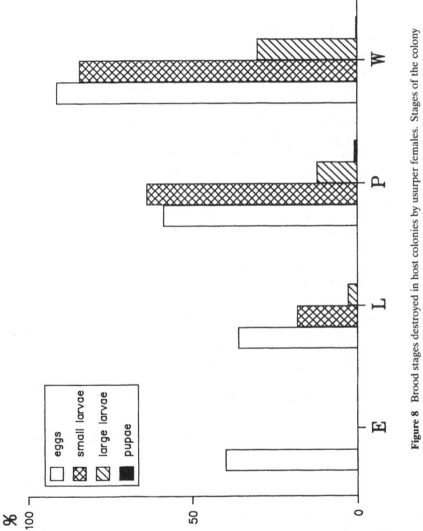

Figure 8 Brood stages destroyed in host colonies by usurper females. Stages of the colony cycle are defined by the oldest immature offspring present in the nests. E = eggs; L = larvae; P = pupae; W = workers (data from Klahn, 1988).

use the eggs to feed herself and her future work force. Finally, destruction of the brood means that cells become available where the usurper can lay her own eggs. Makino (1985) observed that many of the cells which had been emptied by the new female on her arrival were re-used for egg laying.

But, as Litte (1979) observed in usurped colonies of *M. flavitarsis* and Makino (1989) in *P. riparius*, destruction of the immature brood continues long after the usurper's arrival on the nest. Strangely enough, as Makino points out, the main cause of usurper reproductive failure is their own "loss of immatures". On emerging, the workers will themselves destroy most of the eggs they find on their nest which were laid before they were born, plus any young larvae already hatched out (average egg destruction per nest = 80%; ranging from 30% to 100%). The workers responsible for the ensuing brood destruction probably emerged from larvae and pupae left intact by the usurper when she first arrived on the nest, and are therefore the daughters of the former foundress. It is most unlikely that usurper herself is responsible for the carnage, since she would have nothing to gain from destroying her own brood. Usurper-brood destruction is in fact negligible before the birth of the workers (average per nest = 2%; ranging from 0 to 6%). A substantial number of usurper offspring are already exterminated when the mother is still alive on the nest. Once she is dead, any survivors are quickly eliminated. Litte (1979) also observed low reproductive success rates in usurpers of *M. flavitarsis* (only 5 out of 14 produced offspring) and believed the cause lay in destruction of the immature brood by the workers or by other usurpers following later.

In Makino's opinion (1989), this suggests that workers can tell whether the foundress and/or immature brood is related to them or not, even if their mother had been ousted by the alien foundress when they were still immature. If workers are able to recognize a foundress as an alien, they can be expected to destroy any immature brood they find on emerging, since in all probability it belongs to their usurper.

Interspecific Nest Usurpation and Brood Destruction and Predation

Taylor (1939) stated that intraspecific usurpation represents the initial stages of evolution towards true permanent and obligatory social parasitism, *i.e.* inquilinism. Inquilism is the most evolved form of social parasitism found in the social insects. The female parasite is characterized by her inability to found a nest and by loss of the worker caste, which means she is obliged to usurp a colony of another species (the host species) and

must depend on the workers of the parasitized species to raise her offspring (only reproductives).

Among the polistine wasps, only three species of obligate and permanent social parasites belonging to the genus *Polistes* are known (formerly included in the genus *Sulcopolistes*), whilst 4 species of obligate parasites are known to exist in the respine wasps.

In recent studies on social parasitism in *Polistes*, we found that immediately following her arrival on the host colony, the parasite female will destroy vast numbers of the immature brood she finds on the nest (Cervo *et al.* 1990a; Cervo, 1990). As in cases of intraspecific usurpation, the parasite destroys the youngest individuals (Figure 9 a and b). Most of the eggs and small larvae are devoured, whilst the older and operculated larvae suffer to a lesser extent. Furthermore, the number of destroyed eggs we observed is probably an underestimate, as the females may have laid new eggs in cells where eggs had been removed.

In nests monitored during the first hours of invasion and on the days immediately following usurpation, the female parasite was found to be responsible for the disappearance of the immature brood. It was not uncommon to observe the parasite devouring eggs and eating larvae in various stages of development after extracting them from their cells (Cervo *et al.* 1990). On one occasion, a female *P. sulcifer* was seen to lay an egg in a cell immediately after she had removed the small larvae it contained and eaten it.

Occasionally host workers will destroy some of the parasite's immature brood. Bunn (1989) reports the extraction of immature brood in a nest of *D. sylvestris* usurped by a *D. norwegica* queen (facultative interspecific usurpation) after the workers emerged, and Fisher (1987, 1988) observed a similar phenomenon in host workers of *Bombus* colonies after usurpation by a female *Psithyrus*, their permanent and obligate social parasite.

Behavioral patterns associated with brood destruction in polygynic associations, as well as inter and intra specific usurpation arise in competitive situations in the colony related to the assertion of individual fitness. As Klahn (1988) suggests, the pattern of brood destruction may have evolved as a behavior specifically within the context of intraspecific usurpation. However, brood destruction in usurpers is not limited to a genetically distinct subset of the population, and indeed brood destruction can be elicited from foundresses in normal colonies by offering them the comb of a non-sister in place of their own.

Although the suppression of the heterospecific brood by parasitic *Polistes* should be considered as predation, it is possible that this phenomenon originated from intraspecific brood cannibalism since the situational

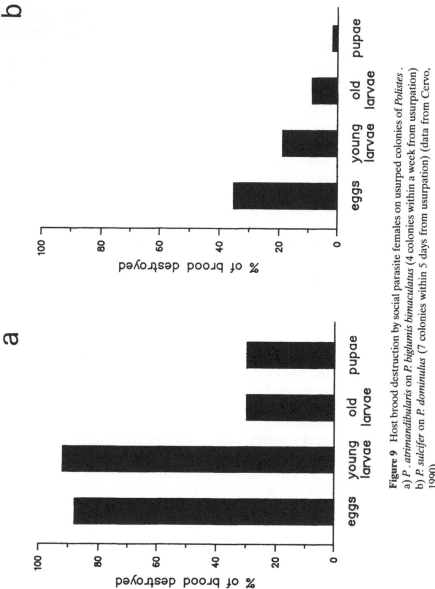

Figure 9 Host brood destruction by social parasite females on usurped colonies of *Polistes*. a) *P. atrimandibularis* on *P. biglumis bimaculatus* (4 colonies within a week from usurpation) b) *P. sulcifer* on *P. dominulus* (7 colonies within 5 days from usurpation) (data from Cervo, 1990).

determinants and functions of this behavior are practically identical. In fact, as we have seen, both the events and the behavioral patterns associated with the destruction of the host brood by a parasite does not substantially differ from those observed in intraspecific usurpation. Heterospecific social parasitism in wasps may well have evolved from species which practiced conspecific nest usurpation as an alternative to founding their own nest, or as a safeguard if their own nests failed early in the cycle. Destruction of a non-related brood in parasitic *Polistes* represents a behavioural pattern inherited before their speciation and now connected to their particular life style (see the "alternative adaptations theory" forwarded by West Eberhard, 1986).

In certain species of *Polistes* (Von Siebold, 1871; Pardi, 1951; Sakagami and Fukushima, 1957; Kasuya *et al.* 1980) females in some colonies are known to occasionally prey on the immature offspring of other females of the same species. We recently observed that *Polistes atrimandibularis*, a social parasite of *P. b. bimaculatus,* often plunders the *brood* in other host colonies and takes the larvae back to the nest where she has her offspring (Cervo *et al.*1990b). This is another case where social parasite behavior finds a parallel in the behavioral repertoire of *Polistes* wasps.

CONCLUSIONS

After our survey on infanticide in social wasp colonies, we realized that brood suppression in these insects probably fits different types of infanticide defined by Hrdy (1979) — although some particular characteristics of eusociality and parental asymmetries found in the Hymenoptera make it difficult to assign some phenomena to one particular category.

Exploitation of immatures is well represented by cannibalism of larvae at the end of the season or in periods of scarce food supplies when they provide a source of protein for the adults. However, as we have seen, a dominant female will also exploit the eggs of her subordinates by eating them to promote her own ovarian development. This is also true of intraspecific usurpers and social parasites. Another type of exploitation of immatures is also practiced by usurper females when they kill the eggs and larvae of defeated females to feed their own larvae. Oophagy by associated females on newly founded nests in spring could also be interpreted as resource competition. When a foundress eats one of her partner's eggs, she not only gains a trophic advantage which can increase her egg-laying potential, she also liberates a cell in which she can lay an egg of her own. As we have already seen, shortage of cell availability at the beginning of

the cycle is a relevant factor in oophagy in *Polistes* associative foundations. The incentive to free cells for deposition may be one of the main factors inducing infanticide of the host brood in intraspecific and interspecific usurper females.

Parental manipulation is widespread and mainly represented by adults which kill some of their own larvae to give as food to others. This can occur in periods of food shortage when a female kills some of own her brood to nourish others. It could be hypothesized that by so doing, the adults are continually enhancing their inclusive fitness, but this aspect has not been investigated. Another phenomenon which may well fit this class is that reported by Archer (1981b), where workers in some colonies of *Vespula vulgaris* kill the male pupae produced by the queen but leave their own sons intact and thus probably promoting their reproductive success. Female foundresses provide a further example of this phenomenon when they improve the chances of reproductive success of their own reproductives by suppressing the eggs and larvae of females they have defeated. If left these immatures would develop into males and non-workers. In both cases the beneficiaries are the direct descendants of the performer of infanticide.

Brood cannibalism observed in captive colonies may be included under the social pathology class. The exact impact of this phenomenon is hard to evaluate. At present the influence of captivity on colonial life in caged colonies (which are commonly used in experiments on social wasps) has not been investigated. However, brood cannibalism certainly occurs if the colony is seriously disturbed. One of us, for example, observed an adult eating a pupa in a colony just collected and put in a plastic bag.

The phenomenon of individuals eliminating immatures to lower competitor reproductive success rate has never been reported in social wasps, but it seems feasible from a theoretical point of view. In fact, males produced by the queen on the same colony are on average 1/2 related to their brothers, but only a 1/4 related to their nephews (*i.e.* sons of any egg laying workers). At the end of the season, the queen's sons and the males born of workers will all compete for the act of mating, thus the queen's sons can possibly be expected to commit infanticide. This, of course, implicates that they are able to perform differential destruction and are not chased away by the workers. The phenomenon of infanticide is obviously well represented in social wasps. This survey does not claim to give an exhaustive coverage of the subject, but we hope it serves to stimulate new interest and further research into the matter.

REFERENCES

ARCHER, M.E. (1981a) Successful and unsuccessful development of colonies of *Vespula vulgaris* (Linn.)(Hymenoptera: Vespidae).*Ecological Entomology*, 6, 1–10.

ARCHER, M.E. (1981b) A simulation model for the colonial development of *Paravespula vulgaris*(Linnaeus) and *Dolichovespula sylvestris* (Scopoli) (Hymenoptera: Vespidae). *Melanderia*, 36, 1–59.

BUNN, D.S. (1989). Usurpation in wasps. *Entomologist's Monthly Magazine*, 171–173.

CARPENTER, J.M. (1982) The phylogenetic relationships and natural classification of the Vespoidea (Hymenoptera).*Systematic Entomology*, 7, 11–38.

CERVO, R. (1990) *Il parassitismo sociale nei Polistes*. Ph.D. Thesis, Università di Firenze.

CERVO, R. and Turillazzi, S. (1989) Nest exchange experiments in *Polistes gallicus* (L.) (Hymenoptera, Vespidae). *Ethology Ecology and Evolution*, 1, 185–193.

CERVO, R., Lorenzi, M.C. and Turillazzi, S. (1990a) Nonaggressive usurpation of the nest of *Polistes biglumis bimaculatus* by the social parasite *Sulcopolistes atrimandibularis* (Hymenoptera Vespidae). *Insectes sociaux,,* 37, 333–347.

CERVO, R., Lorenzi, M.C. and Turillazzi, S. (1990b) *Sulcopolistes atrimandibularis*, social parasite and predator of an alpine *Polistes* (Hymenoptera, Vespidae). Ethology, 86, 71–78.

DANI, F.R. (1990) Caste, variazioni morfometriche stagionali e riproduzione in *Polistes gallicus* (L.) (Hymenoptera, Vespidae). Tesi di laurea, Facolta di Scienze Matematiche fisiche e Naturali, Università di Firenze.

DELEURANCE, Ed. Ph. (1948) Sur le cycle biologique des *Polistes* (Hymenopteres-Vespides). *Compte Rendu de l'Academie des Sciences*, 224, 601–603.

DELEURANCE, Ed. Ph. (1952) Etude du cycle biologique du couvain chez *Polistes*. Les phases "couvain normal" et "couvain abortif". *Behaviour IV*, 2, 104–115.

DELEURANCE, Ed. Ph. (1955) Contribution a l'etude biologique des *Polistes* (Hymenopteres Vespides). II.-Le cycle evolutif du couvain. *Insectes sociaux II,,* 4, 286–302.

DOWNING, H.A. and Jeanne, R.L. (1983) Correlation of season and dominance status with activity of exocrine glands in *Polistes fuscatus* (Hymenoptera: Vespidae). *Journal of the Kansas Entomological Society*, 56, 387–397.

DUNCAN, C.D. (1939) A contribution to the biology of North American Vespine wasps. *Stanford University Publications University Series, Biological Sciences VIII,*, No. 1, 1–272.

FISHER, R.M. (1987) Queen-worker conflict and social parasitism in bumble bees (Hymenoptera: Apidae).*Animal Behaviour*, 35, 1628–1636.

FISHER, R.M. (1988) Observations on the behaviours of three european cuckoo bumble bee species (Psithyrus). *Insectes sociaux*, 35, 341–354.

FREISLING, J.(1942) Zur Psychologie der Feldwespe. *Zeitschrift für Tierpsychologie*, 5, 438–463.

GADAGKAR, R. and Joshi, N.V. (1981) Behaviour of the Indian social wasp *Ropalidia cyathiformis* on a nest of separate combs (Hymenoptera: Vespidae). *Journal of Zoology, London*, 198, 27–37.

GERVET, J. (1964) Le comportement d'oophagie differentielle chez *Polistes gallicus* L. (Hymen., Vesp.). *Insectes sociaux*, 11, 343–382.

HANSELL, M.H. (1981) Nest construction in the subsocial wasp *Parischogaster mellyi* (Saussure) Stenogastrinae (Hymenoptera). *Insectes sociaux*, 28, 208–216.

HELDMANN, G. (1936) Ueber die Entwicklung der polygynen Wabe von *Polistes gellica* L. *Arbeiten uber physiologische und angewandte Entomologie aus Berlin- Dahlem*, **3**, 257–259.

HRDY, S.B. (1979) Infanticide among animals: a rewiew, classification and examination of the implications for reproductive strategies of females. *Ethology and Sociobiology*, **1**, 13–40.

HRDY, S.B. and Hausfater, G. (1984) Comparative and evolutionary perspectives on infanticide: Introduction and overview. In: G. Hausfater & S.B. Hrdy (eds). *Infanticide — Comparative and Evolutionary Perspectives*. Aldine, New York.

ISHAY, J., Bytinski-Salz, H. and Shulev, A. (1968) Contributions to the bionomics of the oriental hornet (*Vespa orientalis F.*).*Israel Journal of Entomology*, **2**, 45–106.

JANET, C. (1895) Etudes sur les Fourmis, les Guêpes et les Abeilles. IX Note: sur *Vespa crabro* L. Histoire d'un nid depuis son origine.*Memoires de la Societe Zoologique de France*, **8**, 1–140.

JEANNE, R.L. (1972) Social biology of the neotropical wasp *Mischocyttarus drewseni*. *Bulletin of the Museum of Comparative Zoology, Harvard*,, **144** (3), 63–150.

KASUYA, E., Hibino, Y. and Ito, Y. (1980) On "intercolonial" cannibalism in japanese paper wasps, *Polistes chinensis antennalis* Perez and *P. jadwigae* Dalla Torre (Hymenoptera: Vespidae). *Researches on Population Ecology*, **22**, 255–262.

KLAHN, J.E. (1988) Intraspecific comb usurpation in the social wasp *Polistes fuscatus*. *Behavioral Ecology and Sociobiology*, **23**, 1–8.

KLAHN, J.E. and Gamboa, G.J. (1983) Social wasps: discrimination between kin and nonkin brood.*Science*, **221**, 482–484.

LITTE, M. (1979) *Mischocyttarus flavitarsis* in Arizona: Social and nesting biology of a polistine wasp. *Zeitschrift für Tierpsychologie*, **50**, 282–312.

LORENZI, M.C. and Cervo, R. (1990) Usurpazione intraspecifica in *Polistes biglumis bimaculatus* (Hymenoptera, Vespidae). *Riassunti del XIV Convegno della Società Italiana di Etologia*.

MAKINO, S. (1985) Foundress-replacement on nest of the monogynic paper wasp *Polistes biglumis* in Japan (Hymenoptera: Vespidae). *Kontyû, Tokyo*, **53**(1), 143–149.

MAKINO, S. (1989) Usurpation and nest rebuilding in *Polistes riparius*: two ways to reproduce after the loss of the original nest (Hymenoptera: Vespidae). *Insectes sociaux*, **36**, 116–128.

MAIDL, F. (1934) *Die Lebensgewohnheiten und Instinkte der staatenbilden den Insekten*. Wien, F. Wagner Verlag, 1–823.

MARINO Piccioli, M.T. and Pardi, L. (1970) Studi sulla biologia di *Belonogaster*. *Monitore zoologico italiano (N.S.) Supplemento*, **3**, 197–225.

MASCHWITZ, U. (1966) Das Speichelsekret der Wespenlarven und seine biologische Bedeutung. *Zeitshrift für vergleiche de Physiologie*, **53**, 228–252.

MATSUURA, M. (1985) Comparative biology of the five japanese species of the genus *Vespa* (Hymenoptera, Vespidae). *Bulletin of the Faculty of Agriculture, Mie University*,, **69**, 1–131.

MIYANO, S. (1986) Colony development, worker behavior and male production in orphan colonies of a japanese paper wasp,*Polistes chinensis antennalis* Perez (Hymenoptera: Vespidae). *Researches on Population Ecology*, **28**, 347–361.

MONTAGNER, M.H. (1966) Sur le determinisme du couvain abortif dans les nids de Guêpes du genre *Vespa*. *Compte Rendu de l'Academie des Science*, **263**, 826–829.

PARDI, L. (1942) Ricerche sui Polistini. V. La poliginia iniziale di *Polistes gallicus* (L.). *Bollettino dell' Istituto di Entomolologia di Bologna*, **14**, 1–106.

PARDI, L. (1948) Ricerche sui Polistini. 11. Sulla durata della permanenza delle femmine nel nido e sull'accrescimento della societá in *Polistes gallicus* (L.). *Atti della societá toscana di scienze naturali* Volume **LV – serie B**, 3–15.

PARDI, L. (1951) Studio delle attivita e della divisione di lavoro in una societá di *Polistes gallicus* (L.) dopo la comparsa delle operaie. *Archivio zoologico italiano*, Volume, **XXXVI**, 363–431.

POLIS, G.A. (1984) Interspecific predation and "infant killing" among invertebrates. In: Hausfater, G. and Hrdy, S.B. (eds). *Infanticide – Comparative and Evolutionary Perspectives*. Aldine, New York.

RAU, P. (1929) Orphan nests of *Polistes* (Hym.- Vesp.). *Entomological News, Philadelphia,*, **40**, 256–259.

REAUMUR, A. F. de (1748) *Memoires pour servir a l'histoire des Insectes*. VI, Part 1.Imp. Royale, Paris.

ROSSI, P. (1794) *Mantissa Insectorum*, Tom. II.

ROUBAUD, (1916) Recherches biologiques sur les guêpes solitaries et sociales d'Afrique. La genese de la vie sociale et l'evolution de l'instinct maternel chez les vespides. *Annales des science Naturelles (Zoologie) (Ser. 10)*, **1**, 1–160.

SAKAGAMI, S.F. and Fukushima, K. (1957) *Vespa dybowskii* André as a facultative temporary social parasite. *Insectes sociaux*, **4**, 1–12.

SEKIJIMA, M., Sugiura M., and Matsuura, M. (1980). Nesting habits and brood development of *Parapolybia indica* Saussure (Hymenoptera: Vespidae). *The Bullettin of the Faculty of Agriculture, Mie University, N.*, **61**, 11–23.

SPRADBERY, J.P. (1971) Seasonal changes in the population structure of wasp colonies (Hymenoptera: Vespidae). *Journal of Entomology (A)*, **47**, 61–69.

TAYLOR, L.H. (1939) Observations on social parasitism in the genus *Vespula* Thomson. *Annals of the Entomological Society of America*, **32**, 304–315.

TRIVERS, R.L. and Hare, H. (1976) Haplodiploidy and the evolution of the social insect. *Science*, **191**, 249–263.

TURILLAZZI, S. and Pardi, L. (1981). Social behavior of *Parischnogaster nigricans serrei* (du Buysson) (Hymenoptera: Stenogastrinae) in Java. *Annals of the Entomological Society of America*, **75**, 657–664.

TURILLAZZI, S. and Francescato, E. (1989) Observations on the behavior of male stenogastrine wasps (Hymenoptera, Vespidae, Stenogastrinae). *Actes des Colloques Insectes Sociaux*, **5**, 181–187.

TURNER, C. H. (1912) An orphan colony of *Polistes pallipes* Lepel.*Psyche, Cambridge*, **19**, 184–190.

VON SIEBOLD, C. Th. (1871) *Beitrage zur Parthenogenesis der Artropoden*. Leipzig, W. Engelmann, pp. 1–238.

WEST, M.J. (1969) The social biology of polistine wasps. *Miscellaneous Publications of the Museum of Zoology, University of Michigan, Ann Arbor N.*, **140**, 101 pp.

WEST EBERHARD, M.J. (1982) The nature and evolution of swarming in tropical social wasps (Vespinae, Polistinae, Polybiini). In: Jaisson, P. (ed.), *Social Insects in the tropics, Vol.1* Universite Paris -Nord, Paris, pp. 97–128.

WEST EBERHARD, M.J. (1986) Alternative adaptations, speciation, and philogeny (a Review). *Proceedings of the National Academy of Sciences of the U.S.A.*, **83**, 1388–1392.

WILSON, E.O. (1971) *The Insect Societies*. Harvard University Press, Cambridge, MA, USA.
WILSON, E.O. (1975) *Sociobiology: The New Synthesis*. Harvard University Press, Cambridge, MA, USA.
ZIKAN, J.F. (1951) Polymorphismus und Ethologie der sozialen Faltenwespen (Vespidae, Diploptera). *Acta zoologica Lilloana,*, 11, 5–51.

CHAPTER 10

PROXIMATE AND ULTIMATE DETERMINANTS OF AVIAN BROOD REDUCTION

L. SCOTT FORBES[1] and DOUGLAS W. MOCK[2]

[1]*Department of Biology, University of Winnipeg, 515 Portage Avenue, Winnipeg, Manitoba R3B 2E9.*

[2]*Department of Zoology, University of Oklahoma, Norman, Oklahoma, USA*

> Take it from somebody who has been around for a million years: when you get right down to it, food is practically the whole story every time.
>
> Kurt Vonnegut, Galapagos

As Darwin learned from the population essay of Malthus, parents in an astonishing variety of life forms produce more offspring than can possibly survive to adulthood. In species with no postnatal parental investment, this can be extreme: ocean sunfish (*Mola mola*) may produce as many as 300,000,000 eggs in a single clutch (Hart, 1973). A giant sequoia (*Sequoiadendron giganteum*) may shed 300,000 seeds annually and live for thirty centuries (Harvey *et al.*, 1980). These taxa, where each offspring is tiny, are playing in an evolutionary lottery; the more tickets they hold, the better their chance of winning. In birds, mammals and some viviparous fish, which invest relatively colossal amounts of parental care in each offspring, the unit costs rise sharply and numbers fall accordingly. Nevertheless, the general parental habit of overproduction is widespread, if more modest, and generates dramas of increasing behavioral complexity for all parties involved. Grizzly bear (*Ursus arctos*) mothers with twins may desert one cub and rear the other (Tait, 1980). A pronghorn (*Antilocapra americana*) embryo may actually kill a sibling *in utero* (O'Gara, 1969), perhaps avoiding post-natal sibling competition for milk through the expedient of

237

a very early 'pre-emptive' strike. Similarly, embryonic sand tiger sharks (*Odontaspis taurus*) hunt for sibs in utero thereby avoiding later intrauterine competition for food (Gilmore *et al.,* 1983).

The behavioral and ecological dimensions of such brood reduction processes have been studied closely for many birds. They are well suited for such studies: being largely diurnal and open-nesting, young birds are much easier to watch than newborn mammals. As well, many conveniently nest in dense colonies, enabling an observer to follow several broods at once.

Our discussion will focus on birds as we address five main questions: (1) Why are surplus offspring created in the first place? (2) Given that too many young are created, how is brood size culled to match available resources? (3) What factors promote siblings to participate actively in the killing of nestmates? (4) Does siblicide (fatal sibling aggression) have anything to do with brood sex ratio? and (5) What, if anything, should parents do to stop siblicide?

WHY DO PARENTS CREATE TOO MANY OFFSPRING?

Parent birds often initiate clutches larger than they normally rear to independence, and do this for at least four reasons. First, the surplus offspring may allow parents to track uncertain resources (Lack, 1947, 1954; Ricklefs, 1965; Temme and Charnov, 1987; Kozlowski and Stearns, 1989). In proposing the original brood-reduction hypothesis, David Lack (1947) recognized the importance of variable food supplies. According to Lack, parents initiate clutches larger than they normally expect to rear and, if food is insufficient, allow brood size to be reduced to match prevailing food conditions. If food is abundant, the full brood may be reared.

Second, surplus offspring may allow for developmental selection (= progeny choice hypothesis of Kozlowski and Stearns, 1989): parents place more offspring into an arena (*e.g.* a nest) than could be expected to survive there, identify the offspring with the highest fitness expectations and eliminate (*e.g.* by parental infanticide), or allow to be eliminated (*e.g.* by siblicide) those with the poorest prospects (Buchholz, 1922; Lloyd, 1980; Simmons, 1988; Kozlowski and Stearns, 1989). Parents can either kill or abandon those offspring with lower fitness in order to concentrate investment in those with the best prospects.

Third, surplus offspring may serve as a food-cache for parents or siblings (Springer, 1948; Ingram, 1959; Eickwort, 1973; Polis, 1984) which Hrdy (1979) referred to as the "exploitation" hypothesis. Offspring may be used as food during periods of food scarcity, or parents may provide 'trophic

offspring' to feed other brood members (*e.g.* the trophic eggs of some snails and sharks, Lyons and Spight, 1973; Gilmore *et al.*, 1983). In birds, cannibalism of kin by both parents and offspring is known in a number of species, primarily raptors and owls (reviewed in Mock, 1984a; Bortolotti *et al.*, 1991). In most cases, such cannibalism appears to be a response to extreme food stress (Bortolotti *et al.*, 1991). It seems unlikely that exploitation is the principal reason for creating surplus offspring in any bird, as there are easier ways of finding a meal, although it may be an adjunct benefit as parents or offspring may recycle nutrients invested in redundant offspring (*sensu* Perrigo, 1987).

Fourth, surplus offspring may serve an insurance function. If a dependent offspring fails unexpectedly due to accident and/or congenital defect, the insurance offspring replaces it; otherwise the insurance offspring is redundant. The insurance hypothesis has been most often invoked to explain the presence of the second egg in birds exhibiting 'obligate' brood reduction (where one chick always dies, or nearly so; certain cranes, boobies, pelicans and penguins: Dorward, 1962; Meyburg, 1974; Warham, 1975; Nelson, 1978; Mock, 1984a; Cash and Evans, 1986; Drummond, 1989; Anderson, 1990). However it is becoming increasingly clear that the insurance hypothesis is also broadly applicable to species practicing 'facultative' brood reduction (where brood reduction occurs in some years but not others: Nisbet and Cohen, 1975; Mock and Parker, 1986; Forbes, 1990a; 1991a). The insurance hypothesis has great intuitive appeal: parents can purchase insurance against the accidental failure of offspring for a small investment at the time of clutch initiation, thereby avoiding breeding delay if an offspring fails. To date, the best demonstration of the insurance hypothesis is the experimental study of Cash and Evans (1986).

Mock and Parker (1986) have suggested that surplus chicks in birds that practice facultative brood reduction may serve both resource tracking and insurance functions simultaneously. Kozlowski and Stearns (1989) similarly note that the resource tracking and developmental selection hypotheses are not mutually exclusive. Indeed, none of the four functions (the resource tracking, developmental selection, food storage or insurance hypotheses) necessarily preclude each other (Forbes, 1990a). Conceivably all could operate simultaneously in species practicing facultative brood reduction (and obligate brood reduction precludes only the resource tracking function). We note that the developmental selection and insurance hypotheses are similar but not identical. In the former, surplus offspring are created to allow quality control. In the latter, surplus offspring not only insure against defective offspring (quality control) but against the accidental failure of offspring as well (*e.g.* due to predation, disease, parasites, injury).

Clearly insurance and developmental selection functions can be satisfied simultaneously.

BROOD REDUCTION SYSTEMS

Having established why parent birds often create surplus offspring, we now describe how parents (and offspring) go about eliminating offspring that are surplus to their needs (*i.e.* brood reduction). Avian brood reduction is usually the lethal outcome of sibling rivalry, although such behavior may serve parental interests, and may occur with or without the use of aggression among siblings (Lamey and Mock, 1991). Brood reduction is facilitated by establishing competitive asymmetries among the offspring prior to hatching via embryo asynchrony or differences in egg size or both, such that the last hatched chick is usually the first to succumb to selective starvation and/or siblicidal aggression.

Evolutionary Significance of Hatching Asynchrony

Hatching asynchrony occurs when incubation begins before the final egg is laid, and is an integral component of brood reduction systems. Aside from the resource tracking and insurance advantages discussed above, there may be other reasons for the evolution of avian hatching asynchrony, only some of which concern the direct effects on sibling rivalry (*e.g.* reduced nest failure: Hussell, 1972; Clark and Wilson, 1981; reduced peak foraging demands on parents: Hussell, 1972; Lessels and Avery, 1989; Mock and Schwagmeyer, 1990; reduced costs of sibling competition: Hahn, 1981; Mock and Ploger, 1987; earlier fledging of oldest chicks: Hussell, 1972; Clark and Wilson, 1981; Slagsvold, 1986; inter-parent conflict: Slagsvold and Lijfeld, 1989; sex-ratio adjustment: Slagsvold, 1990; non-adaptive hormonal effects: Mead and Morton, 1985). However, all hatching asynchrony, even that which evolved for completely different reasons, is likely to influence profoundly the nature of sibling competition and the potential for brood reduction.

Lack's resource tracking hypothesis has received the greatest attention of these, and several of the alternatives are a direct result of perceived deficiencies in Lack's hypothesis. Although it is probably true that the resource tracking hypothesis was accepted too easily for a long while (Clark and Wilson, 1985), the pendulum has clearly swung in the other direction; Lack's resource tracking hypothesis now seems to be dismissed too easily. Lack's hypothesis is usually tested through manipulation of hatching asynchrony: the success of artificially synchronous broods is compared to

normally (and/or artificially) asynchronous broods, the rationale being that if Lack's hypothesis is correct, synchronous broods should exhibit greater nestling mortality and/or lower fledging mass (parents are less able to satisfy the demands of a synchronous brood during periods of food shortage) than asynchronous broods (reviewed in Magrath 1990, Amundsen and Slagsvold 1991). Based upon a perceived lack of congruence between theory and data, Amundsen and Stokland (1988) claimed that Lack's brood reduction hypothesis cannot provide a general explanation for hatching asynchrony. Perhaps this is true, but the empirical shortcomings of the studies used to support this claim are numerous.

First, Lack's hypothesis is usually rejected when no difference is found between synchronous and asynchronous broods. However, when positive conclusions are drawn from negative results, the statistical power of the test (the probability of detecting a real difference if it existed) *must* be evaluated (Toft and Shea, 1983; Forbes, 1990b). To our knowledge, this has never been an explicitly stated component of any hatching asynchrony study.

Second, the success of experimentally synchronized broods is generally compared with unmanipulated natural broods; the act of experimental manipulation could itself confound the results (Magrath, 1990).

Third, the number and (less often) mass of nestlings when they leave the nest is the usual criterion for evaluating the performance of alternative strategies (but see Magrath 1989 who followed nestlings until they were recruited into the breeding population). Assessing the reproductive fitness of parents by the number of chicks fledged may be an inadequate measure (Figure 1); heavier chicks generally exhibit higher post-fledging survival (Perrins, 1963; Coulson and Porter, 1985; Krementz et al., 1989; Magrath, 1991) and fledging mass often varies inversely with brood size (reviewed in Smith et al., 1989). As well, if fledging is delayed, as might occur in synchronous broods, post-fledging survival of offspring may decline (Røskaft and Slagsvold, 1985; Poole, 1989; Hochachka, 1990). Even monitoring fledging mass may not be a satisfactory measure of future success in many species where parental care continues well beyond fledging, as the state of the chick at fledging may be correlated only weakly with subsequent offspring survival. Thus, studies that do not measure post-fledging survival of offspring directly provide at best a weak test of Lack's argument.

Fourth, variability in food supply is central to Lack's hypothesis. If food is abundant during breeding, no difference in synchronous and asynchronous broods is expected, yet food is rarely measured directly. Only Magrath (1989), who manipulated hatching asynchrony and food supplies simultaneously, has adequately addressed this issue. A related question

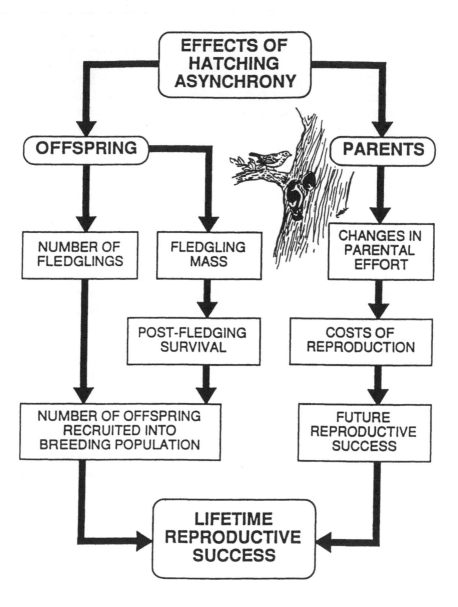

Figure 1 Potential reproductive consequences of hatching asynchrony for parents and off-spring in altricial birds.

is how variable does food have to be in order for resource tracking to be valuable? Temme and Charnov (1987) have addressed this issue in a formal theoretical context, their principal finding being that it depends upon the costs of surplus offspring when food is short (see also Kozlowski and Stearns, 1989; and Pijanowski, 1992); as yet no empirical measurements of the cost of brood reduction are available.

Fifth, studies of the adaptive significance of hatching asynchrony generally ignore the potential implications for parents (Figure 1) – *e.g.* the effects of manipulating hatching asynchrony on the future success of parents have never been measured directly. Examining how hard parents work to provision offspring addresses this question, albeit indirectly. Parents brought more food to artificially synchronous broods of cattle egrets (*Bubulcus ibis*) than to broods with normal asynchrony, and interestingly, chicks in synchronous broods grew as fast as the senior-most sib in normally asynchronous broods (Fujioka, 1985a). Similarly, in a Texas colony, parent cattle egrets brought more food to artificially synchronous and asynchronous broods than to broods with normal asynchrony, but fledging success was highest in broods with normal asynchrony (Mock and Ploger, 1987). They suggested that the natural 1.5 day hatching interval promoted higher reproductive efficiency (fledged chicks per unit of food) for the parents than either a doubled interval or complete hatching synchrony. Gibbons (1987) also found that hatching asynchrony resulted in reduced parental effort in jackdaws (*Corvus monedula*).

What are the consequences of greater foraging effort for these parents? If it exacts a reproductive cost (*e.g.* reduced future survivorship or fecundity; see Nur, 1988), the implications of hatching asynchrony obviously carry beyond the current brood (Figure 1). Once again, this question has yet to be addressed empirically.

Thus we believe that rumors of the death of Lack's hypothesis have been greatly exaggerated: whether Lack's hypothesis can provide a general explanation for avian brood reduction remains an open question. However, it may well be that hatching asynchrony often facilitates maladaptive brood reduction (Amundsen and Slagsvold, 1991).

It is becoming increasingly clear that Lack's original brood reduction hypothesis will be inadequate until it is addressed from a life history perspective. As noted above, the role of parental costs of reproduction (Nur, 1988) must be embedded within a 'neo-Lackian' framework. Furthermore, the role of reproductive variance needs to be addressed. Strategies with lower variance are likely to be favored in temporally varying environments (Cohen, 1966; Gillespie, 1977; Temme and Charnov, 1987; Boyce and Perrins, 1987; Seger and Brockmann, 1987), which is exactly the type of

environment to which Lack originally envisioned brood reduction as being adapted. As Temme and Charnov (1987) and Pijanowski (1992) note, whether a brood reduction strategy will be favored depends upon the frequency of good and bad years, and the cost of the brood reduction mechanism (*e.g.* the cost of creating surplus offspring when food is short). Where brood reduction is efficient, as in siblicidal species, it seems likely these costs will be low. However, where brood reduction is inefficient, as is perhaps the case in many passerines where brood reduction occurs via selective starvation, the costs of a brood reduction system may be substantial (Forbes, 1990a; 1991a; Lamey and Mock, 1991).

Magrath (1990) made one further and important point. Hatching asynchrony lies amid a network of coadapted traits, and several of the hypotheses advanced to explain it are complementary. To presuppose that any single factor operates to the exclusion of others is too simplistic; similarly, to dismiss any single factor because it does not account for all observed variation is dangerous. Rather, the degree of hatching asynchrony will reflect a balance between an array of benefits (*e.g.* reduced nest predation or food demands on parents; early fledging of chicks; the ability to track a variable environment through brood size adjustment), and costs (*e.g.* exaggerated competitive asymmetries among offspring perhaps leading to unwanted brood reduction and/or uneven distribution of food among chicks; increased hatching failure).

Non-Aggressive Brood Reduction

Brood reduction takes both aggressive and non-aggressive forms. The latter is the pattern presumably exhibited by most passerines, and generally involves selective starvation of junior sibs (reviewed in Lamey and Mock, 1991). Larger sibs simply consume more food and leave a smaller residual share for others. Thus, the potential exists for senior siblings to cause the death of junior siblings, albeit indirectly, simply by being more selfish (see Parker *et al.*, 1989 and Forbes and Ydenberg, 1992 for theoretical discussions of such behavior). Although not as spectacular as aggressive brood reduction, the smaller sib is just as dead.

Aggressive Brood Reduction — Siblicide

Siblicide (aggressive infanticide carried out by full or half sibs, *sensu* Mock, 1984a) occurs in a variety of forms. As with other types of brood reduction it can be divided for convenience into the two broad categories of obligate (where it virtually always occurs) and facultative (where its occurrence is less certain) siblicide.

Obligate siblicide

Obligate siblicide is best known in various eagles, boobies, and pelicans. All lay two eggs, hatching is asynchronous, and usually the elder sib kills its younger sib soon after hatch. From the parent's perspective, the second egg/chick serves as insurance against the unexpected failure of the first (Cash and Evans, 1986; Anderson, 1990). Interestingly, obligate siblicide seems to be independent of prevailing food at the time of the junior siblings demise. This is graphically illustrated in Gargett's (1978) description of a four-day-old black eagle (*Aquila verreauxi*) chick, weighing 163 g, pummelling its newly hatched (and doomed) sib while more than 5 kg of prey remained uneaten in the nest. Stinson (1979) proposed that "pending competition" can explain this paradoxical behavior. Simply, food might not be limiting at the time of the junior sib's demise, but may be likely to become so (see also Forbes, 1991b). Moreover, older, larger siblings are likely more difficult to kill later on. Thus, preemptive siblicide is favored. Forbes and Ydenberg (1992), using a simple inclusive fitness model, have further shown that senior sibs should be risk averse when stochastic variation in food levels occurs across years. As a consequence, obligate siblicide may even be favored when sufficient food exists to rear both chicks in most years.

For nearly all such 'obligately siblicidal' species, instances where both chicks survive are reported, albeit rarely (Gargett, 1968; Brown, 1974; Brown et al., 1977; Nelson, 1978; Knopf, 1979; Hustler and Howells, 1986). Clearly 'obligate' siblicide is not truly obligatory. Why this is so is unclear but at least three possibilities exist: (1) these species represent the extreme of the facultative siblicide continuum - *e.g.* chick provisioning rarely reaches the levels necessary to inhibit siblicide, (2) siblicide in these species is polymorphic: a strategy of 'pure' obligate siblicide exists, as does a strategy of facultative siblicide, or (3) the brood reduction mechanism is inefficient, as the senior sib sometimes fails to eliminate its junior sib. Present information does not allow us to discriminate between these alternatives, and experimental alteration of provisioning schedules in these species will be needed to establish the independence of current food and siblicide.

Facultative siblicide

Facultative siblicide, where the survival of the junior sibling is in some way conditional upon the prevailing levels of food, may occur with or without a proximate link between food, hunger, and sibling aggression. Where a link between food and aggression exists, chicks seem to obey the simple behav-

ioral rule: fight if hungry. Food shortage triggers aggression resulting in the demise of the junior-most member(s) of the brood hierarchy.

Where a proximate link between food and aggression does not exist, the aggression per se is obligatory, but chick survival depends upon prevailing food levels. When food is abundant, junior sibs remain sufficiently strong to withstand the beatings from their senior sibs and survive to fledge, whereas if food is short, junior sibs weaken and die. Such a system appears to operate in herons and egrets (Mock *et al.,* 1987a).

The key difference between these two forms of facultative siblicide appears to be within-season variation in food availability (Mock *et al.,* 1987a). When the food a chick receives today is a reliable indicator of what it will receive tomorrow, facultative aggression may be favored. Under such conditions chick hunger accurately portends pending food shortages, and is thus an appropriate trigger for aggression. Similarly, if food is plentiful and likely to remain so, then aggression is superfluous. However, where the food a chick receives today is not a reliable indicator of what it will get tomorrow, obligatory aggression may be favored; senior sibs guarantee priority access to parental resources such that if an unpredictable food shortage does occur, junior sibs are burdened with the shortfall.

FOOD AMOUNT AND FACULTATIVE SIBLICIDE

Aggression is an important determinant of food allocation among nestlings of siblicidal birds, and the role of food amount in governing sibling aggression has received much recent attention (Procter, 1975; Brown *et al.*, 1977; Nelson, 1978; Stinson, 1979, 1980; Poole, 1979, 1982, 1984; Nuechterlein, 1981; Safriel, 1981; Braun and Hunt, 1983; Graves *et al.*, 1984; Mock, 1984a, b, 1985; Fujioka, 1985b; Ploger and Mock, 1986; Mock *et al.*, 1987a,b; Cash and Evans, 1987; Simmons, 1988; Anderson, 1989; Drummond and Garcia Chavelas, 1989; Parker *et al.*, 1989; Forbes, 1991c). Summarizing previous literature, Mock, *et al.* (1987a) articulated a proximate control system for such aggression, which they labelled the Food Amount Hypothesis (FAH): current food shortage increases hunger levels of chicks, which in turn triggers increased (sometimes fatally) levels of sibling aggression. However, an uneven pattern in the proximate relationship between hunger and aggression has begun to emerge: aggression appears to be food-mediated in some species (*e.g.* blue-footed boobies (*Sula nebouxii*), Drummond and Garcia Chavelas, 1989; skuas (*Catharacta maccormicki*), Procter, 1975; oystercatchers (*Haematopus ostralegus*), Safriel, 1981; os-

preys (*Pandion haliaetus*), Poole, 1979, 1984, 1989; Forbes, 1991c), but not in others (*e.g.* cattle and great egrets (*Casmerodius albus*), Mock *et al.*, 1987a; great blue herons (*Ardea herodias*), Mock *et al.*, 1987b).

Experimental manipulations of food amount give us our best insight into the proximate relationship between food and aggression. Such manipulations have generally taken one of three forms: (1) food addition, (2) food deprivation, and (3) brood size enlargement/ reduction. Mock *et al.* (1987a) added a single 'lump sum' of food once daily, approximately doubling the amount of food fed to egret broods. This had no discernible effect on mean daily rates of sibling aggression. However, recently fed egret chicks exhibited a temporary disinclination to attack (Fujioka, 1985a; Mock *et al.*, 1987a) as has been reported for grebes (Nuechterlein, 1981) and kittiwakes (Braun and Hunt, 1983). Why this is so is unclear.

Drummond and Garcia Chavelas (1989) reported increased aggression with artificial food deprivation in boobies; Procter (1975) reported similar observations though serious flaws in experimental design leave these results open to question (Mock *et al.*, 1987a; Drummond and Garcia Chavelas, 1989). Poole (1984) enlarged broods in six osprey nests and found increased aggression among chicks; male parents elevated rates of prey delivery, but junior sibs grew more slowly anyway, as dominant sibs took disproportionate shares of the food.

INFLUENCE OF SIBLING COMPETITION ON SEX RATIOS

Fisher (1930) argued that the ratio of male to female offspring at the end of the period of parental care should be inversely related to the cost of rearing each sex. Many birds are sexually dimorphic, and it seems likely that costs of rearing each sex may differ considerably (Fiala and Congdon, 1983; Bortolotti, 1986a; Slagsvold *et al.*, 1986; Teather and Weatherhead, 1988). Differential rearing costs of males and females sometimes results in differential mortality between the sexes during the brood-rearing period (Howe, 1976; Cronmiller and Thompson, 1981; Fiala, 1981; Bancroft, 1986; Slagsvold *et al.*, 1986). In these cases, the larger males were more likely to perish during periods of food shortage. Apparently members of the smaller sex are not always disadvantaged in sibling competition (Stamps, 1990).

If males and females are not equally affected by competition, or not equally effective at competing with a sibling of the opposite sex, then the cost of rearing sexually dimorphic offspring may not be independent of the

sex composition of the brood (Bortolotti, 1986a). Parents may therefore manipulate the sex ratio of their brood to achieve an optimal combination (order) of the two sexes. Bortolotti determined that the probability of brood reduction is not equal among bald eagle (*Haliaeetus leucocephalus*) broods of different sex composition. The hatching sequence of male first and female second was predicted to have the highest probability of nestling mortality (females being the larger sex, thus are likely to 'catch up' to first hatched males). Interestingly, this type of brood is rare in bald eagles, as there is a sex-dependent hatching sequence in mixed sex broods (93% of the first hatched chicks are female, Bortolotti 1986a). Similarly, there is some suggestion of a post-natal sex bias in golden eagles (*Aquila chrysaetos*) that may favor females in years of food scarcity (Edwards *et al.*, 1988; but see Arnold, 1989; Bortolotti, 1989; Forbes, 1990b).

PARENT-OFFSPRING RELATIONS

Parent-Offspring Conflict in Siblicidal Birds

In siblicidal birds, parents establish competitive asymmetries among offspring at hatching, and in doing so, appear to relinquish at least partial control over (1) how resources are shared among offspring, and (2) eventual brood size. Through size and motor-skill advantages and the use of aggression, senior siblings command a disproportionate share of parental resources, and are capable of eliminating junior siblings. Due to the differing genetic interests of parents and offspring (Trivers, 1974), we expect evolutionary conflict over resource allocation and brood size. Yet behavioral evidence of such conflict in birds is slim. Parents rarely intervene in sibling aggression, and generally do not feed junior sibs preferentially, although some exceptions exist (see Mock, 1984a, 1987; Drummond *et al.*, 1986; Stamps *et al.*, 1985; Drummond, 1989). In effect, natural selection has armed the offspring with the necessary weaponry (talons, sharp bills), and the parents, rather than interfere as peace-keepers, fortify the seniormost sibs via hatching asynchrony. Why do parents seem to collude with elder offspring to facilitate brood reduction? Mock (1987) suggested two alternatives to resolve this seeming paradox. First, it may be that significant parent-offspring conflict does exist and the offspring are "winning". If this is the case, two further questions arise: (1) Why has natural selection not favored opposing parental behavior, and (2) Why do parents establish the initial competitive asymmetries (*e.g.* via hatch asynchrony) among offspring? Alternatively, perhaps no significant evolutionary conflict exists: offspring perform siblicide according to their own selfish tendencies, and

this behavior also serves the parent's interests. That is parents create the initial asymmetries among offspring, but do not intervene further. Such a "laissez-faire" policy (*sensu* Mock, 1987), where resources are shared unequally among offspring, may be favored if it allows parents to adjust brood size to reflect variation in food supplies. Drummond *et al.* (1986) have described such congruency between parent and offspring interests as 'parent-offspring cooperation'. In effect, reduced control over brood size and food sharing may be the evolutionary cost parents pay for a system of adaptive brood reduction.

A third alternative exists however: a parental counter-strategy to curtail investment in selfish offspring may constrain the occurrence of siblicide. O'Connor (1978), using a simple kin selection model, made a simple but important point: senior sibs should prefer to reduce the size of the brood before the parents, *ipso facto* parent-offspring conflict over the timing and/or extent of brood reduction. O'Connor's argument, however, contains two implicit assumptions: parental resources are (1) finite, and (2) fixed. Of course if resources are not limiting, we should expect no phenotypic conflict between parents and offspring over how resources are shared among sibs: everyone gets plenty of food. But such utopian food conditions seem likely to be rare (otherwise parents should increase brood size another notch), and we expect generally that resources are limited. When resources are limited, parents must not only decide how to allocate such resources among present offspring, but how to allocate resources between present and future offspring as well (Winkler and Wallin, 1987). O'Connor's model assumes that only the former is important – *e.g.* the elimination of a brood-mate results in enhanced survival probabilities for the remaining brood members, presumably because fewer mouths are competing for the same amount of food. If, however, parents are optimizing their investment in current and future broods, parents should reduce provisioning to the current brood in favor of increased investment in future broods if siblicide occurs, in accordance with William's principle (Sargent and Gross, 1986). (Of course this presumes that siblicide is not also in the parent's interest as might often be the case). Thus by killing a junior sib, a senior sib may usurp a bigger piece of the parental resource pie (as O'Connor implied), but the size of the pie may shrink. Parents in effect 'tax' the selfishness of current brood members in favor of future broods. Forbes (1993) developed a simple inclusive fitness model of brood reduction incorporating parental costs of reproduction, and showed through numerical simulation that the prospect of such parental retaliation may sometimes (but not always) restrain the siblicidal tendencies of senior sibs.

The Evolutionary Limits of Parental Favoritism

Parents might under some circumstances choose to circumvent the dominance hierarchy and feed junior sibs preferentially. This tactic would seem
to be a more effective means for parents to control resource allocation
among sibs than by working through the dominance hierarchy, but they
may be constrained from doing so by the prospect of evolutionary retaliation by the sibs (Parker and MacNair, 1979; Drummond, 1989). Senior
sibs enjoy priority access to food and hence superior prospects for growth
and survival. Presumably, they would be selected to oppose preferential
feeding of junior sibs since this might upset the status quo (*i.e.* junior sibs
may displace senior sibs in the hierarchy; Ploger and Mock, 1986). As well
junior sibs may take resources that might otherwise go to the senior sib.
In order to minimize these risks, senior sibs have the option of eliminating junior sibs long before they could become a threat (obligate siblicide),
thereby eliminating the benefits of a system of facultative brood reduction to parents. In non-siblicidal species, where parents face no risk of
preemptive siblicide, parents might get away with preferential feeding of
junior sibs. Such may be the case in non-siblicidal budgerigars (*Melopsit-
tacus undulatus*) (Stamps *et al.*, 1987).

SUMMARY

Parents often create surplus offspring to allow for tracking variable food,
insurance against accidental failure of offspring, developmental selection,
and/or exploitation as food during periods of food stress. A system of
brood reduction, enabled by establishing competitive asymmetries among
offspring at hatch, facilitates removal of 'redundant' offspring. Although
hatching asynchrony may be favored by selective forces unrelated to brood
reduction per se, it nonetheless affects profoundly the nature of sibling
competition and the potential for brood reduction. Avian brood reduction is usually the lethal consequence of sibling rivalry (aggressive or otherwise), but such behavior may also serve parental interests. Siblicide
may be obligatory or facultative, the latter being at least partly conditional
upon prevailing food. However sibling aggression in facultatively siblicidal species need not be proximately related to prevailing food. We expect
genotypic conflict between parents and offspring over resource allocation
among siblings, and brood size, yet behavioral evidence of such conflict in
siblicidal birds is slim. Either the offspring are winning, or no phenotypic
conflict exists. Reduced parental control over food sharing and brood size
in the latter case might be the evolutionary price for parents of a system of

adaptive brood reduction. Alternatively, the prospect of parental retaliation (reduced parental investment in the current brood in favor of greater investment in future broods), may constrain the siblicidal tendencies of senior sibs.

REFERENCES

AMUNDSEN, T. and Slagsvold, T. (1991) Hatching asynchrony: facilitating adaptive or maladaptive brood reduction? *Acta XX Congressus Internationalis Ornithologici*, 1707–1719.

AMUNDSEN, T. and Stokland, J. N. (1988) Adaptive significance of asynchronous hatching in the shag: a test of the brood reduction hypothesis. *Journal of Animal Ecology*, **57**, 329–344.

ANDERSON, D. J. (1989). The role of hatching asynchrony in siblicidal brood reduction of two booby species. *Behavioural Ecology and Sociobiology*, **25**: 363–368.

ANDERSON, D. J. (1990) Evolution of obligate siblicide in boobies. 1. A test of the insurance-egg hypothesis. *American Naturalist*, **135**, 334–350.

ARNOLD, T. W. (1989) Sex ratios of fledgling golden eagles and jackrabbit densities. *Auk*, **106**, 521–522.

BANCROFT, G. T. (1986) Nesting success and mortality of the boat-tailed grackle. *Auk*, **103**, 86–99.

BORTOLOTTI, G. R. (1986a) Influence of sibling competition on nestling sex ratios of sexually dimorphic birds. *American Naturalist*, **127**, 495–507.

BORTOLOTTI, G. R. (1986b) Evolution of growth rates in eagles: sibling competition vs. energy considerations. *Ecology*, **67**, 182–194.

BORTOLOTTI, G. R. (1989) Sex ratios of fledgling golden eagles. *Auk*, **106**, 520–521.

BORTOLOTTI, G. R., Wiebe, K. L. and Iko, W. M. (1991) Cannibalism of nestling American kestrels by their parents and siblings. *Canadian Journal of Zoology*, **69**, 1447–1453.

BOYCE, M. S. and Perrins, C. M. (1987) Optimizing great tit clutch size in a fluctuating environment. *Ecology*, **68**, 142–153.

BRAUN, B. M. (1981) *Siblicide, the mechanism of brood reduction in the black-legged kittiwake*. MS Thesis, Univ. of California, Irvine.

BRAUN, B. M. and Hunt, G. L. Jr. (1983) Brood reduction in black–legged kittiwakes. *Auk*, **100**, 469–476.

BROWN, L. H. (1974) A record of two young reared by Verreaux's eagle. *Ostrich*, **45**, 146–147.

BROWN, L.H., Gargett, V. and Steyn, P. (1977) Breeding success in some African eagles related to theories about sibling aggression and its effects. *Ostrich 48*, 65–71.

BUCHHOLZ, J. T. (1922) Developmental selection in vascular plants. *Botanical Gazette*, **73**, 249–286.

CASH, K. J. and Evans, R. M. (1986) Brood reduction in the American white pelican. *Behavioural Ecology and Sociobiology*, **18**, 413–418.

CASH, K. J. and Evans, R. M. (1987) The occurrence, context and functional significance of aggressive begging behaviours in young American white pelicans. *Behaviour*, **102**, 119–128.

CLARK, A. B. and Wilson, D. S. (1981) Avian breeding adaptations: hatching asynchrony, brood reduction, and nest failure. *Quarterly Review of Biology*, **52**, 253–273.

CLARK, A. B. and Wilson, D.S. (1985) The onset of incubation in birds. *American Naturalist*, **125**, 603–611.

COHEN, D. (1966) Optimizing reproduction in a randomly varying environment. *Journal of Theoretical Biology*, **12**, 119–129.

COULSON, J. C. and Porter, J. M. (1985) Reproductive success of the kittiwake *Rissa tridactyla*: the roles of clutch size, chick growth rates and parental quality. *Ibis*, **127**, 450–466.

CRONMILLER, J. R. and Thompson, C. F. (1981) Sex ratio adjustment of malnourished red-winged blackbird broods. *Journal of Field Ornithology*, **52**, 65–67.

DORWARD, E. F. (1962) Comparative biology of the white booby and brown booby Sula spp. at Ascension. *Ibis*, **103b**, 174–220.

DRUMMOND, H. (1989) Parent-offspring conflict and siblicidal brood reduction in boobies. *Acta XIX Congressus Internationalis Ornithologici*, 1244–1253.

DRUMMOND, H., Gonzalez, E. and Osorno, J. L. (1986) Parent-offspring cooperation in the blue-footed booby (*Sula nebouxii*): social roles in infanticidal brood reduction.*Behavioural Ecology Sociobiology*, **19**, 365–373.

DRUMMOND, H. and Garcia Chavelas, C. (1989) Food shortage influences sibling aggression in the blue-footed booby. *Animal Behaviour*, **37**, 806–819.

EDWARDS, T. C. Jr., Collopy, M. W., Steenof, K. and Kochert, M. N. (1988) Sex ratios of fledgling golden eagles. *Auk*, **105**, 793–796.

EICKWORT, K. R. (1973) Cannibalism and kin selection in *Labidomera clivicollis* (Coleoptera: Chrysomelidae). *American Naturalist*, **107**, 452–453.

FIALA, K. L. (1981) Sex ratio constancy in the red-winged blackbird. *Evolution*, **35**, 898–910.

FIALA, K. L. and Congdon, J. D. (1983) Energetic consequences of sexual size dimorphism in nestling red-winged blackbirds. *Ecology*, **64**, 642–647.

FISHER, R. A. (1930) *The Genetical Theory of Natural Selection*. Oxford Univ.Press, Oxford.

FORBES, L. S. (1990a) Insurance offspring and the evolution of avian clutch size. *Journal of Theoretical Biology*, **147**, 345–359.

FORBES, L. S. (1990b) A note on statistical power. *Auk*, **107**, 438–439.

FORBES, L. S. (1991a) Insurance offspring and brood reduction in a variable environment: the costs and benefits of pessimism. *Oikos*, **62**, 325–332.

FORBES, L. S. (1991b) Burgers or brothers: food shortage and the threshold for brood reduction. *Acta XX Congressus Internationalis Ornithologici*, 1720–1726.

FORBES, L. S. (1991c) Hunger and food allocation among broods of facultatively siblicidal ospreys. *Behavioural Ecology and Sociobiology*, **29**, 189–195.

FORBES, L. S. (1993) Avian blood reduction and parent-offspring "conflict". *American Naturalist*, **142**, 82–117.

FORBES, L. S. and Ydenberg, R. C. (1992) Sibling rivalry in a variable environment. *Theoretical Population Biology*, **41**, 000–000.

FUJIOKA, M. (1985a) Food delivery and sibling competition in experimentally even-aged broods of the cattle egret. *Behavioural Ecology and Sociobiology*, **17**, 67–74.

FUJIOKA, M. (1985b) Sibling competition and siblicide in asynchronously-hatching broods of the cattle egret *Bubulcus ibis*. *Animal Behaviour*, **33**, 1228–1242.

GARGETT, V. (1968) Two Wahlberg's eagle chicks – a one in forty eight chance. *Honeyguide*, **56**, 24.

GARGETT, V. (1978) Sibling aggression in the black eagle in the Matopos, Rhodesia. *Ostrich*, **49**, 57–63.

GIBBONS, D. W. (1987) Hatching asynchrony reduces parental investment in the jackdaw. *Journal of Animal Ecology*, **56**, 403–414.

GILLESPIE, J. H. (1977) Natural selection for variance in offspring numbers: a new evolutionary principle. *American Naturalist*, **111**, 1010–1014.

GILMORE, R. G., Dodrill, J. W. and Linley, P. A. (1983) Reproduction and development of the sand tiger shark, *Odontaspis taurus* (Rafinesque). *Fishery Bulletin*, **81**, 201–225.

GRAVES, J., Whiten, A. and Heinzi, P. (1984) Why does the herring gull lay three eggs? *Animal Behaviour*, **32**, 798–805.

HAHN, D. C. (1981) Asynchronous hatching in the laughing gull: cutting losses and reducing sibling rivalry. *Animal Behaviour*, **29**, 421–427.

HART, J. L. (1973) Pacific fishes of Canada. *Fisheries Research Board of Canada Bulletin No. 180*.

HARVEY, H. T., Shellhammer, H. S. and Stecker, R. E. (1980) Giant sequoia ecology. U.S. Dept. Int., Nat. Park Serv., Sci. Monogr. Series No. 12.

HOCHACHKA, W. (1990) Seasonal decline in reproductive performance of song sparrows. *Ecology*, **71**, 1279–1288.

HOWE, H. F. (1976) Egg size, hatching asynchrony, sex and brood reduction in the common grackle. *Ecology*, **57**, 1195–1207.

HRDY, S. B. (1979) Infanticide among animals: A review, classification, and examination of the implications for the reproductive strategies of females. *Ethology and Sociobiology*, **1**, 13–40.

HUSSELL, D. J. T. (1972) Factors affecting clutch size in arctic passerines. *Ecological Monographs*, **42**, 317–364.

HUSTLER, C. W. and Howells, W. W. (1986) A population study of tawny eagles in the Hwange National Park. *Ostrich*, **57**, 101–106.

INGRAM, C. (1959) The importance of juvenile cannibalism in the breeding biology of certain birds of prey. *Auk*, **76**, 218–226.

KNOPF, F. L. (1979) Spatial and temporal aspects of colonial nesting of white pelicans. *Condor*, **81**, 353–363.

KOZLOWSKI, J. and Stearns, S. C. (1989) Hypotheses for the production of excess zygotes: models of bet-hedging and selective abortion. *Evolution*, **43**, 1369–1377.

KREMENTZ, D. G., Nichols, J. D. and Hines, J. (1989) Postfledging survival of European starlings. *Ecology*, **70**, 646–655.

LACK, D. (1947) The significance of clutch size. *Ibis*, **89**, 302–352.

LACK, D. (1954) *The Natural Regulation of Animal Numbers*. Clarendon Press, Oxford.

LAMEY, T. C. and Mock, D. W. (1991) Nonaggressive brood reduction in birds. *Acta XX Congressus Internationalis Ornithologici*, 1741–1751.

LESSELLS, C. M. and Avery, M. I. (1989) Hatching asynchrony in European bee-eaters *Merops apiaster*. *Journal of Animal Ecology*, **58**, 815–835.

LLOYD, D. G. (1980) Sexual strategies in plants. I. An hypothesis of serial adjustment of maternal investment during one reproductive session. *New Phytologist*, **86**, 69–79.

LYONS, A. and Spight, T. M. (1973) Diversity of feeding mechanisms among embryos of Pacific northwest Thais. *Veliger*, **16**, 184–194.

MAGRATH, R. D. (1989) Hatching asynchrony and reproductive success in the blackbird. *Nature*, **339**, 536–538.

MAGRATH, R. D. (1990) Hatching asynchrony in altricial birds. *Biological Reviews*, **65**, 587–622.

MAGRATH, R. D. (1991) Nestling weight and juvenile survival in the blackbird, *Turdus merula*. *Journal of Animal Ecology*, **60**, 335–351.

MEAD, P. S. and Morton, P. I. (1985) Hatching asynchrony in the mountain white-crowned sparrow (*Zonotrichia leucophrys oriantha*): a selected or incidental trait? *Auk*, **102**, 781–792.

MEYBURG, B.-U. (1974) Sibling aggression and mortality among nestling eagles. *Ibis*, **116**, 224–228.

MILLER, R. S. (1973) The brood size of cranes. *Wilson Bulletin*, **85**, 436–441.

MOCK, D. W. (1984a) Infanticide, siblicide, and avian nestling mortality. In: Hausfater, G. and Hrdy, S. B. (eds.), *Infanticide: Comparative and Evolutionary Perspectives*. Aldine, New York, pp. 2–30.

MOCK, D. W. (1984b) Siblicidal aggression and resource monopolization in birds. *Science*, **225**, 731–733.

MOCK, D. W. (1985) Siblicidal brood reduction: the prey size hypothesis. *American Naturalist*, **125**, 327–343.

MOCK, D. W. (1987) Siblicide, parent-offspring conflict, and unequal parental investment by egrets and herons. *Behavioural Ecology and Sociobiology*, **20**, 247–256.

MOCK D. W., Lamey, T. C. and Ploger, B. J. (1987a) Proximate and ultimate roles of food amount in regulating egret sibling aggression. *Ecology*, **68**, 1760–1772.

MOCK D. W., Lamey, T. C., Williams, C. F. and Pelletier, A. (1987b) Flexibility in the development of heron sibling aggression: an intraspecific test of the prey size hypothesis. *Animal Behaviour*, **35**, 1386–1393.

MOCK, D. W. and Parker, G. A. (1986) Advantages and disadvantages of egret and heron brood reduction. *Evolution*, **40**, 459–470.

MOCK, D. W. and Ploger, B. J. (1987) Parental manipulation of optimal hatch asynchrony in cattle egrets: an experimental study. *Animal Behaviour*, **35**, 150–160.

MOCK, D. W. and Schwagmeyer, P. L. (1990) The peak load reduction hypothesis for avian hatching asynchrony. *Evolutionary Ecology*, **4**, 249–260.

NELSON, J. B. (1978) *The Sulidae: the Gannets and Boobies*. Oxford, London.

NISBET, I. C. T. and Cohen, M. E. (1975) Asynchronous hatching in common and roseate terns, *Sterna hirundo* and *S. dougallii*. *Ibis*, **117**, 374–379.

NUECHTERLEIN, G. L. (1981) Asynchronous hatching and sibling competition in western grebes. *Canadian Journal of Zoology*, **59**, 994–998.

NUR, N. (1988) The cost of reproduction in birds: an examination of the evidence. *Ardea*, **76**, 155–168.

O'CONNOR, R. J. (1978) Brood reduction in birds: selection for fratricide, infanticide, and suicide? *Animal Behaviour*, **26**, 79–96.

O'GARA, B. W. (1969) Unique aspects of reproduction in female pronghorn (*Antilocapra americana Ord*). *American Journal of Anatomy*, **125**, 217–231.

PARKER, G. A. and MacNair, M. (1979) Models of parent-offspring conflict. IV. Suppression: evolutionary retaliation by the parent. *Animal Behaviour*, **27**, 1210–1235.

PARKER, G. A., Mock, D. C. and Lamey, T. C. (1989) How selfish should stronger sibs be? *American Naturalist*, **133**, 846–868.

PERRIGO, G. (1987) Breeding and feeding strategies in deer mice and house mice when females are challenged to work for their food. *Animal Behaviour*, **35**, 1298–1316.

PERRINS, C. M. (1963) Survival in the great tit *Parus major*. *Acta XIII Congressus Internationalis Ornithologici*, 717–728.

PIJANOWSKI, B. (1992). A revision of Lack's brood reduction hypothesis. *American Naturalist*, 139: 1270–1292.

PLOGER, B. J. and Mock, D. W. (1986) Role of sibling aggression in the distribution of food to nestling cattle egrets (*Bubulcus ibis*). *Auk*, **103**, 768–776.

POLIS, G. A. (1984) Intraspecific predation and "infant killing" among invertebrates. In: Hausfater, G and Hrdy S. B (eds.), *Infanticide: Comparative and Evolutionary Perspectives*. Aldine Publ. Co., New York, pp. 87–104.

POOLE, A. F. (1979) Sibling aggression among nestling ospreys in Florida Bay. *Auk*, **96**, 415–416.

POOLE, A. F. (1982) Brood reduction in temperate and sub-tropical ospreys. *Oecologia*, **53**, 111–119.

POOLE, A. F. (1984) *Reproductive limitation in coastal ospreys: an ecological and evolutionary perspective*. Ph.D. Thesis, Boston University, Boston, MA.

POOLE, A. F. (1989) *Ospreys*. Cambridge Univ. Press, New York.

PROCTER, D. L. C. (1975) The problem of chick loss in the south polar skua *Catharacta maccormicki*. *Ibis*, **117**, 517–520.

RICKLEFS, R. E. (1965) Brood reduction in the curve-billed thrasher. *Condor*, **67**, 505–510.

RØSKAFT, E. and Slagsvold, T. (1985) Differential mortality of male and female offspring in experimentally manipulated broods of the rook. *Journal of Animal Ecology*, **54**, 261–266.

SAFRIEL, U. N. (1981) Social hierarchy among siblings in broods of the oystercatcher (*Haematopus ostralegus*). *Behavioural Ecology and Sociobiology*, **9**, 59–63.

SARGENT, C. and Gross, M. R. (1986) Williams' principle: an explanation of parental care in teleost fishes. In: T. J. Pitcher (ed.), *The Behavior of Teleost Fishes*, Croom Helm, London, pp. 275–293.

SEGER, J. and Brockmann, H. J. (1987) What is bet-hedging? *Oxford Surveys in Evolutionary Biology*, **4**, 182–211.

SHAW, P. (1985) Brood reduction in the blue-eyed shag *Phalacrocorax atriceps*. *Ibis*, **127**, 476–494.

SIMMONS, R. (1988) Offspring quality and the evolution of Cainism. *Ibis*, **130**, 339–357.

SLAGSVOLD, T. (1985) Asynchronous hatching in passerine birds: influence of hatching failure and brood reduction. *Ornis Scandinavica*, **16**, 81–87.

SLAGSVOLD, T. (1986) Asynchronous versus synchronous hatching in birds: experiments with the pied flycatcher. *Journal of Animal Ecology*, **55**, 1115–1134.

SLAGSVOLD, T. (1990) Fisher's sex ratio theory may explain hatching patterns in birds. *Evolution*, **44**, 1009–1017.

SLAGSVOLD, T. and Lifjeld, J. T. (1989) Hatching asynchrony in birds: the hypothesis of sexual conflict over parental investment. *American Naturalist*, **134**, 239–253.

SLAGSVOLD, T., Røskaft, E. and Engen, S. (1986) Sex ratio, differential cost of rearing young, and differential mortality between the sexes during the period of parental care: Fisher's theory applied to birds. *Ornis Scandinavica*, **17**, 117–125.

SMITH, H. G., Kallander H. and Nilsson, J.-A. (1989) The trade-off between offspring number and quality in the great tit *Parus major*. *Journal of Animal Ecology*, **58**, 383–401.

SPRINGER, S. (1948) Oviphagous embryos of the sand shark, *Carcharias taurus*. *Copeia*, **1948**, 153–157.

256   L.S. FORBES and D.W. MOCK

STAMPS, J. A. (1990) When should avian parents differentially provision sons and daughters. *American Naturalist*, **135**, 671–685.

STAMPS, J. A, Clark, A. B., Arrowood, P. and Kus, B. (1985) The effects of parent and offspring gender on food allocation in budgerigars. *Behaviour*, **101**, 177–199.

STINSON, C. H. (1979) On the selective advantage of fratricide in raptors. *Evolution*, **33**, 1219–1225.

STINSON, C. H. (1980) Weather dependent foraging success and sibling aggression in red-tailed hawks in central Washington. *Condor*, **82**, 76–80.

TAIT, D. E. N. (1980) Abandonment as a tactic in grizzly bears. *American Naturalist*, **115**, 800–808.

TEATHER, K. L. and Weatherhead, P. J. (1988) Sex-specific energy requirements of great-tailed grackle (*Quiscalus mexicanus*) nestlings. *Journal of Animal Ecology*, **57**, 659–668.

TEMME, D. H. and Charnov, E. L. (1987) Brood size adjustment in birds: Economical tracking in a temporally varying environment. *Journal of Theoretical Biology*, **126**, 137–147.

TOFT, C. A. and Shea, P. J. (1983) Detecting community-wide patterns: estimating power strengthens statistical inference. *American Naturalist*, **122**, 618–625.

TRIVERS, R. L. (1974) Parent-offspring conflict. *American Zoologist*, **14**, 249–264.

WARHAM, J. (1975) The crested penguins. In: B. Stonehouse (ed.), *The Biology of Penguins*, University Park Press, Baltimore, pp. 189–269.

WINKLER, D. W. and Wallin, K. (1987) Offspring size and number: a life history model linking effort per offspring and total effort. *American Naturalist*, **129**, 708–720.

CHAPTER 11

PROTECTION AND ABUSE
OF YOUNG IN PINNIPEDS

BURNEY J. LE BOEUF[1] and CLAUDIO CAMPAGNA[2]

[1]Department of Biology and Institute for Marine Sciences,
University of California, Santa Cruz,
California 95064, USA

[2]Centro Nacional Patagonico, Puerto Madryn, Chubut, Argentina

ABSTRACT

The aim of this paper is to describe nurturing and abusive behavior to young in selected species of two families of pinnipeds, the phocids and otariids, attempt to explain the phenomena observed from a functional or proximal perspective, and compare the behaviors seen in pinnipeds to those observed in other large mammals. We focus on the most thoroughly studied species, such as the elephant seals, *Mirounga angustirostris*, and *M. leonina*, and the southern sea lion, *Otaria byronia*, and incorporate observations from other species when possible.

I. NATURAL HISTORY OVERVIEW

A brief review of the social systems and context in which young pinnipeds are produced and reared helps in understanding the treatment of young by adults of both sexes.

All 15 species of eared seals or, *Otariidae*, the sea lions and fur seals, are polygynous and sexually dimorphic with males being larger than females (King, 1983; Riedman, 1990). Adult males of all species are territorial, defending terrestrial or aquatic areas containing females and their pups

257

against other males. All mating and nursing of young occurs on territories. Non-territorial males, most of whom are younger, group together in the general vicinity of territories. The 19 species of phocids, or true seals, vary from moderate polygyny to extreme polygyny, from reversed dimorphism with the female being larger, to monomorphism, to extreme sexual dimorphism with males being larger. Breeding on pack or fast ice is associated with small "family" groups (a female and her pup attended by a male), monomorphism, and a low level of polygyny. In the more social and polygynous species, such as elephant seals, males compete in dominance hierarchies. As a rule, nursing females are less often disturbed by male behavior in territories than when males compete in a dominance hierarchy. In the former, only one male moves among the females and their pups, while in the latter male traffic is great and a threat to female and pup safety. The walrus, *Odobenus rosmarus*, breeds in small social groups on floating ice (Fay 1982). Males are larger than females and the species appears to be moderately polygynous; the mating system around an ice floe where estrous females gather resembles a lek. Males make vocal and visual displays in the water to attract estrous females.

With respect to protection and nurturance, the mother does everything and the males do nothing in all species. Females give birth to a single pup annually in all species except the walrus, and the Australian sea lion, *Neophoca cinerea*, where the time between births exceeds one year (Marlow, 1975). Gestation lasts less than a year in all species except the walrus; gestation in this species lasts 15-16 months (Fay, 1982).

There are many differences in the lactation pattern of phocids and otariids (Costa, 1990; see Table I) and these have important implications for pup health and safety. Lactation is distributed from four months to over a year among the eared seals (Gentry *et al.*, 1986). After giving birth, a female nurses her pup for about one week then goes to sea to feed for 2-7 days. For the next few months the mother alternates nursing the pup on land for about two days with feeding trips to sea that last less than seven days. When the pup is about four months old, the pair may leave the rookery and forage together until the following year when most females wean their pups just before giving birth again. In some species (Galapagos fur seal, *Arctocephalus galapagoensis*, Gentry *et al.*, 1986, and Steller sea lion, *Eumetopias jubatus*, Pitcher and Calkins, 1981), intermittent suckling may last up to three years. In contrast, lactation is concentrated into a brief period among phocids, lasting only four days in hooded seals, *Cystophora cristata*, and up to six weeks in weddell seals, *Leptonychotes weddelli*, and the Hawaiian monk seals, *Monachus schauinslandi*. The brevity of the nursing period of phocids is made possible by fasting. In several well

Table I Differences in the nursing pattern and maternal behavior of seals (phocids) and sea lions (otariids). Source: see references in text.

	Seals	Sea Lions
Nursing		
Periodicity	Concentrated (nurse daily while fasting)	Distributed (nurse 2 days, forage 2–7 days)
Duration	4–45 days	4 mo. to 1 year
Weaning	abrupt	gradual
Pup recognition	poor	good
Adoption	common	rare
Milk stealing	common	rare

studied species, females fast completely while nursing, remaining on land with their pups feeding them daily from body reserves (Le Boeuf et al., 1972; Reiter et al., 1981).

Because phocids are capable of storing all energy that is required for the entire lactation period and otariids must feed during the lactation interval, mother-pup behavior in the two families differs. Phocid mothers remain near their pups and provide protection for them until weaning. On dangerous shifting pack ice, the best thing a hooded seal mother can do is to provide her pup with fat-rich milk and wean it quickly before the ice breaks up and endangers them. Elephant seal pups that become separated from their mothers during the 4-week nursing period may starve because reunions are unlikely due to the crowded conditions in many harems and the fact that individual recognition is poorly developed in both the mother and the pup (Petrinovich, 1974). Without a source of milk, the neonate dies. It may steal milk from other mothers but this is dangerous. Since all milk is derived from the mother's body stores (Ortiz et al., 1984; Costa et al., 1986), a mother that allows an alien pup to suckle is "taking food out of the mouth" of her own pup. Most females viciously attack alien milk thieves (Le Boeuf and Briggs, 1977). Weaning is abrupt in most phocids. When the mother leaves, the pup is approximately 50% fat and in good physical condition but it is inexperienced socially, and alone, it must learn to cope with the environment and to make a living at sea.

In eared seals, the pup is left alone and is vulnerable when the mother goes to sea to feed; this becomes increasingly less important as the pup

matures. When the mother returns from a feeding trip, she calls and singles out her own pup from the rest. The keen recognition of the mothers for their own pups nearly eliminates milk stealing in these animals. Because the mother alternates feeding with nursing, she can modulate nutrient transfer according to her food intake. For example, in some species the mother transfers more milk to sons than daughters (Costa and Gentry, 1986), a behavior consistent with some predictions from sex ratio theory (*e.g.* Trivers and Willard, 1973; Maynard Smith, 1980; Le Boeuf *et al.*, 1989). In the fall, mothers and pups depart the rookery together. It is assumed that the mother provides some protection to her pup from predators and perhaps, helps the pup to feed on its own. In any case, weaning is gradual and the pup begins to feed on its own months before it is weaned from mother's milk.

II. PROTECTING AND NURTURING BEHAVIOR

In all social pinnipeds, one of the greatest threats to a pup is from neighboring females. Consequently, mothers spend much time protecting their pups against perceived threats. In most species, the movement of an individual female elicits threat vocalizations from neighbors. In Steller sea lions, mothers seek out a reasonably flat surface on the rocks out of the surf zone to give birth, but not too far away from the sea because surf spray provides relief from high temperatures (Gentry, 1970). These sites are in short supply and females compete aggressively for them. They vigorously attempt to keep other females at a distance shortly after birth. With time since parturition, they become increasingly tolerable of female neighbors. In elephant seals, the threat of a neighboring female to a mother's pup is constant and real (Le Boeuf *et al.*, 1972; Reiter *et al.*, 1981; McCann, 1982). At any moment a neighbor may turn and bite her pup; bites to the head and nose may be severe and fatal. The older the mother, the larger her size, and the better she is at defending her pup; pups of the most aggressive mothers are least frequently injured (Christenson and Le Boeuf, 1978). The main concern of the elephant seal mother is to keep her pup near her and to adopt the suckling position when the pup calls to signal its hunger. Good mothers are responsive to the pup at all times. They call to their pups if they stray and they pursue them. Mothers are nearly helpless when a stampeding male tramples, crushes and comes to rest on her pup. Nevertheless, a female will bite and vocally threaten the offending bull, a behavior that has little effect on him and appears to be no more than a nuisance to him.

One of the most interesting nurturing behaviors in pinnipeds involves the permanent adoption or temporary fostering of an orphan or a neighboring pup (Riedman, 1982). This phenomenon is most widespread in the elephant seals, grey seals, Weddells seals, Hawaiian monk seals, and the walrus, all species that breed in moderate to large groups where mother-pup separations are frequent and give many opportunities for adoption (Riedman, 1990). D. Boness (pers. comm.) estimates that 25–75% of female grey seals engage in some form of fostering in certain rookeries. In one study of Weddell seals, 7.8 percent of the females in the sample nursed pups different from the ones with which they were first seen (Stirling, 1975). Among monk seals breeding in the Hawaiian leeward islands, the majority of mothers foster alien pups at some time (Johnson and Johnson, 1978, 1984; Alcorn, 1984; Boness, 1989). Fostering behavior has also been observed in eared seals, especially the Australian sea lion, and the Antarctic fur seal, *A. gazella*, but it is rare, perhaps owing to the fact that females have little difficulty recognizing their own pups.

Studies of northern elephant seals at Año Nuevo Island, California (Reiter *et al.*, 1981; Riedman and Le Boeuf, 1982; Riedman, 1983), have revealed a variety of forms that fostering can take (Table II). Most cases of adoption involve females that have recently lost their own pups. Approximately 50% of these adopt a single pup and treat it as they would their own pup for the remainder of the nursing period, about 3–4 weeks. In most cases, the adopted orphan closely resembles the foster mother's own lost pup in age. Just about every other possible combination occurs between a female that has lost her pup and an adoptee but in small frequencies. A female may adopt two pups, usually one in addition to her own, despite the fact that all but the largest females (Fedak *et al.*, 1989) have only sufficient resources to feed one pup and wean it at the mean weaning mass (126 kg for females and 137 kg for males, Le Boeuf *et al.*, 1989). A few females attract a crowd of 5–6 orphans by allowing all of them to nurse at will. In an apparent case of mistaken identity, some females aggressively compete to nurse a neighbor's pup. After a few days of this, the neighbor usually acquiesces and the two females take turns nursing the pup. On occasion, a female who has just lost her pup will adopt as her own a pup that has just been weaned.

The key variable that affects the incidence of adoption and fostering among elephant seals in a given breeding season is the degree of mother-pup separation (Le Boeuf and Briggs, 1977). The latter varies as a function of the interaction of tides, storms and high surf, crowding, and whether the harem site offers high ground from which mothers and pup can retreat from high water. The frequency of permanent adoptions and temporary

Table II The number of females exhibiting various types of fostering behavior on the Point Beach harem of Año Nuevo Island during three breeding seasons. The number of adoption cases is in parentheses. Adapted from Riedman and Le Boeuf (1982).

Fostering behavior	Frequency			
	1977	1978	1979	Total
Female loses her own pup, then:				
Adopts single pup	10	25	13	48
Adopts two pups	0	0	1(2)	1 (2)
Adopts weaned pup	1	4	1	6
Adopts orphan with another female	1(2)	1(2)	0	2(4)
Shares care of pup with its mother	5	5	1	11
Female nurses her own pup, and:				
Adopts single pup	3	4	6	13
Adopts two pups	1(3)	0	0	1(2)

fostering is a function of the number of orphans in the rookery. Since young females are most likely to lose their pups, it is young females that most frequently adopt and foster orphans. These chaotic times offer some pups the opportunity to obtain more milk than they would from their mothers alone. Pups that are nursed by two mothers reach a weaning mass that is twice that of normal weaned pups (Reiter, *et al.*, 1978).

Among all pinnipeds, the incidence of fostering is highest among species where mother-pup separation is most frequent. This statement is supported by data from northern and southern elephant seals, grey seals (Anderson, 1979), Weddell seals (Kaufman *et al.*, 1975), the walrus (Fay, 1982), and even in a few species of otariids (McCann, 1987). One exception to the rule is the Hawaiian monk seal, a species that breeds in small numbers on broad, sandy beaches where the possibility of mother-pup separation due to crowding is remote. Nevertheless, the incidence of fostering can be as high as 87% (Boness, 1989).

Aside from having the opportunity to foster pups, what explanations for the behavior have been used? On the proximal level, some instances of fostering behavior appear to be due to "reproductive errors", a situation where the female misdirects maternal behavior to an alien pup she mistakes for her own (Le Boeuf and Briggs, 1977). Two functional explanations address the benefits a female may derive by adopting. One pertains specifically to elephant seals and involves the continuance of the regular

reproductive cycle after losing a pup (Le Boeuf *et al.,* 1972). It was observed that females that lost their pups and did not nurse, did not copulate and may have missed giving birth the following year. Hence, lactation may be necessary to induce ovulation. Females that lose a pup and adopt or nurse alien pups come into estrus, copulate, and give birth the next year. It is possible to test this hypothesis but it is difficult. The second explanation, which addresses adoption in general, is that a foster mother increases her inclusive fitness if she adopts a genetically related pup. As yet, there is no evidence that this is the case but it is now possible to test this hypothesis using "DNA fingerprinting" (Jeffries *et al.*, 1985). Lastly, it is possible that females receive no benefit from fostering. Since phocids like elephant seals feed pups from body stores laid down during months at sea before parturition, and they are set to spend about one month on land whether they nurse or not, it matters little that they invest this energy in another pup. The cost is "prepaid" and at the end of lactation the female makes up the lost of body reserves quickly (Le Boeuf *et al.*, 1989).

III. ABUSE OF YOUNG

The incidence of pup abuse is highest in social species where individuals form large groups during the breeding season. A variety of forms of pup abuse have been observed.

The dangerous plight of the walrus mother is unique. As nursing mothers migrate with their pups they haul periodically to rest along with other walruses. The consequences of bringing calves ashore into a mixed herd of walruses are severe. The female and her pup must climb over or around other bodies to find a resting place. Walruses jab with their formidable tusks to protect their place and can cause lethal injury to an adult or pup. Taggart (1987) writes that of 30 walruses which were observed jabbing calves, 14 were adult females, 12 were adult males, and 4 were immature animals. Of the 14 females that jabbed calves, two were nursing their own calves. He notes that on one day during his study the entire group of walruses departed the study site (Punuk Island). Left behind were 119 fresh dead walruses. Of these, 40% were calves, neonates and suckling pups. In elephant seals, breeding age males and nursing females account for 60–95% of the pup deaths on the rookery prior to weaning (Le Boeuf *et al.*, 1972; Le Boeuf and Briggs 1977). The pup mortality rate prior to weaning at the Año Nuevo rookery in California varied from 13–26% of pups born over a nine-year period.

Mother-pup separation leading to the trauma-starvation syndrome is a key factor in pup abuse in elephant seals (Le Boeuf and Briggs, 1977).

The main causes of mother-pup separation are the movement of bulls in the harem, weather and tidal conditions (Le Boeuf and Condit, 1983). About 16% of separations per year are due to the behavior of the mothers (Riedman and Le Boeuf, 1982). Some females abandon their pups soon after parturition or confuse them with a pup nearby. Young females, in particular, become embroiled in fights with neighbors and when they are over their pups are gone or are injured. Most orphans occur within a few days after birth. If pups do not reunite with their mothers, or fail to get adopted, they wander about the harem and steal milk from nursing females or starve. "Good" mothers, vigilant and protective of their limited resources which they reserve for their own pups, kill over half of the orphans produced. These wary females give a variety of responses to milk thieves that escalate from vocal threat, nipping the nose, head, or rump, to vicious biting, shaking and throwing the pup. When one female bites an orphan, neighboring females may join in, and in a frenzy, compete to bite, shake and toss the victim and kill it on the spot. More frequently the pup is seriously injured. Once injured, the orphan is less apt to reunite with its mother, less adept at stealing milk, and therefore, is more apt to sustain additional injuries. The downward spiral is rapid and by the mean age of 14 days the pup is dead. During its last few days, the orphaned pup may have a score of injuries ranging from a broken jaw to internal bleeding. It may be blind due to edematous swelling of wounds about the head and it is starving. Barely able to move, it tries to suckle anything: weaned pups, the penile opening of a sleeping male, the eye sockets of a decomposing sea lion carcass. Western gulls begin to peck at its eyes. Even the successful milk thieves that survive to weaning age bear numerous scars and weigh less than pups weaned by their mothers.

The other main source of injury and death to suckling elephant seal pups is being run over by bulls (Le Boeuf and Briggs, 1977). In their effort to obtain mates and exert dominance over each other, males crash through the harem, impervious to young pups in their path. Both the pursuer and the pursued move over pups rather than around them. Pups less than one week old are most vulnerable to being trampled and injured seriously; the itinerant orphans are more than twice as vulnerable as filial pups that remain near their mother's side. Pups trampled by bulls die of internal injuries (ruptured organs and osseous trauma) whereas pups bitten by females die of head wounds and lacerations, complicated by infection and starvation. Adult males do not bite neonates.

Similar behavior has been observed in other social phocids. Most deaths on the rookery in southern elephant seals and grey seals (Bonner and Hickling, 1971; Anderson et al., 1979) are due to starvation resulting from

permanent separation from the mother and injuries caused by adult males and females.

Neonate abuse is far less prevalent among the eared seals than among phocids (*e.g.* several species of fur seals, the Steller sea lion, and the southern sea lion) and females rarely injure pups. However, there are exceptional circumstances and species. It is common for females of several species (*e.g.* the California sea lion, *Zalophus californianus*, C. Heath, pers. comm.) to become aggressive to other females and pups in the area just before and after giving birth. Among northern fur seals, females will occasionally dispute over a pup, pull it in opposite directions and injure it, but this is rare (Bartholomew, 1959, Francis, 1987). A significant exception to the rule that female sea lions do not fatally injure pups occurs in the Australian sea lion, *Neophoca cinerea*. Adult females and males at Kangaroo Island, Australia, are responsible for most neonate deaths on the rookery. Females have an unusually low tolerance for strange pups and will attack, bite and toss them (Marlow, 1972). The results of these attacks vary from slight to serious with some of them leading to death. As in elephant seals, the females that are most aggressive to strange pups are the best mothers, being extremely solicitous of their own pups and guarding them carefully. Marlow (1975) reports that out of 20 attacks on pups that involved serious shaking and tossing, 11 were performed by females, the rest were performed by males. Also unusual in this species is the report that some females, having just lost their pups, will abduct and adopt strange pups. Males, are an even more dangerous threat to pups than females (Marlow, 1972, 1975; Higgins and Tedman, 1990). They injure and abuse pups in three ways: 1) In their constant effort to herd females, males cause substantial mother-pup separation (Marlow, 1975). 2) Like females, males will attack, bite and toss pups; Marlow (1975) observed that males were responsible for nine of 20 attacks on pups; Higgins and Tedman (1990) documented four pup deaths due to males. 3) Bulls also attempt to copulate with small pups which may result in lethal injuries (see also Hooker's sea lion, *Phocarctos hookeri*, Marlow, 1975).

The context in which male southern sea lions in Uruguay (Vaz-Ferreira, 1975) and Patagonia, Argentina (Campagna *et al.*, 1988a), abuse, injure and kill neonates is most unusual. The largest males defend territories and females along the high tide line along beaches (Campagna *et al.*, 1988a, 1988b). Two to three times daily during the peak of the breeding season, adult and subadult males without territories or females raid the breeding unit *en masse*, the adult raiders attempt to secure females or to carry one out. Younger, pubertal subadult males abduct pups that are left behind in the melee and confusion, or who wander about searching for

their mothers following a raid (Campagna *et al.*, 1988b). With the pup held in his jaws, the male retreats from the breeding group, comes to a stop and drops the pup (Figure 1). Sometimes it is allowed to escape and return to the breeding females. More often, the male drops the pup and ignores it so long as it does not wander. If it starts to escape, he pins it down on the substrate or grabs it in his jaws, shakes it from side to side and tosses it into the air. Pups are injured and sometimes killed in this manner. Males may alternate this behavior with sexual mounts including pelvic thrusting (in 9% of abductions). However, intromission has never been observed. In some cases, the abductor takes the pup far out into shallow water, often with several males in pursuit. The males compete to control the pup, grabbing it, biting it, tossing it and repeatedly submerging it. Some of these episodes last up to 30 min and in some cases the pup was not seen again and it was assumed that it drowned. The behavior of controlling, blocking, biting and tossing pups is analogous to the behaviors that adult males direct to breeding females. One gets the impression that the subadults are treating the pups like prospective mates.

The incidence of pup abductions in southern sea lions at Punta Norte, Argentina, is quite high. During four breeding seasons, 21% of 400 pups born were seized by subadult males; 57% of the seized pups were held captive for 10 min to 2 hrs. It was estimated that 5.6% of the pups seized died of bite wounds to the head and neck; 50% taken to the water were drowned.

Weanlings and Juveniles

Because sea lion and fur seal pups are not weaned for a year or more, and after this time they come and go from rookeries at a different time than breeding males, there is little opportunity for males to interact with newly weaned pups. In contrast, among the elephant seals, weaned pups begin to emerge from harems towards the end of the breeding season when estrous females are on the decline and returning to sea, and most of the breeding age males are still present (Figure 2). Weanlings remain on the rookery for 2 1/2 months before going to sea for the first time (Reiter *et al.*, 1978). In this context, males mount weanlings, attempt to mate with them, and in the process, injure and kill some of them. Weekly surveys of weanlings during three breeding seasons (Rose *et al.*, 1991) revealed that by the third week in March, 34–50% of the pups sampled (250–400) showed signs of having been mounted by a male (Figure 2). Mounting of weanlings declined after this date because the males returned to sea. Signs of mounting were caused by neck bites and ranged from missing

Figure 1 A subadult male southern sea lion at Punta Norte, Patagonia, Argentina, abducts a suckling pup (a), a behavior which resembles that of an adult male abducting an estrous female (b).

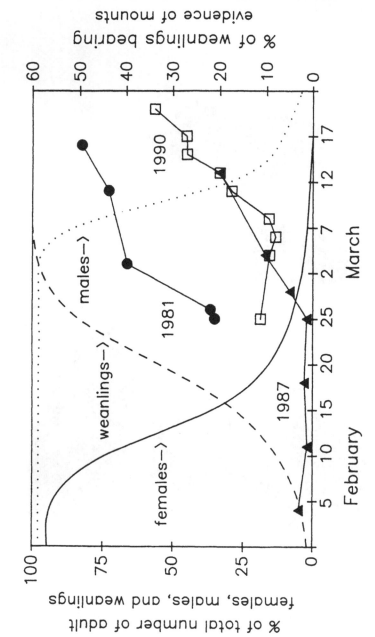

Figure 2 The percentage of weanlings on the rookery bearing evidence of having been mounted by males (tooth marks, injuries or scars) as reflected by successive samples during the breeding season (right axis). The relative frequency (percentage of total that were present) of the total number of adult females and weanlings on the rookery as a function of date in the breeding season (left axis). Adapted from Rose, Deutsch and Le Boeuf (1991).

fur, scraps, and surface cuts to shallow gashes and puncture wounds to deep gashes, exposing the blubber, and profuse bleeding (Figure 3). Two weanlings were observed being killed directly by males mounting them. The age group that most frequently mounted weanlings were 6 year old pubertal males, a group that rarely copulated with females because they were denied access to them. Adult males rarely mounted weanlings. Males that mounted weanlings did not distinguish between the sexes.

IV. EXPLANATIONS FOR ABUSE OF YOUNG AND COMPARISONS WITH OTHER ANIMALS

It is evident from this brief survey that adult male and adult female pinnipeds abuse, injure and kill young under certain circumstances. The circumstances and incidence vary greatly between the two major families in the suborder and are linked to different life history patterns. There is also considerable variation between species in the same family. Nevertheless, abuse of young can be put into three general categories: trampling and crushing, attacking and biting, and sexual assault. Female abuse falls into two categories: 1) females bite strange pups to protect a space around their own pup or to protect resources intended for their own pup, and 2) abandoning young.

Explanations at both the ultimate and proximate level are summarized. Male trampling and crushing of young occurs in species where males are large relative to pups and male movements are frequent in the vicinity of neonates. Abuse of this type is most obvious in the elephant seals. Le Boeuf and Briggs (1977) argue that pups are trampled because breeding males are selected to maximize their reproductive success. It is to a male's advantage to run over a neonate, rather than avoid it, if this means he is more effective in chasing competitors away from females and evading dominant males, and in the end, inseminates more females by this tactic. The fact that the neonate is not likely to be his own, since it was sired during the previous breeding season, makes this argument even more plausible.

Males attacking and biting pups alone, such as occurs in the Australian sea lion, is difficult to explain. Higgins and Tedman (1990) state that misdirected aggression is likely and that pups may be seen as a threat to territorial males. Pup killing by any means might be construed as spiteful behavior that evolved because it decreases the reproductive success of others at a low cost to the actor (Trivers, 1985). Young males, who do most of the killing, are not apt to be killing a pup they sired because they are young and sexually inexperienced. However, it is unclear how a male would benefit

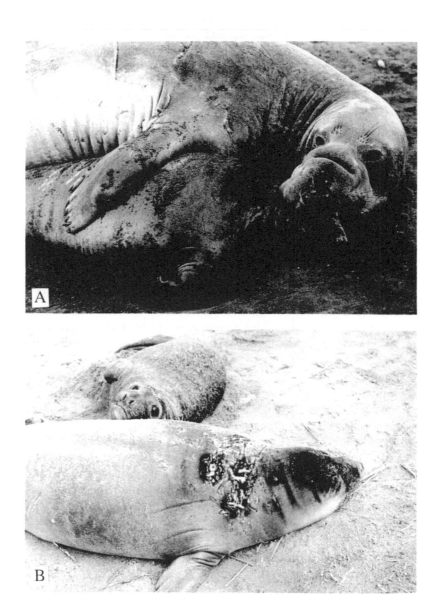

Figure 3 A subadult male northern elephant seal (six years of age) bites and mounts a newly weaned pup on the neck (a). Neckbites during male mounting attempts leave evidence of this behavior in the form of depressions in the pelage, skinscrapes, and open wounds (b).

from spiteful killing because, while reducing the fitness of others, he would also be benefiting many other unaffected individuals, diluting or eliminating his potential relative gain from interference competition (Campagna *et al.*, 1988b).

Males of a number of species respond sexually to neonates, weanlings or juveniles by attempting to mate with them. This behavior has been observed in southern sea lions, Australian sea lions, Hooker's sea lion, northern fur seals, and the two species of elephant seals. This behavior may simply reflect strong selection for high libido in males of these species. Regarding the behavior of southern sea lion males, Campagna *et al.*, (1988b) concluded that the resulting infanticide was of no adaptive value to the perpetrators and that the behavior was a low-cost byproduct of pups being treated as adult females. Males with little or no sexual experience, and a high motivation to mate, generalize their sexual behavior from unobtainable estrous females (protected by territorial males) to the more easily obtained but unsuitable pups. Males do not get obvious, immediate reproductive benefits from seizing, holding and mounting pups but, they gain experience which might make them more effective with adult females later on. A similar explanation applies to male elephant seals mounting weanlings and juveniles (Rose *et al.*, 1991). Sexually inexperienced, sexually mature subadult males are the major perpetrators, and the incidence of the behavior is highest at the end of the breeding season just after the last adult females have left the rookery.

Females biting strange pups is best understood in terms of the mother protecting her own young or a resource (milk) destined for her own young. This stems in large part from the tradeoff of breeding socially. The benefits of group living, such as breeding with a male of demonstrated fitness (Trivers, 1985) or relying on others to signal the approach of predators, are offset by problems such as competition for birthing sites and milk stealing by alien pups. Selfish, aggressive mothers do the best job of rearing their own young by punishing those that threaten their pups or their resources. In northern elephant seals, size and dominance increase with age so it is the prime-age and oldest females that best protect their offspring and injure those of others (Reiter *et al.*, 1981). It is the pups of the youngest females, who get orphaned or wander, that are most often injured and killed.

The abandonment of pups is also a function of age in elephant seals (Le Boeuf *et al.*, 1972; Reiter *et al.*, 1981). Young primiparous females sometimes give birth and never turn around to face the pup but merely orphan it at birth. This seems to be largely a matter of inexperience. Abandonment in this species and in others is often difficult to distinguish from pup "misplacement" and misidentification. Bad weather, crowding,

aggressive female neighbors, and fighting or courting males in the vicinity at the time of parturition can confuse even experienced mothers (Le Boeuf and Briggs, 1977; see Coulson and Hickling, 1964 re Grey seals). For example, if a young elephant seal mother is separated from her pup shortly after birth by a dominant aggressive female, she may not be able to reach her pup for hours or days. By that time, she may not recognize it. Along similar lines, a mother southern sea lion and her pup may be separated shortly after birth (Campagna *et al.*, 1988a). One adult male holds the female and five meters away, another male holds the pup. The mother cannot go to the pup because she is bitten and tossed back to where the male can guard her. The same is true of the pup. In about five days the pup dies of starvation.

This brief review summarizes positive and negative treatment of young by adults in some pinnipeds. Too few species have been studied intensively from this perspective. We have emphasized social species that breed on islands in temperate zones because they have been well studied. Far less is known of the natural history of most phocids that breed on pack and fast ice at low circumpolar latitudes. Moreover, and this is the most telling irony, most information we have about the treatment of young refers to terrestrial behavior. We know woefully little about behavioral interactions at sea where these marine mammals forage and spend much of their time. For example, adult female northern elephant seals spend only 16% of the year on land; the rest of the time they are at sea.

Are the pinnipeds unique in the behaviors displayed to young? No, in our opinion, no new behavior patterns, particularly with respect to abuse of young, emerge. Males injure and kill young in much the same way and for the same reasons, or lack of reasons, in many terrestrial species (Hausfater and Hrdy, 1984). Treating pups as a sexual stimulus is widespread among mammals. Females killing the young of other females occurs in a variety of terrestrial mammals, *e.g.* wolves, African hunting dogs, prairie dogs, ground squirrels (Sherman, 1981). The abandonment of young by primiparous mothers has been documented extensively in a number of mammals, including humans.

Some explanations for abuse of infants and for infanticide in terrestrial mammals do not apply to pinnipeds. For example, infants are not used as "passports" to gain access to resources or as "agonistic buffers" to turn off the aggression of an adversary as has been observed in some primates (Hrdy, 1976; Strum, 1984). Pinnipeds do not cannibalize pups as some other infanticidal mammals do (Struhsaker, 1977; Hoogland, 1985; Takahata, 1985). Killing pups does not cause the mother to resume ovulating sooner, an explanation for infanticide in African lions and hanuman lan-

gurs (Hrdy, 1974, 1977; Packer and Pusey, 1982, 1983), and in any case, the killer is not more apt to mate with the mother. In elephant seals, at least, females whose pups are killed from any cause do not have a higher probability of weaning a pup in the next breeding season (Le Boeuf, *et al.*, 1989), as occurs in prairie dogs, *Cynomis ludovicianus* (Hoogland, 1985).

One question that emerges from this exercise is why is there such variation in abusive treatment of young by adults from one species to the next? Why is it that southern sea lions of both sexes are so abusive to their young and the other four sea lion species are not? Why are elephant seals so different from harbor seals, Weddell seals and harp seals? Is it because of the spotty research coverage with some species being well known and others virtually unstudied? We suspect that there is no simple answer to this question. Life history patterns are important in some cases. For example, some phocids like elephant seals fast while nursing and hence, mothers are strongly selected to prevent milk thievery. Others, like harbor seals, *Phoca vitulina*, do not appear to fast during lactation. Crowding, weather and individual variables, like the age and history of adults involved, influence the frequency of pup harassment but the degree to which these are important remains to be determined.

REFERENCES

ALCORN, D.J. (1984) The Hawaiian monk seal on Laysan Island, 1982. NOAA Technical Memorandum, NMFS.

ANDERSON, S.S. (1979) Cave breeding in another phocid seal, *Halichoerus grypus*. In: *The Mediterranean Monk Seal*, (ed.) K. Ronald and R. Duguy, 151–155. Oxford: Pergammon.

ANDERSON, S.S., Baker, J.R., Prime, J.H. and A. Baird, A. (1979) Mortality in Grey seal pups: incidence and causes. *Journal of Zoology, London*, **189**, 407–417.

BARTHOLOMEW, G.A. (1959) Mother-young relations and the maturation of pup behavior in the Alaska fur seal. *Animal Behaviour*, **7**, 163– 171.

BONNER, W.N. and Hickling, G. (1971) The grey seals of the Farne Islands: Report for the period October 1969–July 1971. *Transactions of the Natural History Society of Northumbria*, **17**, 141–162.

BONESS, D.G. (1989) Fostering in Hawaiian monk seals: Is there a reproductive cost? *8th Biennial Conference on the Biology of Marine Mammals,* Pacific Grove, California, December 7–11, 1989 (abstract).

CAMPAGNA, C., Le Boeuf, B.J. and Cappozzo, H.L. (1988a) Group raids: a mating strategy of male southern sea lions. *Behaviour*, **105**, 224–249.

CAMPAGNA, C., Le Boeuf, B.J. and Cappozzo, H.L. (1988b) Pup abduction and infanticide in southern sea lions. *Behaviour*, **107**, 44–60.

CHRISTENSON, T.E. and Le Boeuf, B.J. (1978) Aggression in the female northern elephant seal, *Mirounga angustirostris. Behaviour,*, **64**, 158–172.

COSTA, D.P. (1990) Reproductive and foraging energetics of pinnipeds: Implications for life history patterns. In: D. Renouf (ed.), *Behaviour of Pinnipeds*. Chapman and Hall, London (in press).

COSTA, D.P. and Gentry, R.L. (1986) Free-ranging energetics of northern fur seals. In: *Fur Seals: Maternal Strategies on Land and at Sea*. Princeton University Press, Princeton, Pp. 79–101.

COSTA, D.P., Le Boeuf, B.J., Huntley, A.C. and Ortiz, C.L.. (1986) The energetics of lactation in the northern elephant seal, *Mirounga angustirostris*. *Journal of Zoology, London*, **209**, 21–33.

COULSON, J.C. and Hickling, G. (1964) The breeding biology of the grey seal, *Halichoerus grypus* (Fab)., on the Farne Islands, Northumberland. *Journal of Animal Ecology*, **33**, 485–512.

FAY, F.H. (1982) Ecology and biology of the Pacific walrus, *Odobenus rosmarus divergens Illiger*. *North American Fauna*, no., **74**. U.S. Dept. of the Interior, Fish and Wildlife Service, Washington, D.C.

FEDAK, M.A., Boyd, I.L., Arnbom, T. and McCann, T.S. (1989) The energetics of lactation in southern elephant seals *Mirounga leonina* in relation to the mother's size. *8th Biennial Conference on the Biology of Marine Mammals*, Pacific Grove, California, December 7–11, 1989 (abstract).

FRANCIS, J.M. (1987) Interfemale aggression and spacing in the northern fur seal, *Callorhinus ursinus*, and the California sea lion, *Zalophus californianus*. Ph.D. thesis, University of California at Santa Cruz.

GENTRY, R.L. (1970) Social behavior of the Steller sea lion. Ph.D. thesis, Univ. of California, Santa Cruz.

GENTRY, R.L. and Holt, J.R. (1986) Attendance behavior of northern fur seals. In: *Fur seals: Maternal Strategies on Land and at Sea*. Princeton University Press, Princeton, pp 41–60.

HAUSFATER, G. and Hrdy, S. (eds.) (1984) *Infanticide: Comparative and evolutionary perspectives*. Aldine, New York.

HIGGINS, L.V. and Tedman, R.A. (1990) Attacks on pups by male Australian sea lions, *Neophoca cinerea*, and the effect on pup mortality. *Marine Mammal Science* (in press).

HOOGLAND, J.L. (1985) Infanticide in prairie dogs: lactating females kill offspring of close kin. *Science*, **230**, 1037–1040.

HRDY, S.B. (1974) Male-male competition and infanticide among the langurs (*Presbytis entellus*) of Abu, Rajasthan. *Folia Primatologica*, **22**, 19–58.

HRDY, S.B. (1976) Care and exploitation of non human primate infants by conspecifics other than the mother. *Advances in the Study of Behavior*, **6**, 101–158.

JEFFRIES, A.J., Brookfield, J.F.Y., and Semeonoff, R. (1985) Positive identification of an immigration test-case using human DNA fingerprints. *Nature*, **317**, 818–819.

JOHNSON, B.W., and Johnson, P.A. (1978) The Hawaiian monk seal on Laysan Island, 1977. Final Report to the UK.S. Marine Mammal Commission, Washington, D.C., for Contract MM7AC009. U.S. Dept. of Commerce, National Technical Information Service, Springfield, Va. Report no. MMC–77/05.

JOHNSON, B.W. and Johnson P.A. (1984) Observations of the Hawaiian monk seal on Laysan Island from 1977–1980. NOAA Technical Memorandum, NMFS, Southwest Fisheries Center no. 49.

KAUFMAN, G.W., Siniff, D.B. and Reichle, R. (1975) Colony behavior of Weddell seals, *Leptonychotes weddelli,* at Hutton Cliffs, Antarctica. *Rapports et Proces-vergaux des Reunions Conseil International Pour L'Exploration de la Mer,* **169**, 228–246.

KING, J.E. (1983) *Seals of the World.* British Museum (Nat. Hist.), Comstock, Cornell Univ. Press, Ithaca.

LE BOEUF, B.J. and Briggs, K.T. (1977) The cost of living in a seal harem. *Mammalia,* **41**, 167–195.

LE BOEUF, B.J. and Condit, R. (1983) The high cost of living on the beach. *Pacific Discovery,* **36**, 12–14.

LE BOEUF, B.J., Condit, R. and Reiter, J. (1989) Parental investment and the secondary sex ratio in northern elephant seals. *Behavioral Ecology and Sociobiology,* **25**, 109–117.

LE BOEUF, B.J., Whiting, R.J. and Gantt, R.F. (1972) Perinatal behavior of northern elephant seal females and their young. *Behaviour* 43, 121–156.

MARLOW, B.J. (1972) Pup abduction in the Australian sea lion, *Neophoca cinerea. Mammalia,* **36**, 161–165.

MARLOW, B.J. (1975) The comparative behaviour of the Australasian sea lions, *Neophoca cinerea* and *Phocarctos hookeri* (Pinnipedia: Otariidae). *Mammalia,* **39**, 159–230.

MAYNARD Smith, J. (1980) A new theory of sexual investment. *Behavioral Ecology and Sociobiology,* **7**, 247–241.

MCCANN, T.S. (1982) Aggressive and maternal activities of female southern elephant seals (*Mirounga leonina*). *Animal Behaviour,* **30**, 268–276.

MCCANN, T.S. (1987) Female fur seal attendance behavior. In: *Status, Biology and Ecology of Fur Seals,* (ed.) J. Croxall and R.L. Gentry, pp.199–200. Proc. of an International Symposium and Workshop, Cambridge, England, Apr 23–27, 1984, NOAA Technical Report NMFS 51.

ORTIZ, C.L., Le Boeuf, B.J. and Costa, D.P. (1984) Milk intake of elephant seal pups: an index of parental investment. *American Naturalist,* **124**, 416–422.

PACKER, C. and Pusey, A.E. (1982) Cooperation and competition within coalitions of male lions: kin selection or game theory? *Nature,* **296**, 740–742.

PACKER, C. and Pusey, A.E. (1983) Adaptations of female lions to infanticide by incoming males. *American Naturalist,* **121**, 716–728.

PETRINOVICH, L. (1974) Individual recognition of pup vocalization by northern elephant seal mothers. *Zeitschrift Tierpsychologie,* **34**, 308–312.

PITCHER, K.W. and Calkins, D.G. (1981) Reproductive biology of Steller sea lions in the Gulf of Alaska. *Journal of Mammalogy,* **62**, 599–605.

REITER, J., Stinson, N.L. and Le Boeuf, B.J. (1978) Northern elephant seal development: the transition from weaning to nutritional independence. *Behavioral Ecology and Sociobiology,* **3**, 337–367.

REITER, J., Panken, K.J. and Le Boeuf, B.J. (1981) Female competition and reproductive success in northern elephant seals. *Animal Behaviour,* **29**, 670–687.

RIEDMAN, M. (1982) The evolution of alloparental care and adoption in mammals and birds. *Quarterly Review of Biology,* **57**, 405–435.

RIEDMAN, M. (1983) The difficult art of mothering in an elephant seal rookery. *Pacific Discovery,* **36**, 4–11.

RIEDMAN, M. (1990) *The Pinnipeds: Seals, Sea Lions, and Walruses.* University of California Press, Berkeley.

RIEDMAN, M. and Le Boeuf, B.J. (1982) Mother-pup separation and adoption in northern elephant seals. *Behavioral Ecology and Sociobiology*, **11**, 203–215.

ROSE, N., Deutsch, C.J. and Le Boeuf, B.J. (1991) Sexual behavior of male northern elephant seals: III. The mounting of weaned pups. *Behaviour*, **119**, 171–192.

SHERMAN, P.W. (1981) In: R.D. Alexander and D.W. Tinkle (eds.) *Natural Selection and Social Behaviour*. Chiron Press.

STIRLING, I. (1975) Factors affecting the evolution of social behavior in the pinnipedia. *Rapport et Proces-vergaux des Reunions Conseil International Pour L'Exploration de la Mer*, **169**, 205–212.

STRUM, S.C. (1984) Why males use infants. In: Taub, D.M., (ed.), *Primate Paternalism*, Van Nostrand Reinhold, New York, pp. 146–185.

STRUHSAKER, T.T. (1977) Infanticide and social organization in the redtail monkey (*Cercopithecus ascanius schmidti*) in the Kibale Forest, Uganda. *Zeitschrift Tierpsychologie*, **45**, 75–84.

TAGGART, S.J. (1987) Grouping behavior of Pacific walruses *(Odobenus rosmarus divergens Illiger)*, an evolutionary perspective. Ph.D. Thesis, University of California, Santa Cruz.

TAKAHATA, Y. (1985) Adult male chimpanzees kill and eat a male newborn infant: newly observed intragroup infanticide and cannibalism in Mahale National Park, Tanzania. *Folia Primatologica*, **44**, 161–170.

TRIVERS, R.L. (1985) *Social Evolution*. Benjamin Cummings, Menlo Park, California.

TRIVERS, R.L. and Willard, D.E. (1973) Natural selection of parental ability to vary the sex ratio of offspring. *Science*, **179**, 90–92.

VAZ-FERREIRA, R. (1975) Factors affecting numbers of sea lions and fur seals on the Uruguayan islands. *Rapports et Proces-verbaux des Reunions Conseil International Pour L'Exploration de la Mer*, **169**, 257–262.

CHAPTER 12

INFANTICIDE IN LIONS: CONSEQUENCES AND COUNTERSTRATEGIES

ANNE E. PUSEY and CRAIG PACKER

Department of Ecology, Evolution and Behavior,
University of Minnesota, St. Paul, Minnesota 55108, USA

INTRODUCTION

The regular occurrence of infanticide by males that have just entered a new social group or area has now been described in a wide variety of mammals (*e.g.* primates, Hiraiwa Hasegawa, this volume, Sommer, this volume; carnivores, Packer and Pusey, 1984; rodents, Parmigiani, this volume). It has become widely accepted that this behavior gives the males a reproductive advantage because they thereby speed up the females' reproduction and sire their own infants more quickly (see Hrdy, 1974, 1979). However, rigorous evidence in support of this theory (see Boggess, 1984) has only been obtained in a subset of these species (*e.g.* Crockett and Sekulic, 1984; Packer and Pusey, 1984; Sommer, this volume). Perhaps the most extensive data from a single species come from the 24 year study of lions in the Serengeti ecosystem (Schaller, 1972; Bertram, 1975; Packer and Pusey, 1984; Pusey and Packer 1987a, Packer *et al.,* 1988). In this chapter we review data on the incidence of male infanticide in lions, and the advantages that males gain from infanticide. We discuss the counterstrategies that females and males show to infanticide, and show how the threat of male infanticide has had far reaching effects on lion social structure and behavior.

STUDY POPULATION AND METHODS

The study areas consist of a 2000 km^2 area of the Serengeti National Park, and the floor of the Ngorongoro crater, Tanzania. The Serengeti ecosystem covers about 25,000 km^2 and contains a population of 2000–3000 lions. Our Serengeti study area contains about 200 individuals at any one time. The Ngorongoro crater is an extinct volcanic caldera that contains a closed population of about 100 lions. All individuals can be recognized by natural markings (see Pennycuick and Rudnai, 1970). Demographic records of individually identified lions have been maintained continuously on all the lions in the two study areas since 1974 or 1975 (Hanby and Bygott, 1979, Packer et al., 1988), and the descendants of two social groups ("prides") in the Serengeti have been continuously studied since 1966 (Schaller, 1972; Bertram, 1975). Since 1984 we and our associates have used radio telemetry to locate study prides in the Serengeti. Each resident lion is located about once every two months and its location, group composition and reproductive state are recorded.

LION SOCIAL STRUCTURE

Lions live in permanent social groups, prides, that occupy stable territories (Schaller, 1972, Packer et al., 1988). Prides typically consist of 2–9 adult females (range: 1–18) and their offspring, and a coalition of 2–6 adult males (range: 1–9). All the females in a pride were born in the same natal pride, while all resident males are immigrants (Pusey and Packer, 1987a). DNA-fingerprinting of 78 cubs and their parents has shown that all cubs are fathered by males of the resident coalition (Gilbert et al., 1991). Females start breeding at about 3 years of age and males at about 4 years of age (Packer et al., 1988). Cohorts of cubs are often born to several females in the pride at about the same time. Most subadult females are recruited into the pride but some leave with companions and form a new pride nearby. All subadult males leave their natal pride with other males of their cohort before breeding and most remain together as a coalition (Pusey and Packer, 1987a). Male coalitions compete to take over prides. When a new male coalition takes over a pride, the males evict all the previous resident males, all the subadult males, and the subadult females that are too young to mate (Hanby and Bygott, 1987; Pusey and Packer, 1987a).

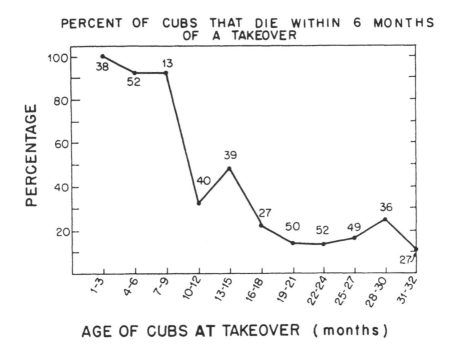

Figure 1 Age-specific cub mortality at takeovers. The percentage of cubs of each age that died within six months of a male takeover (excluding the two cases where males took over their natal pride, see text). Numbers give sample sizes for each age. (Reprinted from Pusey and Packer 1987a).

INCIDENCE OF INFANTICIDE BY MALES

In addition to evicting adult males and subadults, new males cause the death of virtually all small cubs that were present in the pride at the time of the takeover. Bertram (1975) demonstrated that an increase in cub mortality was a regular feature of male replacements and suggested that infanticide accounted for about a quarter of all cub deaths. With more extensive data, we confirmed that virtually all small cubs die after takeovers (Packer and Pusey, 1983a,b, 1984; Pusey and Packer, 1987a) (Figure 1), and calculated that death following male takeovers accounts for 27% of

all cub deaths before the age of 12 months (Packer *et al.*, 1988). Figure 1 does not include two takeovers where males were returning to breed in their natal pride. In these cases the mothers of the cubs were close relatives of the males and the cubs survived (Pusey and Packer, 1987a).

Despite the high incidence of cub deaths following male takeovers in the study areas, the precise cause of death is usually unknown. Dead cubs are likely to be consumed quickly and completely by scavengers, so the chances of finding bodies are low. Because most behavioral observations are made in daylight and lions are primarily active at night, infanticide, when it occurs, is unlikely to be observed. Nevertheless, direct observations of infanticide by males have been made 10 times in the Serengeti (Schaller, 1972; Bertram, 1975; Packer and Pusey, 1984; Packer *et al.*, 1988; Caro and Borgerhoff Mulder, 1989; Matthews and Purdy, pers. comm.) and once in the Masai Mara National Reserve, Kenya, at the northern edge of the Serengeti ecosystem (Jackman and Scott, 1982). Most of these cases involved incidental observations of extra-pride males with wounded or dead cubs (Schaller, 1972; Packer and Pusey, 1984; Caro and Borgerhoff Mulder, 1989).

The most complete observation was made by a film team who had set out specifically to film infanticide. After several weeks of monitoring females that we knew were vulnerable to a male takeover, they observed a mother with three cubs aged 3 months being chased by another pride of females. The cubs remained alone in the initial location. A single resident male from the invading pride followed behind the females and stopped in the vicinity of the cubs. After an hour or so, the male noticed the cubs and immediately ran to them, bit the first in the head, dropped it, and then successively picked up and bit each of the remaining two in the abdomen. He then carried one cub to the shade and ate some of it. The killing was over in less than two minutes. During the attacks, each cub screeched, and one rolled on its back with teeth bared and claws extended. The male growled aggressively in a manner that is completely unlike lions' behavior while killing their prey (Matthews and Purdy, pers. comm. see also the film, Queen of the Beasts, Survival, Anglia).

Besides losing their cubs by infanticide, it is also possible that females sometimes abandon their cubs at a takeover, or are kept from them by the new males until the cubs die of starvation (Bertram, 1975), but we have no evidence of this (Packer and Pusey, 1984). From our observations of attempted takeovers, females appear more likely to defend their cubs than abandon them (see below).

ADVANTAGES OF INFANTICIDE TO MALES

How do new males benefit from infanticide? The average tenure of male coalitions in prides is only 24 months before they are ousted or move on voluntarily to new prides (Pusey and Packer, 1987a) (Figure 2). Because cubs are vulnerable to death following takeovers until they are at least 9 months old (Figure 1), males only have a good chance of fathering surviving cubs if they inseminate females soon after they have taken over a pride. After females have given birth to surviving cubs, they do not come into estrus again until their cubs are at least 18 months old; the average interbirth interval between surviving cubs is 24 months (Pusey and Packer, 1987a) (Figure 2). However, if they lose their cubs, females resume mating activity within days or weeks, regardless of season (Schaller, 1972; Bertram, 1975; Packer and Pusey, 1983a,b), and they mate with which ever males are resident in the pride. Females that lose small cubs (less than 4 months of age) at a takeover conceive again about 4.4 months after the loss (Packer and Pusey, 1983b). In calculating the precise gains males make by killing small cubs rather than waiting for females to come into estrus, it is necessary to consider both females whose dependent cubs died in other circumstances than at a takeover and those whose cubs survived, because cub mortality is high even in the absence of takeovers (Packer et al., 1988). Considering all these females, median postpartum amenorrhea in the absence of a takeover is 11.3 months, and conception occurs about one month later (Packer and Pusey, 1983b). Thus, by killing small cubs when they first take over a pride, males sire cubs about 8 months sooner, on average, than they would if they spared the cubs of the previous males (Packer and Pusey, 1984). Besides speeding up the females' reproduction, another possible advantage of infanticide to males is that the cubs will not provide feeding competition for younger cubs fathered by the males. Bertram (1975) provided evidence that the presence of older cubs depressed the survival of younger cubs within the same cohort, but we have not been able to confirm this between successive cohorts (Pusey and Packer, 1987a).

A final advantage that invading males may gain by killing cubs is that they thereby weaken the alliance between females and their resident males, as has also been suggested in red howlers (Crockett and Sekulic, 1984). Mothers of small cubs appear to cooperate with their resident males to keep out new males. Once they have lost their cubs their current maternal investment is terminated and this may terminate their interest in maintaining the status quo.

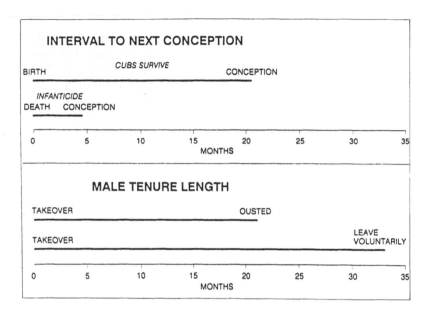

Figure 2 The advantages males gain from infanticide. Females do not conceive until a median of 20.3 months after the birth of surviving cubs (Pusey and Packer, 1987a), but they conceive a median of 4.4 months after their cubs die at takeovers (Packer and Pusey, 1983b). Median tenure length is 21 months for male coalitions that are eventually ousted by other coalitions, and 33 months for coalitions that leave voluntarily (Pusey and Packer, 1987a).

COUNTER-STRATEGIES BY FEMALES

Because infanticide by males is highly detrimental to female reproductive success, females are likely to have evolved counterstrategies that reduce the occurrence or impact of infanticide (reviewed by Hrdy, 1979). Do females behave in ways that reduce the incidence of infanticide? We divide their behavior into immediate responses to alien males, and behavior that may serve to forestall future takeovers.

1. Immediate Responses To Alien Males

a. Responses to roars of alien males

Because they are active at night, the responses of females to extra-pride males are rarely observed. However, in a series of experiments, recorded

male roars were played back to females from loudspeakers placed in their territories. It was thus possible to simulate the presence of various classes of males in the close vicinity of a pride and monitor the responses of females that had cubs (McComb *et al.*, 1993).

Lions of both sexes roar regularly, and males roar more than females (Schaller, 1972). Roaring is mostly nocturnal, and roars can be heard from up to 5 km away; thus roaring lions are usually out of sight from other lions. Male roars were played to females with cubs at times when the resident males of their pride were not present. Roars of two classes of males were played: those of males that were resident in the pride and therefore the fathers of the cubs, and roars of males from elsewhere in the study areas that had not been observed in the vicinity of the females. Females showed a striking difference in response to their own resident males and to alien males (McComb *et al.*, 1993). On hearing the roars of resident males they would usually look up, but remain relaxed, and generally did not move from their initial location. In contrast, when they heard the roars of alien males, all the mothers became highly agitated. They always stood or sat up, usually grimaced or snarled, and stared in the direction of the roars. Sometimes they took a few steps in the direction of the roars in a threatening posture. If several females were present, they usually bunched together. Most females then turned and ran or walked with their cubs in the opposite direction from the roars. Females are thus clearly able to distinguish the roars of resident and alien males, and appear to be aware of the dangers of alien males.

b. Response to direct encounters with alien males

Whereas the playback experiments show that females with cubs will often run away from the roars of alien males that are out of sight, what happens when females come face to face with alien males? Over the study period, we have evidence of the outcome of 11 face-to-face encounters between alien males and females with cubs that were unaccompanied by resident males (Packer *et al.*, 1990). In several cases we actually saw females threatening or attacking the males, who fought back; in other cases we found females to be wounded or even dead (two cases) directly after the encounter (Packer and Pusey, 1984). In these encounters, there was a significant effect of female group size on the mortality of cubs (Table I). The cubs of solitary females were significantly more likely to be killed than the cubs of groups of two or more females. Therefore, grouping by females is an effective defensive strategy.

Table I Aggressive encounters between females and extra-pride males.

Number of defending females:	some cubs survive	all cubs die
≥ two:	5	0
one:	1	5
		p < 0.05, Fisher test

Data collected between 1978 and 1988. Note that no defending males were present in any of these encounters. (Reprinted from Packer et al., 1990).

c. Avoidance of alien males

Although virtually all cubs up to the age of 9 months at the time of a takeover subsequently die, cubs over 10 months old have a better chance of surviving (Figure 1). The mothers of these older cubs sometimes stay with them, and together they successfully avoid the new males by going to new areas or by avoiding the rest of the pride and occupying a peripheral portion of their former range (Packer and Pusey, 1984). Sometimes the females return to the rest of the pride after their cubs have become independent, but sometimes this behavior leads to a permanent pride split (Pusey and Packer, 1987a). Takeovers rarely occur in prides in which the majority of cubs are 6–17 months of age (Packer and Pusey 1983a). This may be because cubs of this age can successfully avoid alien males with their mothers, thereby preventing a takeover. Smaller cubs are much less mobile, and thus more vulnerable to infanticide, while cubs of 18 months or older are better able to survive on their own (Pusey and Packer, 1987a), and their mothers are less likely to accompany them.

2. Behavior That Minimizes The Chances Of Future Takeovers and Infanticide

a. Grouping by females

Because females are better able to defend their cubs from alien males when they are in groups (see above), it might be expected that females would group together when they have cubs. Grouping can occur at two levels: the formation of subgroups within the pride, and the formation of prides of a particular size.

i. sub-grouping within the pride

Although prides are permanent social units consisting of several individuals that associate peacefully together, they are fission-fusion units. Individuals within prides spend considerable periods alone and the rest of the time in temporary sub-groups of varying combinations of individuals (Schaller, 1972; Packer, 1986). However, when females have cubs they pool their cubs in a creche with the cubs of other females in the pride, and the mothers associate together almost constantly. Figure 3 shows that when females have cubs they are much more likely to be found together in a group consisting of all the mothers than when they do not have cubs.

This grouping of females when they have cubs is consistent with female defence of cubs, but other advantages of such grouping must also be considered. It has often been suggested that females group to obtain nutritional advantages from cooperative capture of large prey (reviewed in Packer *et al.*, 1990), but recent data show that solitary females gain as much food as females in any sized group, and that females in moderate sized groups of 2-4 actually suffer nutritionally during periods of food scarcity (Packer *et al.*, 1990). In addition, in an earlier analysis, mothers in groups of three and four were found to have thinner bellies indicative of lower levels of recent food intake than mothers in groups of two or single mothers (Packer, 1986) (Figure 4). We conclude, therefore, that females suffer nutritionally from constant association with other mothers, and that group defence is the best explanation for creche formation by mothers.

ii. optimal Pride Size

Pride sizes vary from 1–18 adult females, and females living in prides of 3–10 have significantly higher reproductive success than females living in smaller or larger prides (Packer *et al.*, 1988). Two important correlates of these differences are female mortality and the rate of male takeovers (Figure 5). Females in prides of one or two suffer higher mortality than those in larger prides, and takeover rates are significantly lower for prides of 2–7 than smaller or larger prides. The higher mortality rates of females in small prides are probably at least partly due to aggressive encounters with infanticidal males, although aggression from female neighbors may also be important (Packer *et al.*, 1990). Small prides are probably more vulnerable to male takeovers both because solitary or pairs of females are less able to repel males, and because they are less likely to have resident males that are always present to protect them (Packer *et al.*, 1988). The reasons for the high takeover rates of large prides are unknown, but it may be that large prides present a particularly attractive target for males

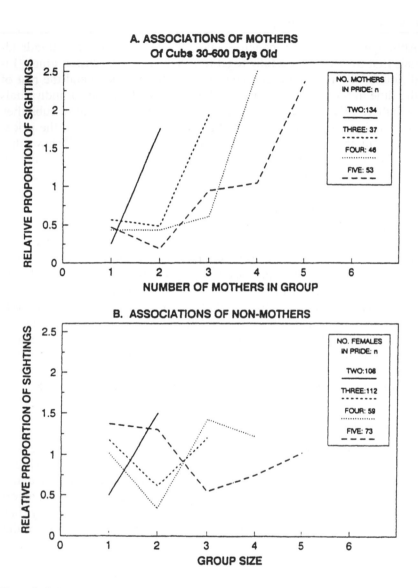

Figure 3 Relative proportion of sightings in which (A) mothers, and (B) non-mothers, were found in groups of each size. The relative proportion is the observed number of sightings of each group size multiplied by x/n, where x is the number of mothers (or non-mothers) in the pride (the number of possible group sizes for mothers in that pride) and n is the total number of sightings of that sized group. Thus, females found equally often in each group size would have a relative proportion of 1.0 for each, regardless of numbers of mothers in the pride. (Redrawn from Packer, *et al.*, 1990).

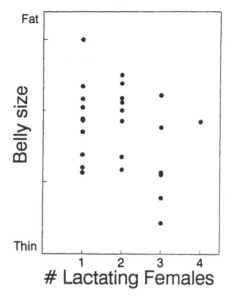

Figure 4 Average belly size of lactating females rearing their cubs alone or communally. Each point is the mean across females of a set when all were lactating simultaneously (minimum of three measurements per female, range = 3–16). Age of cub does not have a significant effect on mother's average belly size. Singletons and pairs had significantly larger belly sizes than sets of three and four. (Reprinted from Packer, 1986).

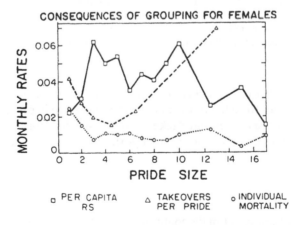

Figure 5 Pride-size-specific reproductive rates, mortality, and frequency of male takeovers. Females in prides of 3–10 have significantly higher reproductive rates than those in smaller prides; mortality rates of solitaries and pairs of females are also significantly higher than those of females in larger prides, and takeover rates are significantly higher for solitary females and females in large prides. (Reprinted from Packer *et al.*, 1988).

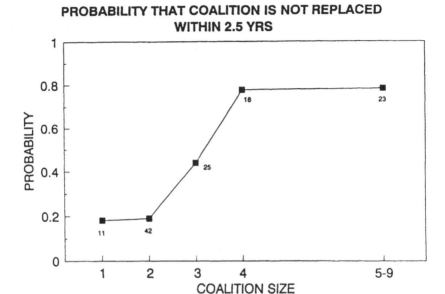

Figure 6 Relationship between male tenure length in prides and male coalition size. For each coalition that gained residence between 1966–1988, the duration of its tenure in each pride was measured. Where the same coalition was resident in more than pride, its tenure in each pride was measured separately. We then calculated the proportion of these tenure lengths for each coalition size that were greater than 2.5 yrs (numbers refer to the number of tenure lengths).

seeking prides to take over. Females can regulate pride size by dispersal of subadult females from the pride, and most females live in prides within the optimal size range (Pusey and Packer, 1987a). We therefore conclude that the risk of infanticide by males is an important factor in the determination of pride size.

b. Behavior that attracts a large male coalition

The length of tenure of male coalitions in prides increases with coalition size (Bygott *et al.*, 1979, Packer *et al.*, 1988) (Figure 6). Because their cubs remain vulnerable to infanticide for many months, females will benefit from having a large male coalition resident in their pride so that their cubs can be protected until they are past the vulnerable stage. Certain aspects

of female reproductive behavior appear to result in the attraction of large male coalitions. Females show a period of heightened sexual activity but reduced fecundity following a takeover.

Females without cubs show regular estrus periods of about 4 days every 16 days (Packer and Pusey, 1983b). Usually a single male forms a consortship with a potentially estrous female during which he stays close to her and guards her from other males. He sniffs her urine frequently throughout the consortship. Males usually guard females for one or two days before they start mating. They then mate about every 20 minutes for two to six days and the male continues to guard the female for a day or more after mating has ceased until he finally loses interest in her. After the first male has stopped guarding her, the female sometimes has a brief consortship with a second male during which mating starts immediately and ceases after one day (Packer and Pusey, 1983a).

Once a new coalition of males has taken over a pride, the females soon start to mate with the males. Their sexual behavior with new males differs from that with familiar males (who fathered their previous cubs) in several ways (Packer and Pusey, 1983a). Although copulation rates are the same with new males and familiar males, females initiate a higher proportion of copulations with new males. They are also more likely to mate with a succession of males during a single estrus in the first few months after a takeover than when mating with males that had fathered their previous litter. These subsequent males are usually from the same coalition as the first male, but occasionally they are from different coalitions.

Despite their active sexual behavior, females show reduced fecundity in the first three months after a takeover (Packer and Pusey, 1983a). This is most obvious when the time from loss of small cubs to next birth for females that were lactating at the time of the takeover is compared with the time from loss of cubs to next birth for lactating females that lost their cubs under other circumstances. Lactating females that lose their cubs at a takeover and then mate with the incoming males give birth a median of 110 days later than females that lose their unweaned cubs then mate with the same males that fathered their previous litters (Figure 7). We estimate that females that lose their cubs at takeovers take 6–9 estrus cycles to become pregnant, whereas most females that lose their cubs under other circumstances conceive in the first or second estrus period (Packer and Pusey, 1983b). This delay in becoming pregnant for the first three months following a takeover is also shown by females that have older, weaned cubs or no cubs at the takeover (Packer and Pusey, 1983a).

The result of this period of reduced fecundity but heightened sexual behavior is that following a takeover there are more estrous females present

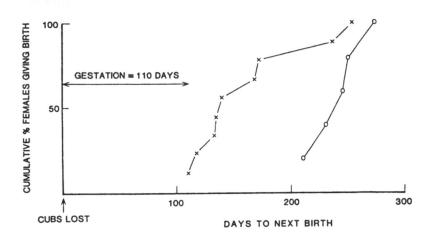

Figure 7 Interval from loss of unweaned cubs to next birth when females mated with males that fathered the previous cubs (median interval = 134 days), and when females mated with incoming males (median = 244 days). The difference is statistically significant (Packer and Pusey 1983a). The intervals from loss of cubs to next conception can be estimated by subtracting gestation length from day of birth. (Reprinted from Packer and Pusey, 1983a).

in the pride and more sexual behavior taking place than at any other time (Packer and Pusey, 1983b). Frequently, males from more than one coalition are attracted to the pride and mate with the females during this period. The coalitions then compete and the largest coalition eventually becomes resident in the pride (Packer and Pusey, 1983a). Often, prides that have been abandoned by their resident males (Pusey and Packer, 1987a) are first taken over by nomadic coalitions of pairs or single males who have little chance of remaining resident for long (Figure 6). In these circumstances, females will do better not to conceive straight away and suffer another takeover, but rather to attract a larger coalition before conceiving. We therefore suggested that instead of decreasing the reproductive success of females as it appears to do at first sight the reduction in fecundity following a takeover is actually an adaptation which leads to the attraction of the largest coalition in the area. This increases the chances that the

pride is taken over by a coalition that can remain resident long enough to protect the females' subsequent cubs (Packer and Pusey, 1983a). We have calculated that as long as the delay in having cubs increases the chances of attracting a large coalition by 30% it will lead to higher lifetime reproductive success of the females (Packer and Pusey, 1983a).

3. Other Possible Female Counterstrategies To Infanticide

Female langurs that are pregnant at the time of male takeovers sometimes solicit copulations with the new male, and Hrdy (1974) suggested that by doing so they might confuse paternity and persuade the male to spare their young once born, although Sommer (this volume) found no evidence that the infants of pregnant females that mated with the new male were any more likely to be spared than those of pregnant females that did not mate with him. Estrus behavior by pregnant females has only rarely been observed in lions (Schaller, 1972; Packer and Pusey, 1983b) and the cubs of females that are pregnant at takeovers do not usually survive (Packer and Pusey 1983a). So pseudoestrus does not seem to be a counterstrategy to infanticide in lions.

Abortion at takeovers has been described in several species (*e.g.* Bruce, 1960; Berger, 1983; Pereira, 1983; Mori and Dunbar, 1985; Sommer this volume), and in some cases may be a reproductive strategy of females in which they terminate investment in young that would later be killed (Hrdy, 1979; Schwagmeyer, 1979; Labov, 1981). In lions, however, there is no evidence of abortion following takeovers. Bertram (1975) speculated that abortion might take place in lions because he observed few births in the first few months after a takeover. However, we looked for direct evidence of abortion but found that females that were obviously pregnant at the time of the takeover all carried their young to term (Packer and Pusey, 1983a). Because of the difficulties of detecting early pregnancy in lions, we have no means of discovering whether abortion occurs at this stage (Packer and Pusey, 1983a).

MALE COUNTERSTRATEGIES

The increase in deaths due to infanticide at takeovers suggests that the presence of resident males deters infanticide by alien males. Resident males regularly patrol the pride range, roar, and keep out alien males (Schaller, 1972; Bygott *et al.*, 1979).

The voluntary movements of males between prides appear to be sensitive to the vulnerability of their cubs. Large coalitions often make

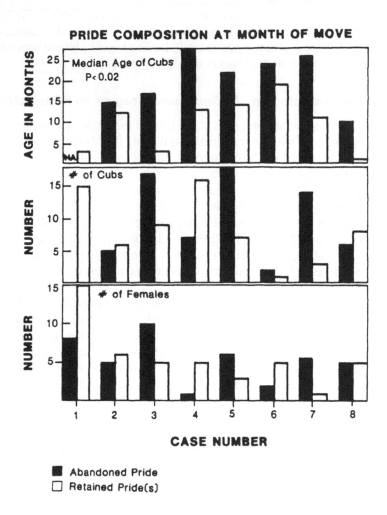

Figure 8 The median age of cubs, number of cubs, and number of females in abandoned prides compared to prides that were retained by male coalitions. Coalitions showed a significant tendency to retain prides with younger cubs, but showed no tendency to retain prides with either more cubs or females (Pusey and Packer, 1987a). (Reprinted from Pusey and Packer, 1987a).

voluntary movements between prides that are not due to eviction (Pusey and Packer, 1987a). These are of three types: (1) a coalition may annex new prides without surrendering previous prides; (2) a coalition that has

been simultaneously resident in more than one pride and is nearing the end of its reproductive lifespan may relinquish one pride while maintaining residence in the other(s); (3) a coalition may completely abandon one pride and become resident in another. In the second case males always maintain residence in the pride that has the youngest and therefore most vulnerable cubs, rather than the pride that has the most females or most cubs (Figure 8). In the third case the movements of males seem to be sensitive both to the vulnerability of cubs, and the number of females in each pride.

We have modelled the conditions under which males should abandon one pride for another, as in (3) above, in order to maximize reproductive success. This model plots a curve of the minimum number of extra females the males should gain at each cub age in order for movement between prides to be profitable. It is based on the survivorship of cubs, their risk of mortality once they have been abandoned, and the interbirth interval of females with surviving cubs (Pusey and Packer, 1987a) (Figure 9). In the first few months after the birth of their cubs, the cubs are very vulnerable and males should only be willing to move on if they thereby gain access to a much greater number of females. When their cubs are 10–28 months, males should be willing to abandon that pride for a new one even if the new pride has fewer females. Cubs of 10–28 months have a good chance of surviving a subsequent takeover (Figure 1), but most of the mothers have not yet resumed mating. After 28 months, almost all the mothers will be breeding again and the males will do well to stay. We only had 5 cases of this type of movement between prides, but the male coalitions with young cubs conformed to the model and moved on to prides with much larger numbers of females. The two coalitions in which the cubs were older moved on to prides with smaller than expected numbers of females. Both these coalitions had maturing daughters in the pride, suggesting that inbreeding avoidance may also be a factor in their movements.

EFFECT OF INFANTICIDE ON THE SEX RATIO

The regularity of male takeovers and subsequent infant death in lions creates a situation in which it may be advantageous for females to modify the sex ratio of their cubs (Packer and Pusey, 1987). Because the reproductive success of individual males depends critically on the number of males in their cohort, while the reproductive success of individual females does not depend on the number of females in their cohort, mothers will benefit by producing more sons than daughters when the chances of these males

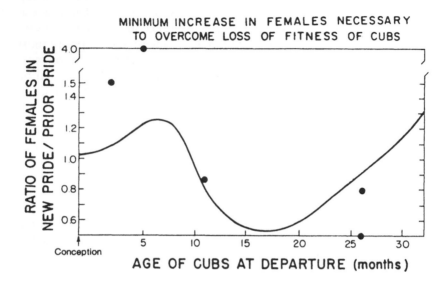

Figure 9 The relative increase in numbers of females necessary to overcome the loss of fitness of cubs at each age. The line represents the minimum ratio of females in a new pride compared to a prior pride that males would have to gain to achieve an overall increase in fitness if the cubs in the previous pride were consequently exposed to a takeover; observed values should all lie above that line (see Pusey and Packer, 1987a for details). The solid dots show the change in numbers of females that five male coalitions experienced when they abandoned one pride for another. (Reprinted from Pusey and Packer, 1987a).

having other male companions in the pride are high. Births are synchronized by male takeovers. Births are more synchronous in the first 300 days following a takeover than afterwards (Packer and Pusey, 1983b), and the sex ratio is indeed significantly biased towards males in the first 300 days after a takeover (Table II). This response by females increases the chances of their sons being able to invade new prides when they mature, thus helping to perpetuate the infanticidal trait.

DISCUSSION AND CONCLUSIONS

Although infanticide by males has only been observed a few times in lions, it appears to be a deliberate and efficient act, and is probably the

Table II Sex of cubs and timing of male takeover

Days since most recent takeover	Male	Female	Proportion Male	Deviation from 0.50
≤ 300	129	96	0.57	$\chi^2 = 4.84, p < 0.05$
> 300	268	281	0.48	$\chi^2 = 0.31$, NS
		Two-by-two $\chi^2 = 4.63, p < 0.05$		

Note: Gestation takes 110 days, and the synchronous births associated with male takeovers occur in the first 300 days after the takeover (see Packer and Pusey, 1983a,b). Data are from 1974–1985 and exclude cubs for which we do not know the timing of birth relative to the preceding male takeover. Note that when all data are included, the overall sex ratio does not deviate significantly from equality: 50.9% male, N = 874, $\chi^2 = 0.29$ (Reprinted from Packer and Pusey, 1987).

usual fate of cubs that die following takeovers. There is no doubt that male lions gain reproductive advantages from infanticide. Following a successful takeover, females mate with the infanticidal males, and DNA-fingerprinting shows that all cubs born in a pride are sired by the males resident in the pride (Gilbert *et al.*, 1991). By killing small cubs, males sire their own cubs an average of 8 months sooner than if they had spared them.

Whereas almost 100% of cubs aged up to 9 months die after takeovers, the percentage falls after this age. However, males evict all older cubs that have not been killed (Hanby and Bygott, 1987). Although some evicted cubs leave the pride with their mothers, the usual effect of eviction is to separate the cubs from their mothers and hence bring the mothers into estrus more quickly. As long as their cubs are present, females do not resume estrus behavior until their cubs are at least 18 months old.

In primates such as red howlers (Crockett and Sekulic, 1984) and langurs (Sommer, this volume) there is a similar relationship between infant age and the proportion that are killed by males after takeovers. In these species, however, more infants of all ages are spared than in lions. Possibly this difference reflects the greater gains male lions make in terms of speeding up the females' reproduction, or lower costs of the behavior. While male lions will speed up females' reproduction if they kill or evict cubs aged up to 18 months, the equivalent age is only 6 months in langurs and 9 months in red howlers. Because of their large size and weaponry, and the fact that lion cubs are sometimes left unprotected, male lions may also find it easier than male primates to kill young.

The behavior of female lions seems to be influenced by the threat of male infanticide in several ways. They are able to recognize the roars of alien males and retreat from them. They form more stable subgroups when they have cubs and these are almost certainly defensive in nature. It is possible, however, that these groups are not only necessary for defence of cubs against incoming males, but also against females from other prides. Females are also known to kill cubs in other prides (Schaller, 1972; Hanby, 1983; Packer and Pusey, 1984). Mothers with mobile cubs sometimes leave the rest of the pride to avoid new males, as has also been described in primates (reviewed in Pusey and Packer, 1987b). Female lions also show heightened sexual behavior but a delay in reproduction following takeovers that may serve to attract larger coalitions that can best protect their next batch of cubs. There is some evidence of a similar delay in conception following male changes in red howlers (Crockett and Sekulic, 1984), but langurs do not show such a delay (Sommer *et al.*, in prep.). This difference may be due to species differences in the likelihood that the first male to enter the group is able to remain there. Perhaps, in langurs, males that are able to enter a new group and start killing infants have a good chance of staying for a long time. In this case females would not benefit from delaying conception. Finally, females appear to respond to takeovers by synchronously producing male-biased litters. This increases the chances that their sons will have numerous coalition partners with whom they can eventually cooperate in taking over a pride elsewhere.

While females live in prides of the size in which takeover rates are lowest, the risk of infanticide is probably not the only factor causing lions to live in groups (Packer, 1986; Packer *et al.*, 1990). Other large cats such as tigers, leopards and cougars also show infanticide but are nevertheless solitary. One way that lions differ from these species is that they live at much higher densities (Packer, 1986). This may increase the costs of territoriality such that group defence of territories becomes advantageous (Brown, 1987; Davies and Houston, 1981). We have recently found that females in very small prides are unable to defend a permanent territory (Packer *et al.*, 1990).

When they encounter alien males, females with cubs sometimes fight with them. In two instances such encounters resulted in the deaths of females. This is in contrast to other species, such as langurs (Sommer, this volume), and rodents (Parmigiani, this volume), where males apparently never wound females that aggressively defend their infants. Clearly the killing of female lions by males is disadvantageous both to the female and the males involved, so why should it ever occur? Because of the rarity of observations of such encounters, we have no data on the factors that

influence a female's readiness to fight (cf. Parmigiani, this volume). It may be that whereas solitary females always run away without risking fighting, and females in very large groups are always able to overcome or drive off one or a few males, in cases of a more even match (*e.g.* between several females and one or two large males), the decision of the females and males to avoid combat is not clear cut. Once attacked, the males may have to fight for their lives and consequently risk wounding or killing a potential mate because of the lethal weaponry involved (see also Packer and Pusey, 1982). There is evidence that female lions can kill fully grown males (Packer and Pusey, unpublished data).

Resident male lions also behave in ways that protect their cubs. Males frequently remain with a pride for many months after all the females have cubs and there are no estrous females present. While resident, they regularly patrol the pride range, roar and keep out alien males. The voluntary movements of males between prides are also sensitive to the vulnerability of their cubs. If they have to abandon one pride out of several, they leave the pride where the cubs are least vulnerable, and they only abandon prides with small cubs if they thereby gain access to much larger numbers of females in another pride.

Although there is tremendous conflict between female lions and extra-pride males, there is a period when females have cubs during which they and their resident males share a common interest in protecting the cubs. The extent to which they cooperate in this endeavor is a topic of current research.

ACKNOWLEDGEMENTS

We thank the Director of Tanzania National Parks, the Coordinator of the Serengeti Wildlife Research Institute, and the Tanzanian National Scientific Research Council for permission and facilities. Over the past 13 years our work has been generously supported by the H.F. Guggenheim Foundation, the National Geographic Society, the Royal Society of Great Britain, the Eppley Foundation, the American Philosophical Society, Sigma Xi, Hewlett-Packard, Klipsch Corporation, NIMH grant no. MH15181, the Graduate School of the University of Minnesota, and NSF grant nos. BSR 8406935 and 8507087. Many people have assisted in the collection of demographic data including Sara Cairns, Larry Herbst, Tony Collins, David Bygott, Jeannette Hanby, Tim Caro, Monique Borgerhoff Mulder, John Fanshawe, David Scheel, Jon Grinnell, Karen McComb, Richard Matthews, Samantha Purdy, Markus Borner and Charlie Trout. Barbie

Allen provided invaluable logistic support throughout the study. We thank the organizers of this conference for inviting us to participate, and Karen McComb for comments on the manuscript. We were supported by J.S. Guggenheim Fellowships while writing this chapter.

REFERENCES

BERGER, Joel. (1983) Induced abortion and social factors in wild horses. *Nature*, **303**, 59–61.

BERTRAM, B.C.R. (1975) Social factors influencing reproduction in wild lions. *Journal of Zoology*, **177**, 463–482.

BOGGESS, J.E. (1984) Infant killing and male reproductive strategies in langurs (*Presbytis entellus*). In: G.Hausfater and S.B. Hrdy (eds), *Infanticide*. Aldine, New York, pp. 283–310.

BROWN, J.L. (1987) *Helping and Communal Breeding in Birds*. Princeton University Press, Princeton, New Jersey.

BRUCE, H.M. (1960) A block to pregnancy in the house mouse caused by proximity to strange males. *Journal of Reproduction and Fertility* 196–103.

BYGOTT, J.D., Bertram, B.C.R. and Hanby, J.P. (1979) Male lions in large coalitions gain reproductive advantages. *Nature*, **282**, 838–840.

CARO, T. and Borgerhoff Mulder, M. (1989) Infanticidal lions caught in the act. *Swara*, **12**, 17–18.

CROCKETT, C. and Sekulic, R. (1984) Infanticide in red howler monkeys. In: G. Hausfater and S.B. Hrdy (eds), *Infanticide*, Aldine, New York, pp.173–191.

DAVIES, N.B. and Houston, A.I. (1981) Owners and satellites: the economics of territory defence in the pied wagtail,*Motacilla alba*. *Journal of Animal Ecology*, **50**, 157–180.

GILBERT, D., Packer, C., Pusey, A.E., Stephens, J.C. and O'Brien, S.J. (1991) Analytical DNA fingerprinting in lions: parentage, genetic diversity and kinship. *Journal of Heredity*, **82**, 378–386.

HANBY, J.P. (1983) *Lions' Share*. Collins, London.

HANBY, J.P. and Bygott, J.D. (1979) Population changes in lions and other predators. In A.R.E. Sinclair and M. Norton-Griffiths (eds), *Serengeti: Dynamics of an Ecosystem*, University of Chicago Press, Chicago, pp. 249–262.

HANBY, J.P. and Bygott, J.D. (1987) Emigration of subadult lions. *Animal Behaviour*, **35**, 161–169.

HRDY, S.B. (1974) Male-male competition and infanticide among the langurs (*Presbytis entellus*) of Abu, Rajasthan. *Folia Primatologica*, **22**, 19–58.

HRDY, S.B. (1979) Infanticide among animals: a review, classification, and examination of the implications for the reproductive strategies of females. *Journal of Ethology and Sociobiology*, **1**, 13–40.

JACKMAN, B. and Scott, J. (1982) *The Marsh Lions*. Elm Tree Books, Hamish Hamilton, London.

LABOV, J.B. (1981) Pregnancy blocking in rodents: adaptive advantages for females. *American Naturalist*, **118**, 361–371.

MCCOMB, K., Pusey, A.E., Packer, C. and Grinnell, J. (1993) Female lions can detect potentially infanticidal males from their roars. *Proceedings of the Royal Society, Series B.*, **252**, 59–64.

MORI, U. and Dunbar, R.I.M. (1985) Changes in the reproductive condition of female gelada baboons following the takeover of one-male units. *Zietschrift für Tierpsychologie*, **67**, 215–224.

PACKER, C. (1986) The ecology of sociality in felids. In: D.I. Rubenstein and R.W. Wrangham (eds), *Ecological Aspects of Social Evolution*. Princeton University Press, Princeton, New Jersey, pp.429–451.

PACKER, C. and Pusey, A.E. (1982) Cooperation and competition within coalitions of male lions: kin selection or game theory? *Nature*, **296**, 740–742.

PACKER, C. and Pusey, A.E. (1983a) Adaptations of female lions to infanticide by incoming males. *American Naturalist*, **121**, 716–728.

PACKER, C. and Pusey, A.E. (1983b) Male takeovers and female reproductive parameters: a simulation of oestrous synchrony in lions (*Panthera leo*). *Animal Behaviour*, **31**, 334–340.

PACKER, C. and Pusey, A.E. (1984) Infanticide in carnivores. In: G. Hausfater and S.B. Hrdy (eds), *Infanticide*, Aldine, New York, pp.31–42.

PACKER, C. and Pusey, A.E. (1987) Intrasexual cooperation and the sex ratio in African lions. *American Naturalist*, **130**, 636–642.

PACKER, C., Herbst, L., Pusey, A.E., Bygott, J.D., Cairns, S.J., Hanby, J.P. and Borgerhoff-Mulder, M. (1988) Reproductive success of lions. In T.H. Chitton-Brock (ed), *Reproductive Success*. University of Chicago Press, Chicago, pp. 363–383.

PACKER, C., Scheel, D. and Pusey, A.E. (1990) Why lions form groups: food is not enough. *American Naturalist*, **136**, 1–19.

PENNYCUICK, C. and Rudnai, J. (1970) A method of identifying individual lions, *Panthera leo*, with an analysis of the reliability of the identification. *Journal of Zoology*, **160**, 497–508.

PEREIRA, M.E. (1983) Abortion following the immigration of an adult male baboon (*Papio cynocephalus*). *American Journal of Primatology*, 493–98.

PUSEY, A.E. and Packer, C. (1987a) The evolution of sex-biased dispersal in lions. *Behaviour*, **101**, 275–310.

PUSEY, A.E. and Packer, C. (1987b) Dispersal and philopatry. In B.B. Smuts, D.L. Cheney, R.M. Seyfarth, T.T. Struhsaker and R.W.Wrangham (eds), *Primate Societies*, University of Chicago Press, Chicago, pp. 250–266.

SCHALLER, G.B. (1972) *The Serengeti Lion*. University of Chicago Press, Chicago.

SCHWAGMEYER, P.L. (1979) The Bruce effect: An evaluation of male/female advantages. *American Naturalist*, **114**, 932–938.

CHAPTER 13

ALTRUISTIC INFANT CARE OR INFANTICIDE: THE DWARF MONGOOSES' DILEMMA

O. ANNE E. RASA

Abt. Ethologie, Zoologisches Institut, Universität Bonn, 5300 Bonn 1, Germany

INTRODUCTION

Allomaternal behavior, or nourishment and protection of the young by individuals other than the mother, is not widespread amongst mammals compared to other vertebrate groups. This is probably due to the fact that mammalian young, in contrast to those of fish and birds, for example, where alloparenting is a common phenomenon, are fed on milk. This mode of nutrition necessitates a particular physiological state in the parturient female which can only rarely be duplicated by other individuals (Creel *et al.* 1991), thus reducing the latter's chances of performing maternal functions. Since care of the young, at least in their earlier stages of development, devolves on the female, there has also been little selection pressure for helping behavior in the male parent. Amongst mammals, alloparental care appears to have had an evolutionary base from which to evolve primarily in cases where there is prolonged infant dependence. Of all the mammalian families, it is particularly common amongst carnivores, where, after weaning, young must progress through a phase of behavioral maturation during which prey capture is learned, before they are capable of fending for themselves. Since the carnivorous mode of nutrition extends the period of infant dependence it thus allows scope for individuals other than the mother to perform nutritive or protective roles.

The widest variety of social structures, incorporating practically every type of alloparental care at present described for mammals, can be found in the carnivore subfamily Viverridae or mongooses. This mammalian group

301

will be considered here as a model to illustrate the selective advantages of sociality and helper systems in small carnivores.

THE BASES OF GROUP LIFE IN VIVERRIDS

Viverrid Social Structures

The evolution of group life amongst carnivores in general has been attributed to a variety of ecological and behavioral factors. These are mainly associated with the improved capture and protection of large prey items capable of feeding several individuals (see Kruuk and McDonald, 1985 for a review). Although this hypothesis may hold for the larger canids, felids and hyaenids, it cannot explain why group life has evolved in small predators feeding almost entirely on small invertebrates and which do not share their prey, such as the viverrids, or the European badger *Meles meles* (Kruuk, 1978). In these cases, protection of a group feeding territory has been postulated as the main factor predisposing towards sociality (Kruuk, 1978; Rood, 1986).

As mentioned previously, viverrids exhibit practically all the forms of social structure found in mammals. Table I gives a synopsis of the information available in the literature regarding social groupings, habitat and feeding preferences for those species where information is available. These data are taken primarily from Smithers (1983), Rosevear (1974) and Albingac (1973). By far the majority of viverrid species belong to the two genera *Herpestes* and *Galerella*, which have only recently been separated from one another taxonomically. *Herpestes*, with the exception of *H. ichneumon*, which is practically pan-African in distribution, is reserved for Asian species, *Galerella* for African species. The other genera listed are considered, taxonomically speaking, to be almost without exception monospecific and will be referred to in this paper as "species".

The species belonging to both the genera *Herpestes* and *Galerella* – are regarded as solitary, except in areas of high food density where loose groups have been reported for *H. ichneumon* (Ben-Yaakov and Yom-Tov, 1983). The social structure, as typified by *H. auropunctatus* (Baldwin et. al., 1952) and *Galerella pulverulenta* (Cavellini and Nel, 1990) is reminiscent of that of many mustelids, a single male holding a territory overlapping with that of one or more females with which he mates. The young remain with the mother until approximately 6 months old, on which they disperse. Species exhibiting the type of social structure described above are classed here as solitary, since the only extended social period is between the mother and her young of the year.

GENUS	GROUP SIZE	ACTIVITY	BIOTOPE	PREY PREFERENCE
HERPESTES auropunctatus	SOLITARY	DIURNAL	WOODLAND/SCRUB	RODENTS/BIRDS/AMPHIBS-REPT
GALERELLA	SOLITARY	DIURNAL	SAVANNAH/WOODLAND	RODENTS-INSECTS/REPTILES
BDEOGALE	SOLITARY	NOCTURNAL	WOODLAND/ROCKS	INSECTS (ANTS)/REPTILES
RHYNCHOGALE	SOLITARY	NOCTURNAL	SAVANNAH/WOODLAND	INSECTS (TERMITES)S.V.
PARACYNICTIS	SOLITARY	NOCTURNAL	SAVANNAH	INSECTS/S.V.
HERPESTES ichneumon	SOLITARY/PAIR	DIURNAL	SAVANNAH/WOODLAND	RODENTS/BIRDS
ICHNEUMIA	SOLITARY/PAIR	NOCTURNAL	SAVANNAH/WOODLAND	INSECTS/S.V./ AMPHIBIANS
ATILAX	SOLITARY/PAIR	DIURNAL/CREP	WATERWAYS	CRUSTACEA/AMPHIBIANS RODENTS
GALIDIA	PAIR/FAMILY (2-4)	DIURNAL	FOREST	S.V./AMPHIBIANS
CYNICTIS	COLONY (4-15)	DIURNAL	ARID SCRUB	INSECTS/REPTILES
SURICATA	FAMILY (2-25)	DIURNAL	ARID SCRUB	INSECTS/REPTILES
HELOGALE	FAMILY (2-35)	DIURNAL	BUSH SAVANNAH	INSECTS/ARTH./S.V.
CROSSARCHUS	GROUP (2-20)	DIURNAL	WOODLAND/FOREST	INSECTS/ARTH./ S.V.
MUNGOS	GROUP(6-40)	DIURNAL	SAVANNAH/WOODLAND	INSECTS/ARTH./ S.V.

Table I A synopsis of information regarding social groupings, habitat and feeding preferences for different genera of viverrids.

Some of the species regarded as solitary are listed in Table I as "solitary or pair-living", owing to conflicting reports regarding group constitution. These species have been recorded by certain authors as "living in pairs". The information available, however, does not specify whether these pairs are heterosexual and representative of a prolonged heterosexual pair-bond or not. In *Galerella pulverulenta*, the Cape grey mongoose, for example, a recent study (Cavallini and Nel, 1990) has shown that at least one such "pair" consisted of an adult male and a male yearling that remained together for several months. Many such reports of pair-living mongooses may thus reflect adult/young long-term bonds rather actual mating systems and the records of long term pair formation are therefore questionable.

The majority of social mongooses are either monogamous or live in polygamous groups containing several members of both sexes. These polygamous groups fall into two classes, those where only the dominant female breeds, *e.g. Helogale* (Rasa, 1977) and *Suricata* (Lynch, 1979) and those where several females produce young, *e.g. Mungos* (Thurnheer, 1990) and *Cynictis* (Wenhold, 1990).

Practically all solitary mongoose species are nocturnal, only, certain members of the genera *Herpestes*, *Galerella* and the Water mongoose *Atilax* being the exception. In the social mongooses, this tendency is reversed and, to date, no group living nocturnal mongooses have been reported.

Social Structure and Prey Preference

If improved capture of large prey were associated with the evolution of sociality in mongooses, one would expect a positive correlation between group size and prey size, as in the pack hunting canids and hyaenids. Table I lists prey preferences for various mongoose species or genera where information is available, the most common prey items being listed first, followed by secondary prey in order of frequency of occurrence in stomach contents. Prey spectra, however, are dependent on prey availability, density and catchability, and since all mongooses are opportunists, the data available may only reflect ecological conditions at particular times. According to the prey capture hypothesis, one would expect that species capturing large prey would have a larger group size than those feeding on small prey items and would also have a large body mass to allow efficient capture of large prey.

Figure 1 shows the correlation between mean body mass and mean prey size for species where data are available. There are no strong trends visible. Some of the larger species, such as *Herpestes ichneumon* and *Atilax* do

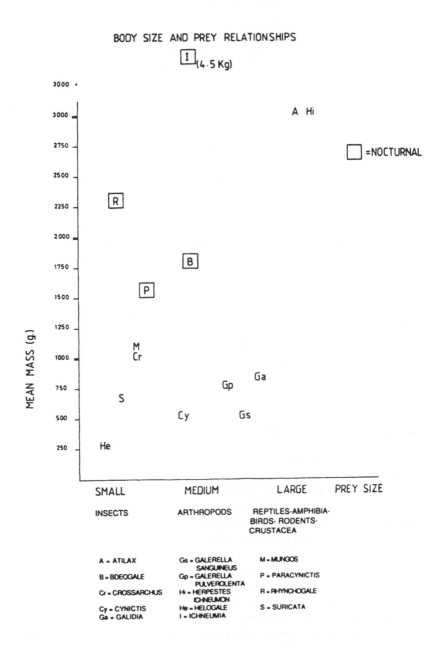

Figure 1 The correlation between mean body mass and mean prey size for different mongoose genera.

take large prey as do some of the smaller *Galerella* species, but all are solitary. Large mongooses, such as *Dologale*, *Bdeogale* and *Paracynictis*, in contrast, feed almost exclusively on small arthropods and the largest mongoose of all, *Ichneumia*, feeds on medium sized arthropods. None of these, however, are social. All the social mongooses have relatively low body mass and feed on small prey. There thus appears to be no relationship between prey size and body mass itself. This finding may be due to the fact that mongooses kill prey by means of a stereotyped skull-crush bite directed to the front of moving prey (Rasa, 1973a) which restricts the size of prey capable of being taken to the size of the mongooses' gape.

When group size is correlated with prey size, however (Figure 2), there is a tendency for species which feed almost entirely on small insects and arthropods to live and forage in groups, those living on larger prey items tending to be solitary. The one exception is the Yellow mongoose, *Cynictis* which, although it lives in a group, forages alone. Amongst the solitary mongooses there is no relationship between group size and prey size although a trend for nocturnal species to feed on smaller prey is present.

These trends are the opposite of what the hypothesis, that group life is associated with improved hunting of larger prey, would predict. Sociality in mongooses does not therefore appear to have evolved as a means of improved capture and defense of large prey items. The tendency for social mongooses to feed primarily on prey at the lower end of the size spectrum indicates that resource renewability is probably more important than prey size itself.

Social Structure and Habitat

If the key to the evolution of sociality in viverrids does not lie in improved capture of large prey, an alternative hypothesis is that it may be associated with protection of the individual, an expression of the "selfish herd" effect (Hamilton, 1971). It would therefore be expected that species living in open habitats would tend to form groups as a protection against predation, especially against aerial predators (Rasa, 1983, 1989a) while those living in denser biotopes would tend to be solitary. As can be seen from Table I, superficially this hypothesis appears to hold true since the majority of social mongooses inhabit relatively open biotopes such as arid scrub and bush savannah. However, of the six known social species, one, *Crossarchus*, lives in thick woodland, another, *Cynictis*, which inhabits the most open habitat — arid scrub — forages alone and several of the solitary mongoose species are diurnal and live in just as open habitats as do the social ones. Group life does not, therefore, appear to correlate closely with the degree

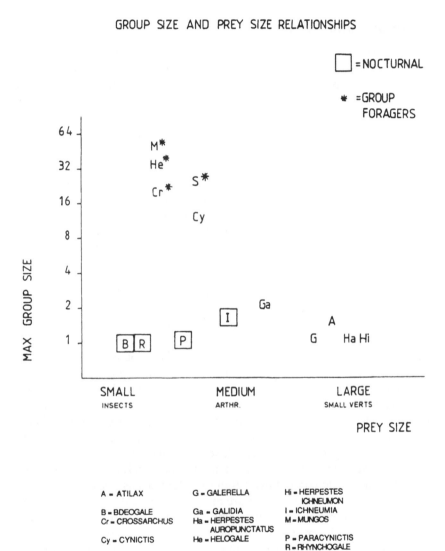

Figure 2 The correlation between mean group size and mean prey size for various mongoose genera.

of exposure while foraging and its main selective advantage is therefore not liable to be the reduction of predation on adults.

Litter Size and Habitat

When, however, the amount of cover offered by the habitat is correlated with mean litter size (Figure 3), a close correlation emerges. Species with small litters or producing only one young at a time inhabit thick cover and litter size increases almost linearly with the degree of openness of the habitat. The only two social species where this tendency does not pertain are *Cynictis* and *Suricata*, both of which are known to produce two litters per year in all areas of their range (Lynch, 1979; Rasa *et al.*, 1992), in contrast to what is known at present for the other genera. Some species, such as *Helogale*, are also capable of producing more than one litter when food is abundant (Rasa, 1979; Creel *et al.*, 1991). The finding that litter size correlates closely with sociality and cover suggests an alternative hypothesis for the evolution of sociality in mongooses, that it evolved as a means of protection of the young rather than the adults.

BREEDING BIOLOGY AND ALLOPARENTING

The present information on breeding biology of various mongoose species indicates that the social mongooses exhibit a high degree of altruism, especially alloparenting. There is no evidence for alloparenting in solitary species but, amongst social ones, almost all types are represented, from allosuckling and babysitting to food bringing at the den (Table II). In order to explain this tendency, the mating systems, where known, provide information as to its probable evolutionary basis. These are shown in Table III. Both *Herpestes* and *Galerella* are known to be polygynous with competition between males for mates. These are also the only genera showing clear sexual dimorphism, males being larger than females, as would be expected in a male competitive system for females as a scarce resource. The majority of the other species are monogamous, either with mating occurring exclusively between pair members or between predominantly the dominants of each sex in group-living species. In these species there is no sexual dimorphism. Only *Cynictis* and *Mungos* are polygamous. In *Mungos* (Thurnheer, 1990), females have synchronous estrus cycles and, although the dominant male attempts to copulate with all females, his main mating activity is concentrated on the alpha female and subordinate males form pairs with "preferred" females with which they mate. In *Cynictis*, females come into estrus approximately 4–5 days apart and are first mated by the

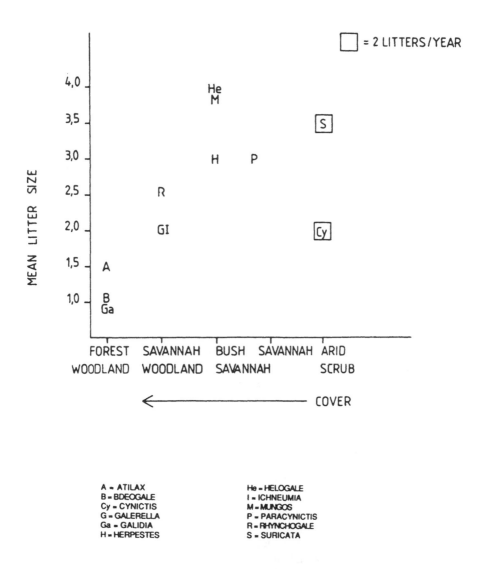

Figure 3 The correlation between degree of cover afforded by the habitat and mean litter size for different mongoose genera.

dominant male but allow subsequent matings by subordinates, bringing the paternity of the offspring into question, since mate guarding by the dominant cannot take place efficiently for the entire estrus period. The dominant male's attention is directed sequentially to the other estrus females in the group (Wenhold, 1990, pers. obs.). For the monogamous group-living species *i.e. Helogale* and *Suricata*, mate guarding is continuous until estrus is past (Ewer, 1963; Rasa, 1977). In the case of *Cynictis* and *Mungos*, therefore, alloparenting by subordinate males may be associated with possible paternity of the offspring, while for *Helogale* and *Suricata* it is more likely to be based on kin-selective processes.

Since the social mongooses show a high degree of altruistic infant care and, amongst viverrids, a close correlation between litter size and degree of exposure of the habitat has been demonstrated, an alternative hypothesis for the evolution of sociality in this sub-family is that it functions primarily as a means of infant protection rather than mutual protection between adults. This hypothesis will be explored here using data on the dwarf mongoose as a model.

THE DWARF MONGOOSE

General Biology

The dwarf mongoose (*Helogale undulata rufula*) is found throughout the savannah and dry bush areas of Africa (Smithers, 1983). Groups have a mean size of 12.3 individuals and consist mainly of close kin, a dominant and breeding alpha pair and their offspring over several years (Rasa, 1972, 1977). The group structure is a matriarchy and only the alpha female produces young regularly and raises them successfully (Rasa, 1987a). Dominant animals are dominant for life (up to at least 14 years) and there is no evident competition for alpha status within groups. The coefficient of relatedness between individuals is high since the majority of group members are usually full siblings, although unrelated immigrant individuals may also be present and, if attaining dominance, breed. Acceptance of an immigrant into a group, however, is not directly equated with its reproductive success within the group, *i.e.* gene transfer, since such immigrants do not automatically gain breeding status. Inbred matings also occur (*i.e.* father/daughter, mother/son) on loss of the mate of the appropriate sex (Rasa, 1987a), therefore the coefficient of relatedness of individuals within a group can vary between 0 and 0.5 or more.

As previously mentioned, Dwarf mongooses are group foragers, feeding mainly on small arthropods, and there is no prey sharing or pack hunting

BREEDING BIOLOGY

GENUS	LITTERS/YR	YOUNG/LITTERS	ALLOPARENTING AT DEN		
			ALLOSUCKLING	BABYSITTING	FOOD BRINGING
HERPESTES	1?	2 - 4	-	-	-
BDEOGALE	1	1	-	-	-
GALIDIA	1	1	-	-	-
GALERELLA	2?	2	-	-	-
RHYNCHOGALE	1	2 - 3	-	-	-
ICHNEUMIA	1	1 - 3	-	-	-
ATILAX	1	1 - 3	-	-	-
PARACYNICTIS	1	2 - 4	-	-	-
CYNICTIS	2	2	Y	Y(♀♂)	Y(♀♂)
HELOGALE	1(2)	2 - 6	Y	Y(♀)	-
SURICATA	2	2 - 5	?	Y(♀)	-
MUNGOS	1	2 - 6	Y	Y(♂♀)	-
CROSSARCHUS	1	2 - 4	?	?	?

Y = Yes

Table II Alloparenting, number of litters produced per year and number of young per litter in different genera of viverrids.

SEXUAL DIMORPHISM AND MATING SYSTEMS

GENUS	MATING SYSTEM	BREEDERS	MEAN MASS (kg)		MASS RATIO	
			♂♂	♀♀	♂♂	♀♀
HERPESTES	POLYGYNY	A♂ A♀♀	3.41	3.12	1 :	0.91
GALERELLA pulverulentus	POLYGYNY	A♂ A♀♀	0.91	0.68	1 :	0.75
G sanguineus	POLYGYNY	A♂ A♀♀	0.64	0.46	1 :	0.72
GALIDIA	MONOGAMY	PAIR ♂ + ♀				
ICHNEUMIA	MONOGAMY	PAIR♂ + ♀	4.5	4.1	1 :	0.91
CYNICTIS	SERIAL POLYGYNY	DOM ♂ A♀♀	0.59	0.55	1 :	0.93
HELOGALE	MONOGAMY	DOM♂ DOM♀	0.27	0.27	1 :	1.00
SURICATA	MONOGAMY	DOM♂DOM♀	0.73	0.72	1 :	0.99
MUNGOS	POLYGAMY (PAIRS)	A♂♂A♀♀	1.3	1.4	1 :	1.08

Table III Mating systems in different genera of viverrids. A = adult. DOM = dominant.

of larger prey. Each group defends a territory in which they are nomadic, wandering daily over distances up to 1 km and sleeping in different dens each night (Rasa, 1986a). Large groups forage over significantly longer distances daily than small groups (Rasa, 1987a). These nomadic tendencies are maintained when young are present, the young being carried at regular intervals between dens.

Alloparental Behavior

Within dwarf mongoose groups there is a high degree of division of labor which is sex based. Subordinate males are involved in a sophisticated vigilance system with exchange of guards along the group's foraging path and a complicated series of warning vocalizations indicating predator type and distance (Rasa, 1986b; Rasa, 1989b; Beynon and Rasa, 1990). Group member loss through predation has been shown to correlate with group size and the number of subordinate males available to perform guarding (Rasa, 1989a). This is attributable to a breakdown in the efficiency of the vigilance system. Individual males are capable of investing up to approximately 30% of their possible foraging time in vigilance (Rasa, 1989b). This appears to present a maximum investment level beyond which individuals are not capable of sustaining further energetic loss. In large groups where several (5 or more) subordinate males are present, an overlap between guards occurs, vigilance is efficient and continuous during foraging. In groups with fewer subordinate males, total vigilance time is only slightly increased per individual, probably in compensation for the lack of males performing this role, and gaps appear between successive guards. These periods where no vigilance is maintained are especially taken advantage of by raptors (Rasa, 1989b).

During the period in which the young are still blind and helpless, they are guarded in the transitory dens by subordinate females, termed babysitters. As in the case of guarding, females also play this role successively, one female remaining behind in the morning when the group goes foraging and being replaced by another in the afternoon, after the group returns to the mound during the midday hours. The major investment by these females in care of the young is, however, not just during their nest-bound phase but also in the three week period subsequent to this when the young are first mobile and move with the group. Babysitters are predominantly responsible for the feeding of the young during this phase, approximately 80% of the prey they capture being given to the youngster accompanying them (Rasa, 1989c). Each youngster accompanies a single adult which feeds it. During this period, considering the amount of

food provided by individual babysitters, it would not be possible for a single adult to care competently for two young simultaneously. Since the mean litter size at the start of the mobile phase is 3.08, at least 3 adults are therefore engaged in the leading and provisioning of the young during at least the first three weeks after they leave the den. If subordinates of either sex are present, the mother does not lead and feed her young when they emerge and the dominant male takes little part in this activity. In the absence of subordinate females, subordinate males takeover the babysitting role (Rasa, 1989c). This reduces the number of individuals available to perform guarding and therefore reduces the efficiency of the protection afforded the group during the start of the youngs' mobile phase.

Factors Involved in Infant Mortality

Predation

During the period the young remain in the den, protection is afforded them by babysitters. At this time they are especially vulnerable to ground predators, especially snakes and larger mongoose species. Guarding by babysitters, however, is relatively efficient, except in the case of large cobras, where the babysitter itself may also be killed by the predator (Rasa, 1989a). Once they emerge from the den, at the start of the mobile phase, they are especially susceptible to attack by raptors, having both a greater vulnerability and wider predator spectrum than the adults. Offspring loss in groups containing 5 or fewer adults is especially high, occurring predominantly in the first three weeks following emergence, and may be total (Rasa, 1987b). This is due to the fact that the vigilance system in such small groups becomes highly inefficient since subordinate males, which normally perform vigilance, now lead and feed the babies, leaving few, if any animals available to guard the group while foraging.

The importance of such vigilance for the survivorship of the young can be inferred from the data on offspring loss. With insufficient guards present, groups are unable to raise offspring (Rasa, 1987b). Whether offspring can survive without babysitters, however, could not be observed under field conditions since males automatically take over this role when no females are present. In groups of more than two adults a babysitter is therefore always available. A single case of a pair of animals attempting to raise two young without helpers was observed. Here male and female alternated in the roles of guard and babysitter (Rasa, 1989b). Both offspring were lost within three weeks of emergence from the den.

Table IV Reproductive success of female dwarf mongooses over 30 pack years (captivity data).

	DOMINANTS	SUBORDINATES
Number of Animals	3	17
Number of Estruses	29	8
Mean estrus/individual	9.7	0.7
% Resulting in Pregnancy	100	62.5
Number of Litters Surviving	27	1 (2 of 3 young dead by Day 26)

Infanticide

Subordinate females in dwarf mongoose groups come into estrus rarely in comparison to dominant females. Data covering all oestruses and pregnancies in 3 captive groups over 30 pack years are shown in Table IV. Behavioral mechanisms are present which inhibit subordinate females being mated by subordinate males (Rasa, 1973b) and they are not mated by the dominant male. Such mechanisms, however, are not fully effective and, as shown in Table IV, a small percentage of the subordinate females become pregnant. When their young are born, however, they are killed and eaten directly after birth by the alpha female. Only one case of a subordinate female successfully raising her young under unusual circumstances has been recorded in 30 pack years of observation on captive groups (Rasa, 1987a).

DISCUSSION

When the causes of infant mortality are taken into account, there appears to be a conflict of interests. On the one hand, the roles performed by subordinates are primarily geared to prevent offspring predation, on the other, the alpha female kills offspring that are not her own. Since the data have indicated that a minimum group size of 6 adults is necessary for the successful rearing of young, one could hypothesize that, at small group sizes, killing of infants from subordinate females would be a disadvantage to the alpha female, since such infants could serve to increase group size rapidly and thus ensure a better survival rate for her own subsequent offspring. This hypothesis would hold especially for small groups. It has been shown

that the optimum group size for ensuring maximal infant survival lies at 9 adults (Rasa, 1989a). This can be regarded, however, as the optimum group size from the point of view of the dominant pair. Mean group size, however, lies at 12.3 adults. This excess over the number necessary for optimal rearing of young has been interpreted as representing the interests of the subordinate group members, a point at which investment costs to subordinates are reduced and the probability of helpers being killed while performing their group-protective roles minimized (Rasa, 1989a). It would therefore also be to the advantage of subordinate animals in small groups if litters produced by subordinate females were to be raised since, in this way, an increase in group size could be achieved rapidly, thus reducing costs to individual subordinates. From a sociobiological point of view, the genetic gain is equal. Since the majority of helpers are full siblings, the raising of young from a brother/sister mating would be genetically equivalent to raising the parents' young. For the parents, however, young from son/daughter matings would also be genetically equivalent to their own offspring but, owing to the fact that groups can contain subordinate immigrants which can also breed, young from such matings would represent a reduction in own gene complement transfer ($r = 0.25$). Investment in such young would therefore not be to the advantage of either the dominant pair or full siblings. Sociobiologically speaking, the question remains why does infanticide occur when it presents a genetic disadvantage only in special cases?

In dwarf mongooses, estrus is synchronized (Rasa, 1977). As a result, the young of both dominant and subordinate females would therefore be born at approximately the same time and even in the same den (Rasa, 1987a). In this case, costs to babysitters during the youngs' non-mobile phase would be approximately equal. Whether a babysitter remains in the den protecting 3 – 4 or 6 – 8 young (taking the mean litter size of 3.08 at emergence as a measure), is, from the point of view of energy loss to the adult individual, irrelevant, since it is the time in which the adult must forgo foraging which is considered the cost of babysitting during this phase. When, however, the young become mobile, severe problems arise.

Table V shows the number of adults necessary per group for optimal protection and feeding of the young once they emerge from the den and the projection of these figures to encompass a case in which two litters are raised simultaneously.

As can be seen from these data, the number of subordinates necessary to bypass the bottleneck which occurs predominantly in the first week after the young emerge from the den, when they not only require food from the babysitters but must also be carried by them (Rasa, 1987a), far exceeds the

Table V Number of animals available to perform group protective roles in groups of varying size after the birth of a single litter and in the case of two females giving birth simultaneously. (Mean group size = 12.3 adults, mean litter size on emergence = 3.08 young.) Dominant animals do not guard or babysit when subordinates are present..

| | | SINGLE LITTER | | |
GROUP SIZE	DOMINANTS	AVAILABLE BABYSITTERS	OVERLAP AVAILABLE GUARDS	BETWEEN GUARDS*
12	2	3	7	Y
9	2	3	5	Y
5	2	3	0	N
3	2	0	0	N
		TWO SIMULTANEOUS LITTERS		
12	2	6	4	N
9	2	6	1	N
5	2	3	0	N

*Based on data from Rasa (1989a). Y = Yes, N = No

mean group size. A typical group would have insufficient animals available to provide adequate protection from predators (guards) and the likelihood of offspring survival would be low. The question then arises, why do dwarf mongooses not live in larger groups?

One could argue that a larger group size would be of advantage, since it reduces costs on individuals performing group protective roles. Dwarf mongoose territories, however, appear to be traditional and expansionist tendencies have not been recorded (Rasa, 1987a). This means that there is a limit on the number of individuals such a territory can support and group growth cannot continue unchecked without severe depletion of resources. Apart from this factor, the time spent foraging per day has been shown to be significantly longer for large groups than for small ones (Rasa, 1987a). The costs and risks of food finding therefore also increase as group size increases. Only in cases where these costs and risks no longer apply, where food is abundant and the risks of predation are drastically reduced, for example in the vicinity of human habitation, have group sizes of 20 or more adults been regularly recorded. Group size therefore appears to be limited at its upper end by ecological pressures.

If the killing of subordinates' young did not take place, group size could reach a critical level where the territory would not contain sufficient resources to support the group within one year. For example, should two females breed simultaneously in a group of 12.3 adults, the mean group size, this would result in a mean increase of $3.08 \times 2 = 6.16$ young yearly (based on the data for litter size at emergence from the den and the production of one litter/year) and group size would increase to 19 individuals. At this group size, costs to subordinates outweigh benefits (Rasa, 1989a). More importantly, with 6 babysitters leading young, only 3 adults would be available for vigilance. With only 3 guards present, guarding becomes inefficient with no overlap between guards and no guard present for up to 22.5% of the foraging time (Rasa, 1989b). Under such conditions a high infant mortality can be expected. The presence of a large number of young in the group at the same time would therefore be a disadvantage, especially for subordinates, which must increase their already high investment and also result in high infant loss. At large group sizes, pack splits also take place. The disadvantages accruing to splitting packs have been discussed elsewhere (Rasa 1987a).

In conclusion, the hypothesis put forward for the evolution of sociality in mongooses, that it was selected for as a means of protection for the young, appears to be substantiated. The helping behavior described is directed primarily towards infant care and geared towards a high survival rate of dependent offspring, not of adults. This can be inferred from the finding that, of the 20 animals observed killed over 2184 hours field observations, 70% were young of 4 months of age or less, although such young only comprised 14.9% of the total population. Dependent young are therefore by far the most vulnerable group members. Without the aid of at least 4 helpers, dwarf mongooses are unable to raise litters. Since the investment in the young by subordinates is so high, an increase in the number of young produced per year would result in an energetic bottleneck in the first three weeks after the youngs' emergence from the den. At this time, each youngster is fully dependent on a helper for food and transport and the number of helpers present must equal the number of young plus sufficient individuals (5) to ensure efficient vigilance. This situation could not come about unless group size was drastically increased. An increase in group size could, however, not be supported by the resources available and would also increase the costs and risks to helpers. Dwarf mongooses are therefore faced with a dilemma: if group size is too small, there is no reproductive success and associated fitness gains, if it becomes too large, resources are depleted and costs to individuals increase. To compensate for the latter situation, reproduction by subordinates is suppressed by a variety

of means. In the former case, natural selection processes come into play and unsuccessful groups are eliminated.

ACKNOWLEDGEMENTS

I should like to thank the Deutsche Forschungsgemeinschaft for supporting the research on the dwarf mongoose over a period of 18 years and the Foundation for Research and Development for their support of the yellow mongoose research over the past 5 years. I should also like to thank Mr. Ray Mayers for allowing me to conduct the dwarf mongoose studies on his ranch in Kenya and the National Parks Board of South Africa for their continuing support of the yellow mongoose studies in the Kalahari Gemsbok National Park. My thanks are also due to my students, especially B. Wenhold for their input and to the Ettore Majorana Centre for Scientific Culture for the invitation to take part in the International School of Ethology meeting at Erice where this paper was given.

REFERENCES

ALBIGNAC, R. (1973) Faune de Madagascar, Tome 36, Mammifere Carnivores. O.R.S.T.-O.M., C.N.R.S. Paris.

BALDWIN P.H., Schwartz, G.W. and Schwartz, E.R. (1952) Life history and economic status of the mongoose in Hawaii. *Journal of Mammalogy*, **33**, 335–356.

BEN-YAAKOV, R. and Yom-Tov, Y. 1983 On the biology of the Egyptian mongoose, *Herpestes ichneumon* in Israel. *Zeitschrift für Saugetierkunde*, **48**, 34–45.

BEYNON P. and Rasa, O.A.E. (1989) Do dwarf mongooses have a language?: Warning vocalizations transmit complex information. *South African Journal of Science*, **85**, 447–450.

CAVALLINI, P. and Nel J.A.J. (1990) Ranging behavior of the Cape grey mongoose *Galerella pulverulenta* in a coastal area. *Journal of Zoology London*, **222**, 352–362.

CREEL, S.R., Monfort, S.L., Wildt, D.E. and Waser, P.M. (1991) Spontaneous lactation is an adaptive result of pseudopregnancy. *Nature*,, **351**, 660–662.

EWER, R.F. (1963) The behavior of the meerkat *Suricata suricatta*. *Zeitschrift für Tierpsychologie*, **20**, 570–607.

HAMILTON, W.D. (1971) Geometry for the selfish herd. *Journal of Theoretical Biology*, **31**, 295–311.

KRUUK, H. (1978) Spatial organization and territorial behavior of the European badger *Meles meles*. *Journal of Zoology London*, **184**, 1–19.

KRUUK, H. and Macdonald, D. (1985) Group territories of carnivores: empires and enclaves. In: R.M.Sibley and R.H.Smith (eds.), *Behavioral Ecology: Ecological Consequences of Adaptive Behavior*. Oxford, Blackwell Scientific. pp. 521–536.

LYNCH, C.D. (1969) Ecology of the suricate *Suricata suricatta* and yellow mongoose *Cynictis penicillata* with special reference to their reproduction. D.Sc. thesis, University of Pretoria.

RASA O.A.E. (1972) Aspects of social organization in captive dwarf mongooses. *Journal of Mammalogy*, **53**, 181–185.

RASA, O.A.E. (1973a) Prey capture, feeding techniques and their ontogeny in the dwarf mongoose, *Helogale undulata rufula*. *Zeitschrift für Tierpsychologie*, **32**, 449–488.

RASA, O.A.E. (1973b) Intra-familial sexual repression in the Dwarf mongoose *Helogale parvula*. *Naturwissenschaften*, **60**, 303–304.

RASA, O.A.E. (1977) The ethology and sociology of the dwarf mongoose *Helogale undulata rufula*. *Zeitschrift für Tierpsychologie*, **43**, 337–340.

RASA, O.A.E. (1979) The effects of crowding on the social relationships and behavior of the dwarf mongoose, *Helogale undulata rufula*. *Zeitschrift für Tierpsychologie*, **49**, 317–329.

RASA, O.A.E. (1983) Dwarf mongoose and hornbill mutualism in the Taru desert, Kenya. *Behavioral Ecology and Sociobiology*, **12**, 181–190.

RASA, O.A.E. (1986a) Ecological factors and their relationship to group size, mortality and behavior in the dwarf mongoose (*Helogale undulata rufula*). *Cimbebasia*, **8**, 16–21.

RASA, O.A.E. (1986b) Coordinated vigilance in dwarf mongoose family groups: the Watchman's song hypothesis and the costs of guarding. *Ethology*, **71**, 340–344.

RASA, O.A.E. (1987a) The dwarf mongoose: a study of behavior and social structure in relation to ecology in a small social carnivore. In: J. Rosenblatt (ed.), *Advances in the Study of Behavior,* Vol. 17, California, Academic Press, pp. 121–163.

RASA, O.A.E. (1987b) Sociability for survival: why dwarf mongooses live in groups. In: L.C. Drickamer (ed.), *Readings from the 19th Int. Ethology Conference*, Vol., **4**, Behavioral Ecology and Population Biology. Toulouse, Privat Press, pp. 35–39.

RASA, O.A.E. (1989a) The costs and effectiveness of vigilance behavior in the dwarf mongoose: implications for fitness and optimal group size. *Ethology, Ecology and Evolution*. **1**, 265–282.

RASA, O.A.E. (1989b). Behavioral parameters of vigilance in the dwarf mongoose: social acquisition of a sex-biased role. *Behavior*, **110**, 125–145.

RASA O.A.E. (1989c) Helping in dwarf mongoose societies: an alternative reproductive strategy. In: O.A.E. Rasa, C. Vogel, and E. Voland (eds), *The Sociobiology of Sexual and Reproductive Strategies*. London, Chapman and Hall, pp. 61–73.

RASA, O.A.E., Wenhold, B.A., Beynon, P., Marais, A. and Pallett, J. (1992) Reproduction in the yellow mongoose revisited. *South African Journal of Zoology*, **27**, 192–195.

ROOD, J.P. (1986) Ecology and social evolution in the mongooses. In: D.I. Rubenstein and R.W. Wrangham (eds.), *Ecological Aspects of Social Evolution. Birds and Mammals*. Princeton NJ, Princeton University Press, pp. 131–152.

ROSEVEAR, D.R. (1974) *The Carnivores of West Africa*. Trustees of the British Museum of Natural History, London.

SMITHERS, R.H. (1983) *The Mammals of the Southern African Subregion*. Pretoria, University of Pretoria Press.

THURNHEER S. (1990) Aspekte der sozialen Organisation einer Mangustengruppe in Gefangenschaft. Diplomarbeit, Universität Zürich.

WENHOLD B.A. (1990) The ethology and social structure of the yellow mongoose *Cynictis penicillata*. M.Sc. Thesis, University of Pretoria.

CHAPTER 14

NEPOTISM AND INFANTICIDE AMONG PRAIRIE DOGS

JOHN L. HOOGLAND

*The University of Maryland, Appalachian Environmental Laboratory,
Frostburg, Maryland, 21532, USA*

INTRODUCTION

Black-tailed prairie dogs (*Cynomys ludovicianus*) excel in amicable conduct towards conspecific juveniles. For example, adults and yearlings improve the survivorship of their own and others' juveniles via antipredator calls, communal nursing, allogrooming, communal sleeping during cold weather, and reduced aggression. Curiously, prairie dogs also excel in their hostile conduct towards juveniles. Most notably, adults and yearlings commonly kill juveniles of kin and nonkin, and such infanticide accounts for the partial or total demise of 39% of all litters born.

Why so much variation in treatment of conspecific juveniles by older prairie dogs? In this chapter I marshal evidence from a 14-year study (1975 through 1988) and investigate answers to this problem.

THE STUDY ANIMAL

Black-tailed prairie dogs are large (400–1,100 grams for adults, who are 2 years old or older), diurnal, burrowing, herbivorous, colonial, harem-polygynous, cooperatively breeding rodents of the squirrel family (Sciuridae) (King, 1955; Koford, 1958; Smith, 1967; Wilson, 1975; Hoogland 1979, 1981a, 1985, 1994). Colonies contain contiguous territorial family groups called coteries, which typically contain 1 adult male, 2-3 adult females, and several yearling and juvenile offspring of both sexes (King, 1955; Hoogland, 1986). My study colony in Wind Cave National Park,

South Dakota, USA, was 6.6 hectares in size, and in late spring of each year contained a mean + SD of 125 + 15.3 adults and yearlings and 86.4 + 26.9 juveniles (Hoogland, 1981a, Hoogland *et al.*, 1988). Prairie dogs in South Dakota copulate in February or March, and gestation lasts for 34 or 35 days (Hoogland and Foltz, 1982). Mothers isolate their pre-emergent juveniles in a separate burrow that they vigorously defend (King, 1955; Hoogland, 1986). Nearly weaned young first emerge from the natal burrow in May or June, approximately 6 weeks after parturition.

Because they are highly philopatric, females in the same coterie are invariably close kin (Hoogland, 1986). Males, however, are like other sciurid males (Holekamp, 1984; Holekamp and Sherman, 1989) and disperse from the natal area before sexual maturity (King, 1955; Hoogland, 1982). Using four different mechanisms, prairie dogs systematically avoid copulations with close kin such as parents, offspring, and siblings (Hoogland, 1982; Foltz and Hoogland, 1983). However, individuals do not avoid copulations with more distant kin. Moderate inbreeding with relatives such as first and second cousins is therefore common (Hoogland, 1992).

Mortality during the first year is approximately 50% for both sexes. Females that survive the first year sometimes live as long as 8 years, but males never live longer than 5 years (Hoogland *et al.*, 1988). Individuals sometimes breed as yearlings, but the usual age of sexual maturity is 2 years (King, 1955; Hoogland and Foltz, 1982). Probably in response to the polygynous mating system (Clutton-Brock *et al.*, 1977; Alexander *et al.*, 1979; Hoogland and Foltz, 1982), prairie dog males are 10%-15% larger than females.

HOSTILE TREATMENT OF JUVENILE PRAIRIE DOGS: THE FOUR TYPES OF INFANTICIDE

As outlined below, infanticide within prairie dog colonies is of four types.

Type I: Immigrant Female Kills Unrelated Juvenile Kin

From 1975 through 1988, only 5 females immigrated into the study colony and successfully reproduced. Three of these 5 females set up territories at the periphery of the study colony. The other 2 immigrant females invaded coterie territories containing recently-emergent juveniles, evicted the resident females from the territory, and then promptly killed all juveniles there. Type I infanticide is rare, simply because female immigration into coteries with juveniles is rare (Table I).

Table I Frequencies of the four types of infanticide among prairie dogs. The numerator indicates the number of litters affected by the particular type of infanticide; the denominator indicates the number of litters monitored for that type of infanticide. For the types of infanticide that are easy to detect, reliable data were available for as many as 11 years (1978 through 1988). For the more difficult types, reliable data were available for fewer years (see Hoogland, 1985).

Type I.	Invading female immigrant kills juvenile nonkin (1978 through 1988)	2/591 = 0.3%
Type II.	Invading male kills	
	(a) juvenile nonkin (1978 through 1988)	33/591 = 5.6%
	(b) juvenile half siblings (1978 through 1988)	6/591 = 1.0%
Type III.	Mother and other coterie members kill abandoned juvenile kin (1983 through 1988)	24/253 = 9.5%
Type IV.	Lactating female kills juveniles of close kin (1981 through 1988)	65/294 = 22.1%
Cumulative percentage of litters affected by infanticide:		38.5%

Why is Type I infanticide limited to female immigrants from another colony? What about females who disperse from the home coterie territory and move into another coterie territory within the same home colony? For reasons that remain unclear, females almost never disperse locally within the home colony to another coterie territory (Knowles, 1985; Garrett and Franklin, 1988; Hoogland, 1992). Consequently, infanticide by locally-dispersing females does not occur.

An infanticidal female immigrant removes juveniles that would grow up and compete with herself and her offspring in future years (Hrdy, 1979; Sherman, 1981). Such removal of competitors might be the ultimate explanation for infanticide by female immigrants, but a sample size of 2 precludes any serious speculation.

Type II: Invading Male Kills Unrelated Juvenile Kin

Male prairie dogs frequently change coteries — either as yearlings from the natal coterie to a breeding coterie or as adults from one breeding coterie to another. If the new coterie contains juveniles sired by the previous male, the invading male usually kills them within a few days after arriving. Type II infanticide accounted for the demise of 7% of the litters born at the study colony in 1978 through 1988 (Table I).

Infanticide after invasion by a new male also occurs in other harem-polygynous mammals (Hrdy, 1974, 1977; Bertram, 1975; Packer and Pusey, 1983; Brooks, 1984; Crocket and Sekulic, 1984; Fossey, 1984; Hausfater, 1984; Huck, 1984; Labov, 1984; Leland *et al.,* 1984; vom Saal, 1984; Labov *et al.,* 1985). The payoff for males in these species is that victimized females come into estrus and conceive more quickly than spared females who continue to lactate (Hrdy, 1974; Bertram, 1975). Invading male prairie dogs, however, usually kill juveniles after the termination of lactation. Further and more important, prairie dog females come into estrus only once each year in February or March (Hoogland and Foltz, 1982). Consequently, infanticide by an invading male cannot easily reduce the time until the next estrus of the victimized female.

Earlier I hypothesized that infanticidal males might gain by reducing the number of individuals who would become competitive yearlings the next year (Hoogland, 1985). But this hypothesis now seems unlikely, because additional data since 1985 indicate that the presence of yearlings does not reduce male reproductive success. Perhaps infanticidal males gain through effects on maternal condition (Clutton-Brock *et al.*, 1989). Females freed from maternal duties devote more time to feeding. Consequently, they gain weight and prepare for the next breeding season better than females whose emergent offspring continue to survive (Hoogland, 1994).

Dispersing adult and yearling males sometimes move into an adjacent coterie territory. Occasionally a yearling male disperses into the same adjacent territory into which his father had dispersed 1–2 years earlier. If the dispersing yearling is infanticidal, as 6 of them were at the study colony, then his victims will be his father's offspring (his own half-siblings) (Table I).

For all 6 invading males who killed juvenile kin, the father was still alive at the time of invasion. Although the fathers resisted invasions by other males, they allowed the invasions by their sons. Why did the fathers allow their yearling sons to invade and kill their new offspring? I obtained no answer for this troubling question.

What happens when males invade at a later stage when the emergent juveniles are larger and less vulnerable? I observed several invasions into coterie territories where the juveniles were large (more than 300 grams) and mobile. The new male consistently interacted aggressively with the large juveniles, but I observed no infanticides. In response to the harassment, the juveniles assiduously avoided the new male. Almost none of the harassed larger juveniles survived until the next breeding season. Invading males evidently can eliminate unwanted juveniles by means other than infanticide.

Type III: Mother Allows Infanticide and Cannibalism of Her Own Offspring

Soon after copulating, a prairie dog female begins to show several maternal behaviors: building an underground nest, defense of the burrow with the nest, sleeping alone in the burrow with the nest, etc. (King, 1955; Hoogland and Foltz, 1982). When females who copulated did not show these maternal behaviors, I assumed in the early years of my research that these females — similar to females under both laboratory (Foreman, 1962) and natural (Knowles, 1987) conditions — either failed to conceive or conceived and then aborted and resorbed their litters. More recently, I learned that almost all females who copulate later give birth. The females who copulate and show maternal behaviors attempt to rear their litters to weaning. Those females who copulate and do not show maternal behaviors, however, usually abandon their offspring within a day or two after parturition. Usually with the mother at parturition are other coterie members, who kill and cannibalize the abandoned neonates. Unlike victims of the other types of infanticide, nonmaternally-acting females make no systematic effort to prevent killing and cannibalism of their newborn offspring. Type III infanticide accounted for the elimination of 10% of the litters born in 1983 through 1988 (Table I).

Are females in poor condition more likely to abandon their neonates? To investigate this possibility, I used body weight during the breeding season as an estimate of overall condition. Whereas the mean + SD weight (grams) for females who abandoned was 640 + 57.4 (N = 15), the weight for females who did not abandon was 707 + 85.4 (N = 24) (P = 0.020, Mann-Whitney U test). Perhaps abandonment is an adaptive response by females in poor condition who are unlikely to wean offspring anyway.

Type IV: Lactating Female Kills the Offspring of Close Kin

Prairie dog mothers defend the home burrow containing pre-emergent juveniles from all other prairie dogs. However, mothers also try to enter the home burrows of other mothers. When one mother eludes another mother's defense, she stays in the strange burrow for 30 minutes or more and then emerges with a bloody face. Immediately the victimized mother stops acting maternally, presumably because her juveniles are dead. Early in my research I wondered whether the "marauding" females actually kill the juveniles of other females. Perhaps they just eat already-dead juveniles that are no longer worth the mother's defense. Two lines of evidence indicate that marauders are the killers. (a) An excavation after a case of presumed infanticide yielded decapitated juveniles. The mother's stomach

was empty shortly after the marauding, but the marauder's stomach was full of flesh and bones. (b) Two different marauders brought live juveniles aboveground before killing them. Killing via Type IV infanticide affected 22% of all litters born in 1981 through 1988 (Table I). Marauding by lactating mothers is thus the most common, and the most puzzling, of the four types of infanticide within prairie dog colonies.

In all but 2 of the 65 cases (3%) of infanticide perpetrated by marauding mothers, the marauder killed the juveniles of another female within the home coterie. Because females within a coterie are invariably close kin (Hoogland, 1983, 1986), the marauder's victims are also kin (Hoogland, 1985). Why would a female kill the juvenile offspring of kin in the home coterie rather than the juvenile offspring of nonkin in a different coterie? The answer might be accessibility. Prairie dog mothers not only defend the home burrow against other coterie members, but they also defend the boundary of the home coterie territory against intruders from other coteries. To kill juvenile kin within the home coterie, then, a marauding female only must elude the defense of the mother. To kill juvenile nonkin in a different coterie, however, a marauder must first elude the defense of the home coterie's territorial boundary by all the resident females, and then must elude the additional defense of the chosen mother's home burrow.

A mother leaves her own offspring undefended while marauding, sometimes at great cost. Specifically, $6/65 = 9\%$ of infanticides by lactating mothers occurred while the victimized mother was herself marauding. Marauding might have other costs. For example, killing the juvenile offspring of close kin might depress a female's "inclusive fitness" (Hamilton, 1964) — unless such killing significantly increases the probability that the killer will successfully wean her own litter.

To this point, I have identified five possible benefits realized by killers, as outlined below. These benefits might explain how increased survivorship of the killer's own juvenile offspring might offset the costs of eliminating the offspring of close kin.

Benefit #1: Removal of future competitors

Hrdy (1979) and Sherman (1981) both pointed out that, in a species where individuals are highly philopatric, infanticide leads to the removal of future competitors. Marauding prairie dog females remove future competitors both from themselves and their offspring, so that both killer and her offspring might realize increased lifetime reproductive success.

Prairie dog females restrict over 99% of all their foraging and other activities to the home coterie territory. Consequently, females compete more

with genetically related females of the home coterie than with unrelated or only distantly related females of other coteries. Rather than from differences in accessibility (see above), perhaps the usual restriction of marauding to the home coterie results more from attempts to remove future competitors. Why should mothers bother to kill juveniles in other coteries with whom they and their offspring will only compete minimally?

Benefit #2: Increased sustenance

Marauders could easily kill underground juveniles without eating them. But they don't (Hoogland, 1985). Through cannibalism, marauders obtain sustenance (protein, rare minerals, etc.) that might be crucial for getting through stressful lactation (Hrdy, 1979; Sherman, 1981; Clutton-Brock *et al.*, 1982, 1989; Hausfater and Hrdy, 1984).

Benefit #3: Increased foraging area

Mothers only defend a burrow and the surrounding territory when their offspring are still alive (King, 1955; Hoogland, 1985, 1986). Upon losing her litter for any reason, a female abruptly terminates maternal defense. By killing the juveniles of another female in the home coterie territory, a marauder thus creates a larger area in which she can easily forage.

Benefit #4: Victimized mothers become better helpers

After losing her litter, a female — liberated from maternal duties — devotes more time to scanning for predators, refurbishing burrows in the home coterie territory other than the single burrow that contains her juveniles, patrolling the coterie's territorial boundary for possible invaders, etc. Regarding these latter behaviors, then, females without offspring are more useful to mothers with offspring. Thus, killers induce victimized females to become better helpers.

Benefit #5: Victimized mothers are less likely to kill

Without her own living juveniles, a female does not maraud. Eliminating a female's litter thus reduces the probability that the killer will later lose her own litter to that female's marauding.

328   J.L. HOOGLAND

PREFERENTIAL TREATMENT OF JUVENILE PRAIRIE DOGS: PARENTAL AND ALLOPARENTAL CARE

Despite the prevalence of infanticide within colonies, prairie dogs are among the most social, parental, alloparental (i.e, showing parental-like behavior towards juveniles other than one's own offspring), and cooperative of all animals, as outlined below.

The Antipredator Call

Upon detecting a predator, a prairie dog commonly gives an antipredator call (King, 1955; Waring, 1970; Smith et al., 1977). Like other sciurid antipredator calls (Dunford, 1977; Sherman, 1977, 1985; Smith, 1978; Yahner, 1978; Noyes and Holmes, 1979; Leger and Owings, 1978; Leger et al., 1980; Schwagmeyer, 1980; Davis, 1984; Owings et al., 1986), the prairie dog call functions to warn genetic relatives and is therefore nepotistic (Figure 1). However, the prairie dog antipredator call is unique for two reasons. First, males are just as likely to call as females (Figure 1). Second,

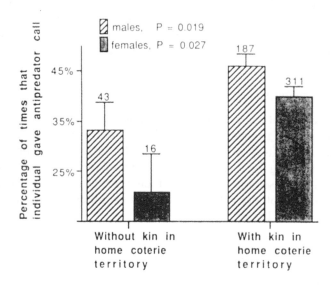

Figure 1 Antipredator calls by prairie dogs that did and did not have living genetic relatives in the home coterie territory. The number above each standard error line indicates the number of different individuals observed (each approximately 7 times) during 928 experimental runs with a stuffed specimen of a natural predator, the badger (*Taxidea taxus*). The p-value is from the Mann-Whitney U test. Males called as often as females (p = 0.178, Mann-Whitney U test). For experimental details, see Hoogland, 1983.

individuals that have only collateral kin (siblings, aunts, uncles, nieces, nephews, etc.) within earshot call as often as individuals with nearby offspring (Hoogland, 1983,1994).

Nepotism and Behavioral Interactions

Prairie dog interactions are either hostile or amicable. Hostile interactions include fights, chases, and long territorial disputes, while amicable interactions include allogrooming, anal sniffing, and oral contacts known as "kisses" (King, 1955; Hoogland, 1979). As expected, individuals consistently interact more amicably with genetically related members of the home coterie than with unrelated or only distantly related individuals of other coteries (Hoogland, 1981b, 1986). However, contrary to expectation, individuals do not discriminate between close and distant kin within the home coterie. Specifically, prairie dogs interact just as amicably with more distant kin such as nieces and cousins as with close kin such as offspring and full siblings. Similarly, prairie dogs spend the night underground during cold weather only with members of the home coterie — but are just as likely to do so with distant kin as with close kin (Hoogland, 1981b, unpublished).

Communal Nursing

The epitome of nepotism and alloparental behavior in prairie dogs is perhaps communal nursing. Shortly after first juvenile emergences, mothers suckle not only their own offspring, but also the offspring of other mothers within the home coterie — the very same offspring they tried to kill during the previous six weeks of lactation (Hoogland et al., 1989)! In some years over 50% of juveniles receive additional milk from foster mothers.

DISCUSSION

Historically, biologists have usually regarded infanticide as a pathological and maladaptive response to overcrowding. Resistance was intense when Hrdy (1974, 1977, 1979, 1984; see also Hausfater, 1984; Sugiyama, 1984) argued that infanticide in Hanuman langurs (*Presbytis entellus*) is an evolved response resulting from sexual selection (Curtin and Dolhinow, 1978; see also Boggess, 1984).

Three lines of evidence indicate that infanticide among prairie dogs is adaptive rather than pathological. (a) In addition to the frequent infanticides at the uncrowded study colony (Table I), killings also occurred at two

other nearby uncrowded colonies (Hoogland, 1985). Poor female reproduction in other studies further suggests the ubiquity of infanticide (King, 1955; Tileston and Lechleitner, 1966; Knowles, 1985, 1987; Halpin, 1987; Garrett *et al.,* 1982; Garrett and Franklin, 1988; Stockrahm and Seabloom, 1988). (b) Infanticide at the study colony was not merely the work of a few idiosyncratic prairie dogs. Rather, 84 different prairie dogs killed: 2 immigrant females, 24 females that abandoned their juveniles shortly after parturition, 38 females who killed the offspring of close kin, and 20 invading males. (c) Infanticide probably enhances the killer's reproductive success (Hoogland, 1985,1994).

In some circumstances, yearlings and adults kill juveniles. In other circumstances, adults treat others' juveniles as they treat their own. Why so much variation? Here I will investigate possible proximate (mechanistic) and ultimate (evolutionary) explanations (Mayr, 1961; Tinbergen, 1963; Williams, 1966; Sherman, 1988) for this question.

In colonial birds, the initiation of parental recognition of offspring usually coincides with the first mixing of young from different clutches — *i.e.,* at the point when the probability of misdirected parental care dramatically increases (Cullen, 1957; Beer, 1970; Hoogland and Sherman, 1976; Beecher, 1982, 1990; Holmes and Sherman, 1983). Before first mixing, parents seem only able to recognize the nest site — they do not discriminate between their own young and young artificially introduced from other nests. Before the inter-litter mixings, prairie dog females resemble parents in colonial avian species: they distinguish between the home burrow containing their own offspring (as targets for nursing and other forms of parental care) and other home burrows (as targets for infanticide). After first mixings, however, prairie dog females are different from avian parents: they seem unable to capitalize on sophisticated mechanisms of parent-offspring recognition. Consequently, infanticide ceases and communal nursing begins.

In proximate terms, then, the switch from infanticide to communal nursing might couple with mechanisms of parent-offspring recognition. Specifically, mothers might kill when others' pre-emergent juveniles are recognizable by their distinctive home burrows. Following first emergences, infanticide ceases so that mothers will not accidentally kill their own, apparently unrecognizable, offspring. Consistent with this proximate interpretation, the only females known to kill emergent juveniles were invading immigrants, who could not have made any mistakes because they did not have any of their own emergent juveniles in the invaded coterie territories.

What is the ultimate explanation for the switch from infanticide to communal nursing? Three related possibilities come to mind. First,

perhaps discrimination and killing are only advantageous when the victims are small, immobile, and defenseless as they are prior to first emergence.

Reproductive value, an individual's expected lifetime reproductive output (Fisher, 1958; Emlen, 1970; Pianka and Parker, 1975; Goodman, 1982; Rubenstein, 1982), might be a second possible ultimate explanation for the switch from infanticide to communal nursing. The probability that a pre-emergent juvenile prairie dog will survive to reproduce is approximately 23%. However, the probability that an emergent juvenile will survive to reproduce is about 35% (Hoogland, 1985, 1994). Natural selection might favor mothers who discriminate against the juvenile offspring of close kin via infanticide when the probability of survivorship and later reproduction is low, but who help such offspring via communal nursing when the probability is higher. One problem with this explanation is that infanticide is the main reason that the reproductive value of pre-emergent juveniles is lower than that of emergent juveniles.

A third ultimate explanation for the shift from infanticide to communal nursing concerns variation in the usefulness of others' offspring. Early in lactation when herbaceous food is scarce (King, 1955), others' offspring might be most useful as sources of sustenance via cannibalism. Following first emergences when herbaceous food is abundant, others' offspring might be most useful for the formation of multi-litter groupings, which probably reduce the probability that one's own offspring will be captured by a predator (Hoogland et al., 1989). Communal nursing encourages the formation of such multi-litter groupings.

What about variation among males in amicability towards juveniles? Like males of other rodent species (Huck et al., 1982; McLean, 1983; vom Saal, 1984; Labov et al., 1985; Parmigiani, 1989; Wolff and Cicerello, 1989), prairie dog males only seem to distinguish between territories where (or time intervals when) they copulated and thus might have sired offspring (as targets for amicable paternal behavior) and territories where they did not copulate (as targets for infanticide). Though I looked, I found no evidence that cuckolded males can discriminate between their own and other males' offspring within the home coterie territory.

Natural selection can work through either the reproduction of individuals themselves ("direct selection;" see West-Eberhard, 1975; Brown and Brown, 1981; Brown, 1987; Armitage, 1987, 1988) or the reproduction of an individual's kin ("indirect selection;" see Hamilton, 1964; Maynard Smith 1964; Alexander, 1974; Brown, 1987). Both direct selection and indirect selection evidently have been important in the evolution of amicable and hostile treatment of juvenile prairie dogs by older conspecifics. Regarding direct selection, for example, both males and females consistently

guard and care for their own offspring but regularly kill the offspring of competing conspecifics. Regarding indirect selection, both males and females that have only nondescendant kin (but no offspring) within earshot regularly give an antipredator call when a predator approaches (Hoogland, 1983). Indirect selection also might ultimately help explain the communal nursing of nondescendant juveniles (Hoogland et al., 1989).

In closing I emphasize that I have not demonstrated that prairie dog males and females are unable to discriminate between their own and others' emergent offspring within the home coterie territory. I can only conclude that such discriminative nepotism is not evident from observations of infanticide and behavioral interactions. Perhaps an investigation of prairie dogs under laboratory conditions (cf. Holmes and Sherman, 1982; Holmes, 1984) would lead to a better understanding of parent-offspring recognition and other forms of "kin recognition" (Waldman et al., 1988; Sherman, 1991) among these fascinating rodents.

ACKNOWLEDGEMENTS

I thank the 114 field assistants who have helped over the years, and I especially thank D. Angell, P. Hardison, J. G. Hoogland, M. A. Hoogland, M. V. Hoogland, S. T. Hoogland, S. Kain, J. Loughry, and P. Walsh. I also thank the staff at Wind Cave National Park, especially L. Butts, R. Klukas, L. McClanahan, E. Ortega, and J. Randall. For financial assistance, I thank The National Science Foundation, The National Geographic Society, The American Philosophical Society, The Center for Field Research, The Max McGraw Wildlife Foundation, The Universities of Michigan, Minnesota, and Maryland, Princeton University, and The Harry Frank Guggenheim Foundation. For discussion and help with the manuscript, I thank Eric van den Berghe, Frank Rohwer, and Paul Sherman.

REFERENCES

ALEXANDER, R.D. 1974 The evolution of social behavior. *Annual Review of Ecology and Systematics*, **5**, 325–383.

ALEXANDER, R.D., Hoogland, J.L., Howard, R.D., Noonan, K.M. and Sherman, P.W. 1979 Sexual dimorphisms and breeding systems in pinnipeds, ungulates, primates, and humans. In: Chagnon, N.A. and Irons, W. eds. *Evolutionary Biology and Human Social Behavior*. Duxbury Press, North Scituate, Massachusetts. Pages 402–435.

ARMITAGE, K.B. 1987 Social dynamics of mammals: reproductive success, kinship, and individual fitness. *Trends in Ecology and Evolution*, **2**, 279–284.

ARMITAGE, K.B. 1988 Social organization of ground-dwelling squirrels. In: C.N. Slobod-chikoff, ed. *The Ecology of Social Behavior.* Academic Press, New York. Pages 132–155.

BEECHER, M.D. 1982 Signature systems and kin recognition. *American Zoologist,* 22, 477–490.

BEECHER, M.D. 1990 The evolution of parent-offspring recognition in swallows. In: Dewsbury, D.A., ed. *Contemporary Issues in Comparative Psychology.* Sinauer Associates, Sunderland, Massachusetts. Pages 360–380.

BEER, C.G. 1970 Individual recognition of voice in the social behavior of birds. In: Lehrman, D.S., Hinde, R.A. and Shaw, E., eds. *Advances in the Study of Behavior,* Volume 3. Academic Press, New York. Pages 27–74.

BERTRAM, B.C.R. 1975 Social factors influencing reproduction in wild lions. *Journal of Zoology,* 177, 463–482.

BOGGESS, J. 1984 Infant killing and male reproductive strategies in langurs (*Presbytis entellus*). In: Hausfater, G. and Hrdy, S.B. eds. *Infanticide: Comparative and Evolutionary Perspectives.* Aldine, New York. Pages 283–310.

BROOKS, R.J. 1984 Causes and consequences of infanticide in populations of rodents. In: Hausfater, G. and Hrdy, S.B. eds. *Infanticide: Comparative and Evolutionary Perspectives.* Aldine, New York. Pages 331–348.

BROWN, J.L. 1987 *Helping and Communal Breeding in Birds.* Princeton University Press, Princeton.

BROWN, J.L. and Brown, E.R. 1981 Kin selection and individual selection in babblers. In: Alexander, R.D. and Tinkle, D.W. eds. *Natural Selection and Social Behavior.* Chiron Press, New York. Pages 244–256.

CLUTTON-BROCK, T.H., Albon, S.D. and Guinness, F.E. 1989 Fitness costs of gestation and lactation in wild mammals. *Nature,* 337, 260–262.

CLUTTON-BROCK, T.H., Guinness, F.E. and Albon, S.D. 1982 *Red Deer: Behavior and Ecology of Two Sexes.* Chicago University Press, Chicago.

CLUTTON-BROCK, T.H., Harvey, P.H. and Rudder, B. 1977 Sexual dimorphism, socionomic sex ratio and body weight in primates. *Nature,* 269, 797–799.

CROCKETT, C.M., and Sekulic, R. 1984 Infanticide in red howler monkeys (*Alouatta seniculus*). In: Hausfater, G. and Hrdy, S.B., eds. *Infanticide: Comparative and Evolutionary Perspectives.* Aldine, New York. Pages 173–191.

CULLEN, E. 1957 Adaptations in the kittiwake to cliff-nesting. *Ibis,* 99, 275–302.

CURTIN, R., and Dolhinow, P. 1978 Primate social behavior in a changing world. *American Scientist,* 66, 468–475.

DAVIS, L.S. 1984 Alarm calling in Richardson's ground squirrels (*Spermophilus richardsonii*). *Zeitschrift fur Tierpsychologie,* 66, 152–164.

DUNFORD, C. 1977 Kin selection for ground squirrel alarm calls. *American Naturalist,* 111, 782–785.

EMLEN, J.M. 1970 Age specificity and ecological theory. *Ecology,* 51, 588–601.

FISHER, R.A. 1958 *The Genetical Theory of Natural Selection.* Second edition. Dover, New York.

FOLTZ, D.W. and Hoogland, J.L. 1983 Genetic evidence of outbreeding in the black-tailed prairie dog (*Cynomys ludovicianus*). *Evolution,* 37, 273–281.

FOREMAN, D. 1962 The normal reproductive cycle of the female prairie dog and the effects of light. *Anatomical Record,* 142, 391–405.

FOSSEY, D. 1984 Infanticide in mountain gorillas (*Gorilla gorilla beringei*) with comparative notes on chimpanzees. In: Hausfater, G. and Hrdy, S.B. eds. *Infanticide: Comparative and Evolutionary Perspectives*. Aldine, New York. Pages 217–235.

GARRETT, M.G. and Franklin, W.L. 1988 Behavioral ecology of dispersal in the black-tailed prairie dog. *Journal of Mammalogy*, **69**, 236–250.

GARRETT, M.G., Hoogland, J.L. and Franklin, W.L. 1982 Demographic differences between an old and a new colony of black-tailed prairie dogs (*Cynomys ludovicianus*). *American Midland Naturalist*, **108**, 51–59.

GOODMAN, D. 1982 Optimal life histories, optimal notation, and the value of reproductive value. *American Naturalist*, **119**, 803–823.

HALPIN, Z.T. 1987 Natal dispersal and the formation of new social groups in a newly established town of black-tailed prairie dogs (*Cynomys ludovicianus*). In: Chepko-Sade, B.D. and Halpin, Z.T., eds. *Mammalian Dispersal Patterns. The Effects of Social Structure on Population Genetics*. University of Chicago Press, Chicago. Pages 104–118.

HAMILTON, W.D. 1964 The genetical evolution of social behavior. I & II. *Journal of Theoretical Biology*, **7**, 1–52.

HAUSFATER, G. 1984 Infanticide in langurs: strategies, counterstrategies, and parameter values. In: Hausfater, G. and Hrdy, S.B., eds. *Infanticide: Comparative and Evolutionary Perspectives*. Aldine, New York. Pages 257–281.

HAUSFATER, G. and Hrdy, S.B., eds. 1984 *Infanticide: Comparative and Evolutionary Perspectives*. Aldine, New York.

HOLEKAMP, K.E. 1984 Dispersal in ground-dwelling sciurids. In: Murie, J.O. and Michener, G.R., eds. *The Biology of Ground-dwelling Squirrels*. University of Nebraska Press, Lincoln. Pages 297–320.

HOLEKAMP, K.E., and Sherman, P.W. 1989 Why male ground squirrels disperse. *American Scientist*, **77**, 232–239.

HOLMES, W.G. 1984 Sibling recognition in thirteen-lined ground squirrels: effects of genetic relatedness, rearing association, and olfaction. *Behavioral Ecology and Sociobiology*, **14**, 225–233.

HOLMES, W.G. and Sherman, P.W. 1982 The ontogeny of kin recognition in two species of ground squirrels. *American Zoologist*, **22**, 491–517.

HOLMES, W.G. and Sherman, P.W. 1983 Kin recognition in animals. *American Scientist*, **71**, 46–55.

HOOGLAND, J.L. 1979 Aggression, ectoparasitism, and other possible costs of prairie dog (Sciuridae: *Cynomys* spp.) coloniality. *Behaviour*, **69**, 1–35.

HOOGLAND, J.L. 1981a The evolution of coloniality in white-tailed and black-tailed prairie dogs (Sciuridae: *Cynomys leucurus* and *C. ludovicianus*). *Ecology*, **62**, 252–272.

HOOGLAND, J.L. 1981b Nepotism and cooperative breeding in the black-tailed prairie dog (Sciuridae: *Cynomys ludovicianus*). In: Alexander, R.D. and Tinkle, D.W., eds. *Natural Selection and Social Behavior*. Chiron Press, New York. Pages 283–310.

HOOGLAND, J.L. 1982 Prairie dogs avoid extreme inbreeding. *Science*, **215**, 1639–1641.

HOOGLAND, J.L. 1983 Nepotism and alarm calling in the black-tailed prairie dog (*Cynomys ludovicianus*). *Animal Behaviour*, **31**, 472–479.

HOOGLAND, J.L. 1985 Infanticide in prairie dogs: lactating females kill offspring of close kin. *Science*, **230**, 1037–1040.

HOOGLAND, J.L. 1986 Nepotism in prairie dogs varies with competition but not with kinship. *Animal Behaviour*, **34**, 263–270.

HOOGLAND, J.L. 1992 Levels of inbreeding among prairie dogs. *American Naturalist*, **139**, 591–602.

HOOGLAND, J.L. 1994 *The Black-tailed Prairie Dog: Social Life of a Burrowing Mammal.* Chicago University Press, Chicago.

HOOGLAND, J.L., Angell, D.K., Daley, J.G. and Radcliffe, M.C. 1988 Demography and population dynamics of prairie dogs. In: Uresk, D.W., Schenbeck, G.L. and Cefkin, R., technical coordinators. Eighth Great Plains Wildlife Damage Control Workshop Proceedings. *Great Plains Agricultural Council Publication Number*, **121**, 1–231. Lincoln, Nebraska. Pages 18–22.

HOOGLAND, J.L. and Foltz, D.W. 1982 Variance in male and female reproductive success in a harem-polygynous mammal, the black-tailed prairie dog (Sciuridae: *Cynomys ludovicianus*). *Behavioral Ecology and Sociobiology*, **11**, 155–163.

HOOGLAND, J.L. and Sherman, P.W. 1976 Advantages and disadvantages of Bank Swallow (*Riparia riparia*) coloniality. *Ecological Monographs*, **46**, 33–58.

HOOGLAND, J.L., Tamarin, R.H. and Levy, C.K. 1989 Communal nursing in prairie dogs. *Behavioral Ecology and Sociobiology*, **24**, 91–95.

HRDY, S.B. 1974 Male-male competition and infanticide among the langurs (*Presbytis entellus*) of Abu, Rajasthan. *Folia Primatologia*, **22**, 19–58.

HRDY, S.B. 1977 Infanticide as a primate reproductive strategy. *American Scientist*, **65**, 40–49.

HRDY, S.B. 1979 Infanticide among animals: a review, classification, and examination of the implications for the reproductive strategies of females. *Ethology and Sociobiology*, **1**, 13–40.

HRDY, S.B. 1984 Assumptions and evidence regarding the sexual selection hypothesis: a reply to Boggess. In: Hausfater, G. and Hrdy, S.B., eds. *Infanticide: Comparative and Evolutionary Perspectives*. Aldine, New York. Pages 315–319.

HUCK, U.W. 1984 Infanticide and the evolution of pregnancy block in rodents. In: Hausfater, G. and Hrdy, S.B., eds. *Infanticide: Comparative and Evolutionary Perspectives*. Aldine, New York. Pages 249–365.

HUCK, U.W., Soltis, R.L. and Coopersmith, C.B. 1982 Infanticide in male laboratory mice: effects of social status, prior sexual experience, and basis for discrimination between related and unrelated young. *Animal Behaviour*, **30**, 1158–1165.

KING, J.A. 1955 Social behavior, social organization, and population dynamics in a black-tailed prairiedog town in the Black Hills of South Dakota. *Contributions from the Laboratory of Vertebrate Biology, The University of Michigan 67*, 1–123.

KNOWLES, C.J. 1985 Observations on prairie dog dispersal in Montana. *Prairie Naturalist*, **17**, 33–40.

KNOWLES, C.J. 1987 Reproductive ecology of black–tailed prairie dogs in Montana. *Great Basin Naturalist*, **47**, 202–206.

KOFORD, C.B. 1958 Prairie dogs, whitefaces, and blue grama. *Wildlife Monographs*, **3**, 1–78.

LABOV, J.B. 1984 Infanticidal behavior in male and female rodents: sectional introduction and directions for future research. In: Hausfater, G. and Hrdy, S.B., eds. *Infanticide: Comparative and Evolutionary Perspectives*. Aldine, New York. Pages 323–329.

LABOV, J.B., Huck, U.W., Elwood, R.W. and Brooks, R.J. 1985 Current problems in the study of infanticidal behavior of rodents. *Quarterly Review of Biology*, **60**, 1–20.

LEGER, D.W. and Owings, D.H. 1978 Responses to alarm calls by California ground squirrels: effects of call structure and maternal status. *Behavioral Ecology and Sociobiology*, **3**, 177–186.

LEGER, D.W., Owings, D.H. and Gelfand, D.L. 1980 Single-note vocalizations of California ground squirrels: graded signals and situation- specificity of predator and socially evoked calls. *Zeitschrift fur Tierpsychologie*, **52**, 227–246.

LELAND, L., Struhsaker, T.T and Butynski, T.M. 1984 Infanticide by adult males in three primate species of the Kibale Forest, Uganda: a test of hypotheses. In: Hausfater, G. and Hrdy, S.B., eds., *Infanticide: Comparative and Evolutionary Perspectives*. Aldine, New York. Pages 151–172.

MAYNARD SMITH, J. 1964 Group selection and kin selection. *Nature*, **201**, 1145–1147.

MAYR, E. 1961 Cause and effect in biology. *Science*, **134**, 1501–1506.

MCLEAN, I.G. 1983 Paternal behaviour and killing of young in arctic ground squirrels. *Animal Behaviour*, **31**, 32–44.

NOYES, D.H. and Holmes, W.G. 1979 Behavioral responses of free-living hoary marmots to a model golden eagle. *Journal of Mammalogy*, **60**, 408–411.

OWINGS, D.H., Hennessy, D.F., Leger, D.W. and Gladney, A.B. 1986 Different functions of "alarm" calling for different time scales: a preliminary report on ground squirrels. *Behaviour*, **99**, 101–116.

PACKER, C. and Pusey, A.E. 1983 Adaptations of female lions to infanticide by incoming males. *American Naturalist*, **121**, 716–728.

PARMIGIANI, S. 1989 Inhibition of infanticide in male mice (*Mus musculus*): Is kin recognition involved? *Ethology, Ecology, and Evolution*, **1**, 93–98.

PIANKA, E.R. and Parker, W.S. 1975 Age-specific reproductive tactics. *American Naturalist*, **109**, 453–464.

RUBENSTEIN, D.I. 1982 Reproductive tactics and behavioral strategies: coming of age in monkeys and horses. In: Bateson, P.P.G. and Klopfer, P.H., eds. *Perspectives in Ethology*. Plenum Press, New York. Pages 469–487.

SCHWAGMEYER, P.L. 1980 Alarm calling behavior of the thirteen-lined ground squirrel, *Spermophilus tridecemlineatus*. *Behavioral Ecology and Sociobiology*, **7**, 195–200.

SHERMAN, P.W. 1977 Nepotism and the evolution of alarm calls. *Science*, **197**, 1246–1253.

SHERMAN, P.W. 1981 Reproductive competition and infanticide in Belding's ground squirrels and other animals. In: Alexander, R.D. and Tinkle, D.W., eds. *Natural Selection and Social Behavior*. Chiron Press, New York. Pages 311–331.

SHERMAN, P.W. 1985 Alarm calls of Belding's ground squirrels to aerial predators: nepotism or self-preservation? *Behavioral Ecology and Sociobiology*, **17**, 313–323.

SHERMAN, P.W. 1988 The levels of analysis. *Animal Behaviour*, **36**, 616–619.

SHERMAN, P.W. 1991 Multiple mating and kin recognition by self-inspection. *Ethology and Sociobiology*, **12**, 377–386.

SMITH, R.E. 1967 Natural history of the prairie dog in Kansas. *Miscellaneous Publications of the Museum of Natural History, The University of Kansas*, **49**(1), 1–39.

SMITH, S.F. 1978 Alarm calls, their origin and use in *Eutamias sonomae*. *Journal of Mammalogy*, **59**, 888–893.

SMITH, W.J., Smith, S.L., Oppenheimer, E.C. and deVilla, J.G. 1977 Vocalizations of the black-tailed prairie dog, *Cynomys ludovicianus*. *Animal Behavior*, **25**, 152–164.

STOCKRAHM, D.M.B. and Seabloom, R.W. 1988 Comparative reproductive performance of black-tailed prairie dog populations in North Dakota. *Journal of Mammalogy*, **69**, 160–164.

SUGIYAMA, Y. 1984 Proximate factors of infanticide among langurs at Dharwar: a reply to Boggess. In: Hausfater, G. and Hrdy, S.B., eds. *Infanticide: Comparative and Evolutionary Perspectives*. Aldine, New York. Pages 311–314.

TILESTON, J.V. and Lechleitner, R.R. 1966 Some comparisons of the black- tailed and white-tailed prairie dogs in north-central Colorado. *American Midland Naturalist*, **75**, 292–316.

TINBERGEN, N. 1963 On the aims and methods in ethology. *Zeitschrift fur Tierpsychologie*, **20**, 410–433.

VOM SAAL, F.S. 1984 Proximate and ultimate causes of infanticide and parental behavior in male house mice. In: Hausfater, G. and Hrdy, S.B., eds. *Infanticide: Comparative and Evolutionary Perspectives*. Aldine, New York. Pages 401–424.

WALDMAN, B., Frumhoff, P.C. and Sherman, P.W. 1988 Problems in kin recognition. *Trends in Ecology and Evolution*, **3**, 8–13.

WARING, G.H. 1970 Sound communications of black-tailed, white-tailed, and Gunnison's prairie dogs. *American Midland Naturalist*, **83**, 167–185.

WEST-EBERHARD, M.J. 1975 The evolution of social behavior by kin selection. *The Quarterly Review of Biology*, **58**, 155–183.

WILLIAMS, G.C. 1966 *Adaptation and Natural Selection*. Princeton University Press, Princeton, New Jersey.

WILSON, E.O. 1975 *Sociobiology*. Harvard University Press, Cambridge, Massachusetts.

WOLFF, J.O. and Cicerello, D.M. 1989 Field evidence for sexual selection and resource competition infanticide in white-footed mice. *Animal Behaviour*, **38**, 637–642.

YAHNER, R.H. 1978 Seasonal rates of vocalizations in eastern chipmunks. *Ohio Journal of Science*, **78**, 301–303.

PART 4: EXPERIMENTAL STUDIES OF INFANTICIDE AND PARENTAL CARE IN RODENTS

CHAPTER 15

INFANTICIDE AND PROTECTION OF YOUNG IN HOUSE MICE (*Mus domesticus*): FEMALE AND MALE STRATEGIES

STEFANO PARMIGIANI, PAOLA PALANZA, DANILO MAINARDI and
PAUL F. BRAIN

*Dipartimento di Biologia e Fisiologia Generali,
Università di Parma, 43100 Parma, Italy.*

*Biomedical and Physiological Research Group, Biological Sciences,
University College of Swansea, Swansea, U.K.*

> "There is a time to kill and a time to care"
> (Qoheleth's Book, 3,3)

INTRODUCTION

Initial ethological observations strongly suggested that animals rarely kill conspecifics when they fight in resource competition (Lorenz, 1966). It was consequently maintained that an "inhibition to kill" evolved for the "good of the species". Indeed, using this argument, and noting the power of so-called "infantile signals" to potently inhibit attack, the young (especially those of social species) were for long considered to be almost immune from deadly attack by adult conspecifics. On this basis, any behavior which resulted in death or injuries to offspring, either by parents or other conspecifics, was considered "pathological" or due to "abnormal" conditions (e.g. crowding or disturbance in rodents, Calhoun, 1962). Only recently has the phenomenon of infanticide (here defined as the killing of preweanling young by conspecifics) been viewed within the context of evolutionary

341

biology (Hrdy, 1974). In fact, relatively recent detailed ethological studies of several species of vertebrates have produced evidence that the killing of young (especially of subjects unrelated to the perpetrator) may be reproductively advantageous to the killer (Hrdy, 1979). One of the most common forms of infant killing by mammals involves intrasexual competition to gain access to mates. In this scenario, the capacity of parents to defend their offspring from infanticidal conspecifics is an important determinant of their reproductive success. Such behaviors have been extensively observed in house mice (*Mus domesticus and M. musculus*) providing a useful model for studying the roles of infanticide and protection of young in shaping reproductive biology and social dynamics. This present chapter describes laboratory experiments designed to elucidate the proximal and ultimate causes of infanticide and protection of young in male and female Swiss mice.

FEMALE STRATEGIES

As for most mammals, reproductive effort and parental investment of female house mice is generally more costly than for male counterparts. Females not only expend more energy producing gametes but also incur the direct costs of pregnancy and lactation (Trivers, 1972), as well as the costs associated with offspring defense. In this chapter a series of experiments on strategies adopted by female mice to maximize their reproductive fitness are described.

Lactating Female Aggression: Proximate and Ultimate Causes

During lactation female house mice become intensely aggressive towards unfamiliar conspecifics (especially males) that intrude into the nest area (see also chapter by Svare and Boechler); this behavior is variously referred to as maternal, lactational or postpartum aggression. Maternal aggression serves the function of protecting a female's litter from being killed by conspecifics (Svare, 1977; Ostermeyer, 1983). However, prior to the Hrdy's (1979) hypothesis of sexually selected infanticide, maternal aggression was viewed as a defense against predation by conspecifics who were presumed to be attempting to engage in cannibalism for the purpose of using the pups as a food resource.

In initial studies, a postpartum female's attack behavior was described as consisting of bites being directed to the most vulnerable regions (i.e. head, ventral and inguinal areas) of the intruder's body (Brain, 1981). As a

consequence, aggression in lactating mice was termed "defensive" in order to distinguish it from "offensive" attack behavior between males in which the topography of attack rarely involves biting vulnerable body regions; the back, flanks and tail are primary targets (Brain, 1981). More recently, findings from our laboratory (Parmigiani, 1986; Parmigiani *et al.*, 1988b) have suggested that maternal aggression is a heterogeneous phenomenon ranging in form from offensive to defensive attack, according to the context and sex of the conspecific intruders. Specifically, different patterns of attack are generated when lactating females confront sexually naive male and virgin female intruders. Sexually naive males are severely attacked by lactating females, with bites primarily directed toward the head and ventral body surface, whereas virgin females are comparatively rarely (three-five times less than in the case of males) bitten on these vulnerable regions of the body. Furthermore, female intruders elicit more social investigation by lactating residents than do males, whereas males evoke more responses consistent with "fearfulness" (i.e. loud squeaking even while attacking) in lactating females than do intruder females (Parmigiani *et al.*, 1988b; 1990). Importantly, the majority of male intruders counter-attack and dominate the lactating female, thereafter killing her pups, a behavior rarely shown by female intruders (Parmigiani, 1986; Parmigiani *et al.*, 1988a; 1989b). Studies with the opiate antagonist naloxone (Parmigiani *et al.*, 1988c) and the antiaggressive drug fluprazine (Parmigiani *et al.*, 1989c) also support the view that there are differences in the neurobiological substrates of the two maternal attack patterns. This may reflect a functional distinction between these behaviors.

In essence, maternal aggression appears to be a dichotomous phenomenon subserving different functions based on characteristics of the intruder. The more damaging and less ritualized (defensive) type of attack directed towards the male (i.e. the sex which is more likely to kill pups) may primarily be a counterstrategy to infanticide (Ostermeyer, 1983; Paul, 1986; Parmigiani, 1986) as suggested for other species of rodents (Wolff, 1985). Conversely, the "offensive" attack directed towards female intruders may be a form of competitive aggression (Brain, 1981; Archer, 1988; also see section on territorial behavior and infanticide by females).

Is Maternal Aggression a Counter-Strategy to Infanticide?

Several studies support the hypothesis that attack by lactating female mice toward males is a deterrent to infanticide. Infanticide in the Swiss albino line of mice is a sexually dimorphic trait (see Figure 1). For example, the majority (typically 70–80%) of 60-day-old sexually naive males

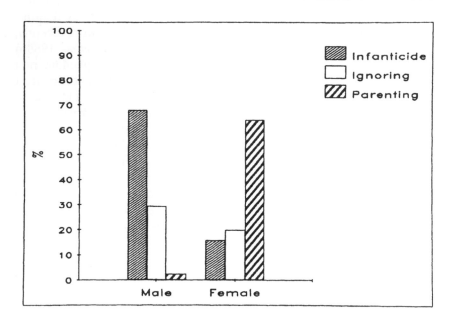

Figure 1 Percentage of males (N=200) and females (N=50) showing infanticide, ignoring and parenting behavior towards 2-4-day-old mouse pups.

exhibit infanticide when exposed to a pup in their home cage, whereas similarly aged virgin females in the same situation display parental behavior in 62 % of cases (Palanza and Parmigiani, 1990). Lactating females simultaneously confronted with a sexually naive male and a virgin female attacked only the male (Parmigiani *et al.*, 1989b).

The reproductive history of an intruder male influences the likelihood of his being attacked by a lactating female. When lactating females are confronted with either virgin, sexually experienced or parentally experienced unfamiliar conspecific males, maternal attack is mainly elicited by sexually naive males and males which had mated 24 hr before testing; these males were most likely to kill pups (Parmigiani *et al.*, 1988a). Conversely, very few of the males (15%) which were in a paternal state (*i.e.* mated subjects that were cohabiting with a nursing female and assisting in rearing their own pups) killed the unrelated pups, and these

males were hardly attacked by the resident lactating females. The data support the view that lactating female mice are capable of assessing the infanticidal potential of conspecific intruders and of adopting an appropriate behavioral strategy. More recently, Elwood *et al.* (1990b) have provided evidence that CS1 lactating female mice are more likely to attack infanticidal than non-infanticidal sexually naive males. What remains to be elucidated are the proximate mechanisms of female discrimination between males with different infanticidal tendencies. One possibility is that infanticidal, non-infanticidal (those ignoring the pup) and paternal males may differ in the chemical composition of their urine (Elwood *et al.*, 1990a).

The intensity of maternal aggression increases with the number of pups in the litter (Maestripieri and Alleva, 1990). According to parental investment theory (Trivers, 1972), this finding lends support to the hypothesis that maternal aggression is related to offspring defense. Importantly, the defense of the litter was successful when male intruders were incapable of defeating the female. These males were best characterized by their low levels of social aggression in previous inter-male aggression tests (Parmigiani *et al.*, 1989b). Altogether, the reported data support the view that the defensive attack generated when lactating female mice encounter sexually naive male intruders serves as a deterrent to infanticide and may be interpreted as a female counterstrategy to such behavior (Ostermeyer, 1983; Parmigiani, 1986; Paul, 1986).

While the above hypothesis seems logical, in a majority of cases maternal attack does not prevent the killing of a litter by sexually naive male intruders. In fact, Parmigiani *et al.* (1989b) have shown that the proportion of males exhibiting infanticide either in the presence or absence of the mother does not differ, and attack by the mother merely delayed the killing of the pups. Two factors may account for the apparent failure of litter defense in our experimental situation: 1) in natural conditions, nests are often located within defensible areas such as burrows with a narrow entrance, which facilitates the defense of the litter (Wolff, 1985), and 2) maternal attack might attract the attention of the resident stud male. The male that sired the litter is usually absent in laboratory tests, but in the natural environment he would likely intervene and assist the female in defending the litter. In the latter case, female attack may prevent infanticide until the temporarily absent mate arrives. Consequently, the effects of providing protected nests and of the presence of the stud male on the outcome of encounters between lactating females and sexually naive conspecifics were investigated.

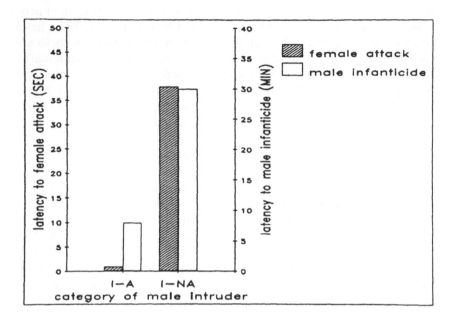

Figure 2 Female attack on and male infanticide by selected infanticidal males of high (A, N=10) and low (NA, N=10) intermale aggressiveness.

Effects of Protected Nests

We removed the stud male approximately one week before parturition. Pregnant females were provided with transparent plastic bottles in which to deliver their litter and cotton wool for nest material. The bottles had a narrow entrance measuring 3 cm. in diameter to facilitate nest defense. When the pups were three days old, the lactating females were tested for maternal aggression against sexually naive male intruders. Before being used as intruders, the males were individually housed and tested in their home cages both for infanticide and aggressive behavior toward anosmic males (these males elicit attack but do not fight back). As a result of these screening tests, two categories of intruders were obtained: 1) Infanticidal and aggressive males and 2) Infanticidal and non-aggressive males (non-infanticidal males were not included in this study). The results in Figure

2 show that the latency to attack by the lactating females was shorter toward the aggressive than the non-aggressive male intruders. However, the aggressive males entered the nest and killed the pups significantly faster than the non-aggressive males. Two out of 10 non-aggressive males did not exhibit infanticide while all aggressive males were infanticidal. This experiment suggests that females can discriminate between males of high or low aggressiveness (toward other males) and that the level of aggression of the male intruder (*i.e.* his "fighting ability") influences the latency of maternal attack, even though all males had exhibited infanticide on the screening test. Results showed that, in comparison to previous experimental conditions with unprotected nests, the protected location of the nest did not improve the defense of the litter by the lactating female.

There is an association between social status and tendency to exhibit infanticide in male mice: virtually all highly aggressive, dominant males kill pups, whereas, completely subordinated males which have been repeatedly defeated during encounters with other males do not exhibit infanticide (they ignore the pups; vom Saal and Howard, 1982; Huck, *et al.*, 1982). However, among males which show variable levels of aggression (and thus possibly different social status), such as those used in our study described above, there appears to be no clear relationship between the males' fighting ability and the likelihood of killing pups sired by another male.

Effects of the Presence of the Stud Male

Lactating females were tested towards sexually naive male and female intruders in the presence of the stud male, who also had the opportunity to assist in the defense of the litter. Figure 3 shows that in this situation, the resident stud male attacked virtually all of the male intruders but none of the female intruders. In contrast, male and female intruders were equally likely to be attacked by lactating females. This finding is surprising in that when a lactating female is confronted with an intruder male or an intruder female without the presence of the stud male, the intruder male is attacked at much higher intensity than is the intruder female (Parmigiani *et al.*, 1988b). In all tests in which the resident male defeated the intruder male (10/13), all pups were unharmed, while in the three cases where the intruder defeated the resident male, the entire litter was destroyed. It is important to note that all of these studies were conducted with a stock of mice in which the incidence of infanticide by the virgin females used as intruders is very low. However, the marked difference in the behavior of the resident stud male and lactating female in their attack behavior toward

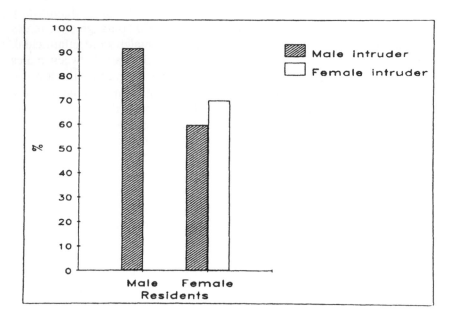

Figure 3 Percentage of stud male and lactating female residents (N = 13) attacking intruders of either sex.

intruders of different sex may reflect the different role played by male and female mice in social organization.

Cost-benefit considerations suggest that it is adaptive for a lactating female to allow the stud male to defend the territory (and her litter) against an interloper male whenever the stud male is present. It follows that if maternal aggression is indeed a counterstrategy to infanticide, the lactating female will increase her rate of attack toward an interloper (and potentially infanticidal) male when he is not attacked by her territorial mate. To simulate this situation the stud male's aggressive behavior was pharmacologically depressed by injection of 2 mg/kg of the antiaggressive drug, fluprazine, whose defining characteristic is to decrease male competitive aggression without impairing other components of the behavioral repertoire (Bradford *et al.*, 1984; Flannelly *et al.*, 1985). None of the lactating females (N = 10) housed with a fluprazine-treated (non-aggressive) male

attacked the male intruder. Conversely, 61% (8/13) of the lactating residents whose saline-treated male was aggressive attacked the intruders (the short duration of the tests precluded obtaining information about the frequency of infanticide). These findings suggest that the lactating female's aggressive response towards male intruders is modulated by the aggressive behavior of her stud, but the results were opposite to the predicted direction of change in aggressiveness by females. We described above the observation that the lactating females' response to an intruder male is modulated by the aggressive tendency of the intruder male, even though all of the males were infanticidal. Thus, maternal attack appears to be modulated by the aggressive characteristics of both the resident and intruder males rather than being a response to the potential risk to the litter by the intruder. These observations cast doubt on the view that maternal attack on interloper males is only a counterstrategy to infanticide.

One may speculate as to the ultimate causation of maternal aggression toward male intruders. Since females whose litters are destroyed quickly come into estrus and can be inseminated by the usurper male, female attack might also serve to assess the fighting ability, and indirectly the genetic quality, of the interloper male in terms of resources holding potential (RHP; Parker, 1974; Maynard-Smith, 1982). This male is likely to become her future mate (*i.e.* the father of the next litter). Indeed, when the stud male is absent (*i.e.* the territory is vacant), the cost of defending her current litter may be outweighed by the gain of a new sexual mate characterized by good RHP. In this scenario, maternal aggression may prevent reproduction with males of low aggressive potential.

The assumption that maternal aggression serves the function of testing the quality of males is supported by previously reported experimental data showing that lactating females do not engage in costly fights (*i.e.* in terms of energy expenditure) with an interloper male when her mate (*i.e.* the father of the litter) is not aggressive and submissive to the intruder (Parmigiani, 1989b). Moreover, it is known that social status of male mice influences whether they successfully mate with females, which is considered to reflect female choice. Courtship in this species resembles an agonistic interaction with the male subduing the female prior the copulation, suggesting that only vigorous males can successfully copulate (Parmigiani *et al.*, 1982; Hurst, 1986). Aggression by a lactating female toward an intruder male may be primarily a method of the female assessing the male's future potential to assist in nest defense, and thus may represent a category of intersexual selection by females for males with "good" genes (Krebs and Davies, 1981; Trivers, 1985). It is premature to suggest that maternal aggression is not effective in defending her litter against infanticidal conspecific

intruders in the natural environment. However, rather than being mutually exclusive, both functions (intersexual selection and counterstrategy to infanticide) could have operated synergistically in the evolution of maternal aggression.

Indeed, it is possible that there is an evolutionary continuum between the functions of maternal attack and pregnancy block (abortion) when a female mouse comes in contact with a strange male during the first week of pregnancy, which is referred to as the Bruce effect (Bruce, 1960). In fact, two hypotheses have been recently proposed to explain the evolution of the Bruce effect in terms of female advantage: 1) as a counterstrategy by which a pregnant female minimizes parental investment in embryos that are certain to be susceptible to infanticide by the interloper male as soon as they are delivered; Elwood and Kennedy (1990) have shown that infanticidal male mice are more effective than non-infanticidal counterparts in inducing pregnancy block, and 2) as a method of mate selection. Pregnancy block may be a strategy by which a female may increase the likelihood that her offspring will be sired by dominant (*i.e.* aggressive) males (Schwagmeyer, 1979). Indeed, Huck (1984) has shown that fewer pregnant female mice abort when exposed to subordinate rather than dominant strange males. Storey and Snow (1990) have recently observed that Meadow voles (*Microtus pennsylvanicus*) are more likely to reabsorb implanted embryos and become sexually receptive in response to aggressive (*i.e.* subjects that intensely attacked them) rather than non-aggressive male intruders. These data suggest that pregnancy block in females has evolved to be regulated by cues associated with male aggressiveness. Consequently, maternal aggression and the pregnancy block phenomenon can be both viewed as female strategies to maximize reproductive success.

Intra-Female Competition: Territorial Behavior and Infanticide

Female mice have usually been regarded as being "non aggressive" except when lactating(Mackintosh, 1981). An important feature of maternal aggression is, however, the fact that lactating females may use aggression both in competition with same sex conspecifics to establish a social hierarchy (Parmigiani, 1986) and as an intersexual sexual selection mechanism. However, there are now numerous observations showing that female aggression is not restricted to the postpartum period in mice. For example, after females have either cohabited with males or been exposed to male bedding, there is an increase in the proportion of females that attacked a female intruder (Parmigiani *et al.*, 1989). Interfemale aggression may thus

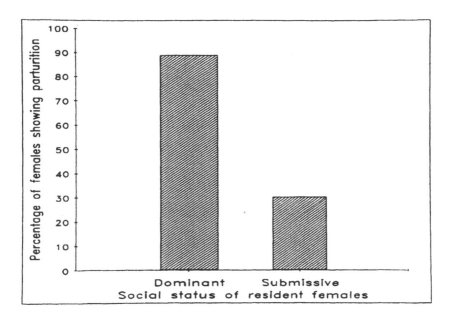

Figure 4 Comparative reproductive successes of dominant and submissive females housed together in artificial territories.

be relevant to population dynamics and be involved in the establishment of a social hierarchy which influences their reproductive success (Lloyd and Christian, 1969; vom Saal, 1983; 1984).   Indeed, some studies of freely growing populations have reported that females exhibit patterns of intrasexual aggressive behavior similar to those seen in males (Yasukawa *et al.*, 1985; Chovnik *et al.*, 1987). Moreover, in certain situations, female mice attack juveniles, which may play a role in dispersal (Gray, 1979), and in wild stocks, they kill alien pups (McCarthy and vom Saal, 1985).

    In order to understand female competition strategies, outside of and during breeding, and their role in shaping social structure, a longitudinal study in artificial territories (consisting of multiple cages connected by runaways) was carried out (Parmigiani *et al.*, 1989a).   Resident females attacked same sex conspecifics and developed stable dominant-subordination relationships. The dominant female had greater reproductive success than females which had been defeated (see Figure 4). When

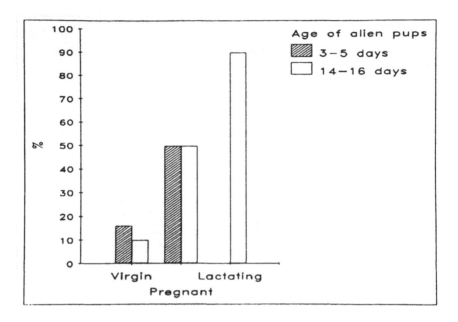

Figure 5 Percent of females exhibiting infanticide toward an alien pup of two different ages. Females were tested as virgins (N=10), during the last week of pregnancy (N=10), and between 3–5 days postpartum while lactating (N=10).

pregnant female intruders were introduced into the test enclosure, they were fiercely attacked by the pregnant or lactating residents, and none of the intruders pups survived. These data confirm previous suggestions that pregnancy increases the likelihood of infanticide in female mice (McCarthy and vom Saal, 1985).

It has been observed that in Swiss albino females the frequency with which infanticide is observed varies with their reproductive state and the age of the alien pup (see Figure 5). In fact, while virgin females are generally parental both towards newborn pups and preweaning young, pregnant females kill pups and attack preweaning young in 50% of cases. Interestingly, lactating females behaved parentally towards stranger pups which were the same age of their own offspring but strongly attacked young which were older (Parmigiani *et al.*, 1989a).

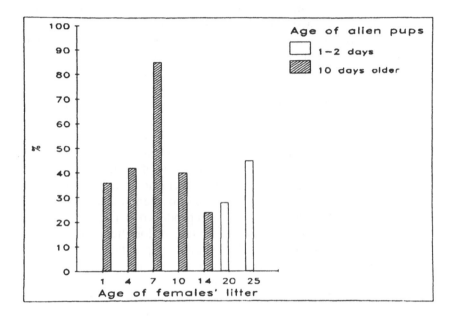

Figure 6 Percent of lactating females whose litters are of differing ages that attacked an alien pup which was either 1–2 days old or 10-days older than the females' own pups. No female exhibited infanticide toward the 1–2 day-old pup when tested on days 1–14. Females tested on days 20 and 25 were not tested for their behavior toward pups which were 10-days older than the females' own litter.

The proximal causes of infanticide in lactating females were investigated taking into account the age differences between their litters and those of unfamiliar pups placed into the lactating females home cage. The results (see Figure 6) suggest that, irrespective of the age of their litters, the lactating females never attacked alien pups when they were the same age as their own offspring. Newborn (1–2 days old) alien pups were generally adopted (*i.e.* retrieved to the nest) by lactating females who had litters ranging in age between 4 and 18 days. However, when the female's litter was at the age of weaning (24 days old), newborn alien pups were ignored (in 65% of cases) or killed (35% of cases), but not parented.

Lactating females generally attacked young which were 10 days older than their own pups, and the highest frequency of attack (90% of cases)

occurred in females with litters of 7-day-old pups. These findings suggest an age-dependent discrimination of unfamiliar pups by lactating females and that this phenomenon is modulated by factors related to the stage of lactation and to the age difference between the alien and related pups. Svare and Gandelman (1973) have suggested that the presence of body hair on the alien young may be one stimulus involved in evoking attack in lactating mice whose young are still hairless. However, other proximal mechanisms such as the body size, olfactory cues and the age-dependent production of ultrasonic vocalizations by the young may contribute to the differential responsiveness of lactating females to pups of different ages. The parental behavior of resident lactating mice toward alien young of the same age as or younger that their litters could be functionally reducing the risk of killing related pups. In a deme, females are often genetically related (Berry, 1981), and, as shown above and in other studies in mice (Yasukawa *et al.*, 1985), the pups of pregnant females who attempt to immigrate into a deme are typically killed. Thus, any pup produced in a deme by a familiar female is likely to be a relative, and kin-selection theory predicts that the lactating female should not kill the younger pup. Alien pups which are older than a females own pups must be immigrants and thus genetically unrelated competitors of the resident lactating females own young, and they are killed; this falls under Hrdy's (1979) definition of resource competition infanticide. When resident females are lactating, the territory-holding male is generally inhibited in his attack and killing of young (see section on male strategies). Consequently, the female intolerance towards alien young during different reproductive phases may prevent immigration of juveniles which would compete for resources with their offspring.

In essence, competition between females (expressed by intrasexual aggression, inhibition of subordinate female reproduction and killing of unrelated young) is an important factor in the regulation both of reproductive potential of a deme-like unit and the number of growing young within the defended area. Thus, female house mice actively contribute to the regulation of population size, a role that has been previously exclusively ascribed to male territorial aggression by most authors.

MALE STRATEGIES

As observed in other species of vertebrates, one of the most common circumstances in which male mice kill unrelated young is when taking over a territory (vom Saal, 1984; Paul, 1986; Parmigiani, 1989a). When introduced into a lactating female's nesting area, the majority of sexually

naive males initiate infanticide after having subdued the female. In the light of Darwin's (1871) sexual selection theory, Hrdy (1979) proposed that this kind of infanticide, rather than being pathological, is a form of post-mating reproductive competition among males. In fact, as seen above for females, the elimination of a competitor's offspring (genes) by the infanticidal male increases his reproductive fitness; the death of the litter and cessation of lactation accelerates the return to estrus by the female and dramatically advances the opportunity for the infanticidal male to sire offspring (vom Saal and Howard, 1982; McCarthy and vom Saal, 1986). This observation, together with the evidence that infanticidal behavior is at least in part genetically mediated (Svare *et al.*, 1984) and thus may be subject to selection pressure, supports the hypothesis that infanticide evolved through sexual selection.

Another important assumption concerning the adaptiveness of "sexually selected" infanticide (Hrdy, 1979; vom Saal and Howard, 1982) is the presumed existence of mechanisms to inhibit infanticidal males from killing their own offspring. Indeed, male mice do not harm their own pups even if they are infanticidal prior to mating. Moreover, colony/intruder tests in artificial territories (see sections of female territorial aggression) indicate that a resident stud male does not kill alien young when the resident females with which he had mated are pregnant or lactating (Parmigiani *et al.*, 1989a). A series of laboratory experiments were consequently designed to determine the factors which lead infanticidal male mice to become parental.

Inhibition of Infanticide: Proximate and Ultimate Causation

Several mechanisms have been proposed to mediate the transition from infanticide to paternal behavior in stud males. Some investigators report reductions in pup killing by males immediately (Huck *et al.*, 1982) or approximately 12–20 days (vom Saal and Howard, 1982) after sexual experience (*i.e.* copulation with ejaculation). Conversely, other authors maintain that post-copulatory cohabitation with the pregnant mate for 18–20 days (*i.e.* the duration of pregnancy) even without contact with pups at birth, inhibits infanticide by males (Elwood and Ostermeyer, 1984; Elwood, 1985). Other studies suggest that the tendency of sexually naive male to kill pups can be inhibited by cohabitation with females that are pregnant by other males (Soroker and Terkel, 1988). Notwithstanding the possibility of strain differences in the mechanisms mediating the inhibition of infanticide in males prior to the birth of their offspring (Labov *et al.*, 1985), the reported

studies confirm that male socio-sexual interactions with females play a crucial role in inhibiting infanticide.

We suggest that in order to understand the evolution of the proximate mechanisms mediating the inhibition of infanticide in male mice, it is necessary to examine the house mouse's social structure; house mice are typically found living commensally with man. This species frequently forms demes (*i.e.* defended areas containing a dominant adult male, a variable number of breeding females and their offspring; Bronson, 1979). Male mice on reaching puberty are normally driven out of the territory by the dominant resident male (*i.e.* the father), and this leads to dispersal and colonization of new territories. In such circumstances, dispersing sexually naive males may exhibit infanticide when taking over the reproductive area of another male. Thus, a potential sequence of events in a reproductively successful male's life should include: acquisition of territory and mates, copulation and cohabitation with females, and the eventual siring of litters. Copulation and subsequent cohabitation with the pregnant mate suppresses infanticide in all previously infanticidal males, which obviously has to occur or else there would not be the possibility that any litter of mice would ever survive (vom Saal and Howard, 1982; Elwood and Ostermeyer, 1984; Elwood, 1985; Parmigiani, 1989a; see chapter by Perrigo and vom Saal). It is relevant that sexually naive males kill pups found either inside or outside of their home cage with similar frequencies, indicating that holding a territory (*i.e.* a urine marked area) without any contact with females does not suppress male infanticide (vom Saal, 1985; Parmigiani, 1989b). Findings concerning which of the factors (*i.e.* holding a territory, copulation, cohabitation with females, or cues emitted by pregnant females) regulate the suppression of infanticide in male mice are also discussed in the chapters by Perrigo and vom Saal as well as Elwood and Kennedy.

Consequently, a follow-up investigation evaluated the effects of variable methods of exposure to females in different reproductive conditions on the male's behavior towards unrelated infants (Palanza and Parmigiani, 1991). Sexually naive Swiss male mice, pre-selected for infanticide, were each housed with a female. Males and females were either in physical contact (*i.e.* male and female cohabited) or sensory contact (*i.e.* male and female were separated by a wire-mesh partition). In the latter case, the interactions were of auditory, olfactory, tactile (through the wire mesh) and visual nature. It was found that physical contact (for 6 days) and mating with virgin females (whose later pregnancy confirmed successful copulation) inhibited pup killing only in a minority of males (20%). A similar physical contact by a virgin male with a pregnant female (*i.e.* impregnated by another male) for any period of time prior to, but not

including, parturition did not reduce male infanticide. In contrast, physical contact with a female (impregnated by another male) during the end of pregnancy and throughout the time of parturition inhibited infanticide in 50% of virgin males. Mating and postcopulatory physical contact (for 18 days) with a female throughout pregnancy but without being present at the delivery of the litter was effective at suppressing infanticide in all males (for similar findings see chapter by Perrigo and vom Saal).

Sensory contact (for 6 days) with virgin, pregnant or parturient females (all impregnated by other males) failed to reduce the incidence of infanticide in virgin males. Post-copulatory sensory contact with the mate for the entire period of pregnancy was sufficient to inhibit pup killing in virtually all males, whereas copulation followed by sensory contact with a non-pregnant mate (reabsorption was induced) did not suppress infanticide in the majority of males. Sensory contact without copulation with a pregnant female did not reduce the incidence of male infanticide.

Taken together, the results of these studies suggest that copulation functions as a "primer", but cues (possibly olfactory) emitted by the pregnant mate prior to and during parturition are also necessary to induce the shift from infanticidal to paternal behavior in the majority of Swiss males. Although copulation followed by physical contact with the female mate induces a non-infanticidal state in all males, a minority of individuals are inhibited from exhibiting infanticide solely by copulation. Physical contact of a virgin male with a female (inseminated by another male) at the time of pup delivery also inhibited infanticide in about one-half of Swiss males. In summary, multiple mechanisms underlying the suppression of infanticide can co-exist in house mouse populations (Palanza and Parmigiani, 1991). The mechanisms involved in inhibiting male infanticide also show heterogeneity in wild mouse populations (Soroker and Terkel, 1988). This behavioral "spectrum" within and between populations of mice may be due to genetic variability coupled with genetic differences in responsiveness to hormonal effects during fetal and adult life (vom Saal, 1984; Svare et al., 1984; Brain and Parmigiani, 1990). These observations may account for the conflicting data obtained from different strains in terms of the inhibition of male infanticide. The variation in proximal mechanisms responsible of the transition from killing to caring for pups among domesticated and wild lines of mice might be due to different reproductive strategies due to different environmental selection pressures.

The events associated with reproduction induce neurophysiological changes in males which account for the induction of parental behavior towards any pups after a male has mated and/or cohabited with a pregnant

female (Perrigo and vom Saal; Elwood and Kennedy, this volume). After mating and cohabitation with the pregnant mate, males do not kill related (their own) or unrelated young, either in the presence or absence of the familiar lactating female (Parmigiani, 1989b). This suggests that there is an indirect (*i.e.* not genetic or maternally mediated) kin recognition of young by males. Selection appears to have operated to suppress infant killing at times when there is a strong probability that a male will encounter his own offspring. In fact, once a male has established a territory and achieved dominance, he is likely to sire the majority of the litters within the territory (Bronson, 1979; vom Saal, 1984). Consequently, the preferential survival of related infants appears to be a by-product of the typical "deme" social organization. This "mechanism", which results in kin selection, does not necessitate the animals having to show kin recognition.

The experimental evidence broadly supports Hrdy's hypothesis of the adaptiveness of sexually selected infanticide. Infanticidal behavior can be considered as a unique form of intraspecific aggression, since, by definition, it always involves the death of a conspecific. Since in the majority of domestic stocks of mice males generally eat any killed pups, a debate has developed whether the killing of young conspecifics by adult males is actually a form of aggression or an expression of intraspecific predation (cannibalism) (Svare and Bartke, 1978). Recent work in our laboratory has provided the first indirect indication that male infanticidal behavior, rather than being predatory, has very similar substrates to those involved in the control of social aggression. Indeed, a series of experiments on the behavior of male mice have shown that administration of fluprazine, which inhibits aggression between adult males, decreases infanticide and intermale aggression to the same degree, but does not significantly alter predation of an insect larva, which is a palatable food source for mice (Parmigiani and Palanza, 1991).

CONCLUSION

Both situational determinants and reproductive life stages strongly affect the likelihood that a male or a female mouse will kill or care for an unrelated conspecific pup. The Biblical quote that "There is a time to kill and a time to care" briefly summarizes the male and female strategies concerning the protection and killing of young in mice. The sex-differences in the display and the timing of these behaviors can generate conflicts (e.g. female's parental defense against an intruder male or blocking of pregnancy by a strange male) but may also be complementary and thus

act together in order to maximize male and female reproductive success within a deme. Indeed, males are most likely to show infanticide when they are sexually naive and during the taking over of a territory, whereas females exhibit the highest propensity of killing unrelated young when they are near parturition and/or lactating and the stud territorial male is thus not infanticidal. In male mice, infanticide occurs due to intraspecific competition for mates. In contrast, in female mice, infanticide appears to be a resource-competition strategy augmenting the probability of access to food and survival of their offspring, which will ultimately increase their inclusive fitness. The killing of older, unrelated young (juvenilicide) may serve the function of territorial defense against juvenile immigrants.

The observation that females become aggressive as soon as they become associated with a male holding a territory suggests that competition for mates and physical resources and infanticide toward alien young are intimately linked. The relationship between infanticide and other types of intraspecific aggression is supported by the observation that the levels of inter-male attack, maternal aggression, and infanticide by males within and between different stocks of mice usually co-vary (Parmigiani, 1989a). For example, a colony-intruder study in the relatively non-aggressive DBA strain of mice showed that territorial aggression toward same sex conspecifics and infanticide toward unrelated pups was less intense in males and virtually absent in resident females (irrespective of their reproductive condition) in comparison to Swiss mice, which is a highly aggressive stock (Parmigiani et al., 1989a). It is likely that the responses of adults other than the mother towards conspecific infants may reflect the kind of demic social organization formed by commensal populations of mice. In this respect, the house mouse is a widely distributed and very divergent group of animals (seven recognized species and several chromosomal races) capable of exploiting a varied set of environments as diverse as fields, cold stores, warehouses, hayricks, pacific atolls and arctic islands (Berry, 1981). Different populations of mice at different locations and at varied times within the same location can exhibit a variety of social organizations, ranging from exclusive male territories to complex hierarchical groups (Brain and Parmigiani, 1990). The actual social organization employed seems to largely reflect differences in the degrees of female and male aggressiveness, which determine the spatial distribution of the species (Lorenz, 1966; Brown, 1975; Brain and Parmigiani, 1990).

The variation in female and male behavioral strategies, as far as intra- and inter-sexual aggression, infanticide, and protection of young might be related to the levels of competition within the population due to different ecological pressures. Indeed, individual variability in behavioral responses

towards young observed in the outbred Swiss albino line, and also reported in wild mice, indicates that different strategies coexist within a genetically variable population of mice. This phenotypic variability (due to genotype, intrauterine position and experience) might "pre-adapt" individuals to cope with the changing environmental conditions; thus accounting for the reported flexibility of this highly opportunistic species. Consequently, it seems heuristic to evaluate social relationships between individuals in wild mice populations living in different environments in order to attempt to understand the evolutionary significance of the proximate mechanisms modulating the behavior of adults towards young and their impact on social structures.

ACKNOWLEDGEMENTS

We would like to thank the CNR (Consiglio Nazionale delle Ricerche) and MURST (Ministero dell'Università e delle Ricerca Scientifica e Tecnologica) for having supported the research described in this chapter. We thank F. vom Saal for suggestions during the preparation of the chapter.

REFERENCES

ARCHER, J. (1988) *The Biology of Aggression.* Cambridge, Cambridge University Press.

BERRY, R.J. (1981) The biology of the house mouse. *Symposium of the Zoological Society of London*, **47**. New York, Academic Press.

BRADFORD, L.D., Olivier, B., van Dalen, D. and Schipper, J. (1984) Serenics: the pharmacology of fluprazine and DU 28412. In: K.A. Miczek., M.R. Kruk and B. Olivier (eds), *Ethopharmacological Aggression Research*, New York, A.R. Liss, pp.191–207.

BRAIN, P.F. (1981) Differentiating types of attack and defense in rodents. In: P.F. Brain and D. Benton (eds), *Multidisciplinary Approaches to Aggression*, Elsevier North Holland, Biomedical Press, pp.53–78.

BRAIN, P.F. and Parmigiani, S. (1990) Variation in aggressiveness in house mouse populations. *Biological Journal of the Linnean Society*, **41**, 257–269.

BRONSON, F.H. (1979) The reproductive ecology of the house mouse. *Quarterly Review of Biology*, **54**, 265–299.

BROWN, J.L. (1975) *The Evolution of Behavior.* New York, Norton.

BRUCE, H.M. (1960) A block to pregnancy in the mouse caused by proximity of strange males. *Journal of Reproductive Fertility*, **1**, 96–103.

CALHOUN, J.B. (1962) Population density and social pathology. *Scientific American*, **206**, 139–148.

CHOVNICK, A., Yasukawa, N.J., Monder H. and Christian, J.J. (1987) Female behavior in populations of mice in the presence or absence of a male hierarchy. *Aggressive Behavior*, **13**, 367–375.

DARWIN, C. (1871) *The Descent of Man and Selection in Relation to Sex*. New York, Modern Library Random House.

ELWOOD, R.W. (1985) The inhibition of infanticide and the onset of paternal care in male mice, *Mus musculus. Journal of Comparative Psychology*, **99**, 457–467.

ELWOOD, R.W. and Kennedy, H.F. (1990) The relationship between infanticide and pregnancy block in mice. *Behavioral and Neural Biology*, **53**, 277–283.

ELWOOD, R.J. and Ostermeyer, M.C. (1984) Does copulation inhibit infanticide in rodents? *Animal Behaviour*, **32**, 293–305.

ELWOOD, R.W., Kennedy, H.F. and Blakely, H.M. (1990a) Responses of infant mice to odors of urine from infanticidal, non infanticidal and paternal mice. *Developmental Psychobiology*, **23**, 309–317.

ELWOOD, R.W., Nesbitt, A.A. and Kennedy, H.F. (1990b) Maternal aggression and the risk of infanticide. *Animal Behaviour*, **40**, 1080–1086.

FLANNELLY, K.J., Murakoa, M.Y., Blanchard, C. and Blanchard, R.J. (1985) Specific anti-aggressive effects of fluprazine hydrochloride. *Psychopharmacology*, **87**, 86–89.

GRAY, L.E. (1979) The effects of the reproductive status and prior housing condition on the aggressiveness of female mice. *Behavioral Neural Biology*, **26**, 508–513.

HRDY, S.B. (1974) Male-male competition and infanticide among the langurs (*Presbytis entellus*) of Abu, Rajastan. *Folia Primatologica*, **22**, 19–58.

HRDY, S.B. (1979) Infanticide among animals: a review, classification, and examinations of the implications for the reproductive strategies of females. *Ethology and Sociobiology*, **1**, 13–40.

HUCK, U.W., Soltis, R.L. and Coopersmith, C.B. (1982) Infanticide in male laboratory mice: Effects of social status, prior sexual experience, and bases for discriminating between related and unrelated young. *Animal Behaviour*, **30**, 1158–1165.

HUCK, U.W. (1984) Infanticide and the evolution of pregnancy block in rodents. In: G. Hausfater and S.B. Hrdy (eds), *Infanticide: Comparative and Evolutionary Perspective*, New York, Aldine Publishing Company, pp. 349–366.

HURST, J.L. (1986) Mating in free living wild house mice (*Mus domesticus Rutty*). *Journal of Zoology*, **210**, 623–628.

KREBS, C.B. and Davies, N.B. (1981) *An Introduction to Behavioral Ecology*. Oxford, Blackwell Scientific Publication.

LABOV, J.B., Huck, U.W., Elwood, R.W. and Brooks, R.J. (1985) Current problems in the study of infanticidal behavior of rodents. *Quarterly Review of Biology*, **61**, 1–20.

LLOYD, J.A. and Christian, J.J. (1969). Reproductive activity of individual females in 3 experimental freely growing populations of house mice (*Mus musculus*). *Journal of Mammology*, **30**, 49–59.

LORENZ, K. (1966) *On Aggression*. New York, Harcourt.

MACKINTOSH, J.H. (1981) Behaviour of the house mouse. *Symposium of The Zoological Society of London*, **47**, 337–335.

MAESTRIPIERI, D. and Alleva, E. (1990) Maternal aggression and litter size in the female house mouse. *Ethology*, **84**, 27–34.

MAYNARD-SMITH, J. (1982) *Evolution and the Theory of Games*. Cambridge, Cambridge University Press.

McCARTHY, M.M. and vom Saal, F.S. (1985) The influence of reproductive state on infanticide by wild female house mice *(Mus musculus)*. *Physiology and Behavior*, **35**, 843–849.

McCARTHY, M.M. and vom Saal, F.S. (1986) Inhibition of infanticide after mating in wild male house mice. *Physiology and Behaviour*, **36**, 203–209.

OSTERMEYER, M.C. (1983) Maternal aggression. In: R.W. Elwood (ed), *Parental Behavior in Rodents*. New York, Wiley & Sons Ltd, pp. 151–179.

PALANZA, P. and Parmigiani S. (1990) Male and female infanticide in Swiss albino mice (*Mus domesticus*). *Ethology Ecology Evolution*, **2**, 319–320.

PALANZA, P. and Parmigiani, S. (1991) Inhibition of infanticide in male Swiss mice: behavioral polymorphism in response to multiple mediating factors. *Physiology and Behavior*, **49**, 797–802.

PARKER, G.A. (1974) Assessment strategy and the evolution of fighting behavior *Journal of Theoretical Biology*, **47**, 223–243.

PARMIGIANI, S. (1986) Rank order in pairs of communally nursing female mice and maternal aggression towards conspecific intruders of differing sex. *Aggressive Behavior*, **12**, 377–386.

PARMIGIANI S. (1989a) Maternal aggression and infanticide in the house mouse: consequences on the social dynamics. In: P.F. Brain, D. Mainardi and S. Parmigiani (eds), *House Mouse Aggression*, London, Harwood Academic Press, pp.161–178.

PARMIGIANI, S. (1989b) Inhibition of infanticide in male house mouse (*Mus domesticus*): is kin recognition involved? *Ethology Ecology Evolution*, **1**, 93–98.

PARMIGIANI, P. and Palanza, P. (1991) Fluprazine inhibits intermale attack and infanticide, but not predation, in male mice. *Neuroscience and Biobehavioral Reviews*, **15**, 511–513.

PARMIGIANI S., Brain, P.F. and Palanza P. (1989a) Ethoexperimental analysis of different forms of intraspecific aggression in the house mouse (*Mus musculus*). In: R.J. Blanchard, P.F. Brain, D.C. Blanchard and S. Parmigiani (eds), *Ethoexperimental Approaches to the Study of Behavior,* Dordrecht, Kluwer Academic Publisher, pp.418–431.

PARMIGIANI, S., Brunoni, W. and Pasquali, A. (1982) Behavioral influences of dominant, isolated and subordinated male mice on female socio-sexual preferences. *Bollettino di Zoologia*, **49**, 73–78.

PARMIGIANI, S., Palanza, P. and Brain, P.F. (1989b) Intraspecific maternal aggression in the house mouse (*Mus domesticus*): a counterstrategy to infanticide by male? *Ethology Ecology Evolution*, **1**, 341–352.

PARMIGIANI, S., Sgoifo, A. and Mainardi D. (1988a) Parental aggression displayed by female mice in relation to the sex, reproductive status and infanticidal potential of conspecific intruders. *Monitore Zoologico Italiano*, **22**, 193–201.

PARMIGIANI, S., Brain, P.F., Mainardi, D. and Brunoni W. (1988b) Different patterns of biting attack generated when lactating female mice (*Mus domesticus*) encounter male and female conspecific intruders. *Journal of Comparative Psychology*, **102**, 287–293.

PARMIGIANI, S., Palanza, P., Mainardi, M. and Mainardi, D. (1990) Fear and defensive components of maternal aggression in mice. In: P.F. Brain, S. Parmigiani, R. Blanchard and D. Mainardi (eds), *Fear and Defence*. London, Harwood Academic Publishers, pp. 109–126.

PARMIGIANI, S., Rodgers, R.J., Palanza, P. and Mainardi, M. (1988c) Naloxone differentially alters parental aggression by female mice towards conspecific intruders of differing sex. *Aggressive Behavior*, **14**, 213–224.

PARMIGIANI, S., Rodgers, R.J., Palanza, P., Mainardi, M. and Brain, P.F. (1989c) The inhibitory effects of Fluprazine on parental aggression in female mice are dependent upon intruder sex. *Physiology and Behavior*, **46**, 455–459.

PAUL, L. (1986) Infanticide and maternal aggression: synchrony of male and female reproductive strategy in mice. *Aggressive Behavior*, **12**, 1–11.

SOROKER, V. and Terkel, J. (1988) Change in incidence of infanticide and parental responses during the reproductive cycle in male and female wild mice *Mus musculus*. *Animal Behaviour*, **36**, 1275–1281.

SCHWAGMEYER, P.L. (1979) The Bruce effect: An evaluation of male/female advantages.
American Naturalist, **114**, 932–938.

STOREY, A.E. and Snow, D.T. (1990) Postimplantation pregnancy disruptions in meadow
voles: relationship to variation in male sexual and aggressive behavior. *Physiology and
Behavior*, **47**, 19–26.

SVARE, B. (1977) Maternal aggression in mice: influence of the young. *Biobehavioral
Review*, **1**, 151–164.

SVARE, B. and Gandelman, R. (1973) Post-partum aggression in mice: experiential and
environmental factors. *Hormones and Behavior*, **4**, 323–334.

SVARE, B. and Bartke, A. (1978) Food deprivation induces conspecific pup-killing in mice.
Aggressive Behavior, **4**, 253–261.

SVARE, B., Broida, J. and Mann, M. (1984) Psychobiological determinants underlying
infanticide in mice. In: G. Hausfater and S.B. Hrdy (eds), *Infanticide: Comparative
and Evolutionary Perspectives*, New York, Aldine Publishing Company, pp. 387–400.

TRIVERS, R.L. (1972) Parental investment and sexual selection. In: B. Campbell (ed),
Sexual Selection and Descent of Man. Chicago, Aldine, pp.136–179.

TRIVERS, R.L. (1985) *Social Evolution*. Menlo Park, The Benjamin/Cumming.

VOM SAAL, F.S. (1983) Models of early hormonal effects on intrasex aggression in mice. In:
B. Svare (ed), *Hormones and Aggressive Behavior*, New York, Plenum Press, pp. 197–
222.

VOM SAAL, F.S. (1984) Proximate and ultimate causes of infanticide and paternal behavior
in male house mice. In: G. Hausfater and S.B. Hrdy (eds), *Infanticide: Comparative and
Evolutionary Perspectives,* New York, Aldine Publishing Company, pp. 401–425.

VOM SAAL, F.S. (1985) Time-contingent change in infanticide and parental behavior induced by ejaculation in male mice. *Physiology and Behavior*, **34**, 7–15.

VOM SAAL, F.S. and Howard, L.S. (1982) The regulation of infanticide and parental
behavior: implication for reproductive success in male mice. *Science*, **215**, 1270–1272.

WOLFF, J.O. (1985) Maternal aggression as a deterrent to infanticide in *Peromyscus leucopus*
and *P. maniculatus*. *Animal Behaviour*, **33**, 117–123.

YASUKAWA, N.J., Harvey, M., Leff, F.L. and Christian, J.J. (1985) Role of female behavior
in controlling population growth in mice. *Aggressive Behavior*, **11**, 49–64.

CHAPTER 16

BEHAVIORAL CYCLES AND THE NEURAL TIMING OF INFANTICIDE AND PARENTAL BEHAVIOR IN MALE HOUSE MICE

GLENN PERRIGO[1,3] and FREDERICK S. VOM SAAL[1,2,3]

Division of Biological Sciences, Department of Psychology and
The John M. Dalton Research Center, University of Missouri-Columbia,
Columbia, Missouri 65211, USA

Infanticide is a unique form of intraspecific aggression. Most forms of intraspecific aggression rarely lead to the death of interacting animals, whereas infanticide — by definition — is the killing of conspecific young. Much of the early research on this controversial subject assumed that such behavior was maladaptive and symptomatic of grossly abnormal social conditions in nature or in the laboratory. In his classic studies of the Norway Rat, Calhoun (1962) observed that sociopathologic conditions, such as populations stressed by excessive crowding, can indeed lead to a severe social breakdown and thus a high incidence of infanticide.

In recent years, however, a prominent new view of infanticide has emerged: it is a violent but adaptive reproductive strategy found in a variety of mammals and other vertebrates (*e.g.* Hrdy, 1979; Hausfater and Hrdy, 1984). Field and laboratory studies have dramatically documented the reproductive advantages that accrue when an infant-killing male usurps the territory of another male (Hrdy, 1979; vom Saal and Howard, 1982; Packer and Pusey, 1984, this book). By killing the offspring of a defeated competitor, an infanticidal male benefits in two ways. First, he eliminates potential reproductive and resource competition with his own offspring, and second, once a female's own young are killed, she rapidly ovulates again and mates with the usurper male.

The male house mouse, (*Mus domesticus* and *M. musculus*), has become the focus of much research concerning the socioecology, evolution and physiology of infanticidal and parental behavior in males. Male house mice routinely attack and kill alien young whenever they encounter them, but an effective infanticidal strategy must allow a male to recognize when his own offspring might be present. Thus, a fundamental issue in the study of infanticide concerns the factors that prevent male mice from harming their own progeny. It is now widely accepted that multiple behavioral mechanisms — and combinations thereof — are responsible for inhibiting pup-killing behavior in male mice (*e.g.* Soroker and Terkel, 1988; Palanza and Parmigiani, 1991; Elwood, this book). The major theme of this chapter, however, involves our investigation of one of these inhibitory mechanisms — specifically, the dramatic changes in behavior toward offspring that are triggered by the act of ejaculation.

THE EJACULATORY PHENOMENON

In male house mice, the act of ejaculation during mating provides a fail-safe neural signal for timing the onset of paternity (vom Saal, 1985). The specific stimulus of ejaculation inhibits infanticide. However, a remarkable aspect of this phenomenon is that a male's pup-killing behavior often does not cease for many days after mating, but nearly always ceases by the time his own sired offspring would be born three weeks later. When infanticide ceases, males react non-aggressively toward pups and often express parental behavior similar to that of a newly lactating female. Furthermore, infanticidal behavior spontaneously re-emerges at a time that coincides with the weaning and dispersal of offspring (vom Saal, 1985). These timed behavioral changes — which result specifically from ejaculation — are clear-cut and remarkably consistent among various house mouse stocks (McCarthy and vom Saal, 1986; Kennedy and Elwood, 1988; Palanza and Parmigiani, 1991; Soroker and Terkel, 1988; Perrigo *et al.*, 1989a, 1990). This phenomenon has also been verified in the Norway rat, *Rattus norvegicus*, (Mennella and Moltz, 1988) so it may occur in other rodents as well. In terms of reproductive advantages, male mice and rats are thus likely to eliminate pups sired by a competitor, whereas copulation ensures that they do not harm their own pups during the period of their mate's lactation.

Our primary interest here is the unique neural timing mechanism that gauges the passage of time between ejaculation and the birth of pups in male house mice. The remainder of this chapter presents an overview of what we have learned from CF-1 stock males (*Mus domesticus*) about the

nature of the ejaculatory trigger and the timing mechanism that regulates time-delayed changes in their behavior toward pups. First, we have examined the role that pituitary and gonadal hormones play in the modulation of these behavioral changes. Second, we have employed experimental strategies where behavioral changes toward pups were carefully monitored when males were maintained and mated at various light/dark regimens, including free-running rhythm conditions of either constant light or constant dark. Third, we have tested the effects of remating and the effects of aging on the timing of pup-killing and parenting strategies. And finally, we have examined individual variation in the neural timing of these behaviors and correlated these results with hormonal events that occur during late fetal development. All in all, our research has suggested a variety of new phenomena to be studied, including the possibility of a circadian-based timing system that facilitates the long-term synchronization of behavioral changes.

The Time-Course of Behavioral Changes Triggered by Ejaculation

Figure 1 is a schematic representation of this phenomenon as observed in CF-1 stock mice from our laboratories. The CF-1 male's behavioral cycle toward pups has four distinct phases:

(1) *Pre-mating.* In virgin males, half of all individuals spontaneously kill pups while the other half do not harm them. These latter males either "Parent" pups (about 40%) or they "Ignore" them (about 10%). By definition, a parental male retrieves a pup to his nest where he incubates it and keeps it warm (vom Saal, 1985; Elwood, 1986). Males who "Ignore" pups neither harm nor parent them.

(2) *Ejaculation and Pregnancy of Mate.* Ejaculation intensifies pup-killing behavior. Virtually all males will attack and kill pups immediately after mating; they will continue to kill pups during part or most of their mate's pregnancy.

(3) *Lactation of Mate.* By the time pups are born, 19-20 days after mating, infanticide ceases and most males behave parentally toward pups. They remain parental throughout their mate's lactation.

(4) *Post-weaning of Offspring.* Between 50 and 60 days after mating, many males spontaneously begin killing pups again. The re-emergence of infanticidal behavior thus coincides with the weaning of pups.

Remarkably, this entire behavioral cycle toward pups occurs even when a CF-1 male is kept totally isolated from his mate and deprived of any

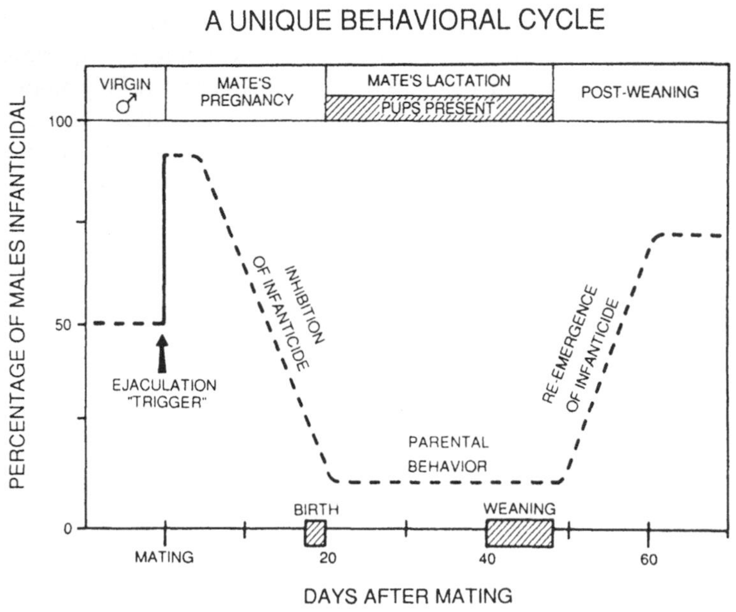

A UNIQUE BEHAVIORAL CYCLE

Figure 1 Schematic representation of temporal changes in behavior toward pups resulting from ejaculation in CF-1 stock male house mice.

female cues whatsoever following ejaculation (vom Saal, 1985; Perrigo and vom Saal, 1989). Furthermore, the inhibition of infanticide occurs after mating even when a male's pituitary and testes are removed (Perrigo *et al.*, 1989a). We are unaware of any other similar neural phenomenon in mammals where such dramatic shifts in adaptive behavior are timed to occur so many days — indeed weeks — after a specific stimulus such as coital ejaculation.

The Measurement of Infanticide and Parental Behavior

When a male house mouse encounters a neonate he either attempts to kill it or he does not harm it. These are clear-cut, unambiguous responses. We can easily assess a male's behavior by placing a 1–3 day old pup in his home cage. If a male is infanticidal, he will typically approach the pup, rattle his tail, and suddenly lunge at and attempt to kill the pup with rapid

bites to the head and back. This is an acute and dramatic response, so we immediately intervene and try to rescue the pup from attack as quickly as possible.

The opposite response from infanticide is parental behavior. A Parental male will typically groom the pup about the head and genitals before retrieving it to his nest. Interestingly, when a parental CF-1 male is allowed to incubate a pup, he appears sedated and is largely inattentive to disturbance. When the small subset of CF-1 males who "Ignore" pups are tested repeatedly, some may become infanticidal while others begin retrieving and incubating pups (vom Saal, 1985). "Ignorer" males thus appear to straddle a neutral behavioral state between infanticide and true parental behavior (Perrigo and vom Saal, 1989).

Recently, however, we have modified the above test procedure so that injuries to live pups are virtually eliminated (Perrigo et al., 1989b). Test pups are now placed within a tube made of 1.5-mm^2 wire mesh screen. A tube 4–5 cm long and 1.5 cm in diameter is large enough to slide a neonate comfortably inside — the screen-encased pup is quiescent, secure and completely buffered from attack. Thus, when an infanticidal CF-1 male encounters a screen-protected pup, he routinely attacks and repeatedly bites at the screen, without injuring the neonate (Perrigo et al., 1989b). If, however, the male does not show any intent to harm the pup, the next step is to introduce an unprotected pup for 30 minutes and determine whether the male ignores the pup, or, if he is truly Parental. While this humane test procedure seems to be a reliable assessment of infanticidal tendencies in CF-1 stock males, a screen-protected pup has not proven to be an effective testing paradigm in either wild stock (unpublished observation) or other laboratory stock house mice (Elwood et al., 1990) .

Finally, it should also be emphasized that a male's reaction toward a newborn pup is a generalized, non-specific response. Neither the sex, age (1-10 days old) nor relatedness of the pup appear to have any discernable influence on a male's tendency to exhibit infanticide or parental behavior (vom Saal and Howard, 1982; Svare et al., 1984; vom Saal, 1985; McCarthy and vom Saal, 1986). Male mice also have spontaneous ejaculations nearly every night (Huber and Bronson, 1980), but this event does not in any way influence their behavior toward pups.

Female Cues *Per Se* Do Not Inhibit Infanticide in Virgin CF-1 Males

As noted in the introduction, there are multiple inhibitory mechanisms that prevent a male mouse from harming his offspring, and cohabitation

with a female will effectively inhibit male infanticidal behavior in some stocks of house mice. With regard to the ejaculatory phenomenon, however, the CF-1 mouse is an ideal choice of study, mainly because virgin CF-1 males appear totally insensitive to female cues *per se* as an inhibitor of infanticide. To illustrate this point, two groups of adult, virgin CF-1 males were identified as infanticidal on a pretest and then exposed to an array of chemical and tactile cues from either pregnant or nonpregnant females.

In the first phase of this experiment, a wire mesh bottom cage containing either three females, 10–11 days pregnant, or three nonpregnant females was suspended over each male's cage; thus, female urine and wastes showered freely on the males below. Pregnant females were changed every seven days so that none would deliver pups while suspended overhead (nonpregnant females were also changed every seven days). Each male was tested with a pup beginning at the time of first exposure to females suspended overhead (Day 0), and then retested every three days thereafter until day 21 (pups would have been born by now if the males had been allowed to mate).

In the second phase of this experiment, males in both groups were now allowed to cohabit with either a pregnant female or a sexually unreceptive, ovariectomized female. Likewise, pregnant and ovariectomized females were also replaced every seven days. During this phase of the experiment, testing with a pup occurred every seven days. Finally, after 21 days of cohabitation, all females in both groups were replaced with a pregnant female scheduled to give birth 10–12 days later. One more test with a pup occurred on Day 28 and each female was allowed to give birth in the male's own cage. When litters were born, cages were checked daily each morning over the next 10 days for any evidence of pup-killing. Throughout both phases of this experiment, a control group of virgin infanticidal males was also tested on the same schedule. The control males were not exposed to any female cues whatsoever.

As shown in Figure 2, which visually summarizes all of the results of this experiment, only the delivery of pups in the male's own cage inhibited infanticidal behavior in the virgin males (P < .0001 as compared with the control males). Only one litter was killed in the virgin male Groups 1 and 2, and previous experiments have established that if a male mouse kills one pup he will also kill the entire litter (vom Saal and Howard, 1982; Palanza and Parmigiani, 1991). All surviving litters appeared healthy (8–14 pups per litter) and there was no evidence of bitten or missing pups over the next 10 days, nor was there any evidence of wounds indicating that a female had attacked the male to defend her litter. Evidence from Parmigiani's

EXPOSURE TO ♀ CUES

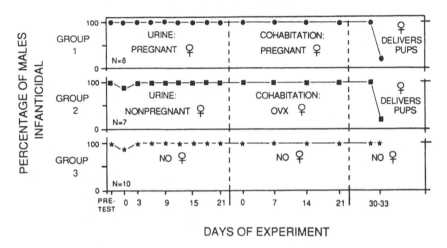

DAYS OF EXPERIMENT

Figure 2 The response toward newborn pups when virgin infanticidal males were exposed to various combinations of chemosensory and tactile cues from female mice (OVX = ovariectomized). A zero to 100 percentage scale is shown for each of the three groups.

laboratory suggests that a direct tactile or chemotactile mechanism is involved here, because infanticide is inhibited only when the male has actual physical contact with the female and/or her pups during birth (see Palanza and Parmigiani, 1991; Parmigiani *et al.*, Chapter 15). Regardless of these results, the act of ejaculation is still the primary stimulus that inhibits infanticide and regulates the timing of parental behavior in CF-1 males. Figure 2 also confirms that even when tested repeatedly with a pup, this procedure does not in any way inhibit infanticide among virgin males.

The Effects of Gonadal and Pituitary Hormones

As will be detailed later, differential exposure to sex steroids during fetal development clearly influences the way adult male mice behave toward young. Some steroid-sensitive behaviors are "organized" during perinatal development and do not require the presence of specific gonadal hormones in order for the behavior to occur in adulthood (Beatty, 1979; vom

Saal, 1983). But other behaviors may be "sensitized" during perinatal development and therefore require the presence of gonadal hormones in adulthood in order for the behavior to occur ("activation").

When virgin CF-1 males are castrated and allowed to interact with a pup for the first time, males behave parentally (Perrigo *et al.*, 1989a). If, however, these same castrated males are implanted with a 1 cm Silastic capsule containing 5 mg of crystalline testosterone (dissolved in .02 cc of sesame oil) and retested for their behavior toward pups several days later, half of the previously parental males will exhibit infanticide. As depicted earlier in Figure 1, this finding mimics the typical 50/50 proportion of spontaneously infanticidal versus non-infanticidal males observed whenever a large random sample of gonadally intact virgin CF-1 males are tested. Removal of the testosterone capsule abolishes infanticide. In summary, concurrent exposure to testosterone is required for a virgin male to exhibit infanticide.

The act of ejaculation causes a dramatic surge in LH (luteinizing hormone) and testosterone in male mice (Coquelin and Bronson, 1980), so we tested whether the hormonal changes triggered by mating might be responsible for mediating changes in a male's behavior toward pups. Twenty-two spontaneously infanticidal males were hypophysectomized (their pituitary gland was removed) and castrated, and, in order to maintain both their ability to mate and exhibit infanticide (vom Saal, 1983), they were also implanted with a 5 mg testosterone capsule (see above). Half of the males were allowed to mate while the other half were not. When tested with a pup at 20 days after mating, only 1 out of 12 mated males exhibited infanticide while 6 out of 10 non-mated males still exhibited infanticide ($P <$ .05; Perrigo *et al.,* 1989a). Males who were hypophysectomized and mated thus showed a post-mating inhibition pattern identical to that observed in previous experiments using intact males (vom Saal, 1985). This finding revealed that the mating-induced inhibition of infanticide is a neurally-timed and mediated response, operating independently from pituitary hormone secretions or changes in gonadal secretions resulting from of mating (Perrigo *et al.*, 1989a).

How Do Mated Males Keep Track of Time?

Since the stimulus of ejaculation results in an unusually prolonged sequence of behavioral changes, a mated male must somehow be able track his mate's pregnancy and thus stop killing pups and behave parentally at the appropriate time. This prompted us to ask how mated males could measure the passage of time after mating. Because CF-1 males do not need female cues or typical hormonal cues in order to exhibit this response, this

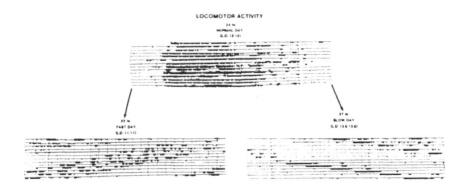

Figure 3 Daily locomotor patterns in a representative *fast day* male (left) and *slow day* male (right); 24 hr strips from an event recorder are pasted consecutively over several days. The dark bars in each strip represent when the animal was active on a running wheel. All individuals began their entrainment at a normal L:D 12:12 cycle; the top picture represents a typical activity pattern of a standard 24 hr day. The stair-step patterns generated in the lower pictures demonstrate that the *fast day* animal and the *slow day* animal were entrained to their respective photocycles.

clearly suggested that the efficacy of their strategy depended on a unique neural timekeeper.

We speculated that mated males could keep track of time either: 1) by measuring the absolute amount of time passing after ejaculation, or 2) by assessing the number of light/dark cycles experienced after ejaculation. Since photoperiodic variation in nature always provides infallible temporal cues for entraining daily (circadian) and seasonal cycles of feeding, breeding, metabolism and movement, we suspected the latter hypothesis. To test both possibilities, we used an experimental paradigm that allowed us to distinguish between absolute time (a standard 24 hr day) versus the number of light/dark cycles experienced: CF-1 males were thus housed at artificially fast (L:D 11:11 = 22 hr.) versus artificially slow (L:D 13.5:13.5 = 27 hr) daylengths (Perrigo *et al.*, 1990).

"Fast" versus "slow" time

One hundred adult CF-1 males were placed in light-tight, coffin-sized boxes illuminated inside with a 15-Watt fluorescent lamp (L:D 12:12 was

their initial light/dark cycle). Fifty males in each group were slowly adapted over a 25 day period to the 22-hour *fast day* cycle or the 27 hr *slow day* cycle by either increasing or decreasing the length of their light and dark exposure by several minutes each day. To verify behavioral entrainment, the locomotor patterns of several randomly chosen males in each group were monitored in cages with a running wheel interfaced to an event recorder (Figure 3). Males were allowed to mate and then screened for infanticide one day after ejaculation. Parental males were discarded from the experiment while the remaining infanticidal males (about 85% in both the *fast* and *slow day* groups) were retested with a pup between 16 and 25 absolute (24 hr) days after mating. The rationale behind our test procedure is illustrated by the timeline diagram in Figure 4. Specifically, half of the *fast day* males were retested at 16.3 absolute days (=18 light/dark cycles) and half were retested at 20 absolute days (= 22 light/dark cycles) after mating, while half of the *slow day* males were retested at 20 absolute days (=18 light/dark cycles) and half were retested at 24.8 absolute days (= 22 light/dark cycles) after mating. Our objective here was to directly compare both groups at 20 absolute days after mating and also control for the number of equivalent light/dark cycles experienced by both groups (18 versus 22 cycles).

Figure 5 shows the post-mating inhibition of infanticide graphed in two complementary perspectives: First, in relation to the number of absolute (24 hr) days experienced after mating, versus second, in relation to the number of light/dark cycles experienced after mating. When viewed side-by-side, the graphs suggest the presence of a unique neural timekeeper. At 20 absolute (24 hr) days after mating there was a significant difference in the frequency of infanticide between the *fast* and *slow day* groups (13% versus 61%, respectively; P < .005), suggesting that mated males did not rely on the amount of absolute time after mating as a cue to inhibit infanticide. Likewise, no differences were noted in the frequency of infanticide with both groups matched for experiencing the same number of light/dark cycles. This suggested that photoperiodic cues synchronized the shift in behavior, since a sudden transition from violent to benevolent behavior toward pups occurred as a function of the number of light/dark cycles experienced after ejaculation rather than the amount of absolute time (24 h days) experienced (Perrigo *et al.*, 1990).

The fact that dramatic shifts from infanticide to parental behavior in male mice parallel the behavioral and temporal dimensions of pregnancy in females is in itself interesting. Infanticide is also a fundamental component of the behavioral repertoire of female house mice — virtually all pregnant wild-stock females kill pups up to the time of parturition, at which

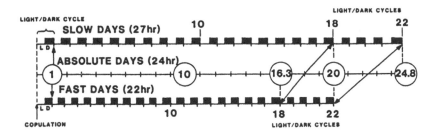

Figure 4 The relationship between absolute time (24 hr days) and the increasing desynchronization of light/dark cycles experienced by both groups during the course of this experiment. The alternating dark bars on the fast and slow time scales represent the dark phase of the repeating light/dark cycle. By 20 absolute days after mating, *fast day* males had experienced 4 more light/dark cycles than *slow day* males.

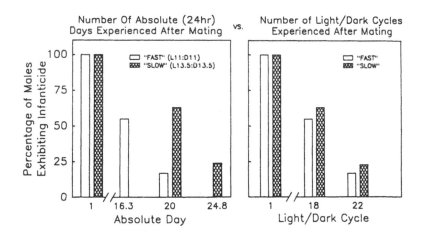

Figure 5 The percentage of male mice maintained at *fast* (=22 hr) and *slow* (=27 hr) days exhibiting infanticide when graphed in relation to absolute (24 hr) days versus the number of light/dark cycles experienced after ejaculation.

time they become parental (McCarthy and vom Saal, 1985; Soroker and Terkel, 1988). Thus, both sexes exhibit infanticide and seem to share a common suite of parental behaviors expressed at the time pups are born. But female house mice rely on the cues from developing fetuses and always seem to gauge the length of pregnancy in absolute time (Lanman and Seidman, 1977). Even when entrained to extreme light/dark cycles mimicking a 20 versus 28 hour day, female house mice still give birth the same number of absolute days after insemination (Davis and Menaker, 1981).

Male mice have apparently evolved a novel timekeeping solution for synchronizing their behavior toward pups with the duration of their mate's pregnancy. Some photoperiodically mediated phenomena in rodents, such as the preovulatory surge in LH, and hence, the organization of estrous cycles, regularly occur at 4–5 day multiples of daily light/dark cycles (Alleva et al., 1968; Fitzgerald and Zucker, 1976). Unlike our present finding, however, these events are mediated by cyclic changes in the secretion of pituitary and gonadal hormones. As described in the previous section, the mating-induced inhibition of infanticide can occur in male mice even in the absence of pituitary hormones or changes in gonadal secretions.

How Does the Neural Timing System Operate in the Presence or Absence of Photoperiodic Cues?

The above results suggested that ejaculation triggers a photoperiodically mediated timing mechanism that can synchronize a male's parental behavior with the presence of his pups (Perrigo et al., 1990). Male Norway rats maintained at L:D 12:12 also show a sharp change from pup-killing to parenting between 18 and 20 days after mating — just before the birth of their offspring at 22 days (Mennella and Moltz, 1988). The most obvious physiological explanation of this result would be a circadian-based timing mechanism, which, in the absence of entraining photoperiodic cues, would exhibit an endogenous rhythm with a period of about 24 hours.

Thus, the following experiment examined the timing of these behaviors when male mice were housed, mated and tested at typical husbandry conditions of L:D 12:12 versus free-running rhythm conditions of constant dark (DD) or constant light (LL). Our hypothesis was that males kept in different free-running conditions might undergo the transition from pup-killing to parenting at different times after mating.

Pretesting and partitioning of virgin males

As noted earlier, half of all virgin CF-1 males spontaneously kill pups whenever they encounter them, while the other half will usually retrieve and parent them. Previous experiments have also shown that whether or not a virgin CF-1 male is infanticidal or parental appears to be programmed by *in utero* variation in sex steroid exposure during late fetal development (vom Saal, 1983; Perrigo and vom Saal, 1989; Perrigo *et al.,* 1989a). These behavioral differences apparently result from testosterone emanating from fetal siblings and thus depends on whether a male fetus develops next to same or opposite sex fetuses (vom Saal and Bronson, 1980; vom Saal, 1989). Thus, we controlled for this physiological variation by identifying whether a virgin male was infanticidal or parental before he was delegated to an experimental treatment. The necessity for doing this will be obvious from our results.

Ninety-one virgin male mice (5 months of age and maintained since birth at L:D 12:12) were, for the first time, pretested for their behavior toward a 1-3 day old pup. The males were classified as follows: 46 were infanticidal (51%), 33 were parental (36%), and 12 males ignored the pup (13%). This replicates previous findings (vom Saal, 1985; Perrigo *et al.,* 1989a). One day later, the pretested males were evenly distributed among three animal rooms maintained at one of three light/dark treatments: constant light (LL), L:D 12:12, or constant darkness (DD). Thirty males were delegated to each light condition and partitioned as follows: 15 Infanticidal males, 11 Parental males and 4 Ignorer males. A partially covered 34-Watt fluorescent light was the illumination source in both the L:D 12:12 and LL rooms (range of 20-30 Lux as measured at the bottom and top, respectively, of the cage rack where males were housed).

After three weeks of accommodation to their light/dark condition, each male had two female CF-1 mice (50 days of age) placed in his cage (beginning at the time of lights on at 1200 hrs in the L:D 12:12 treatment; matings in the DD and LL rooms were also done at the same absolute time). All females were removed after three hours. The presence of a vaginal plug confirmed whether a male had ejaculated. This procedure was repeated for five days until most of the males in each group had mated, resulting in a distribution of 14-15 Infanticidal, 8-10 Parental and 3-4 Ignorer males within each of the three experimental groups. Upon confirmation of ejaculation (Day 0), each male was immediately tested with a pup and then retested between 1400 and 1500 hrs every three days thereafter until Day 30 after mating. Two more retests were done at Day 60 and Day 90 after mating. All observations in DD were done with a 15-Watt red light, which, for control purposes remained on constantly in all three treatment rooms.

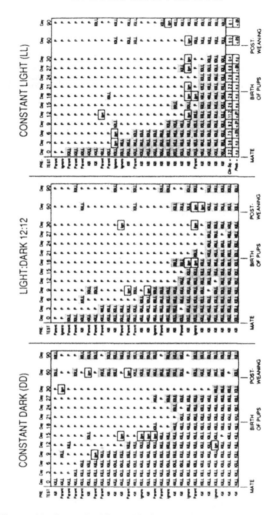

Figure 6A & B   Figure 6A shows the hierarchically sorted response patterns of individual male house mice who were mated and tested for their behavior toward a newborn pup (KILL = infanticidal; P = parental; Ign = ignored pup) at conditions of constant dark (DD), L:D 12:12 and constant light (LL). Males were tested immediately after ejaculation (Day 0) and every three days therafter until Day 30; two more tests occurred at Day 60 and Day 100 after mating. The behavior of each male prior to mating is shown in the Pretest column. Chi-Square values (df = 2) and probabilities (n.s. = not significant) comparing the frequency of infanticidal versus noninfanticidal behavior among all three groups at each test day are reported at the bottom of the columns under the LL group. Figure 6B represents a different perspective of the data in 6A. Specifically, the data from all three groups are summarized and displayed as a single frequency graph; asterisks (*) represent significant differences in the frequency of infanticidal males in the DD group compared with the LL and L:D 12:12 groups.

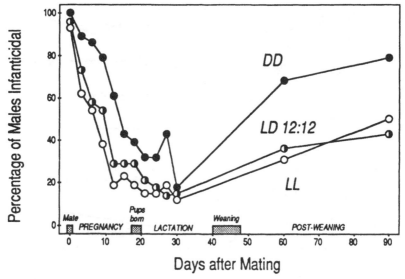

Figure 6B

The time-course of behavioral changes

Figures 6 and 7 show complementary perspectives of how each treatment group and pretest behavior of males behaved toward pups during the entire 90 day test period following mating. The raw data in the Figure 6 matrix are hierarchically sorted for the first 30 days of this experiment to reveal the complete range of behavioral and timing variation for each individual within each light treatment, whereas the Figure 7 graph focuses on the overall pattern of timing variation among males subdivided in relation to their pretest behavior toward pups. The behavioral differences shown in these figures are striking. Statistical differences were assessed with Chi-square analyses comparing the frequency of infanticidal versus noninfanticidal behavior among the groups at each test day (for analysis purposes, parenting and ignoring behaviors were combined as noninfanticidal).

Transitions in behavior after mating

As shown in the Figure 6 matrix, there were significant overall group differences (p = .05 or less) in the frequency of infanticide on Days 6, 9, 12, 27, 60, and 90 after mating. In general though, the group differences

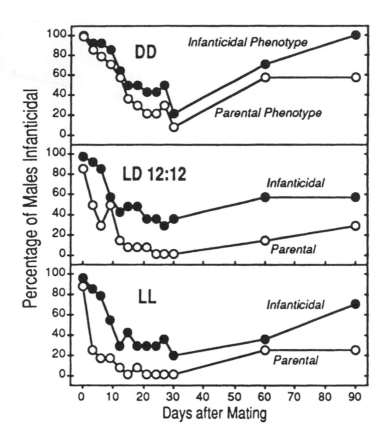

Figure 7 The temporal pattern of behavioral changes observed when males within each treatment group are graphed in relation to their behavioral phenotype as virgins (Infanticidal or Parental), which is correlated with testosterone exposure during late fetal development (see text). Males in each group were ranked according to the day after mating when they expressed their first bout of parental behavior. Mann-Whitney U-tests revealed significant transition time differences between the two phenotypes in LL (p < 0.005) and L:D 12:12 (p < 0.005), whereas in the absence of light (DD), there were no significant differences (p > 0.40). (Redrawn from Perrigo, *et al.*, 1991).

resulted from a higher frequency of infanticide in the DD males. Also obvious in Figure 6 is the irregular pattern of transition from pup-killing to parenting observed in some males. This transition occurred smoothly between consecutive test days in many individuals, but in some males, transient and spontaneous episodes of parenting or pup-killing occurred sporadically throughout the course of testing. Furthermore, within the

first 30 days after mating, the majority of "Ignore" events occurred at the interface between infanticidal to parental transitions.

The frequency results from the L:D 12:12 group in Figure 6 also match previous experiments where CF-1 males were tested for infanticide only one or two times between days 1 and 21 after mating (vom Saal, 1985), instead of every three days as in this experiment. This suggests that repeated testing with a pup does not influence the timing of infanticide inhibition after mating.

Behavior of pretested Parental males

As shown in the Figure 7 graph, which partitions males according to their pretest behavior toward pups, ejaculation immediately triggered pup-killing in virtually all of those males classified as Parental in the pretest. It should also be emphasized here that mere exposure to females, without ejaculation, does not induce pup-killing in virgin Parental males (Perrigo *et al.*, 1991). The transition back to parental behavior occurred rapidly in the LL and L:D 12:12 males, but was significantly prolonged in the DD males. These groups differed (p = .05 or less) on Days 3, 6, 9, 12, and 15. Significant group differences disappeared at Day 18, coinciding with the birth of pups, but reappeared again at Day 27. By 60 Days after mating — pups would have been weaned by this time — many of the parental males, especially those in DD, became pup-killers again.

Behavior of pretested Infanticidal males

Figure 7 also reveals that among those males classified as Infanticidal in the pretest, there were no significant behavioral differences except for Day 60 and Day 90 after mating (100% of the DD males were infanticidal again by Day 90). Nevertheless, the majority of pretested Infanticidal males in each group still expressed parental behavior within 21 days after mating. Some pup-killing males, however, never expressed parental behavior at all.

The effects of light

One further point is visually obvious in Figure 7, namely that males who were categorized as Parental on their pretest underwent transitions to parental behavior more rapidly than their Infanticidal counterparts. Thus, all males were ranked according to the day after mating when they expressed their first bout of parental behavior. Indeed, Mann-Whitney U-tests revealed significant transition time differences between the two

pretest categories of males in both the LL (p < .005) and L:D 12:12 treatments (p < .005). In sharp contrast, no significant transition time differences were noted between the two pretest categories of DD males (p >
.40). This suggests that pretested Noninfanticidal males responded to the
stimulus of light differently than pretested Infanticidal males: The presence of light in the LL and L:D 12:12 treatments accelerated the post-
mating transition to parenting among parental phenotypes, whereas in the
absence of light (DD), there were no significant behavioral differences
among pretested infanticidal and parental males.

*The period of free-running locomotor activity in DD versus LL: Test of the
circadian-linked timing hypothesis*

After the above experiment was completed (90 days after mating), eight
LL males and ten DD males were randomly chosen and placed in cages
with a running wheel interfaced to an Esterline-Angus event recorder.
Free-running activity patterns were recorded for 40 days, at which time we
estimated the period (τ = tau) of each male's activity cycle by eyefitting
a line through his activity onsets during the last 15 days of activity. The
purpose of this experiment was to examine whether any correlation existed
between the period (τ) of a circadian activity/rest cycle and the timing of
post-mating behavioral changes in either DD or LL.

Our results were consistent with "Aschoff's Rule" (Aschoff, 1960; Pittendrigh, 1960); thus, the free-running activity cycle of DD males was considerably shorter, about 1 hr less, than the cycle of LL males. The period (τ)
of the activity/rest cycle in DD males was 24.13 ± .05 hrs (meaan ± sem)
while in the LL males, τ = 25.15 ± .17 hrs (p < .0001). These results, and
the results from the previous experiment, do not in themselves support the
hypothesis of a circadian timing mechanism keeping track of daily cycles
experienced after mating (Perrigo et al., 1990). If this had been the case,
then a post-mating shift to parental behavior should have occurred more
rapidly in DD males because of their shorter free-run period. Or, at the
very least, there should have been no differences between the LL and DD
groups. Figures 6 and 7, however, clearly suggest the opposite result. It
should also be noted that the act of ejaculation has no effect on the phase
(ϕ) nor period (τ) of the free-run when CF-1 males are maintained in DD
(unpublished observation).

Within the eight LL males, however, there was a significant positive
correlation between the period of an individual's activity cycle and the
number of days elapsed between ejaculation and their first expression of
parental behavior (r^2 = .75, p < .05). In contrast, no such correlation
existed in DD animals (r^2 = .16, p > .25). These results are unique, but they

neither support nor refute the involvement of a circadian-based timing mechanism, they only suggest a correlation in constant light. Nor, as will be discussed later, do they eliminate the possibility of "masking" in constant light.

Does testing in light versus dark influence behavior toward pups?

One potential pitfall in the apparent light-mediated differences noted above is that males might be more prone to kill pups depending on whether they are tested in the dark or in the light. Specifically, DD males were always tested in darkness, while the L:D 12:12 and LL males were always tested with the lights on. In a follow-up experiment, 58 virgin males were thus tested for their behavior toward a newborn pup at opposite times of day: 29 males were tested at three hours after lights on while 29 males were tested in darkness three hours after lights off. Males in this experiment were 5 months of age, maintained since birth at L:D 12:12. The results showed there were no differences in pup-killing or parenting at opposite times of day (66% of the males were infanticidal when tested with the lights on, while 55% were infanticidal when tested with the lights out; p >.50). Thus, testing in either light or dark did not seem to influence how a male behaves toward pups.

Observing the transition at 12 hour intervals and the repeatability of the timing phenomenon within individuals

The following experiment asked whether the effect of first mating and the timing of inhibition observed in Figure 6 is repeatable within individuals. Six males who showed rapid transition times from infanticidal to parental behavior were thus selected from each of the three treatments in Figure 6 and mated for the second time at 100 days after their first mating. The first test with a pup occurred immediately after ejaculation was confirmed. Unlike their first mating, however, in which retests occurred at three day intervals, each retest in this experiment occurred instead at 12 hr intervals until the end of Day 12. Thus, all subsequent retests in the L:D 12:12 room occurred at the time of lights off (2400 hrs) and at the time of lights on (1200 hrs). Tests in the DD and LL rooms were done at the same absolute time.

As shown in Figure 8, a clean transition from pup-killing to parenting sometimes occurred in as little as 12 hours, but only in five of the 18 males tested here. In contrast, the majority of males (13 out of 18) showed fluctuations, sometimes periodic, in their behavior toward pups before

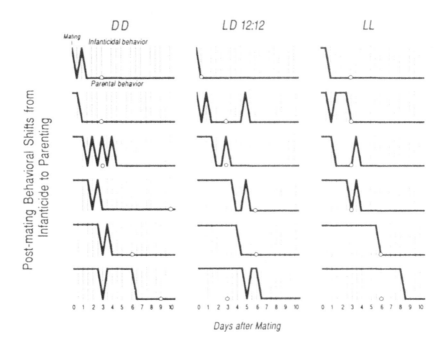

Figure 8 Day-to-day oscillations in behavior toward pups in individual males (six from each treatment) who were re-mated at 9 months of age and tested for infanticide every 12 hours (each division = 12 hrs of real time) for 12 days (the scale ends at 12 days for clarity). For comparison purposes, a white dot indicates the day when parental behavior was first noted (as measured at 3-day intervals) when males were mated for the first time at 6 months of age.(Redrawn from Perrigo, *et al.*, 1991).

"locking in" to consistent parenting. The white dot shown in Figure 8 indicates the test day after their first mating (from Figure 6) when parental behavior was first noted in each male. Since these males were tested at different intervals (every 12 hrs versus every three days), we cannot make a legitimate statistical comparison here. However, with the exception of one of the DD males, the visual results agree remarkably well, timewise, with the pattern of changes observed when the same individuals were tested three months earlier. These data suggest that after a behavioral cycle toward pups is "reset", ejaculation can trigger a new cycle of infanticidal inhibition, one that appears programmed to last about the same length of time within an individual.

Aging: A more prolonged and attenuated response

The males kept in the L:D 12:12 condition (from Figure 6) remained individually housed until 18 months of age, at which time they were again allowed to mate. This provided a direct comparison between the same individuals when tested at one full year after their first mating experience (*i.e.*, 6 months versus 18 months of age). All surviving males (18 out of the original 28) were mated and then tested every three days for 30 days (same procedure as earlier). As shown in Figure 9, the transition to parental behavior after mating was significantly more prolonged when the males were a year older. In fact, five of the older males (28%) never ceased killing pups. With regard to individual variation, the scatterplot in Figure 10 shows there was no individual correlation (as suggested by the previous experiment) between the time elapsed between mating and parenting when each male was a full year older (p > .35). In general then, the neural timing mechanism triggered by ejaculation appears to attenuate with age.

OVERVIEW AND CONCLUSIONS

Both sexes of house mice exhibit infanticide and express parental behavior at the time pups are born, but male mice lack the cues from developing fetuses that precisely regulate the timing of infanticidal and parental behavior in females (Svare, 1981; McCarthy and vom Saal, 1985; Perrigo *et al.*, 1990). While the use of different stocks of house mice among different laboratories was originally a source of considerable debate (and confusion) regarding the behavioral mechanisms that inhibit infanticide (Huck *et al.*, 1982; Labov *et al.*, 1985; vom Saal, 1985; Elwood, 1986), it is now extensively documented that multiple, redundant mechanisms have indeed evolved to inhibit pup-killing behavior in male mice (*e.g.* see Elwood, Parmigiani, this volume). As a general rule, these behavioral polymorphisms stem from the amplification of genetic differences among the myriad of inbred domestic and wild stocks used in the study of infanticide (*e.g.* see Figure 11; Perrigo and vom Saal, 1989; Palanza and Parmigiani,1991: Parmigiani *et al.*, Chapter 15). While house mice originated in the Old World, they can now be found living worldwide in an amazing variety of feral and commensal habitats, mainly because of their enormous behavioral and reproductive flexibility (Bronson, 1979; Berry, 1981; Perrigo, 1990). Thus, the degree to which these multiple inhibitory mechanisms are expressed may reflect subtle differences in the socioecology and deme structures of original founder stocks.

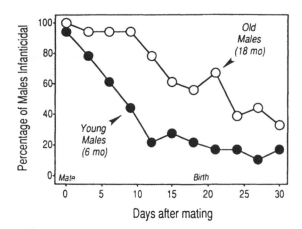

Figure 9 Males kept in L:D 12:12 condition remained individually housed until 18 months of age, at which time the 18 (out of 28) survivors were again mated and re-tested for infanticide every three days until 30 days after mating. The graph demonstrates that the transition time from infanticide to parenting becomes substantially prolonged by aging. (Redrawn from Perrigo, *et al.*, 1991).

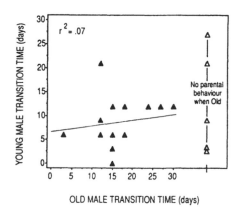

Figure 10 The neural timing mechanism also attenuates with age. There was no correlation in transition time, defined here as the test day after mating when an infanticidal male exhibited his first bout of parental behavior, among individual males (closed triangles) when they were mated when Young (6 months old) versus when they were Old (18 months old); $r^2 = .07$, p > .35. Since five of the Old males did not cease killing pups and, hence, did not exhibit parental behavior after 30 days of testing, they are shown separately (open triangles) on the right side of graph.

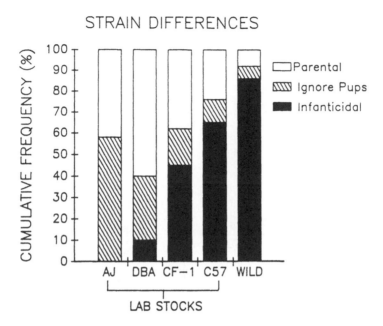

Figure 11 The genetic variation in background frequencies of spontaneous infanticidal and noninfanticidal behavior (parenters and ignorers) among virgin male house mice from four laboratory stocks and one wild stock.

The contrast in multiple, redundant mechanisms is especially evident in wild-trapped stocks of *Mus domesticus* and *M. musculus* (for a discussion of taxonomic differences see: Marshal and Sage 1981). Wild male *M. domesticus* trapped in Alberta, Canada behave similar to CF-1 males. As shown in Figure 12, their pattern of infanticide inhibition after ejaculation in L:D 12:12, including the expression of parental behavior and re-emergence of infanticide 60–90 days later, resembles that shown in Figure 6; likewise, 20+ days of cohabitation with pregnant females does not inhibit infanticide either, although neither does the birth of pups in a wild male's home cage (unpublished observation). But in wild male *M. musculus* from Israel, ejaculation and female cohabitation are independent mechanisms. Either mechanism, by itself, will effectively inhibit infanticide (Soroker and Terkel, 1988). And in another muroid rodent, the Norway rat, ejaculation alone can inhibit infanticide, but so will chemosensory cues if virgin male rats are exposed only to the soiled cage bedding of a pregnant female. The latter results suggests the evolution of female counter-strategies to defend

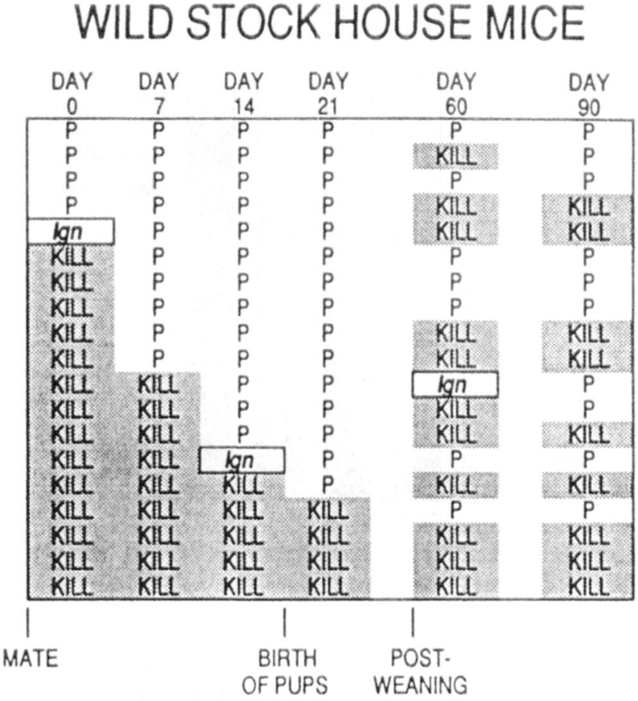

WILD STOCK HOUSE MICE

DAY 0	DAY 7	DAY 14	DAY 21	DAY 60	DAY 90
P	P	P	P	P	P
P	P	P	P	KILL	P
P	P	P	P	P	P
P	P	P	P	KILL	KILL
Ign	P	P	P	KILL	KILL
KILL	P	P	P	P	P
KILL	P	P	P	P	P
KILL	P	P	P	P	P
KILL	P	P	P	KILL	KILL
KILL	P	P	P	KILL	KILL
KILL	KILL	P	P	Ign	P
KILL	KILL	P	P	KILL	P
KILL	KILL	P	P	KILL	KILL
KILL	KILL	Ign	P	P	P
KILL	KILL	KILL	P	KILL	KILL
KILL	KILL	KILL	KILL	P	P
KILL	KILL	KILL	KILL	KILL	KILL
KILL	KILL	KILL	KILL	KILL	KILL
KILL	KILL	KILL	KILL	KILL	KILL

MATE   BIRTH   POST-
                       OF PUPS   WEANING

Figure 12 The response pattern of 19 individual virgin male wild stock house mice (five months of age) who were maintained at LD 12:12 and tested for their behavior toward a newborn pup (KILL=infanticidal; P=parental; Ign=ignored pup) immediately after mating (Day 0), and at Days 7, 14, 21, 60 and 90 after mating.

their litters from infanticidal attack (Mennella and Moltz, 1988; see also Parmigiani *et al.*, 1988; Elwood and Kennedy, 1990).

As demonstrated earlier, however, CF-1 mice are a genetic stock in which virgin males appear insensitive to female cues *per se* as an inhibitor of infanticide. In contrast, the birth of pups in the presence of a male in his home cage clearly seemed to inhibit infanticide (Figure 2). Interestingly, the data from Figure 2 also suggest that when a virgin male is present during the birth of pups, infanticide is inhibited to such a degree that even the ensuing postpartum mating will not trigger his typical pup-killing behavior. It cannot be discerned, however, whether this reflects a female counter-strategy or represents yet another redundant mechanism to prevent a male from accidentally killing his own offspring.

Evidence for Fetal Hormonal Programming of Individual Timing Variation and Responses to the Effects of Light

Since most nocturnal rodents see light only during short, crepuscular periods each day, one might argue that a housing condition of constant light — or even a typical photoperiodic laboratory condition of L:D 12:12 — represents an unnatural environment for a nocturnal species such as the house mouse. On the other hand, house mice in the laboratory routinely show a great deal more flexibility in their daily activity patterns than do other nocturnal rodents (Perrigo, 1987,1990), nor is their reproduction under photoperiodic control (Bronson, 1979; Berry, 1981).

Regardless of these potential arguments, Figures 6 and 7 still revealed that the stimulus of light clearly influenced the timing of behavioral changes following ejaculation in house mice. Photoperiodic (L:D 12:12) and constant light (LL) dramatically accelerated the inhibition of infanticide and emergence of parental behavior following ejaculation, but mainly in those males who had displayed parental behavior when they were pretested as virgins. These results, together with prior findings, suggest a unique link between the way in which timing variation and differential responses to photic stimuli among adults are programmed by hormonal events during late fetal development.

As mentioned earlier, previous experiments have established that CF-1 males who develop between two male fetuses — and are thus exposed to higher concentrations of testosterone — are significantly more likely to exhibit parental behavior both before and after mating than are their male counterparts who developed between two female siblings (vom Saal, 1983; Perrigo et al., 1989a; see also Samuels et al., 1981). This is known as the intrauterine position phenomenon (vom Saal, 1989) and describes the fact that, in mammals such as house mice that produce large litters, fetuses are positioned randomly in the uterine horns and are therefore exposed to differential sex steroid concentrations depending on whether they develop next to same or opposite sex siblings. As a result, an individual's intrauterine position has been correlated with a profound range of variation among reproductive, morphologic, and behavioral characteristics expressed when both sexes are adult, including adult-infant interactions (vom Saal, 1989).

The broad range of timing variation shown among all three treatment groups in Figure 6 matches with the phenotypic variation in infanticide predicted by the intrauterine position model (Perrigo et al., 1989a; Perrigo and vom Saal, 1989). In any large random sample of CF-1 males there are always some individuals who simply do not respond to the stimulus of ejaculation. Some males always remain parental and, as noted above, these individuals most likely underwent fetal development between two

male siblings. In direct contrast, some males always kill pups, regardless of how much time elapses after mating, and these individuals most likely underwent fetal development between two female siblings. Both subsets of CF-1 males represents about 10-15% of the population, and the data from Figure 6 fit these expected proportions remarkably well. Among those few individuals in which mating *per se* does not seem to inhibit pup-killing, the stimulus of ejaculation and female cohabitation are probably both required in order to inhibit infanticide (see also Elwood and Ostermeyer, 1984). In fact, we have obtained preliminary evidence for this by examining the five Old males from Figure 9 who were still infanticidal at 30 days after mating. When cohabited with a pregnant female, four out of five of these mated males did indeed express parental behavior within 14 days of cohabitation (unpublished observation).

A Proposed Neuroethological Model: The Programming of Individual Behavioral Thresholds

Figure 13 is a scheme depicting the transition from infanticide to parental behavior following ejaculation in CF-1 males. As depicted in the top half of Figure 13, the intense sympathetic stimulus of ejaculation triggers a sudden and dramatic behavioral change, which thus intensifies a male's motivation to kill pups. Implicit in this prediction is that the act of ejaculation is a neural "supercharge" that immediately drives most parental CF-1 males well above their threshold for infanticidal behavior (verified in Figures 7 and 8). From an ecological standpoint, it is not surprising that ejaculation triggers such an immediate pup-killing reaction. Female house mice exhibit a strong postpartum estrus within 24 hours after parturition; thus, if a virgin male copulates with a newly lactating female he will increase his reproductive success by immediately seeking out and destroying her litter.

Ejaculation also activates the time-delayed inhibition process. The neural substrate(s) governing infanticide thus appear to undergo an inhibitory decay, which, over the course of time, eventually diminishes a male's motivation to kill pups. Once an individual's threshold for the inhibition of infanticide is reached (*i.e.* the Transition Zone in Figure 13), then parental responses can emerge. When CF-1 males are hypophysectomized, which eliminates the pituitary hormone prolactin, a facilitator of parental behavior in females, they are significantly less likely to display parental behavior after mating and thus tend to ignore pups (Perrigo *et al.,* 1989a). Since hypophysectomy does not prevent the post-ejaculatory inhibition of infanticide, when viewed together, these findings suggest that

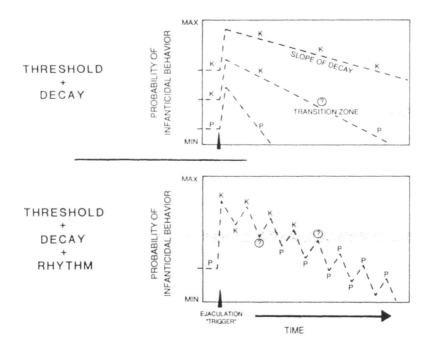

Figure 13 A potential scheme for understanding how the post-ejaculatory inhibition of infanticide might operates in CF-1 stock males. As shown in the top half of the figure (Threshold + Decay), males who behave parentally (P) before they mate are stimulated by the act of ejaculation to exhibit infanticide, while in males who are already infanticidal (K) before they mate, the stimulus of ejaculation simply intensifies their motivation to kill pups (see text). After ejaculation occurs, an inhibitory decay process is activated, which eventually eliminates infanticidal behavior by falling below the Transition Zone threshold, thus allowing parental behavior to occur. This model assumes that any one of three behaviors — Infanticide (K), Ignoring or Parenting (P) — can occur when males are at or near their particular Transition Zone threshold. The bottom half of this figure (Threshold + Decay + Rhythm), while purely speculative, depicts the potential behavioral results if a rhythmic (oscillating) behavioral decay is superimposed on the inhibition process (see text).

the inhibition of infanticide and the occurrence of parental behavior are independent phenomena.

As implied by the previous section, the slope of a CF-1 male's behavioral decay is, to a large extent, programmed by hormonal events related to his fetal position, although light cues, genetic differences among individuals and social factors (*e.g.* synergistic female cues) undoubtedly influence

this decay process too. With regard to aging, the time interval between ejaculation and parenting was relatively constant in individual young males (6 to 9 months of age), while in old males (18 months of age) there was a more prolonged inhibition of infanticide and an increased proportion of males who failed to show parental behavior following ejaculation. This suggests that the functioning of the timing system, shifts in response thresholds, and/or the need for redundant female cues also change during aging.

Also implicit in the Figure 13 model is that when a male "ignores" a pup, he is at or near his behavioral threshold (Transition Zone) during the inhibition process. Thus, if testing with a pup occurs during this transition period, then any one of three behavioral states could be observed: Infanticide (K), ignoring, or parenting (P). The evidence for this comes from Figure 8. By testing males at 12 hr intervals we observed fluctuations among all three behaviors during the transition phase. In fact, several males actually seemed to oscillate between infanticidal and noninfanticidal behaviors during this transition period. While purely speculative, the bottom half of Figure 13 depicts the potential behavioral results when an oscillating behavioral decay is superimposed on the inhibition process. This type of mechanism is one possibility that could explain the unusual behavioral fluctuations shown in Figure 8.

Is There a Circadian Timing Mechanism?: Pros and Cons

How this behavioral transition is neurally timed still remains perplexing. When we compared the timing of the transition from infanticide to parental behavior in CF-1 males entrained to 22 hr daylengths (L:D 11:11) versus 27 hr daylengths (L:D 13.5:13.5), the results clearly suggested that ejaculation triggers a neural timing system that "counts" photocycles. Thus, the objective in testing animals in conditions of constant dark and constant light was to determine how males would undergo the post-mating transition from infanticide to parental behavior in the absence of a light/dark cycle. The males who were given a running wheel in constant dark (DD) showed a free-run of almost 24 hr ($\tau = 24.1$ hr). When tested at 18 days (or activity cycles) after ejaculation, 39% of all males housed in DD exhibited infanticide. This result, however, is not statistically different from the proportion of males housed in 22 hr or 27 hr days that killed pups when tested at 18 activity cycles (Perrigo et al., 1990). On one hand, these findings are not inconsistent with the hypothesis that males can use an endogenous circadian system to "count" cycles following ejaculation.

But on the other hand, males kept in the free-running condition of constant, relatively low intensity light (LL) did indeed appear to measure the passage of time after mating differently than DD males. The inhibition of infanticide after ejaculation occurred more rapidly when males were housed in LL; in fact, the L:D 12:12 and LL groups both showed the same pattern in transition times. It must also be emphasized here that males in the L:D 12:12 condition do indeed exhibit a 24 hr activity/rest cycle under the lighting conditions provided in this experiment (unpublished observation). These findings are thus inconsistent with an entrainable "day-counting" mechanism, since the period (τ) of free-running activity for the males who were given running wheels in LL was 25.2 hrs. This suggests that males in LL should have taken longer to undergo the transition from infanticide to parental behavior after ejaculation, or, at the very least, they should not have differed so dramatically from the DD males. It must be explicitly emphasized, however, that within the context of this experiment, it not possible to eliminate whether a circadian timing system could have been "masked" by exposure to constant conditions (*e.g.* see Aschoff and von Goetz, 1988; Binkley, 1990). The process of masking occurs when an internal rhythm may be present, but is obscured (masked) by other direct effects; in this case, the propensity for nocturnal house mice to run on an activity wheel more often in darkness, and vice versa, to run less frequently when subjected to constant light.

Finally, we have made no attempt to explain how infanticidal behavior spontaneously re-emerges two months after mating. In general though, the presence of light seemed to inhibit infanticide during this phase too. No simple physiological explanation can thus account for these time-delayed responses, and little more can be said here except that our present and past experiments suggest both parallels and paradoxes with widely studied behavioral and reproductive timing processes (*e.g.* Silver and Bittman, 1984; Daan, 1987). Ablation of the Suprachiasmatic Nucleus, exposure to high intensity constant light (both of which can disrupt circadian rhythms by rendering animals arrhythmic), or maintenance at skeleton photoperiods are several potential experiments that could distinguish whether light has a photoperiodic, direct, or some other regulatory role in the timing of this entire behavioral cycle.

In conclusion, and regardless of how such a neural timing system operates, the two to three week timespan often intervening between ejaculation and the inhibition of infanticide — plus the spontaneous re-emergence of infanticide after pups are weaned — clearly seems to redefine the possible temporal and behavioral relationships between a neural stimulus and its response.

ACKNOWLEDGEMENTS

We thank W. Cully Bryant and Lee Belvin for their help. This work was supported by NSF Grants BNS 8813375 to GP and DCB 9004806 to FSvS.

REFERENCES

ALLEVA, J.J., Waleski, M.V., Alleva, F.R. and Umberger, E.J. (1968) Synchronizing effect of photoperiodicity on ovulation in hamsters. *Endocrinology*, **82**, 1227–1235.

ASCHOFF, J. (1960) Exogenous and endogenous components on circadian rhythms. *Cold Spring Harbor Symposium on Quantitative Biology*, **25**, 11–28.

ASCHOFF, J., and von Goetz, C. (1988) Masking of circadian activity rhythms in hamsters by darkness. *Journal of Comparative Physiology*, **162(A)**, 559–562.

BEATTY, W. (1979) Gonadal hormones and sex differences in nonreproductive behaviors in rodents: Organizational and activational influences. *Hormones and Behavior*, **12**, 112–163.

BERRY, R.J. (1981) Town mouse, country mouse: adaptation and adaptability in *Mus domesticus* (*M. musculus domesticus*). *Mammalian Review*, **11**, 91–136.

BINKLEY, S. (1990) *The Clockwork Sparrow: Time, Clocks, and Calendars in Biological Organisms.* Prentice-Hall, Englewood Cliffs, New Jersey.

BRONSON, F.H. (1979) The reproductive ecology of the house mouse. *Quarterly Review of Biology*, **54**, 265–299.

CALHOUN, J.B. (1962) Population density and social pathology. *Scientific American*, **206**, 139–148.

COQUELIN, A. and Bronson, F.H. (1980) Secretion of luteinizing hormone in male mice: Factors that influence release during sexual encounters. *Endocrinology*, **106**, 1224–1229.

DAAN, S. (1987) Clocks and hourglass timers in behavioural cycles. In: T. Hiroshige and K. Honma (eds) *Comparative Aspects of Circadian Clocks.* Hokkaido University Press, Sapporo, Japan, pp 42–54.

DAVIS, F. and Menaker, M. (1981) Development of mouse circadian pacemaker: independence from environmental cycles. *Journal of Comparative Physiology*, **143(A)**, 527–539.

ELWOOD, R. (1986) What makes male mice paternal? *Behavioral and Neural Biology*, **46**, 54–63.

ELWOOD, R. and Ostermeyer, M. (1984) Does copulation inhibit infanticide in male rodents? *Animal Behaviour*, **32**, 293–294.

ELWOOD, R. and Kennedy, H. (1990) The relationship between infanticide and pregnancy block in mice. *Behavioral and Neural Biology*, **53**, 277–283.

ELWOOD, R., Masterson, D. and O'neill, C. (1990) Protecting pups for infanticidal responsiveness in mice, *Mus domesticus. Animal Behaviour*, **40**, 778–780.

FITZGERALD, K. and Zucker, I. (1976) Circadian organization of the estrous cycle of the golden hamster. *Proceedings of the National Academy of Science USA*, **73**, 2923–2927.

HAUSFATER, G. and Hrdy, S. eds (1984) *Infanticide: Comparative and Evolutionary Perspectives.* Aldine Publishing, Chicago.

HRDY, S. (1979) Infanticide among animals: A review, classification, and examination of the implications for the reproductive strategies of females. *Ethology and Sociobiology*, **1**, 13–40.

HUBER, M. and Bronson, F.H. (1980) Social modulation of spontaneous ejaculation in the mouse. *Behavioral and Neural Biology*, **29**, 390–393.

HUCK, U., Soltis, R. and Coopersmith, C. (1982) Infanticide in male laboratory mice: effects of social status, prior sexual experience, and basis for social discrimination between related and unrelated young. *Animal Behaviour*, **30**, 1158–1165.

KENNEDY, H. and Elwood, R. (1988) Strain differences in the inhibition of infanticide in male mice (*Mus musculus*). *Behavioral and Neural Biology*, **50**, 349–353.

LABOV, J., Huck, U., Elwood, R. and Brooks, R. (1985) Current problems in the study of infanticidal behavior of rodents. *Quarterly Review of Biology*, **60**, 1–20.

LANMAN, T.J. and Seidman, L. (1977) Length of gestation in mice under a 21-hour day. *Biology of Reproduction*, **17**, 224–227.

MARSHAL, J.T. and Sage, R.M. (1981) Taxonomy of the house mouse. *Symposium of the Zoological Society of London*, **47**, 15–25.

MCCARTHY, M. and vom Saal, F.S. (1985) The influence of reproductive state on infanticide by wild female house mice (*Mus musculus*). *Physiology and Behavior*, **35**, 843–849.

MCCARTHY, M. and vom Saal, F.S. (1986) Inhibition of infanticide after mating in wild male house mice. *Physiology and Behavior*, **36**, 203–209.

MENNELLA, J. and Moltz, H. (1988) Infanticide in rats: Male strategy and female counter-strategy. *Physiology and Behavior*, **42**, 19–31.

PACKER, C. and Pusey, S. (1984) Infanticide in carnivores. In: G. Hausfater and S. Hrdy (eds) *Infanticide: Comparative and Evolutionary Perspectives*. Aldine Publishing, Chicago, pp 31–42.

PALANZA, P. and Parmigiani, S. (1991) Inhibition of infanticide in Swiss male mice: Behavioral polymorphism in response to multiple mediating factors. *Physiology and Behavior*, **49**, 797–802.

PARMIGIANI, S., Sgoifo, A. and Mainardi, D. (1988) Parental aggression displayed by female mice in relation to the sex, reproductive status and infanticidal potential of conspecific intruders. *Monitore Zoologico Italiano*, **22**, 193–201.

PERRIGO, G. (1987) Breeding and feeding strategies in deer mice and house mice when females are challenged to work for their food. *Animal Behaviour*, **35**, 1298–1316.

PERRIGO, G. (1990) Food, sex, time and effort in a small mammal: Energy allocation strategies for survival and reproduction. *Behaviour*, **114**(1–4),191–205.

PERRIGO, G., and vom Saal, F.S. (1989) Mating-induced regulation of infanticide in male mice: Fetal programming of a unique stimulus-response. In: R.J. Blanchard, P.F. Brain, D.C. Blanchard and S. Parmigiani (eds) *Ethoexperimental Approaches to the Study of Behaviour*. Kluwer, Dordrecht, The Netherlands, pp 320–336.

PERRIGO, G., Bryant, W.C. and vom Saal, F.S. (1989a) Fetal, hormonal and experiential factors influencing the mating-induced regulation of infanticide in male house mice. *Physiology and Behavior*, **46**, 121–128.

PERRIGO, G., Belvin, L., Bryant, W.C. and vom Saal, F.S. (1989b) The use of live pups in a humane, injury-free test for infanticidal behavior in male mice. *Animal Behaviour*, **38**, 897–898.

PERRIGO, G., Bryant, W.C. and vom Saal, F.S. (1990) A unique neural timing mechanism prevents male mice from harming their own offspring. *Animal Behaviour*, **39**, 535–539.

PERRIGO, G., L. Belvin, and F.S. vom Saal (1991). Individual variation in the neural timing of infanticide and parental behavior in male house mice. *Physiology and Behavior*, **50**, 287–296.

PITTENDRIGH, C.S. (1960) Circadian rhythms and the circadian organization of living systems. *Cold Spring Harbor Symposium on Quantitative Biology*, **25**, 159–184.

SAMUELS, O., Jason, G., Mann, M. and Svare, B. (1981) Pup-killing behavior in mice: suppression by early androgen exposure. *Physiology and Behavior*, **26**, 473–477.

SILVER, R and Bittman, E. L. (1984) Reproductive mechanisms: interaction of circadian and interval timing. *Annals of the New York Academy of Sciences*, **423**, 488–514.

SOROKER, V. and Terkel, J. (1988) Changes in incidence of infanticidal and parental responses during the reproductive cycle in male and female wild mice *Mus musculus*. *Animal Behaviour*, **36**, 1275–1281.

SVARE, B. (1981) Maternal Aggression in Mammals. In: D.J. Gubernick and P.H. Klopfer (eds) *Parental Care in Mammals*. Plenum Press, New York, pp 179–210.

SVARE, B., Kinsley, C., Mann, M. and Broida, J. (1984) Infanticide: Accounting for genetic variation. *Physiology and Behavior*, **33**, 137–152.

VOM SAAL, F.S. (1983) Variation in infanticide and parental behavior in male mice due to prior intrauterine proximity to female fetuses: Elimination by prenatal stress. *Physiology and Behavior*, **30**, 675–681.

VOM SAAL, F.S. (1985) Time-contingent change in infanticide and parental behavior induced by ejaculation in male mice. *Physiology and Behavior*, **34**, 7–15.

VOM SAAL, F.S. and Bronson, F.H. (1980) Sexual characteristics of adult female mice are correlated with their blood testosterone levels during perinatal development. *Science*, **208**, 597–599.

VOM SAAL, F.S. (1989) Sexual differentiation in litter–bearing mammals: influence of adjacent fetuses *in utero*. *Journal of Animal Science*, **67**, 1824–1840.

VOM SAAL, F.S. and Howard, L. (1982) The regulation of infanticide and parental behavior: Implications for reproductive success in male mice. *Science*, **215**, 1270–1272.

CHAPTER 17

SELECTIVE ALLOCATION OF PARENTAL AND INFANTICIDAL RESPONSES IN RODENTS: A REVIEW OF MECHANISMS

ROBERT W. ELWOOD and HAZEL F. KENNEDY

Division of Environmental and Evolutionary Biology,
School of Biology and Biochemistry,
The Queen's University of Belfast, Belfast BT7 1NN, Northern Ireland

INTRODUCTION

An adult rodent might benefit from infanticide in a variety of ways (Hrdy, 1979; Sherman, 1981; Elwood and Ostermeyer, 1984a, b). There will be costs, however, in terms of inclusive fitness (Hamilton, 1964), and the precise utility of infanticide will depend on the following:

$$\frac{\text{benefit to killer} + (\text{benefit to surviving kin} \times r)}{\text{cost to killer} + (\text{cost to kin} \times r)}$$

where r is the degree of relatedness. The killer might benefit directly in terms of improved survival and reproductive success and benefit might also accrue to surviving kin in terms of their increased survivorship and future reproduction. For example, a lactating female that engaged in pup-cannibalism may be able to provide better quality milk for her surviving infants (Konig and Markl, 1987; Perrigo, 1987) or for other kin when communal suckling occurs (Hoogland, *et al.*, 1989). In this latter case the level of benefit, in terms of inclusive fitness, depends on the degree of relatedness between the killer and the relatives that gain. The costs of killing include time and effort spent searching for and killing the infants and might also include a cost due to intervention from conspecifics. There

is also a potential cost in terms of the destruction of shared genes and this depends on the degree of relatedness between the killer and the victim. However, if the probability of survival of the victim was low had it not been killed, *e.g.* if the victim was starving, then the cost to the killer will be devalued.

There will be a range of results from this calculation, from cases in which the inclusive fitness of the potential killer would be decreased to cases in which it would be enhanced. In general, however, animals that harm their own offspring will incur the greatest costs in terms of lost shared genes and thus there would have to be substantial gains in the fitness of other close kin to offset this cost (O'Connor, 1978). Parental killing in rodents usually involves cannibalism in times of food shortage or severe disturbance *i.e.* when there is a low probability of survival or of reaching a sufficient size to have a reasonable chance of reproducing (reviewed by Elwood, 1992).

There are fewer constraints on non-parental infanticide because the cost to shared genes is likely to be considerably lower. Thus, when adults have access to unrelated infants, infanticide is likely to be common. However, for infanticide to be used efficiently the perpetrator must have some way of minimizing harm to kin. Mechanisms are expected to evolve that enable infanticide to be directed on a selective basis.

Similar reasoning requires parental care to be allocated selectively (Trivers, 1974). Animals that are indiscriminate in their allocation of care to non-kin will incur costs with little or no benefit in terms of survival of shared genes. Natural selection should provide mechanisms that ensure the allocation of parental care to kin but not to non-kin.

Adult rodents thus face a dilemma when they encounter infants. Should they kill the infants and risk harm to close kin or should they act parentally and risk wasting investment in non-kin? It is the purpose of this chapter to examine mechanisms that enable rodents to make appropriate decisions concerning infanticide and parental care.

SELECTIVITY OF INFANTICIDE AND PARENTAL CARE BY FEMALE RODENTS

With the exception of some laboratory strains of mice (Noirot and Goyens, 1971) virgin female rodents tend not to be maternal. For example, virgin rats tend either to avoid unrelated test pups or to kill them (Jakubowski and Terkel, 1985; Rosenblatt, 1967) and the response of virgin hamsters (Richards, 1966), gerbils (Elwood, 1977) and wild mice (McCarthy, *et al.*

1986) is to cannibalize test pups. This avoidance or cannibalism of unrelated infants stops prior to parturition and a dramatic change in maternal responsiveness occurs in the last few days of gestation. For example, female gerbils tested at this time commonly show a startle response on encountering the test pup but then alternate between sniffing and licking the pup and licking their own genital region, in a manner similar to that seen during parturition (Elwood, 1977). This may be followed by retrieving the pup to the nest, nest-building and even assuming the lactation position. Clearly, the female enters a 'maternal state' prior to parturition at which time she is receptive to both kin and non-kin.

This onset of maternal responsiveness occurs in a wide variety of mammals and appears to be mediated by the hormonal changes seen during pregnancy (Rosenblatt and Siegel, 1981, 1983). In the rat, progesterone levels increase during pregnancy but show a sudden drop in the final 4 days. This contrasts with the sudden increase in levels of both estradiol and prolactin at the same time (Rosenblatt and Siegel, 1983). In wild mice, infanticidal females become inhibited from infanticide within 1 hr of injection with prostaglandin, whereas injection with oxytocin not only inhibits infanticide but also initiates maternal care (McCarthy, *et al.*, 1986) and there is rapidly accumulating evidence that oxytocin plays a role in the mediation of the onset of maternal care (reviewed by Insel, 1990).

The important point to be made here is that several hormones have been implicated in the onset of the maternal state and the precise mechanism is likely to be complex. It is this physiological change that mediates the temporal change in behavior and, for most of the time, ensures that females do not waste parental investment on non-kin. This allows some species to use cannibalism as a means of obtaining food, while ensuring that they avoid harming their own infants when they are present.

Non-pregnant female gerbils that have reared a previous litter also cannibalize test pups but again change to maternal responses in late pregnancy (Elwood, 1977). The maternal state is temporary, its onset being due to the hormonal state but its maintenance being due to stimuli from the pups. Removing the pups at birth results in a more rapid return to cannibalism than if the litter remains with the female in the gerbil (Elwood, 1981), the hamster (Siegel and Greenwald, 1978) and the California mouse (Gubernick and Alberts, 1989). In some species it is the unexpected loss of a litter that seems to stimulate the parent to be cannibalistic towards unrelated infants *e.g.* ground squirrels (Sherman, 1981).

SELECTIVITY OF INFANTICIDE AND PARENTAL CARE BY MALE RODENTS

A number of hypotheses have been proposed to explain how males might avoid harming their own offspring and, instead, give paternal care (reviewed by Elwood, 1985). These are of two main types. One type, previously referred to as "recognition hypotheses" (Elwood, 1985), proposes that the male uses some aspect of the situation in which the infants are encountered to make an 'on the spot' decision. The second type proposes that males are brought into a 'non-infanticidal state' or even a "parental state" at a time when their own young are likely to be born.

Recognition Hypotheses

Recognition of the pup

It is possible that a male might directly recognize a pup as kin by phenotype matching, probably using odor cues (Hepper 1986; Waldman, *et al.*, 1988). Phenotype matching might be due to a) self matching *i.e.* the pup has some genetically determined feature that is shared by the male, b) matching to other known kin *e.g.* siblings of the male or previous offspring, or c) matching to some feature of the female with which the male mated. In this final case the feature might be genetically determined and produced by the pup or might depend on odors from the female that are deposited on her offspring. Regardless of the precise mechanism it is some feature of, or on, the pup that is recognized.

One study indicates that kinship of unfamiliar pups may be recognized by male mice (Paul, 1986). Males of three inbred strains of mice were presented with four unfamiliar 1-day-old pups that were either their own offspring or had been fathered by another male of the same strain. For each of the strains no male harmed his own young but about one third of males killed infants that were not their own. All tests were conducted in the home cage of the male with no other adults present indicating that it was some aspect of, or on, the pup that was recognized. Many other experiments, however, have been conducted in which males were exposed to unrelated pups and whether infanticide or paternal care occurred was due to the prior treatment of the male, not the relatedness of the pups (*e.g.* vom Saal, 1985; Elwood, 1985, 1986; Parmigiani, 1989; Soroker and Terkel, 1988). Further studies have introduced males to infants in the presence of females and generally have failed to show any evidence of pup-recognition. Huck, *et al.* (1982), for example, reported that more males harmed their own pups if encountered in the nest of a strange female than harmed

strange pups that were in the nest of a familiar female, indicating that pups *per se* are not recognized. The data on direct pup recognition(sensu Waldman, *et. al.*, 1988) by males are thus contradictory.

A recent study investigated whether male mice of two strains were able to discriminate between infants on the basis of kinship and determined if infanticidal or paternal responsiveness occurred (Elwood and Kennedy, 1991). To maximize the possibility of kinship being discriminated, three types of pups were offered, those fathered by the male ('own'), those fathered by a different male of the same strain ('other') and those of a different strain ('alien').

The subjects were albino CS1 and brown CBA males that had been weaned prior to the birth of the next litter and had not been exposed to infants other than their litter mates. Unlike previous experiments in this laboratory (*e.g.* Elwood, 1985), there was no preliminary screening for infanticidal tendencies. At 90–120 days of age each male was isolated for 10 days and then placed with a female of his own strain and the female was checked twice each day for the presence of a copulatory plug. Females were removed when a plug was found and housed in separate cages.

Males of each strain were assigned to one of three groups, depending on the type of test pup to be presented. Males were tested on the day that their own litter was born. A single newborn test pup was placed into the center of the male's home cage and the responses of the male recorded onto a check sheet, in 15 sec intervals, for 7 1/2 minutes, using period sampling. Four parental activities were recorded: sniff/lick pup, nest-build, huddle over pup and retrieve and also one non-parental activity, that being activity directed towards the bars on the cage top. A parental behavior score was produced by summing the scores for sniff/lick, huddle, nest-build and retrieval speed (total number of 15 sec observation periods less the number of periods that elapsed prior to retrieval). Recording was terminated if the pup was attacked and intervention by the experimenter ensured that the pup was removed and humanely killed within 10 seconds, a procedure employed in all our experiments. Males that harmed pups were rated as infanticidal and their scores for other activities discarded from further analysis.

There was no evidence of infanticide being selectively directed at non-kin (Table I). This finding is in marked contrast to the study of Paul (1986) in which males discriminated between pups that the male had fathered and others of the same strain even "without noticeable investigation close to the pup's body". Parmigiani (1989) also investigated the ability of males to discriminate on the basis of kinship but, in that study, males and females were permitted to cohabit after copulation. This results in the

Table I Number and type of pups killed by males of both strains in experiment 1.

	CS1	CBA
No. of "own" pups killed	1/41	6/26
No. of "other" pups killed	1/33	6/24
No. of "alien" pups killed	2/35	2/19
Total killed	4/109	14/69
% killed	3.7	20.3

complete elimination of infanticidal responsiveness (Elwood and Oster-meyer, 1984c; Elwood 1985, 1986) and thus no selectivity in infanticide was noted by Parmigiani (1989) because no male harmed pups. The methods used by Paul (1986) and those of Elwood and Kennedy (1991) are similar but why the data are so different is not clear. In the latter study, male mice of the two strains were not able to selectively direct infanticide on the basis of kinship, even between strains, whereas Paul (1986) found discrimination, even within strains.

Male mice were, however, able to be selective in their paternal responsiveness, apparently on the basis of kinship (Elwood and Kennedy, 1991). There was no difference for CS1 males between pup types in the amount of sniff/lick, huddle or latency to retrieve. However, different amounts of nest-building and bar-directed activities were shown. In these cases, males showed least nest-building and most bar activity in response to 'alien' pups. A significant difference also occurred for the parental behavior score (Figure 1) with the least parental behavior being elicited by 'alien' pups and the most by 'other' pups. The scores for CBA males did not differ between the type of pup for sniff/lick, nest-build, huddle, latency to retrieve or for bar-directed activity. A significant difference occurred, however, for the combined parental behavior score (Figure 1) with most parental behavior being directed to 'own' pups and least to 'alien' pups. Thus in both CS1 and CBA mice there were differences in parental behavior towards the three types of pup.

Location-based recognition

Apparent kin recognition might be achieved if paternal responses are given in locations in which kin might be expected, whereas infanticide or ignoring of pups occurs where kin are unlikely to be found (Elwood, 1985). That is location might influence the male's responses to infants when

PARENTAL SCORE

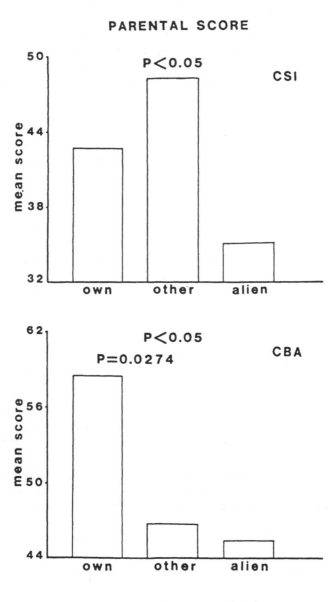

Figure 1 The mean scores for parental behavior shown towards the males' "own" pups, pups sired by another male of the same strain ("other") or sired by a male of a different strain ("alien") are shown for two strains of male mice. The probability values refer to Kruskal-Wallis tests.

location and relatedness tend to correlate. A recent study on *Peromyscus leucopus* suggests that this occurs in the field (Wolff and Cicirello, 1989). Males outside of their normal home range killed infants but those within their home range did not. It has been suggested that house mice (*Mus musculus*) also show different responses to infants located in familiar or unfamiliar locations (Beilharz, 1975). Infants encountered in home cages received paternal care whereas those encountered in novel environments were more likely to be killed (Beilharz, 1975). However, the repeated use of the same animals in an unbalanced experimental design makes the evaluation of these results difficult. Different results were obtained by Parmigiani (1989) using male mice that had copulated and cohabited with their pregnant mates. When these males encountered pups in the home cage or novel area there was no difference in infanticide but observations on paternal responses were not reported. The lack of a difference in infanticide appeared to be due to the onset of a non-infanticidal state that occurs when male mice copulate and cohabit with pregnant females (Elwood and Ostermeyer, 1984c; Soroker and Terkel, 1988) and no male harmed pups in the study of Parmigiani (1989).

The possibility that male mice vary their paternal and infanticidal responses, according to location, was examined in an experiment by El-wood and Kennedy (1991). Males were allowed to copulate and were tested at a time that their own young should be born. They were not, however, allowed to cohabit with the pregnant female and therefore should not have been totally inhibited in infanticide (Elwood, 1985, 1986; Soroker and Terkel, 1988; Kennedy and Elwood, 1988).

CS1 and CBA males were reared, maintained and mated as described previously, females being removed immediately after mating. The day their own litter was born each male was presented with a single unre-lated newborn pup of their own strain. For group 1 the male was removed from its cage, the pup placed in the center of the cage, and the male re-turned. Group 2 males were tested in the cage of a male of the alterna-tive strain, the resident being removed just prior to the test. The males were observed for 10 minutes and categorized as 'paternal' if the male retrieved and/or either huddled over the pup or exhibited nest-building around the pup, 'ignoring' if the male did not harm the pup but failed to reach the criterion to be classed as paternal, or 'infanticidal' if the pup was harmed.

The data indicate that the familiarity of the location in which a pup is encountered does not influence infanticide by male mice (Table II). There were, however, significant differences in paternal responses with more CS1 and CBA males being paternal in their home cage than the cage of

Table II Number of males behaving paternally (P), ignoring (I) or infanticidally (K) when presented with a test pup in their own cage or in the cage of another male.

Location of pup	Response of Male	STRAIN	
		CS1	CBA
Male's Own Cage	P	13	13
	I	6	4
	K	1	2
Total		20	19
Cage of Another Male	P	6	3
	I	12	13
	K	3	5
Total		21	21

a strange male (Fisher exact probability tests, p = 0.021 and p = 0.0006 respectively). It is not clear if this difference in paternal responses is due to an adaptive decision by the male concerning the pup or if it is due to a greater level of exploration of the novel cage with consequent less time for paternal behavior. The data, however, are congruent with the hypothesis that males use location to determine selectivity of paternal responses.

Recognition of previous sexual partners

Another way in which "indirect recognition" (Waldman, *et al.* 1988) might be achieved is by recognizing a third party. Holmes and Sherman (1981) used the term "mediated recognition" to describe the situation in which female ground squirrels behaved more favorably towards unfamiliar maternal half siblings if first encountered in the presence of their common mother (Sherman, 1980). Male mammals might use a similar mechanism to recognize offspring by acting favorably to infants that are associated with a previous sexual partner (Hrdy, 1979). This possibility has been tested for mice in a number of experiments. Male Swiss Webster mice were more likely to kill unrelated infants encountered in the cage of a strange female than they were their own infants with a familiar female (Huck, *et al.* 1982). That it was not the infants per se that were recognized was demonstrated by subsequently presenting males with their own infants, in cages of strange females, or unrelated infants, in cages of familiar females. Pup-killing by males was more common in the former group (13/22 compared to 3/22)

(Huck, *et al.* 1982). McCarthy and vom Saal (1986) investigated the possibility that "wild type" male mice might be more likely to kill pups encountered in the presence of a strange female rather than with a familiar female and found that 71% and 33% respectively killed pups. These males had all previously killed pups in initial screening but a second experiment, which did not employ screening, found no significant difference when only 5% killed pups with a strange female and none killed with a familiar female. A third experiment used screened-infanticidal CF1 males but none of these killed pups, whether encountered with a familiar or strange female (McCarthy and vom Saal, 1986). Several other studies have failed to find any effects of the familiarity of females (vom Saal and Howard, 1982; Brooks and Schwarzkopf, 1983; Parmigiani, 1989).

The experiment described below examined the role of familiarity of the female in determining male responses towards infants and also investigated which cues from the female might be recognized (Elwood and Kennedy, 1991). For example, in the above experiments the males were exposed to odors deposited in the cage as well as those on the female. Furthermore, the males and females were allowed to interact and the behavior of females might have been different towards strange or familiar males (Ostermeyer, 1983). Thus the female might influence the male's responsiveness towards infants without any 'recognition' of familiarity or strangeness by the male (Elwood, 1985).

Male CS1 and CBA mice, were housed singly for 10 days and then mated with a female of their own strain. When a copulatory plug was found the male was placed in a separate cage, the female remaining in the mating cage. The males were tested 20 days after copulation in one of four experimental groups. Groups 1 and 3 were tested in the cage of their recently parturient mate and Groups 2 and 4 in the cage of a recently parturient strange female of the alternative strain. For Groups 1 and 2 the females and their litters were removed just prior to the test whereas in Groups 3 and 4 the litters were removed and the females restrained behind a wire mesh partition at one end of the cage. In all cases a single unrelated pup of the same strain as the male was placed in the center of the cage and observed as in the previous experiment.

The data (Table III) do not support the hypothesis that males alter their infanticidal tendencies depending on the odors of familiar or unfamiliar females (Groups 1 and 2) or of the presence of those females (Groups 3 and 4). Furthermore, because in Groups 1 and 3 the males were returned to their previous home cages, which had not been cleaned since the male had been removed 20 days earlier, there was no support for familiarity of location influencing infanticide.

Table III Number of males responding paternally (P), ignoring (I), or infanticidally (K) when presented with a test pup in the cage of a previous mate or an unfamiliar female. Females were either not present or were present but restrained.

Group	Where pup presented	Female present?	Behavior of male	STRAIN CS1	STRAIN CBA
1	Cage of previous mate	no	P	3	9
			I	23	5
			K	0	1
	Total			26	15
2	Cage of another female	no	P	4	4
			I	11	10
			K	1	1
	Total			16	15
3	Cage of previous mate	yes (restrained)	P	8	8
			I	10	6
			K	0	1
	Total			18	15
4	Cage of another female	yes (restrained)	P	1	7
			I	16	8
			K	0	0
	Total			17	15

The data agree with those of Brooks and Schwarzkopf (1983), Parmigiani (1989) and two experiments of McCarthy and vom Saal (1986). The data do not agree, however, with those reported for one experiment of McCarthy and vom Saal (1986) or of Huck, *et al.* (1982). In those experiments, however, the females were free to interact with the males and it is possible that this physical interaction is required for males to discriminate between females. Alternatively, it might be the females that discriminate between males and this might influence the males' behavior towards pups (Elwood, 1985).

Although there was no evidence of indirect kin recognition with respect to infanticide there was for paternal responsiveness. Males tended to be more paternal towards unrelated pups found with previous mates than with strange females. These data are similar to those of the previous experiment on location-based recognition and might simply be due to males investigating strange cages more than familiar ones. However, males had been removed from their familiar cages 20 days prior to the test and it seems more likely that the discrimination is of cues associated with the females. This discrimination, however, was only significant for CS1 males when the females were present, albeit restrained, but there was a similar trend for CBA males when the females were absent. This seemed to be due to CS1 males showing little paternal care in the empty cage of a familiar female but significantly more when the familiar female was present. There was no such shift in the behavior of CBA males, which tend to be more paternal than CS1 males. As discussed for the infanticidal responses, males might be more discriminating in their paternal responses if they are able to physically interact with familiar or unfamiliar females and this possibility was examined in another experiment (Elwood and Kennedy, 1991).

CS1 and CBA male mice were paired with females of the same strain and when a copulatory plug was found the male was rehoused in a separate cage. Males were tested 20 days after copulation by placing them into a female's cage with one pup of the same strain as the male but unrelated to him. In one condition, the male was returned to the cage of his former mate whereas, in a second condition, he was placed into the cage of a strange female of the alternative strain. All females were newly parturient but their pups were removed just prior to the test. The females were free to interact with the males and test pups. The data enabled each male to be classed as 'paternal' if they huddled over the pup or nest-built or sniff-licked the pup more than 20 times, 'ignore' if it did not harm the pup but failed to reach the criterion for paternal, or 'infanticidal' if the male harmed the pup.

Table IV Number of males responding paternally (P), ignoring (I) or infanticidally (K) when presented with a test pup in the cage of a previous mate or an unfamiliar female. All females were present and unrestrained.

Group	Where pup presented	Female present?	Response of male	STRAIN CS1	CBA
			P	6	10
1	Cage of previous mate	yes (unrestrained)	I	10	6
			K	0	0
	Total			16	16
			P	5	3
2	Cage of another female	yes (unrestrained)	I	11	14
			K	0	0
	Total			16	17

There was no evidence to support the hypothesis that infanticide is mediated by familiarity with the female with which infants are associated (Table IV). In this respect the data are similar to those of other studies (Brooks and Schwarzkopf, 1983; McCarthy and vom Saal, 1986; Parmigiani, 1989). The lack of infanticide found in the experiment by Elwood and Kennedy (1991) was unexpected but cannot be explained by an inhibition due to cohabitation. Possibly the inhibition was because the males were removed and placed in new cages immediately after mating, a procedure not used in some other experiments. The females did not prevent access to the pups and all males sniffed the pup at least once, thus it is doubtful if the females prevented infanticide. This conclusion is supported by the finding that CS1 females are unable to prevent unmated infanticidal males from killing pups (Elwood, *et al.* 1990).

Parental responsiveness of CBA males was influenced by the type of female with which the pup was associated. They were more likely to be paternal and more likely to huddle over the pup when the pup was associated with the former mate than with a strange female. These results, however, might be directly influenced by the behavior of the female rather than by recognition on the part of the male. CBA males were more likely to fight with a strange female than with the previous mate and were more

likely to be chased and bitten by a strange female than by their previous mate. Thus there was more time for paternal responses with the previous mate. CS1 males showed no differences in paternal responses towards the pups in the two situations but there were differences in their interactions with females. CS1 males were more likely to chase-bite their former mate than a strange female and were more likely to receive a lunge from their former mate.

Huck, *et al.* (1982) and McCarthy and vom Saal (1986) suggested that their results demonstrated discrimination of the female by the male, however, the data of Elwood and Kennedy (1991) support the suggestion of Elwood (1985), that the female's responses might influence the male's behavior. Thus, when the female is free to interact with males, it is difficult to ascribe cause and effect concerning the males' responses to pups. When the female was absent or restrained however, the males did show differences in paternal responses depending upon familiarity with the female.

Conclusion on recognition hypotheses

The overall conclusion from these experiments is that they provide no evidence that males use direct or indirect recognition of kin to enable selectivity in infanticidal responses. Infanticide appears, instead, to be controlled by a particular state (Elwood, 1985, 1986; vom Saal, 1985; Soroker and Terkel, 1988). However, paternal responsiveness does appear to be mediated both by direct recognition and indirect kin recognition.

Non-Infanticidal State Hypotheses

The possibility that a non-infanticidal or paternal state might exist was first investigated in the Mongolian gerbil (Elwood, 1975a, 1977), a species that shows considerable paternal care towards its own young (Elwood, 1975b, Elwood and Broom, 1978). The subjects were males that had been housed with females for some weeks but that had not yet produced their first litter. Each female was removed from the cage and a single, unrelated newborn placed in the cage with the male. Males with non-pregnant mates were likely to attempt to cannibalize the test pup whereas those with pregnant mates were more likely to be paternal. Indeed, no male housed with a female in the final 6 days of gestation harmed the test pup. Since a) none of the pups were related to the male, b) all tests were performed in the home cage, and c) all females were removed at the time of testing, none of the three recognition hypotheses can account for this onset of paternal care. A similar onset of paternal care has been reported for the house

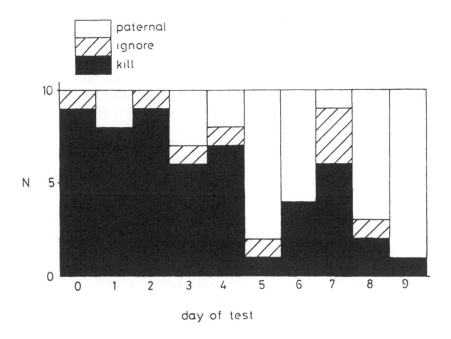

Figure 2 Number of previously infanticidal males that killed or ignored pups or acted paternally when tested at different times after mating. Each male remained with the female and was only tested once (Elwood, 1985).

mouse (Figure 2) (Labov, 1980; vom Saal, 1984, 1985; Elwood, 1985) and the rat (Brown, 1986; Mennella and Moltz, 1988). It thus seems that male rodents are brought into a non-infanticidal or paternal state at a time when their own young are likely to be encountered. Three main hypotheses have been put forward to account for this inhibition of infanticide and onset of paternal care.

Copulation

Male gerbils that have their pregnant mates removed from the cage show a return to pup-cannibalism and thus copulation cannot account for the maintenance of the inhibition of cannibalism (Elwood, 1975a, 1980). Studies on rats have failed to show any effect of copulation on infanticide in some strains (Brown, 1986) but have in others (Mennella and Moltz, 1988), although the methods used in these studies differ. In the mouse there is considerable evidence that copulation has an effect (*e.g.* vom Saal and Howard, 1982; Huck, *et al.* 1982; Elwood, 1985; Soroker and Terkel,

1988). In these studies, copulation has been shown to markedly reduce the proportion of males that are infanticidal, however, this has not been the case in every experiment (Elwood and Ostermeyer, 1984c; McCarthy and vom Saal, 1986; Palanza and Parmigiani, 1991). One problem in these studies has been the separation of effects due to copulation from those possibly due to postcopulatory interactions but this was achieved in a study by vom Saal (1985). In that study male CF1 mice were allowed to a) interact with a non-estrous female, b) mount and intromit with a receptive female, or c) mount, intromit and ejaculate with a female. Only this latter group showed a reduction in infanticide and an onset of paternal care when tested 20 days later, indicating that a change was brought about by ejaculation.

In Elwood (1985), previously infanticidal males were a) allowed to mate, the female being removed on the day of copulation or b) not allowed to mate. When tested at a later date there was a marked difference in paternal responsiveness with those that had copulated showing a greater paternal responsiveness than those that had not (Figure 3, Groups 2 and 3). These data are thus in agreement with those of vom Saal (1985). The not mated group of Elwood (1985) were subsequently mated, the female removed, and the males tested at a later date. Again, copulation appeared to enhance paternal responsiveness but not as much as seen in the other group which had copulated (Figure 3, compare Groups 2 and 3b). It thus appears that differences in experimental procedure between these two groups had influenced their responsiveness even though both groups had copulated. The procedures differed in respect of the period of isolation prior to mating and in the number of infanticidal experiences. Subsequent experiments investigated which of these two factors influenced the effectiveness of copulation in bringing about the onset of paternal care (Elwood, 1986). There were six experimental groups of previously infanticidal males. Groups 1A and 1B were isolated for 7 days prior to the screening for infanticidal tendencies, and then mated, the females being removed when copulatory plugs were located. Groups 2A and 2B were also isolated for 7 days prior to screening but then isolated for a further 14 days prior to mating. Groups 3A and 3B were isolated for 21 days prior to screening. The B groups were offered a pup on the day of copulation and most males killed those pups; the A groups had no such test. The males were tested 18 days after copulation. The data indicated that there was no effect of differing isolation periods on parental behavior or infanticide. Males exposed to pups on the day of copulation, however, were less likely to be paternal and more likely to be infanticidal when tested 18 days later (Figure 4).

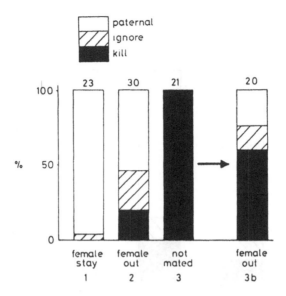

Figure 3 Number of previously infanticidal males that killed or ignored or acted paternally 18 days after mating (Groups 1, 2 and 3b) or at a similar time since initial screening (Group 3). See text for details (Elwood, 1985).

Prior infanticidal experience also appeared to reduce the effectiveness of copulation in wild mice when males were tested by placing them with a female and her litter but not when tested by placing a newborn pup into the male's cage (McCarthy and vom Saal, 1986). It may be concluded from these studies that copulation in mice has an effect in reducing infanticide in males and in initiating paternal care but that the magnitude of this effect varies with the methods used. The magnitude of the effect also varies with the genotype of the mice as CS1 males show an effect but CBA males do not, even when tested with identical methods (Figure 5) (Kennedy and Elwood, 1988). Intrauterine position of the developing embryo also influences the effectiveness of copulation in adulthood in eliminating infanticide (Perrigo and vom Saal, 1989).

What is the role of cohabitation?

It is clear that in many cases copulation has an effect in reducing infanticide and enhancing paternal responsiveness but it is equally clear that cohabitation also has an effect in gerbils (Elwood, 1977, 1980), rats (Brown, 1986; Mennella and Moltz, 1988), and mice (Labov, 1980; Elwood and Ostermeyer, 1984c; Elwood, 1985, 1986; Kennedy and Elwood, 1988; Soroker

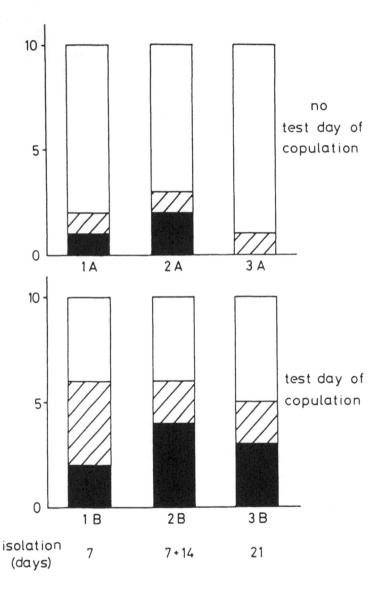

Figure 4 Number of previously infanticidal males that killed (black), ignored (hatched) or acted paternally (white) are shown for an experiment in which isolation period and previous exposure to pups were the variables (Elwood, 1986).

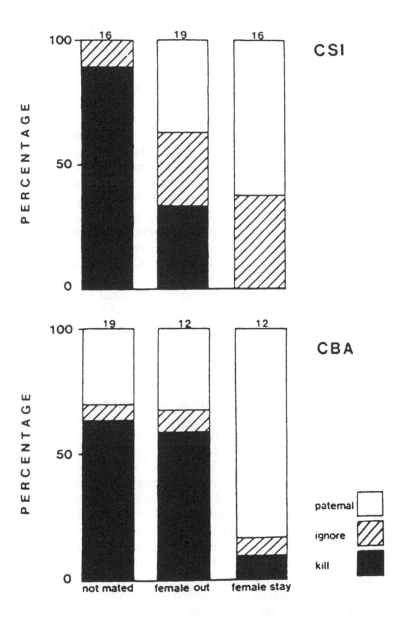

Figure 5 The percentage of previously infanticidal CS1 and CBA males that killed or ignored or acted paternally after different sexual and cohabitation experience (Kennedy and Elwood, 1988).

and Terkel, 1988). These studies have consistently shown inhibition of infanticide and enhanced paternal responsiveness in a greater proportion of males that have copulated and then cohabited with the female (female stay groups) than those that have only copulated (female out groups) (Figures 3 and 5). In another study (Elwood, 1985), infanticidal males were mated and the female either removed or allowed to remain with the male. The males were then tested either day 7, 13 or 18 after mating (Figure 6). There was a significant decline in infanticide for the female stay group with increasing time since copulation and a significant increase in paternal responses. There was no such change for the groups in which the female was removed. Brown (1986) demonstrated that copulation followed by cohabitation with a non-pregnant female was ineffective in inhibiting infanticide in rats but if the female was pregnant inhibition occurred. In mice, cohabitation by sexually naive males with pregnant females caused a slight inhibition of infanticide (Soroker and Terkel, 1988) whereas cohabitation with non-pregnant females did not (McCarthy and vom Saal, 1986; Soroker and Terkel, 1988). Cohabitation has to follow copulation to eliminate infanticide in CS1 males (Elwood, 1985, 1986; Elwood and Kennedy, 1990) and both are essential for elimination of infanticide in CBA males (Figure 5) (Kennedy and Elwood, 1988). The question remains, however, what factor(s) associated with cohabitation inhibit infanticide and influence the onset of the paternal state?

Pheromones

The hypothesis that a pheromone, released by the pregnant female, might cause the inhibition of infanticide in males was first proposed by Elwood (1975a, 1977). Experiments, in which pregnant female gerbils were housed above their mates so that urine could be dripped into the males' cages, however, failed to maintain the inhibition of infanticide in those males (Elwood and Ostermeyer, 1984a). Experiments that investigated the possible role of the midventral gland of pregnant female gerbils also proved negative (Elwood and Ostermeyer, 1984a). Further experiments, in which males were either separated from their pregnant mates by wire mesh, clear plastic or opaque plastic, also found no evidence of a pheromonal effect in the maintenance of the inhibition of infanticide seen in male gerbils during their mates' pregnancies. Gubernick and Alberts (1989), however, using similar techniques to Elwood and Ostermeyer (1984a) found a pheromonal role in the postpartum maintenance of paternal responsiveness in *Peromyscus californicus*. A pheromone has also been implicated in the prepartum inhibition of infanticide in male rats (Mennella and Moltz, 1988). Sexually naive Wistar rats provided with bedding, soiled by

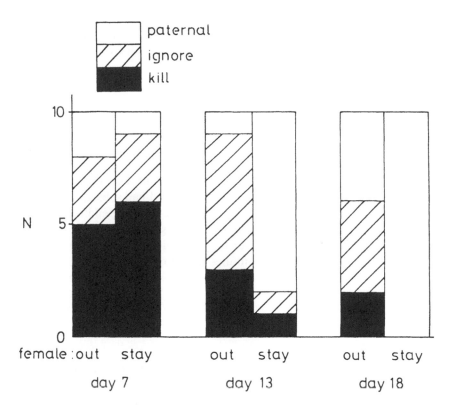

Figure 6 Number of previously infanticidal males that killed or ignored or acted paternally when time since copulation and cohabitation were the experimental variables (Elwood, 1985).

pregnant females, were less likely to be infanticidal (2/10) than were those provided with bedding soiled by non-pregnant females (7/10). Neither the source or the nature of the pheromone has been identified but the authors conclude that it is probably of high molecular weight as there was no evidence of airborne effects in their colony (Mennella and Moltz, 1988).

The possibility of airborne pheromones having an effect on the inhibition of infanticide on onset of paternal care in CS1 mice was tested by placing mated, screened killers in a room devoid of pregnant females (Elwood, 1985). There was, however, no difference in responses to pups when compared to males housed within the main colony room. Pheromonal effects have also been examined in CBA mice (Kennedy, 1989). Screened infanticidal males were mated and then the females removed. In one

Table V Numbers of CBA males that acted parentally, ignored, or responded infanticidally towards test pups after exposure to soiled bedding from either virgin or pregnant females.

	TYPE OF BEDDING	
Response of male	Virgin Female	Pregnant Female
Parental	8	9
Ignoring	1	0
Infanticidal	11	13

group, soiled bedding was provided each day from the cages of females that had copulated at least 12 days previously and had subsequently been housed singly. In a second group, males received soiled bedding from the cages of virgin females. The results (Table V) showed no effect of soiled bedding on the infanticidal or paternal responsiveness of CBA males.

It seems, therefore, that pheromones play no role in the onset of paternal care in gerbils or in mice but they do in rats. Further experimentation on a variety of species is required to determine if the use of pheromones in this context is common.

Social Subordination

The behavior of female gerbils changes during pregnancy and Elwood (1975b) considered the possibility that this might influence the altered responsiveness of the males towards infants. No specific mechanism was proposed, however, until the suggestion that females might subordinate their mates and that subordination might cause the inhibition of infanticide (Elwood and Ostermeyer, 1984a, b, c). Subordinate male mice have been shown to be less infanticidal than dominants (Huck, *et al.* 1982; vom Saal and Howard, 1982) as have subordinate male gerbils (Elwood and Ostermeyer, 1984c). However, if social subordination is the normal mechanism by which infanticide is inhibited a change in responsiveness to pups would be expected when the social status changes.

To test this possibility infanticidal and non-infanticidal CS1 male mice were placed in male:male pairs in cages for 10 min each day for 5 consecutive days, but remained isolated at other times (Elwood, 1986). Social status was assessed by comparing the proportion of dominant and subordinate acts performed by each member of a pair. Eighteen days after the initial pairing the males were again tested for their infanticidal tendencies.

In this final test 15/15 non-infanticidal males that were subordinated by another male remained non-infanticidal (Figure 7) whereas 5/17 that

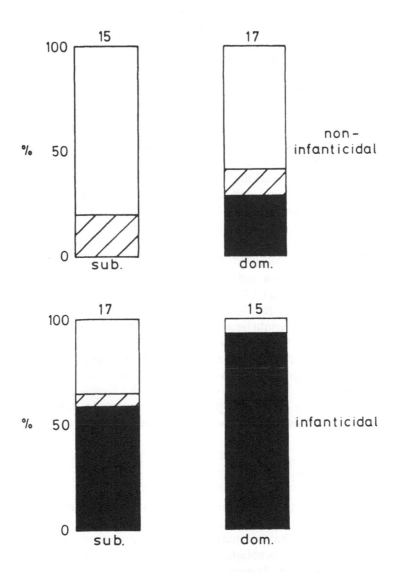

Figure 7 Number of previously non-infanticidal (top two groups) and infanticidal males (bottom two groups) that were infanticidal (black) or ignored (hatched) or paternal (white) after they were subordinated or became dominant in interactions with other males (Elwood, 1986).

became dominant switched to being infanticidal (Fisher exact test p = 0.03). Fourteen out of 15 infanticidal males that became dominant remained infanticidal in the final test compared to only 10/17 that were subordinated (p = 0.026). There were, however, no significant changes in the number that were paternal. Furthermore, the proportion of subordinate acts (*i.e.* degree of subordination) influenced whether previously infanticidal males switched to being non-infanticidal. Those that became non-infanticidal showed a greater degree of subordination than did those that remained infanticidal (p < 0.05 Mann-Whitney).

These data support the subordination hypothesis. However, CS1 males need only to be housed with the female for 24 hrs after copulation for an effect due to cohabitation to be shown (Elwood, 1985). Observations of male:female pairs showed that shortly after copulation the female is aggressive towards the male and that 24 hrs after copulation appears to be dominant to the male (Elwood, 1985). Furthermore, previously infanticidal males with relatively large females are less likely to be infanticidal 13 days after copulation than are those with relatively small females (Elwood, 1986), which again is consistent with the subordination hypothesis.

More recent data have failed to support the subordination hypothesis (Kennedy, 1989). When CS1 and CBA infanticidal males were paired with other males for a single 20 minute encounter there was no change in behavior towards pups that could be ascribed to subordination. Furthermore, 20 min encounters on 3 consecutive days also failed to produce an effect as did three 20 min encounters on a single day. It had been anticipated that encounters with aggressive males on a single day would mimic the effects of a single day of post copulatory cohabitation with a female (Elwood, 1985) but this was not the case. Perhaps in the absence of copulation exposure to dominant animals has to be more prolonged to have an effect.

CONCLUSIONS

There seems to be good agreement that female rodents are brought into a maternal state due to endogenous changes associated with pregnancy and parturition and that this maternal state is subsequently maintained by exogenous stimuli (Rosenblatt, 1990). This results in the elimination of infanticidal tendencies. However, there is less agreement concerning the inhibition of infanticide and onset of paternal care in rodents. This is because various techniques have been employed to test six distinct hypotheses using a number of strains and species. Variation in approaches used by workers imposes certain problems. It can lead to dogmatic statements

concerning the role of particular mechanisms when those statements are based on the outcome of one or two experiments on a single strain of laboratory rodent. The benefit of a varied approach, however, is that it has provided evidence that a number of factors are involved in the regulation of infanticide and paternal care.

The approach in this laboratory has been an attempt to gain benefits from using two different laboratory strains of mice (CS1 and CBA), reared in standard conditions, thus allowing comparison. There is no evidence from these experiments that male mice regulate infanticide by a) recognition of relatedness of infants, b) location-based recognition or c) recognition of cues of a previous sexual partner. There is evidence, however, that the paternal responsiveness of males is influenced by all three mechanisms. The lack of selectivity of infanticide suggests that males are in an infanticidal state or not. The infanticidal state appears to depend on experiences prior to the encounter with the pup and not influenced by factors concerned specifically with the encounter. The term 'paternal state' has been used previously (e.g. Elwood, 1991) but the results presented here suggest that the term 'non-infanticidal state' might be more appropriate. This is because non-infanticidal animals appear to be able to vary their paternal responsiveness dependent upon a variety of cues.

The onset of the 'non-infanticidal state' in mice appears to be influenced by two main factors. It is partially inhibited by copulation in CS1 mice under appropriate conditions (Elwood, 1985) but not under other experimental conditions (Elwood and Ostermeyer, 1984b; Elwood, 1986). Copulation does not influence the non-infanticidal state in CBA mice (Kennedy and Elwood, 1988; Elwood and Kennedy, 1991). Cohabitation after copulation, however, brings about a virtual elimination of infanticide in both strains. Social subordination appears to play a role in CS1 males (Elwood, 1985, 1986) but there is no evidence that pheromones influence infanticide in either CS1 or CBA males.

Events associated with reproduction induce a non-infanticidal state in male mice but this still gives the option to the male of being parental or not. In this way male mice may be certain of not harming their own young in mistaken infanticidal attacks but may avoid giving paternal care to unrelated young. The induction of a temporally based non-infanticidal state is an example of "indirect kinship recognition" (sensu Waldman, et al. 1988) but this means of indirect recognition is usually ignored in reviews of kinship recognition, even those dealing with parent-offspring recognition (e.g. Holmes, 1990). It is clear that temporal changes in responsiveness to pups occur in males and females and that this is a major factor in the selective allocation of parental and infanticidal responses.

REFERENCES

BEILHARZ, R.G. (1975) The aggressive response of male mice (*Mus musculus L.*) to a variety of stimulus animals. *Zeitschrift fur Tierpsychologie*, **39**, 141–149.

BROOKS, R.J. and Schwarzkopf, L. (1983) Factors affecting incidence of infanticide and discrimination of related and unrelated neonates in male *Mus musculus*. *Behavioral and Neural Biology*, **37**, 149–161.

BROWN, R.E. (1986) Social and hormonal factors influencing infanticide and its suppression in adult Long-Evans rats (*Rattus norvegicus*). *Journal of Comparative Psychology*, **100**, 155–161.

ELWOOD, R.W. (1975a) Paternal and maternal behaviour of the Mongolian gerbil, and the development of the young. Unpublished PhD Thesis, University of Reading.

ELWOOD, R.W. (1975b) Paternal and maternal behaviour of the Mongolian gerbil. *Animal Behaviour*, **23**, 766–772.

ELWOOD, R.W. (1977) Changes in the responses of male and female gerbils (*Meriones unguiculatus*) towards test pups during the pregnancy of the female. *Animal Behaviour*, **25**, 46–51.

ELWOOD, R.W. (1980) The development, inhibition and disinhibition of pup-cannibalism in the Mongolian gerbil. *Animal Behaviour*, **28**, 1188–1194.

ELWOOD, R.W. (1981) Postparturitional re-establishment of pup-cannibalism in female gerbils. *Developmental Psychobiology*, **14**, 209–212.

ELWOOD, R.W. (1985) The inhibition of infanticide and the onset of paternal care in male mice, *Mus musculus*. *Journal of Comparative Psychology*, **99**, 457–467.

ELWOOD, R.W. (1986) What makes male mice paternal? *Behavioral and Neural Biology*, **46**, 54–63.

ELWOOD, R.W. (1991) Parental states as mechanisms for kinship recognition and deception about relatedness. In P. Hepper (ed). *Kinship Recognition in Vertebrates*, Cambridge University Press, pp. 289–307.

ELWOOD, R.W. (1992) Pup-cannibalism in rodents. In M. Elgar and B. Crespi (eds). *Cannibalism*, Oxford University Press, pp. 299–322.

ELWOOD, R.W. and Broom, D.M. (1978) The influence of litter size and parental behavior on the development of Mongolian gerbil pups. *Animal Behaviour*, **26**, 438–454.

ELWOOD, R.W. and Kennedy, H.F. (1991) Selectivity in paternal and infanticidal responses by male mice: effects of relatedness, location and previous sexual partners. *Behavioral and Neural Biology.* **56**, 129–147.

ELWOOD, R.W. and Ostermeyer, M.C. (1984a) Infanticide by male and female Mongolian gerbils: ontogeny, causation and function. In: G. Hausfater and S.B. Hrdy (eds). *Infanticide: Comparative and Evolutionary Perspectives*, New York, Aldine, pp.367–386.

ELWOOD, R.W. and Ostermeyer, M.C. (1984b) The effects of food deprivation, aggression and isolation on infanticide in the male Mongolian gerbil. *Aggressive Behavior*, **10**, 293–301.

ELWOOD, R.W. and Ostermeyer, M.C. (1984c) Does copulation inhibit infanticide in rodents? *Animal Behaviour*, **32**, 293–305.

ELWOOD, R.W., Nesbitt, A.A. and Kennedy, H.F. (1990) Maternal aggression and the risk of infanticide in mice. *Animal Behaviour*, **40**, 1080–1086.

GUBERNICK, D.J. and Alberts, J.R. (1989) Postpartum maintenance of paternal behaviour of the biparental California mouse *Peromyscus californicus*. *Animal Behaviour*, **37**, 656–664.

HAMILTON, W.D. (1964) The genetical evolution of social behavior I. *J. Theoretical Biology*, 7, 1–16.

HEPPER, P.G. (1986) Kin recognition: functions and mechanisms a review. *Biology Reviews*, 61, 63–93.

HOLMES, W.G. and Sherman, P.W. (1983) Kin recognition in animals. *American Science*, 71, 46–55.

HOLMES, W.G. (1990) Parent-offspring recognition in mammals: a proximate and ultimate perspective. In: N.A. Krasnegor and R.S. Bridges (eds). *Mammalian Parenting*, New York, Oxford University Press, pp. 441–460.

HOOGLAND, J.L., Tamarin, R.H. and Levy, C.K. (1989) Communal nursing in prairie dogs. *Behavioural Ecology and Sociobiology*, 24, 91–95.

HRDY, S.B. (1979) Infanticide among animals: a review, classification and examination of the implication for the reproductive strategies of females. *Ethology and Sociobiology*, 1, 13–40.

HUCK, U.W., Soltis, R.L. and Coopersmith, C.B. (1982) Infanticide in male laboratory mice: effects of social status, prior sexual experience, and basis for discrimination between related and unrelated young. *Animal Behaviour*, 30, 1158–1165.

INSEL, T.R. (1990) Oxytocin and maternal behaviour. In: N.A. Krasnegor and R.S. Bridges (eds). *Mammalian Parenting*, New York, Oxford University Press, pp. 260–280.

JAKUBOWSKI, M. and Terkel, J. (1982) Infanticide and caretaking in non-lactating *Mus musculus*: influence of genotype, family group and sex. *Animal Behaviour*, 30, 1029–1035.

JAKUBOWSKI, M. and Terkel, J. (1985) Transition from pup killing to parental behavior in male and virgin female albino rats. *Physiology and Behavior*, 34, 683–686.

KENNEDY, H.F. (1989) Father-infant recognition in mice. Unpublished PhD Thesis, The Queen's University of Belfast.

KENNEDY, H.F. and Elwood, R.W. (1988) Strain differences in the inhibition of infanticide in male mice (*Mus musculus*). *Behavior and Neural Biology*, 50, 349–353.

KONIG, B. and Markl, H. (1987) Maternal care in house mice. I. The weaning strategy as a means of parental manipulation of offspring quality. *Behavioral Ecology and Sociobiology*, 20, 1–9.

LABOV, J.B. (1980) Factors influencing infanticidal behavior in wild male house mice (*Mus musculus*). *Behavioral Ecology and Sociobiology*, 6, 297–303.

MCCARTHY, M.M., Bare, J.E. and vom Saal, F. (1986) Infanticide and parental behavior in wild female house mice: effects of ovariectomy, adrenalectomy and administration of oxytocin and prostaglandin F2. *Physiology and Behavior*, 36, 17–23.

MCCARTHY, M. and vom Saal, F. (1986) Inhibition of infanticide after mating by wild male house mice. *Physiology and Behavior*, 36, 203–209.

MENNELLA, J.A. and Moltz, H. (1988) Infanticide in rats: male strategy and female counter strategy. *Physiology and Behavior*, 42, 19–28.

NOIROT, E. and Goyens, J. (1971) Changes in maternal behaviour during gestation in the mouse. *Hormones Behavior*, 2, 207–215.

O'CONNOR, R.J. (1978) Brood reduction in birds: selection for fratricide, infanticide and suicide? *Animal Behaviour*, 26, 79–96.

OSTERMEYER, M.C. (1983) Maternal aggression. In R.W. Elwood (ed). *Parental Behaviour of Rodents*, Chichester, John Wiley and Sons, pp. 151–179.

PALANZA, P. and Parmigiani, S. (1991) Inhibition of infanticide in male Swiss mice: behavioral polymorphism in response to multiple mediating factors. *Physiology and Behavior*, **49**, 797–802.

PARMIGIANI, S. (1989) Inhibition of infanticide in male house mice (*Mus domesticus*): is kin recognition involved? *Ethology, Ecology and Evolution*, **1**, 93–98.

PAUL, L. (1986) Infanticide and maternal aggression: synchrony of male and female reproductive strategies in mice. *Aggressive Behavior*, **12**, 1–11.

PERRIGO, G. (1987) Breeding and feeding strategies in deer mice and house mice when females are challenged to work for their food. *Animal Behaviour*, **35**, 1298–1316.

PERRIGO, G. and vom Saal, F.S. (1989) Mating-induced regulation of infanticide in male mice: fetal programming of a unique stimulus-response. In: R.J. Blanchard, P.F. Brain, D.C. Blanchard and S. Parmigiani (eds). *Ethoexperimental Analysis of Behavior*, Dordrecht, Kluwer, pp. 320–336.

RICHARDS, M.P.M. (1966) Maternal behaviour in the golden hamster: responsiveness to young in virgin, pregnant, and lactating females. *Animal Behaviour*, **14**, 310–313.

ROSENBLATT, J.S. (1967) Nonhormonal basis of maternal behaviour in the rat. *Science*, **156**, 1512–1514.

ROSENBLATT, J.S. (1990) Landmarks in the physiological study of maternal behavior with special reference to the rat. In: N.A. Krasnegor and R.S. Bridges (eds). *Mammalian Parenting*, New York, Oxford University Press.

ROSENBLATT, J.S. and Siegel, H.I. (1981) Factors governing the onset and maintenance of maternal behavior among nonprimate mammals. The role of hormonal and nonhormonal factors. In: D.J. Bubernick and P.H. Klopfer (eds). *Parental Care in Mammals*, New York, Plenum Press, pp. 13–76.

ROSENBLATT, J.S. and Siegel, H.I. (1983) Physiological and behavioral changes during pregnancy and parturition underlying the onset of maternal behaviour in rodents. In: R.W. Elwood (ed). *Parental Behaviour of Rodents*, Chichester, John Wiley and Sons, pp. 23–66.

SHERMAN, P. (1980). The limits of ground squirrel nepotism. In: G.W. Barlow and J. Silverburg (eds). *Sociobiology: Beyond Nature-nurture?*, Boulder, Colorado, Westview Press, pp. 505–544.

SHERMAN, P.W. (1981) Reproductive competition and infanticide in Belding's ground squirrels and other animals. In: R.D. Alexander and D.W. Tinkle (eds). *Natural Selection and Social Behavior*, New York, Chiron Press, pp. 311–331.

SIEGEL, H.I. and Greenwald, G.S. (1978) Effects of mother-litter separation on later maternal responsiveness in the hamster. *Physiology and Behavior*, **21**, 147–149.

SOROKER, V. and Terkel, J. (1988) Changes in incidence of infanticidal and parental responses during the reproductive cycle of male and female wild mice *Mus musculus*. *Animal Behaviour*, **36**, 1275–1281.

TRIVERS, R.L. (1974) Parent-offspring conflict. *American Zoologist*, **14**, 249–264.

VOM SAAL, F.S. (1984) Proximate and ultimate causes of infanticide and parental behavior in male house mice. In: G. Hausfater and S.B. Hrdy (eds). *Infanticide: Comparative and Evolutionary Perspectives*, New York, Aldine, pp. 401–424.

VOM SAAL, F.S. (1985) Time-contingent change in infanticide and parental behavior induced by ejaculation in male mice. *Physiology and Behavior*, **34**, 7–15.

VOM SAAL, F.S. and Howard, L.S. (1982) The regulation of infanticide and parental behaviour: implications for reproductive success in male mice. *Science*, **215**, 1270–1272.

WALDMAN, B., Frumhoff, P.C. and Sherman, P.W. (1988) Problems of kin recognition. *Trends in Ecology and Evolution*, **3**, 8–13.

WOLFF, J.O. and Cicirello, D.M. (1989) Field evidence for sexual selection and resource competition infanticide in white-footed mice. *Animal Behaviour*, **38**, 637–642.

CHAPTER 18

BIPARENTAL CARE AND MALE-FEMALE RELATIONS IN MAMMALS

DAVID J. GUBERNICK

Psychology Department, University of Wisconsin, Madison, Wisconsin, U.S.A., 53706

INTRODUCTION

The evolution and proximate causation of paternal behavior in mammals poses an interesting theoretical problem. Males normally maximize reproductive success by mating with multiple females, whereas female reproductive success is ordinarily limited by time and energy constraints (Williams, 1966; Trivers, 1972). Females are likely to benefit from increased male parental assistance (Wittenberger and Tilson, 1980), but any male investment in one female's young necessarily reduces the male's chances of inseminating other females, which would appear to create an insurmountable evolutionary barrier to male parental care (Kurland and Gaulin, 1984).

Because mammalian infants are completely dependent on the mother for nutritional needs, males are emancipated from care of young (Orians, 1969) compared with other animals, such as birds, in which male care is common (Lack, 1968). Yet paternal behavior does occur in mammals, primarily among rodents, carnivores and primates (Kleiman and Malcolm, 1981). Biparental care in mammals is generally, but not exclusively, associated with monogamy (Kleiman, 1977; Wittenberger and Tilson, 1980; Kleiman and Malcolm, 1981). In order to understand biparental care we need to view it as part of the social, ecological and sexual context from which it emerges.

In this chapter I will focus initially upon the evolution of monogamy in mammals and discuss three separate but interrelated topics: mating

427

exclusivity, pair-bonding and male parental care. Mating exclusivity and pair-bonding are often considered synonymous (Wickler and Seibt, 1982), and both are assumed to be associated with male care of young.

Most evolutionary analyses of monogamy have emphasized either mate fidelity or biparental care. In order to understand mate fidelity or biparental care, however, it is essential to examine the nature of the relationship between the male and female. Thus I will attempt to bring the pair bond to center stage as a logical starting point for investigation of monogamy and biparental care. To do this, I first consider briefly some conceptual issues that have generated confusion and that have made it difficult to advance the study of monogamy. Next I deal with pair bonds in more detail by providing an evolutionary rationale for pair bonds and discussing their proximate causation. Finally, I provide an example of how pair bonds between parents affect biparental behavior in a monogamous rodent. This chapter is not a review of pair bonds, but rather provides a conceptual framework for analysis of pair bonds and their importance for understanding the ultimate and proximate causation of mate fidelity and biparental care. Although my focus is on mammals, the ideas apply more broadly to other taxa as well.

CONCEPTUAL CONFUSION

Despite the renewed interest in the evolution of monogamy (Gowaty and Mock, 1985; Barlow, 1988; Mock and Fujioka, 1990), the topic has been plagued by confusion and ambiguity in terminology (Wickler and Seibt, 1983; Dewsbury, 1988). This has resulted in erroneous interpretations and conclusions concerning observations of apparent monogamy in nature and the selective forces favoring the evolution of monogamy. Part of the difficulty with use of the concept monogamy stems from viewing monogamy as a unitary phenomenon (Mock, 1985), which obscures recognition and appreciation of the variation in monogamous mating systems (Kleiman, 1981; Mock, 1985; Mock and Fujioka, 1990) and the diversity of selection pressures favoring the evolution of various forms of monogamy in different species. Monogamy, as discussed below, is multidimensional. Thus the ambiguity and confusion reflects the failure to recognize the underlying variation in monogamous mating systems (Gowaty, 1983). Monogamy is commonly defined as a prolonged association and essentially exclusive mating relationship between one male and one female (Wittenberger and Tilson, 1980). Paternal care has often been linked with monogamy and with the necessity of biparental care for rearing young

(Brown, 1975; Wilson, 1975), although monogamy in mammals may not imply anything about the extent of paternal investment (Kleiman, 1977). Thus three different concepts of monogamy are often confounded: mating exclusivity (monogamy as a mating pattern), association or pair-bond between a male and female (monogamy as a social relationship) and biparental care (monogamy as a rearing pattern) (Wickler and Seibt, 1983; Dewsbury, 1988).

Mating Exclusivity

Mating exclusivity refers typically to a male and female copulating only with each other (Selander, 1972; Brown, 1975; Kleiman, 1977; Thornhill and Alcock, 1983), although exclusivity sometimes is used to refer to exclusive mating by only one sex (Selander, 1972). Either sex may obtain occasional extrapair copulations (*e.g.* Eastern bluebirds, *Sialia sialis*, Gowaty and Karlin, 1984) and still be considered essentially monogamous (Kleiman, 1977; Wittenberger and Tilson, 1980).

In most cases, mating exclusivity is often inferred from indirect evidence, such as male-female association patterns, absence of unrelated adult conspecifics in the pair's home range or territory, preference for the mate, and breeding by only one adult pair in the family group (Kleiman, 1977). Such indirect evidence alone is not adequate for determining reproductive behavior because it confounds a social relationship with a mating relationship (Kleiman, 1977).

Extra-pair copulations have been documented in many species of apparently monogamous birds (Gowaty and Karlin, 1984; McKinney, Cheng and Bruggers, 1984; Gyllensten *et al.,* 1990), indicating a wide variation in the extent of mating exclusivity. Extra-pair copulations may also be widespread in monogamous mammals, for example, aardwolves, *Proteles cristatus* (Richardson, 1987), Mongolian gerbils, *Meriones unguiculatus* (Agren, Zhou and Zhong, 1989), and old field mice, *Peromyscus polionotus* (Foltz, 1981).

Mating exclusivity is often difficult to determine in the field because it requires continuous monitoring of pairs. An individual may not exhibit "behavioral" mate fidelity (i.e. she or he copulates with someone else), yet exhibit "genetical" mate fidelity if no fertilization occurs. Because extrapair copulations may not result in successful fertilization (Gyllensten, Jakobsson and Temrin, 1990) the use of DNA fingerprinting or genetic markers may overestimate the extent of behavioral mate fidelity. Radioactive labels, used in conjunction with genetic evidence, offer a promising approach for assessing mating exclusivity. A radioisotope injected into a

male labels his semen, and the radioactivity can be detected in a female for varying lengths of time after copulation (Scott and Tan, 1985). Different isotopes can be used to label different males and each separate label can be detected in the same female if she has mated with more than one male (Scott and Tan, 1985). This has yet to be done for any "monogamous" mammal.

PAIR BOND

A pair bond generally refers to a prolonged and preferential or exclusive association between a particular male and female (Wilson, 1975; Wittenberger and Tilson, 1980) and also to an emotional attachment (Wickler, 1976). Different researchers have been inconsistent in their use of pair bond, sometimes meaning association patterns and at other times meaning an emotional attachment. Association patterns of most species are more easy to observe in the field than copulation and paternal behavior. As a result, the presence of a pair bond (association) between a male and female is often used to infer mating exclusivity and the likelihood of paternal care.

However, association patterns exhibited by supposedly monogamous species show extensive variation, ranging from a pair that shares the same home range but rarely is seen together (*e.g.* rufus elephant shrew, *Elephantulus rufescens*, Rathbun, 1979; extreme facultative monogamy, Kleiman 1977, 1981), to a pair usually seen together in a temporary nuclear family in which the pair remains together during the breeding season when rearing the young (*e.g.* aardwolves, *Proteles cristatus*, Richardson, 1987) to a pair seen in permanent nuclear (*e.g.* the California mouse, *Peromyscus californicus*, Ribble, 1990) or extended families (*e.g.* silverback jackal, *Canis mesomelas*, Moehlman, 1986; extreme obligate monogamy, Kleiman, 1977, 1981).

In contrast to association patterns, the pair bond as an emotional attachment related to monogamy has received little systematic investigation. In fact, Kleiman (1977) argued that pair bond as an emotional attachment has little usefulness for understanding monogamy. I will present a very different view later.

Biparental Care

Male care of young can be divided into (1) direct care, where the male directs caregiving behavior (such as holding, grooming, carrying, protection) toward particular infants and (2) indirect care, where the male performs behaviors that could benefit infants (such as defending a territory or chas-

ing away predators) but are not directed at any particular infant (Kleiman and Malcolm, 1981). Although indirect caregiving may sometimes be important, it is in practice difficult to determine when male behavior such as territory defense is designed to benefit infants and when such potential benefits are simply the incidental byproduct of behavior performed for other reasons, *e.g.* to exclude male rivals. Because of this difficulty male care refers here to direct care of young.

Although male care of the young is more common in monogamous mammals than in polygynous mammals, biparental care is not invariably associated with monogamy. Males and females may exhibit mating exclusivity without frequent pair association or male care (elephant shrews, *E. rufescens*, Rathbun, 1979). Pairs may mate exclusively and associate with each other in the absence of direct male care (klipspringers, *Oreotragus oreotragus*, Dunbar, 1984; Alaskan hoary marmots, *Marmota caligata*, Holmes, 1984; gibbons, *Hylobates* spp., Leighton, 1987). On the other hand, some species show extensive male care and association in the absence of mating exclusivity (e.g. saddle-back tamarins, *Saguinus fuscicollis*, Terborgh and Goldizen, 1985).

The presence of male care of infants also varies in polygynous mammals. Whereas some polygynous species exhibit no paternal care (*e.g.* hamadryas baboons, *Papio hamadryas*, Kummer, 1968; elephant seals, *Mirounga angustirostris*, LeBoeuf, 1974), other species may display extensive care of infants (*e.g.* Barbary macaques, *Macaca sylvanus*, Taub, 1984). In species where males care for young, males should invest in their own infants rather than those of another male (Trivers, 1972). This fundamental principle leads to the prediction that the degree of male care should reflect paternity probability (Trivers, 1972). Paternity probability should be high in monogamous species in so far as females typically mate with one male. Thus male care of young has been used to infer mating exclusivity, and hence monogamy.

However, increased probability of paternity is not invariably associated with male care of infants. Paternity certainty is high in monogamous gibbons, *Hylobates* spp., and de Brazza monkeys, *Cercopithecus neglectus*, and in one-male groups of hamadryas baboons, *P. hamadryas*, and gorillas, *Gorilla gorilla*, but male care is low, contrary to prediction (see Snowdon and Suomi, 1982, and Smuts and Gubernick, 1992, for reviews). In addition, in some multi-male primate groups where paternity probability is low (*e.g.* savannah baboons, *Papio cynocephalus*), males care for young contrary to the paternity certainty hypothesis (Smuts, 1985). [Werren, Gross and Shine (1980) provide a theoretical model for the inadequacy of paternity certainty as an explanation for male care.] Male care of young may occur

because it increases the male's chances of mating with the infant's mother in the future (*e.g.* savannah baboons, Smuts, 1985) and not because it is essential for infant survival. Thus, male care can at times be considered a form of mating effort rather than parental effort (Smuts and Gubernick, 1992).

EVOLUTION OF MONOGAMY: WHAT ARE WE TRYING TO EXPLAIN?

Numerous hypotheses have been given for the evolution of monogamy in mammals and other taxa (see Table I). There is no general consensus about the ultimate causes of monogamy, in part because there has been confusion over what it is that we are trying to explain (see above). I have indicated in Table I, based upon each author's statements, whether each hypothesis refers to mating exclusivity, pair bonding (association), or paternal care. It is clear from Table I that the various hypotheses offered to explain the evolution of monogamy do not necessarily deal with the same dimension of monogamy or may apply to more than one feature.

Of the 17 hypotheses listed, 14 apply to mating exclusivity, 8 apply to pair bonding, and 6 to paternal care. Ten of these hypotheses apply to two or more dimensions of monogamy. Only the hypothesis that male care is essential for raising young applies to all 3 features of monogamy. Site attachment and female choice may also apply to each dimension.

Additional confusion may occur because each of the three dimensions has been used as an explanation for the evolution of the other. For example, mating exclusivity is hypothesized to select for pair bonding and male care (Alexander and Noonan, 1979; Rasmussen, 1981). Likewise, a pair bond may select for mating exclusivity and male care (Smuts, 1985). And the need for male care may select for mating exclusivity (Wilson, 1975) and pair bonding (Ember and Ember, 1979). Thus each dimension can be considered as the cause or consequence of the others and it is relatively arbitrary which dimension is chosen as the consequence (outcome) to be explained. In the absence of other information, there is no *a priori* reason to focus on one dimension over the other as the primary selective force initiating the evolution of a monogamous system. The challenge is to understand the interrelationships between mating exclusivity, pair bonds and male care in each species (see Gowaty, 1985 for a related argument about birds).

My intent here is not to further evaluate these or additional hypotheses for the evolution of monogamy (see Wittenberger and Tilson, 1980;

Table I Hypotheses for the evolution of monogamy

	MATING EXCLUSIVITY [A]	PAIR BONDING (ASSOCIATION) [B]	MALE CARE [C]
1. Monopolization of Single Female	+	?	
2. Polygyny Threshold Not Reached	+		
3. Male Care Essential	+	+	+
4. Female Aggression	+	+	
5. Male Less Successful with Two Females	+		
6. Resource Distribution	+	+	
7. Site Attachment	+	?	+
8. Predictable Food		+	
9. Earlier Breeding	+	+	
10. Equal Sex Ratio	+		
11. Predator Vigilance		+	
12. Protection From Infanticide	+	+	
13. Paternity Certainty		+	+
14. Female Choice	+	?	+
15. Low Maternal Investment at Birth	+		+
16. Male Opportunity to Invest	+		+
17. Kin Selection	+		

References: 1A: Wilson, 1975; Emlen and Oring, 1977; Wittenberger, 1979; Wrangham, 1980; Slobodchikoff, 1984. 1B: Wittenberger and Tilson, 1980. 2A: Orians, 1966; Kleiman, 1977; Wittenberger and Tilson, 1980. 3A: Wilson, 1975; Kleiman, 1977. 3B: Ember and Ember, 1979; Wittenberger and Tilson, 1980. 3C: Kleiman, 1977; Snowdon and Suomi, 1982. 4A: Wittenberger and Tilson, 1980. 4B: Wittenberger and Tilson, 1980. 5A: Trivers, 1972. 6A: Emlen and Oring, 1977; Wrangham, 1980. Slobodchikoff, 1984. 6B: Emlen and Oring, 1977. 7A: Wickler and Seibt, 1981. 7C: Baylis, 1981. 8A: Weinrich, 1977. 9A: Wilson, 1975. 9B: Wilson, 1975. 10A: Wittenberger and Tilson, 1980. 11A: Dunbar, 1984. 11B: Dunbar, 1984; van Schaik and Dunbar, 1990. 12A: van Schaik and Dunbar, 1990. 12B: van Schaik and Dunbar, 1990. 13B: Rasmussen, 1981. 13C: Trivers, 1972; Kleiman, 1977. 14A: Wittenberger and Tilson, 1980. 14B: Smuts and Gubernick, 1992. 14C: Trivers, 1972; Smuts and Gubernick, 1992. 15A: Zeveloff and Boyce, 1980. 15C: Zeveloff and Boyce, 1980. 16A: Zeveloff and Boyce, 1980; Knowlton, 1982. 16C: Knowlton, 1982; Elwood, 1983. 17A: Peck and Feldman, 1988.

Murray, 1984) but to illustrate that the various hypotheses deal with different dimensions of monogamy. I assume that mating exclusivity, pair-bonding and paternal care can, in principle, evolve independently (Murray, 1985; Dewsbury, 1988), under different selection pressures, but they can also co-evolve such that one can help drive or constrain the other.

We have looked briefly at the three essential features of monogamy: mating exclusivity, pair bonding and paternal care. It is important to keep these features conceptually separate to reduce confusion (in terminology and in hypothesis generation and testing) but also because these features can co-vary in different ways in different species. The primary goal is to relate these components to the selection pressures responsible for their evolution and to determine how these features affect one another at both the ultimate and proximate levels. For the remainder of the chapter, the term monogamy will refer, at the minimum, to a significant degree of mating exclusivity without specifying the nature of pair bonds or paternal care.

EVOLUTIONARY ANALYSIS OF MONOGAMY

Most analyses have focused upon ecological factors (*e.g.* spatio-temporal distribution of resources; Emlen and Oring, 1977) and different reproductive strategies of males and females (Trivers, 1972) for understanding the evolution of monogamy (see Barlow, 1988).

There is a basic conflict of interest between the reproductive strategies of males and females. Males can enhance their fitness by mating with multiple females and not investing in young, whereas females are less likely to increase their fitness by mating with multiple males (Trivers, 1972; Kurland and Gaulin, 1984), but would often benefit from male parental assistance. The ways in which these reproductive conflicts of interest between males and females are resolved depends in part upon ecological conditions. Certain ecological conditions favor certain mating systems over others (Emlen and Oring, 1977; Wrangham, 1980). For example, monogamy may result when females are widely dispersed since males can not economically defend more than one female (Emlen and Oring, 1977; Holmes, 1984).

The focus on how each sex maximizes its reproductive success and how reproductive strategies of males and females relate to resources is useful, but it ignores a fundamental feature of monogamy: the nature of the relationship between a male and female. Most non-monogamous mammalian mating systems involve ephemeral bonds between males and

females. For example, in elephants, *Loxodonta africana*, fully adult males and females come together only to mate (Moss, 1988). In most non-monogamous mammals, mating partners of at least one sex (polygyny) and sometimes both sexes (promiscuity) are typically shared with others. Prolonged pair-wise associations between males and females are uncommon in non-monogamous mammals (see Smuts, 1985 for an exception).

Monogamy is of special interest because it often implies more of an emphasis on mutualistic or cooperative relations between males and females than is the case in other mating systems. Thus I argue that to understand monogamy we must begin to explore the nature of male-female relationships and why they take the forms that they do (*e.g.* pair bond). The logical starting point for examining monogamy in a given species is the relationship between the male and female, or the pair bond.

PAIR BOND: THE CORE OF MONOGAMY

Although Kleiman (1981) called attention to some of the dramatic differences in pair bonds (associations) in monogamous (exclusive mating) mammals, the nature of these relationships has remained largely unexplored.

By a pair bond I mean a male-female relationship that is characterized by preferential treatment and an emotional tie that differentiates that relationship from others. Pair bonds may be reflected in partner preferences, extent of association or proximity, coordination or synchrony in behavior, provisioning, response to loss of the partner, and the effects of presence or absence of the partner on aspects of the other's behavior, such as parental care and mating exclusivity. Both within and between species, pair bonds may be characterized by varying degrees of reciprocity, cooperation, mutual dependence, and intimacy. Furthermore, a pair bond, as used here, is distinguished from other types of bonds by its functional consequences. A pair bond necessarily involves reproduction to distinguish it from, say, mother-son or brother-sister bonds, or male-female friendships, which in some species can involve each of the elements mentioned above.

Kleiman (1977) has claimed that consideration of a pair bond as an emotional attachment is of limited usefulness for understanding monogamy because "...mating exclusivity has genetic consequences, whereas an emotional bond (or lack of bond) does not" (p.40). Emotional attachments, however, can be considered as powerful proximate mechanisms that facilitate the formation and continued maintenance of a pair bond. Pair bonds have functional consequences in that they promote the reproductive

success of each partner (see below). Therefore, emotional attachments can have genetic consequences through their link with the pair bond. I argue that the emotional component is the very essence of a pair bond and that to understand monogamy we need to study the emotional ties between males and females. The evidence is clearest for humans. The nature of the emotional relationship between husband and wife affects paternal and maternal behavior (Belsky, 1984; Goldberg and Easterbrooks, 1984). Affectionate and communicative marital relations are associated with sensitive mothering (Engfer, 1988) and mother's feeding competence, and marital harmony is positively related to approval and physical affection shown by both parents towards their children (Easterbrooks and Emde, 1988).

Of the three dimensions of monogamy, mating exclusivity, pair bonds and male care, pair bonds are least emphasized in research on monogamy and are less well understood. We need to know more about both the evolution of pair bonds and their proximate causes, and we also need to know how pair bonds affect mating exclusivity and paternal behavior. I consider each of these in turn.

PAIR BOND: ULTIMATE CONSIDERATIONS

Various hypotheses have been offered for the evolution of pair bonds (see Table I). Of particular concern to the present edited volume is the hypothesis that infanticide by conspecifics, especially male conspecifics, has selected for pair bonds between males and females to protect infants against the threat of infanticide from outside the pair (*e.g.* van Schaik and Dunbar, 1990). In addition, males must reduce their infanticidal tendencies to prevent harm to their own offspring and hence pair bonds also may have been favored to reduce the infanticidal tendencies of the paired male (there should be proximate mechanisms associated with the formation of pair bonds that inhibit male infanticide; see below. See chapters in this volume for proximate mechanisms inhibiting male infanticide in non-pairbonded species).

Whatever the ultimate reasons for pair bonds, animals could not evolve stable, long-term, reciprocal, mutualistic, intimate relationships, including pair bonds, with non-kin without also evolving mechanisms to ensure that, on average, each member of the cooperating unit received reproductive benefits greater than s/he would receive if acting alone or in cooperating with others (Trivers, 1971; Wrangham, 1982; Smuts, 1988). Such partnerships entail both competition and cooperation. Because genetic interests of males and females are not identical, conflicts of interest invariably

occur and threaten the survival of the relationship. Yet cooperative relationships may be the most effective way for individuals to enhance reproductive success (Axelrod and Hamilton, 1981). Thus, competition within partner pairs should be inhibited relative to unallied individuals because of benefits each member derives as a result of their joint action (Wrangham, 1982). We might expect more intimate relations between a male and female where there is selection for cooperation, though not necessarily male care of young.

Females are susceptible to desertion because males can enhance their reproductive success by fertilizing multiple females (Trivers, 1972). If males also invest in young (whatever the selective forces) then males are susceptible to cuckoldry (Trivers, 1972). Therefore females should be less concerned with their mate's fidelity (exclusivity) and more concerned with male desertion, unless mate infidelity indicates that a male will leave or take away resources important to the female's reproductive success. Males should be more concerned with fidelity (female exclusivity) and less so with desertion by the female. In species with substantial male care, males should employ a mixed strategy - not only should they assist a single female to raise young but also to mate with other females, should the opportunity arise, whom they will not aid in rearing young (Trivers, 1972).

Natural selection should favor adaptations in each sex to reduce such susceptibilities. Females are expected to be more sensitive to threats of desertion and males to threats of infidelity. Males should therefore be interested in forming a bond that facilitates female mating exclusivity and females should be interested in a bond that facilitates male commitment. The pair bond may be viewed as the outcome of these two different, but partially compatible goals. It represents an adaptive compromise to conflicts of reproductive interest.

Pair bonds do not eliminate these conflicts of interests between males and females, but constrain them. Such reproductive conflicts are evident from observations of extra-pair copulations in otherwise monogamous animals (McKinney et al., 1984). The reason for male extra-pair copulations is obvious, whereas the functional consequences for female extra-pair copulation are less well understood (Gowaty, 1985). Females may gain better genes or more viable offspring by mating with a "superior" male but remaining with a male that will help care for her young. Females could possibly obtain an ally by mating with another male, hedge her bets for better breeding opportunities with that male in the future, reduce male sexual harassment or aggression towards her, reduce infanticidal tendencies of males most likely to come into contact with her young, or possibly reduce direct competition between the extra male and her partner for needed

resources. Whatever the reasons for female infidelity, it would seem to be contrary to her male's reproductive interests.

The focus on paternal care in the evolution of monogamy may have overemphasized male care as the primary benefit to males and females, as evidenced by the variation in male care in monogamous species. Males provide a wide variety of benefits to females other than male care, for example, predator vigilance, protection and resource defense. If male care of young or other male behaviors are critical to female reproductive success, then we might expect females to be more invested than males in maintaining the pair bond, for example, by remaining faithful (except in the case of forced copulations) even in the absence of the male. If females could easily find another male to benefit them (*e.g.* male-biased sex ratio), then males may invest more in pair bond maintenance than they would otherwise. Thus, from an evolutionary perspective, we might expect sex differences in the proximate mechanisms underlying pair bonds.

PAIR BOND: PROXIMATE CONSIDERATIONS

Evolution operates on outcomes (consequences) but what is selected are the proximate mechanisms and processes that produce an adaptive outcome. Monogamy is thus the result of selection pressures on each sex favoring certain proximate mechanisms. I argue that the proximate mechanisms at the core of monogamy are those responsible for the development and maintenance of particular types of male-female relations, or pair bonds. Two questions arise immediately: How are pair bonds formed and maintained? What is the nature of the pair bond?

Pair Bond Formation

Relatively little is known about the proximate mechanisms underlying pair bond formation or maintenance in monogamous mammals. An important component of bond formation, mate choice, has received little study in the field or laboratory. Pair bonds and mate choice have been examined in the prairie vole, *Microtus ochrogaster,* a predominantly monogamous rodent (Getz, Carter and Gavish, 1981; Getz and Hoffman, 1986). Field evidence for monogamy in prairie voles is based on association patterns, joint nesting, and shared home range. There is no evidence yet for mating exclusivity or paternal care in the field (see Dewsbury, 1988). In the laboratory, male prairie voles help care for the young (Hartung and Dewsbury, 1979; McGuire and Novak, 1984; Oliveras and Novak, 1986) but mating is not exclusive (see below).

Carter and her associates are examining behavioral and physiological substrates of pair bond formation in the prairie vole (Carter, Getz and Cohen, 1986; Carter *et al.*, 1988). Pair bonding, as measured by association preferences (physical contact), apparently develops as a result of copulatory interactions (Carter *et al.*, 1988). A social preference may develop after 24 hours of cohabitation, but not after 6 hours of familiarity, whereas an association preference will develop after 6 hours of cohabitation if copulation occurs (Williams, Catania and Carter, 1992). Interestingly, sexual interactions with a familiar mate affected subsequent social preferences for the familiar partner more strongly than sexual preferences; that is, females associated with their familiar sexual partner but also mated with unfamiliar males, thus providing no evidence of mating exclusivity (Carter, Williams and Witt, 1990; see also Fuentes and Dewsbury, 1984 and Shapiro *et al.*, 1986 below). In prairie voles, oxytocin is released during sexual interactions. Because pair bond formation in female prairie voles is facilitated by copulation, Carter hypothesizes that oxytocin may facilitate pair bond formation (Carter *et al.*, 1990). In female prairie voles, oxytocin increases affiliative behavior towards males and reduces aggression (Witt, Carter and Walton, 1990) and more importantly, oxytocin released directly into the lateral ventricles of females, induces partner preferences during just 6 h of cohabitation without copulation (Williams, Carter and Insel, 1992).

Dewsbury and his associates have examined female mate choice in prairie voles and in non-monogamous montane voles, *Microtus montanus*. Female prairie voles spent more time with and copulated more with a familiar male they had previously mated with than with an unfamiliar male (Shapiro *et al.*, 1986). In contrast, female montane voles exhibited no such preferences for males that they had previously copulated with. Mere familiarity, without copulation, was not sufficient to induce a preference in female prairie voles (see also Carter and Getz, 1985). Female prairie voles also display a preference for dominant males over subordinant males, whereas montane voles show no such preference (Shapiro and Dewsbury, 1986).

Male prairie voles prefer unmated females over mated females, they copulate more and spend more time with unmated females, whereas male montane voles show no such preference (Ferguson *et al.*, 1986). In addition, male prairie voles exhibit more fidelity to a single partner than do male montane voles. They copulate selectively with one female when given access to four females simultaneously, whereas montane voles distribute their copulations more widely (Fuentes and Dewsbury, 1984). These data are consistent with field data indicating a more monogamous mating system in prairie voles and a polygynous system in montane voles (Jannett,

1980). As in the Carter studies above there was no evidence of mating exclusivity in prairie voles.

Proximate Nature of Pair Bonds

The proximate nature of pair bonds has received relatively little attention and so far only two types of studies have been used to investigate the nature of the pair bond: separation-reunion and response to strangers.

(a) Separation-reunion

Extensive research on infant-mother attachment demonstrates that separation from the attachment figure elicits distinct behavioral and physiological responses indicative of stress (Mineka and Suomi, 1978; Reite and Fields, 1985) and that reunion with the attachment figure may reduce stress and also reveal aspects of the attachment bond not observable otherwise (Ainsworth et al., 1978). Attachment bonds may be formed throughout life (Ainsworth, 1989) and, in humans, adult male-female attachment (love) shares many features in common with infant-mother attachment (Shaver et al., 1988). Thus, separation and reunion may provide a good paradigm for examining the emotional nature of pair bonds.

Titi monkeys (Callicebus moloch) are monogamous New World primates characterized by pair bonds and male care for young (Wright, 1984; Robinson et al., 1987). When parents were given preference tests between their mate and infant, both parents preferred their mate (Mendoza and Mason, 1986). Separation from the mate elicited elevated plasma cortisol in both males and females but separation from their infant did not. These data indicate a stronger or at least a different kind of bond between mates than between either parent and the offspring. These results for females are particularly interesting, since in mammals the mother-infant bond is the strongest relationship of all, certainly stronger than the female-male bond. Thus these experiments suggest that, at least in some monogamous species, selection has favored dramatic changes in the nature of attachment bonds.

It would be of interest to document the range of variation within and between species in response to separation (and reunion) from the mate in the wild, both naturally occurring separations and experimentally induced separations. In some species emotional ties will be minimal and a separated or lost partner may be more easily replaced than in other species. If male-female cooperation is not essential (e.g. for resource defense, feeding the young) then there may be little or no necessity for an emotional bond. Loss of a partner may result in changes in activity patterns, food

intake (*e.g.* weight loss), sleep patterns, social behavior (Reite and Field, 1985), and immune responsiveness (Coe *et al.*, 1989). Individuals may withdraw from social interactions with others and show signs of behavioral depression (Harlow, *et al.*, 1971). Physiological measurements in response to pair bond disruption offer a promising approach for understanding the emotional nature of pair bonds (see Reite and Field, 1985).

(b) Response to strangers

The response of pair-bonded individuals to strangers has been studied in several monogamous primates. In the titi monkey mates increase spatial proximity, contact and coordination of activities, and vocal duets and they direct agonistic behaviors toward intruder pairs (Anzenberger *et al.*, 1986). Males more often interposed themselves between their mate and the strangers, restrained the female from approaching the strangers, and increased sexual behavior directed towards their mates. Males were also more physiologically aroused as indicated by increased heart rate activity but not cortisol in presence of the strangers.

Common marmosets (*Callithrix jacchus*) are considered monogamous based on laboratory studies and males help raise young (Tardif *et al.*, 1986). Both sexes are highly aggressive toward opposite-sex strangers (Sutcliffe and Poole, 1984) when their own mate is present and strange males receive more aggression than strange females (Evans, 1983). In the absence of their own mate, females are indifferent to novel males, whereas males actively solicit sexual interactions and contact with novel females (Evans, 1983; see also Anzenberger, 1985). This suggests a sex difference in the maintenance of the pair bond and mate fidelity in common marmosets: males are willing to mate with another female (in the absence of their partner), whereas females are more likely to be faithful than males. Female aggression appears to be the major factor in maintaining the pair bond (Anzenberger, 1985).

Response to strangers in the presence of the mate differs in other Callitrichids. In the facultatively polyandrous saddle-back tamarins, *Saguinus fuscicollis* (Terborgh and Goldizen, 1985), both sexes are aggressive towards same-sexed strangers but strange females receive more aggression than strange males (Epple, 1977; Epple and Alveario, 1985). In cotton-top tamarins, *Saguinus oedipus*, females are nonaggressive to intruders of either sex and males are aggressive only to strange males (French and Snowdon, 1981; Snowdon, 1990). Just the opposite occurs in golden lion tamarins, *Leontopithecus rosalia* in the laboratory. Female lion tamarins are highly aggressive to female intruders, whereas males are relatively nonagonistic (French and Inglett, 1989).

These variations in response to intruders are interesting, but are consistent in that aggression is primarily directed to same-sex strangers, suggesting intrasexual competition as a proximate mechanism maintaining pair bonds and mate fidelity (Wittenberger and Tilson, 1980; Snowdon, 1990). It is noteworthy that in each of these species the male shows extensive care of the young and that such help is essential to infant survival (see Snowdon and Soumi, 1982). The Callitrichids are a promising group for examination of the interrelationships between variations in mating exclusivity, pair bonds and male care.

Examination of mate fidelity in the presence or absence of the partner as discussed above will provide important information about the nature of pair bonds and mate fidelity. It would be of value to also examine the other partner's response upon reunion with the unfaithful partner. For example, does the partner behave more aggressively towards a mate who was away and unfaithful than towards one that was faithful while away (Zenone *et al.*, 1979).

(c) Other approaches

Additional approaches and measurements can be used to examine pair bonds, such as the ratio of affiliative/agonistic interactions within the dyad, synchrony in daily routine patterns (*e.g.* foraging), sexual behavior (*e.g.* extent of courtship) (Kleiman, 1981), and microanalysis of the synchrony of interaction patterns within a pair and, where appropriate, analysis of facial expressions (Ekman, 1982).

The extent of agonistic interactions between partners has received little attention. Such agonistic interactions may test the bond (Zahavi, 1976) and the willingness of the partners to invest in the relationship, especially if the individuals are vulnerable to exploitation (*e.g.* cheating, desertion).

Courtship patterns and duration will likely reflect differences in pair bonds between individuals within a species as well as differences between species. What each sex needs to evaluate to determine whether their interests will be met may require prolonged courtship, relative to the reproductive cycle, in monogamous species compared to polygamous species. A similar idea would apply to individual males and females within a species.

Variation in sexual behavior, especially copulatory patterns within pairs (*e.g.* brief versus prolonged copulation; single versus multiple copulations; Dewsbury, 1981) may reflect underlying differences in the nature of the pair bond and, in some species, sexual behavior may be involved in the formation and/or maintenance of a pair bond (Carter *et al.*, 1988).

PAIR BONDS AND BIPARENTAL CARE

Because of the lack of emphasis on pair bonds, we know relatively little about the relationship between pair bonds and parental behavior in mammals, especially male care of young. Male investment in young is affected by the male's relationship with the mother in promiscuous savannah baboons, *Papio cynocephalus anubis*. Males that had a special friendship with a female (spent more time in proximity and groomed with the female more than did other males) were more likely than other males to help care for (e.g. cuddle, protect) their friend's infant (Smuts, 1985) whether or not they were likely fathers of the infant. Male care may also help establish bonds with females. Caring for a female's infant increased the male's likelihood of mating with her in the future (Smuts, 1985).

The evidence for the effects of the pair bond on parental care in monogamous mammals is meager. In my laboratory we have examined how disruption of the pair bond affects biparental care in a monogamous rodent, the California mouse, *Peromyscus californicus*. This research is reviewed below. *Peromyscus californicus* provides a model system for an integrative approach toward understanding the evolution and proximate causation of monogamy in mammals because they exhibit mating exclusivity, pair bonding, and biparental care.

A Monogamous Rodent

The California mouse (Figure 1) is found in coastal California from San Francisco Bay south to Baja Peninsula in chaparral, sage scrub, and oak-woodland habitats (McCabe and Blanchard, 1950; M'Closkey, 1972; Merritt, 1974, 1978; Meserve, 1974). They breed throughout the year but primarily during the rainy season between October and May (McCabe and Blanchard, 1950; Ribble and Salvioni, 1990). They live for 9-18 months or more in the field (Chandler, 1979) and give birth to 1–3 young/litter (Rood, 1966; Drickamer and Vestal, 1973). Juveniles emerge from their nest about 35 days of age and disperse when they are about 75–85 days old (Ribble, 1989). Males are philopatric, *i.e.* they establish a home range close to their natal area, whereas females usually disperse (Ribble, 1992). A male may be in residence from 1 to 8 months before a female settles down in his territory and they may remain together for months before producing young (Ribble, 1990).

Field studies indicate that males and females form persistent bonds and associate together for long periods (McCabe and Blanchard, 1950; Chandler, 1979). The home ranges of a male and female overlap extensively

Figure 1 A male and female *Peromyscus californicus* and their young.

and are virtually exclusive of other pairs' home ranges (Ribble and Salvioni, 1990). Pairs nest together not only throughout the breeding season but also during the nonbreeding season when they are not rearing young (Ribble and Salvioni, 1990), and pairs remain together permanently unless one mate dies (Ribble, 1990). These data are corroborated by laboratory studies in which male-female pairs of *P. californicus* remained paired together rather than exchanging mates with other pairs inhabiting the same enclosure (Eisenberg, 1962, 1963). Intromission latencies during copulation are long in *P. californicus* in the laboratory (Dewsbury, 1974), which is a copulatory pattern typical of rodent species that tend to form mating pairs in their natural habitat (Dewsbury, 1981).

Field data are consistent with the idea that *P. californicus* exhibits biparental care. Radiotelemetry studies indicate that the male spends substantial amounts of time in the nest with the female when she is lactating (Ribble and Salvioni, 1990) and we determined with the aid of fiber optics that the male is in the nest with the female and their young (Gubernick and Ribble, unpublished observations). Biparental care in *P. californicus* is confirmed by laboratory studies (Eisenberg, 1962; Dudley, 1974; Gubernick and Alberts, 1987). Beginning at birth, fathers exhibit all of the components of parental behavior displayed by mothers, except nursing (Gubernick and Alberts, 1987). Mothers and fathers spend substantial and equivalent amounts of time in the nest with their pups and in physical contact with them throughout lactation. Paternal care involves a major investment in time generally equivalent to that of the female. In fact, fathers devote more time than mothers to licking pups. Mothers nurse for at least four weeks, and fathers and mothers both build nests and carry young (Gubernick and Alberts, 1987). Similar results have been obtained in both large enclosures (Gubernick and Alberts, 1987) and 100 sq ft rooms containing several nest boxes (Gubernick, unpublished observations).

P. californicus displays mating exclusivity in the wild. Based on DNA-fingerprinting and paternity exclusion analysis of 82 pups from 22 families in which the mother and presumptive father were known, in every case only the paired male sired each litter (Ribble, 1991).

Proximate Linkages Between Pair Bonds and Biparental Care

The field and laboratory data indicate the *Permyscus californicus* exhibit pair bonds and biparental care. We examined whether pair bonds between a male and female affect the onset and maintenance of paternal behavior.

(a) Onset of paternal behavior

In many rodents, a large percentage of unmated, virgin males are infanticidal but become parental prior to birth of their own young, *i.e.* they respond paternally towards unrelated alien young presented to them (wild house mice, *Mus musculus*, Jakubowski and Terkel, 1982; laboratory mice, *M. domesticus*, vom Saal and Howard, 1982; Elwood, 1986a; Mongolian gerbils,*M. unguiculatus*, Elwood, 1977; white-footed mice, *P. leucopus*, Cicirello and Wolff, 1990; see also chapters in this volume). This transition from infanticidal behavior to parental responsiveness typically occurs during the female's pregnancy (Elwood, 1977; vom Saal and Howard, 1982; Soroker and Terkel, 1988; this volume) and is essential to prevent otherwise infanticidal males from harming their own offspring (Wolff and Cicirello, 1989). The prepartum inhibition of male infanticide and the prepartum onset of paternal behavior in mice is mediated by copulation and/or cohabitation with a female (Elwood and Ostermeyer, 1984; vom Saal, 1985; Elwood, 1986a, 1986b; McCarthy and vom Saal, 1986; Soroker and Terkel, 1988; Cicirello and Wolff, 1990; chapters in this volume).

Virgin male *P. californicus* are frequently infanticidal prior to the formation of a pair bond (Gubernick, Schneider and Jeannote, in press; see also Figure 2), but they are parental when their own young are born (Gubernick and Alberts, 1987). We examined in the laboratory whether male*P. californicus* also inhibit their infanticidal behavior and become parental prior to birth of their offspring. We tested virgin males individually in their home cage for 10 minutes with a 1- to 3-day-old alien pup, then paired each male with a virgin female, and subsequently tested the males individually 1, 15 and 30 days after copulation (gestation is normally 31–33 days, Gubernick, 1988). A male was considered parental if it spent 1 min or more licking the pup or crouched over it in a nursing posture, whereas an animal was considered infanticidal if it attacked the pup. We terminated a test at the moment of attack, hence pups would have been killed if we had not interceded. Males resided with their partner throughout the study and all females gave birth within 35 days after copulation.

In contrast to other species of mice (this volume), the vast majority of male *P. californicus* did not become parental nor inhibit their infanticidal tendencies prior to birth of their young (Figure 2; between 33–50% of males were infanticidal). A small, but significant, percentage of males did become parental by 1 day after copulation and most of these same males remained parental throughout gestation (Figure 2). Thus for most males, sexual experience and cohabitation with their pregnant partner does not inhibit their infanticidal behavior prepartum. However, the pair bond with a female does affect the postpartum maintenance of

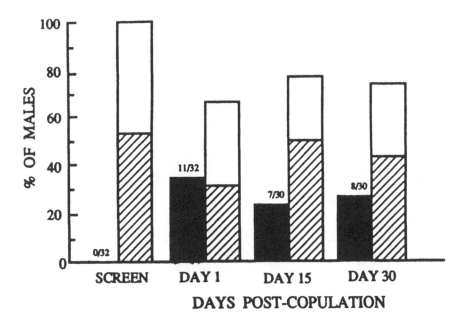

Figure 2 The percentage of males exhibiting parental (solid bar) or non-parental: attack (hatched bar) or ignore (open bar) behavior towards a test pup during their partner's pregnancy. Adult virgin males initially were screened and then paired with a virgin female. Males lived with their partner and males were tested at different times after mating (N = 30–32). [Adapted from Gubernick *et al.*, in press.]

paternal behavior and the postpartum inhibition of infanticide as described next.

(b) Maintenance of paternal behavior

To determine the linkage between pair bonds and parental care postpartum, we disrupted the pair bond by removing the mate and examined its effects on the postpartum maintenance of maternal and paternal behavior. Because parents may be affected by the presence or absence of their pups (Rosenblatt and Siegel, 1981), we first determined whether pup stimulation alone was crucial for the regulation of parental responsiveness.

Mothers and fathers lived together and either their pups remained with them or were removed within 4 hr of parturition. On Day 3 postpartum, parents were tested individually in their home cage for 10 min with a

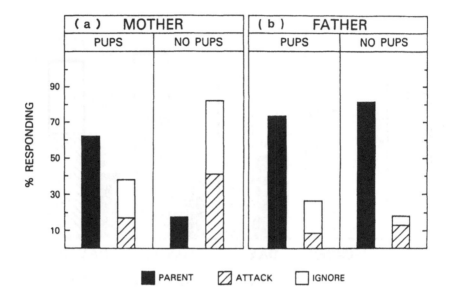

Figure 3 The percentage of mothers (a) and fathers (b) exhibiting parental or non-parental (attack, ignore) behavior towards a test pup on day 3 postpartum. Parents either remained together with their pups until testing (Pups present; N = 29 mothers and 34 fathers) or their pups were removed on the day of birth (Pups absent; N = 22 mothers and 22 fathers). [Adapted from Gubernick and Alberts, 1989].

1- to 3-day-old alien pup. An animal was considered parental or infanticidal as described earlier.

Pup stimulation is indispensable for the postpartum maintenance of maternal behavior. Pup absence reduced the number of mothers acting parentally and increased the number of mothers that acted infanticidally (Figure 3a). Father's absence, however had no effect on maternal responsiveness (Gubernick and Alberts, 1989).

In contrast, fathers continued to act parentally in the absence of pup stimulation (Figure 3b) (Gubernick and Alberts, 1989). The female's presence, however, was essential to maintain male parental care and to inhibit infanticide. In the absence of pup stimulation, fathers were significantly more likely to act parentally if the pair had remained together postpartum than if the mother had been removed (Figure 4).

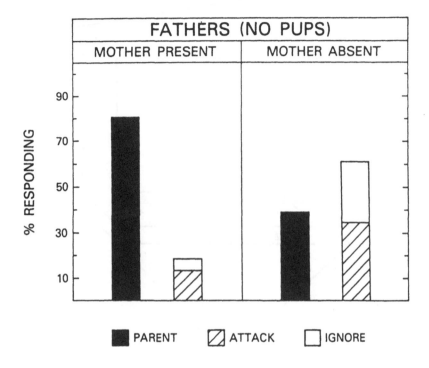

Figure 4 The percentage of fathers exhibiting parental or non-parental (attack, ignore) behavior towards a test pup on day 3 postpartum. The father's own pups were removed on the day of birth and his mate either remained (Mother present; N = 29) or was also removed (Mother absent; N = 26). [Adapted from Gubernick and Alberts, 1989].

We next examined the proximate mechanisms underlying the effects of the female on paternal behavior. Direct physical contact with the mother was not essential for the maintenance of paternal responsiveness. In the absence of pup stimulation, fathers that were separated from their mate on the day of birth by a double wire mesh barrier were as likely to be parental as males that freely interacted with their mate and young, indicating that a nontactile maternal cue(s) influenced paternal responsiveness (Gubernick and Alberts, 1989).

Maternal urine is the critical stimulus sufficient to maintain paternal behavior. Fathers exposed to maternal urine via a two-story townhouse drip cage (Figure 5) were parental compared to unexposed males (Figure 6; Gubernick and Alberts, 1989). Furthermore, maternal urine placed

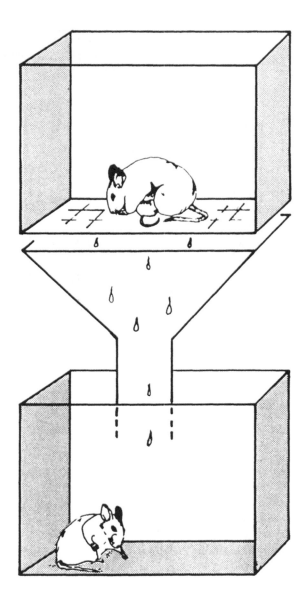

Figure 5 A two-story "townhouse" used to expose males to female urine and other olfactory cues. A hardware cloth bottom in the upper chamber allowed urine to fall into the lower chamber where the male was located.

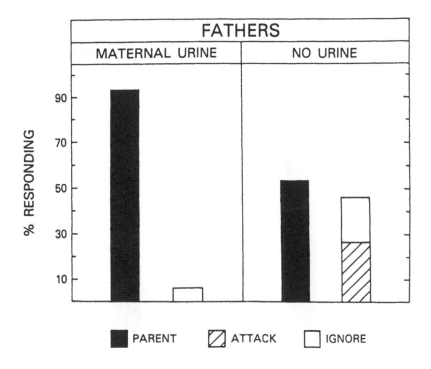

Figure 6 The percentage of fathers exhibiting parental or non-parental (attack, ignore) behavior towards a test pup on day 3 postpartum. Fathers were either exposed to urine from their mate (N = 15) or remained unexposed to maternal urine (N = 15) for three days while fathers were in the "townhouse." [Adapted from Gubernick and Alberts, 1990].

onto the nares of isolated fathers maintained paternal responsiveness compared to fathers similarly treated with distilled water (Figure 7) (Gubernick, 1990).

These data indicate that disruption of the pair bond reduced paternal responsiveness and increased infanticide, whereas a chemosignal in maternal urine reinstated paternal solicitude and reduced male infanticide. Copulation with the female during her postpartum estrus was not implicated because males were removed before they could copulate. In species that form persistent monogamous pair bonds between males and females, such as *P. californicus*, we might predict that fathers would be more responsive to maternal signals from their mate than to cues from other females. And this is precisely what we found.

Fathers were placed in the townhouse and exposed to urine from either their mate, a virgin female, another lactating female that gave birth the

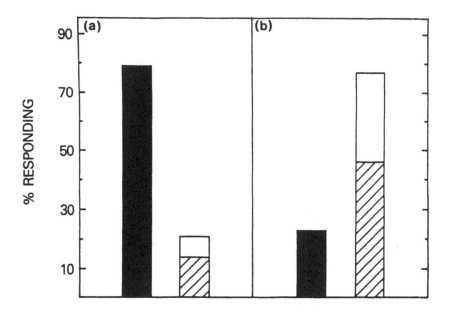

Figure 7 The percentage of fathers exhibiting parental (solid bar) or non-parental: attack (hatched bar) or ignore (open bar) behavior towards a test pup on day 3 postpartum. Fathers were removed from their mate and offspring shortly after birth and were exposed to either 100 ul of maternal urine on their nares twice/day for 3 days (a) (N = 15) or 100 ul of distilled water twice a day for 3 days (b) (N = 15). [Adapted from Gubernick, 1990].

same day as the male's mate, or no female. Exposure to urine from the male's mate maintained paternal responsiveness and inhibited infanticide compared to chemosignals from the other females (Figure 8) (Gubernick, 1990). In other experiments, we determined that mere familiarity with a female and exposure to the familiar female's urine was not sufficient to maintain paternal behavior postpartum and that the active chemosignal is contained in the volatile fraction of maternal urine (Gubernick, 1990).

These experiments clarify the effects of the pair bond on the maintenance of parental behavior and inhibition of infanticide in *P. californicus*. Disruption of the pair bond reduces paternal behavior but not the maintenance of maternal responsiveness. Pup stimulation, and not the male, is essential for maternal behavior. In contrast, a maternal urinary chemosignal is the critical proximate mechanism underlying the maintenance of

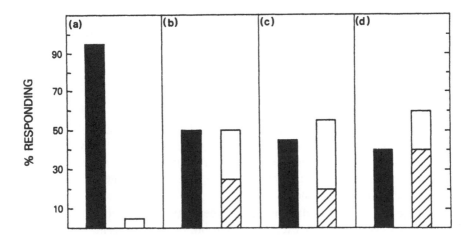

Figure 8 The percentage of fathers exhibiting parental (solid bar) or non-parental: attack (hatched bar) or ignore (open bar) behavior towards a test pup on day 3 postpartum. Fathers were removed from their mate and pups were placed in the "townhouse" and were either exposed to: (a) urine from their mate, (b) no urine, (c) urine from a virgin female, or (d) urine from another lactating female. [Adapted from Gubernick, 1990].

paternal solicitude and inhibition of infanticide during the first 3 days postpartum, and the effect is specific to the male's mate.

Although the vast majority of males remain parental only when exposed to cues from their mate, approximately 25–45% of males in the various control groups continue to be parental even in the absence of cues from their partner. This suggests that for some males, female cues are unnecessary for the maintenance of paternal behavior and the inhibition of infanticide postpartum. It is possible that these males are the same males that become parental prepartum as described in (a) above and continue to be parental into the postpartum period even in the absence of cues from their mate. Furthermore, the males that are infanticidal and do not become parental prepartum may be the same males that require maternal chemosignals to keep them parental and inhibit their infanticidal behavior postpartum. This is indeed the case. (Gubernick *et al.*, in press)

Thus there appears to be multiple proximate mechanisms in *P. californicus* to achieve the same end.

Now that we have established a link between pair bonds and paternal behavior in *P. californicus*, I am currently focusing upon the formation and maintenance of pair bonds, the nature of pair bonds, mate fidelity, and how these are interrelated.

CONCLUSIONS

The study of biparental behavior and the study of male-female relations in mammals have in many ways remained separate enterprises. I have argued for a more integrative approach that incorporates analysis of pair bonds, mating exclusivity, and biparental care for understanding the ultimate and proximate causation of monogamy.

Monogamy is a multidimensional concept, and it is important to keep the dimensions of mating exclusivity, pair bonds, and biparental care *conceptually* separate to reduce confusion in both terminology and explanations for the evolution of monogamy. Most explanations and analyses of monogamy focus upon one dimension at a time, either mating exclusivity or biparental care. I have argued for the importance of investigating pair bonds and their relationships to biparental care and mating exclusivity.

Male-female relations, or pair bonds, are central to monogamy because males and females in many monogamous species are doing something together that in most non-monogamous species males and females do not do — cooperate. The hypothesized functional consequences of monogamy have to do primarily with intersexual cooperation, often in the rearing of young and mate fidelity. Thus, the ultimate pay-off of such a system is likely to be a direct outcome of the nature and extent of male-female cooperation.

Investigation of pair bonds will also provide insight into the selective forces favoring the evolution of monogamy by determining what each sex is most concerned about and what each sex is willing to invest in (*e.g.* mate fidelity, rearing of young, resource defense, predator vigilance). Examination of the emotional nature of pair bonds will be valuable for this endeavor. The study of emotions is a knotty problem (Scherer and Ekman, 1984), but this is no reason to dismiss or ignore the role of emotions in the lives of animals and especially within pair bonds. Our lack of knowledge of precisely how emotions offer selective advantages should not deter us from addressing the evolution and proximate causation of male-female intimacy. As male-female pair bonds change we might expect mate fidelity

and male care of young to be affected. The challenge is to decipher the interrelationships between each of these dimensions. Understanding the proximate mechanisms underlying pair bonding will lead to the formulation of better arguments for the evolution of monogamy, parental care and male-female relations in mammals.

ACKNOWLEDGEMENTS

I wish to extend my warmest appreciation to the following people for their helpful comments and suggestions: Don Dewsbury, Patty Gowaty, Warren Holmes, Douglas Mock, David Ribble, Barbara Smuts, and Chuck Snowdon. My research and the writing of this chapter were supported in part by a NSF grant (BNS-8919302).

FOOTNOTE

1. Ever since Darwin (1872) emotions have been viewed as the product of evolution and as functional adaptations (Daly *et al.*, 1982; Ekman, 1982; Plutchik, 1984; Scherer and Ekman, 1984). "The emotions are specialized modes of operation shaped by natural selection to adjust the physiological, psychological, and behavioral parameters of the organism in ways that increase its capacity and tendency to respond adaptively to the threats and opportunities characteristic of specific kinds of situations. This formulation predicts that each emotion should correspond to a particular kind of adaptively significant situation that has occurred repeatedly in the course of evolution, and that the detailed characteristics of an emotional state can be analyzed as design features that increase the individual's ability to cope with the particular kinds of adaptive challenges that arise in this situation" (Nesse, 1990, p. 68). For applications of this approach see Nesse (1990; fear and anger) and Daly *et al.* (1982, for male sexual jealousy).

REFERENCES

AGREN, G., Zhou, Q., and Zhong, W. (1989) Ecology and social behaviour of Mongolian gerbils, *Meriones unguiculatus*, at Xilinhot, Inner Mongolia, China. *Animal Behaviour*, **37**, 11–27.
AINSWORTH, M. D. S. (1989) Attachments beyond infancy. *American Psychologist*, **44**, 709–716.
AINSWORTH, M. D. S., Blehar, M. C., Waters, E., and Wall, S. (1978) *Patterns of Attachment: A Psychological Study of the Strange Situation*. Hillsdale, N.J., Erlbaum.

ALEXANDER, R. D. and Noonan, K. M. (1979) Concealment of ovulation, parental care, and human social evolution. In: N. A. Chagnon and W. Irons, (eds.), *Evolutionary Biology and Human Social Behavior: An Anthropological Perspective*, North Scituate, MA: Duxbury, pp. 436–453.

ANZENBERGER, G. (1985) How stranger encounters of common marmosets (*Callithrix jacchus jacchus*) are influenced by family members: The quality of behavior. *Folia Primatologica*, **45**, 204–224.

ANZENBERGER, G., Mendoza, S. P., and Mason, W. A. (1986) Comparative studies of social behavior in Callicebus and Saimiri: Behavioral and physiological responses of established pairs to unfamiliar pairs. *American Journal of Primatology*, **11**, 37–51.

AXELROD, R. and Hamilton, W. D. (1981) The evolution of cooperation. *Science*, **211**, 1390–1396.

BARLOW, G. W. (1988) Monogamy in relation to resources. In: C. N. Slobodchikoff (ed.), *The Ecology of Social Behavior*, San Diego, Academic Press, pp. 55–79.

BAYLIS, J. R. (1981) The evolution of parental care in fishes, with reference to Darwin's rule of male sexual selection. *Environmental Biology of Fishes*, **6**, 223–251.

BELSKY, J. (1984) The determinants of parenting: A process model. *Child Development*, **55**, 83–96.

BROWN, J. L. (1975) *The Evolution of Behavior.* New York: Norton Press.

CARTER, C. S. and Getz, L. L. (1985) Social and hormonal determinants of reproductive patterns in the prairie vole. In: R. Gilles and J. Balthazart (eds.), *Neurobiology*, Berlin, Springer-Verlag, pp. 18–36.

CARTER, C. S., Getz, L. L., and Cohen-Parsons, M. (1986) Relationships between social organization and behavioral endocrinology in a monogamous mammal. *Advances in the Study of Behavior*, **16**, 109–145.

CARTER, C. S., Williams, J. R., and Witt, D. M. (1990) The biology of social bonding in a monogamous mammal. In: J. Balthazart (ed.), *Hormones, Brain and Behavior*, Basel, S. Karger, pp. 154–164.

CARTER, C. S., Witt, D. M., Thompson, E. G., and Carlstead, K. (1988) Effects of hormonal, sexual, and social history on mating and pair bonding in prairie voles. *Physiology and Behavior*, **44**, 691–697.

CHANDLER, T. (1979) *Population Biology of Coastal Chaparral Rodents.* Unpublished doctoral dissertation, University of California, Los Angeles.

CICIRELLO, D.M. and Wolff, J.A. (1990) The effects of mating on infanticide and pup discrimination in white-footed mice. *Behavioral Ecology and Sociobiology*, **26**, 275–279.

COE, C. L., Rosenberg, L. T. and Levine, S. 1989 Immunological consequences of psychological disturbance and maternal loss in infancy. In: C. Rovee-Collier and L. P. Lipsitt (eds.), *Advances in Infancy Research*, vol. 5, Norwood, N.J., Ablex Publishers, pp. 97–134.

DALY, M., Wilson, M., and Weghorst, S. J. 1985 Male sexual jealousy. *Ethology and Sociobiology*, **3**, 11–27.

DARWIN, C. (1872) *The Expression of Emotions in Man and Animals.* London, John Murray.

DEWSBURY, D.A. (1974) Copulatory behavior of California mice (*Peromyscus californicus*). *Brain, Behavior and Evolution*, **9**, 95–106.

DEWSBURY, D.A. (1981) An exercise in the prediction of monogamy in the field from laboratory data on 42 species of Muroid rodents. *The Biologist*, **63**, 138–162.

DEWSBURY, D.A. (1988) The comparative psychology of monogamy. In: D.W. Leger (ed.),*Comparative Perspectives in Modern Psychology*, Nebraska Symposium on Motivation, Vol. 35, Lincoln, University of Nebraska Press, pp. 1–50.

DRICKAMER, L.C. and Vestal, B.M. (1973) Patterns of reproduction in a laboratory colony of Peromyscus. *Journal of Mammalogy*, 54, 523–528.

DUDLEY, D. (1974) Paternal behavior in the California mouse, *Peromyscus californicus. Behavioral Biology*, 11, 247–252.

DUNBAR, R.I.M. (1984) The ecology of monogamy. *New Scientist*, 103, 12–15.

EASTERBROOKS, M. A. and Emde, R. N. (1988) Marital and parent-child relationships: The role of affect in the family system. In: R. A. Hinde and J. Stevenson-Hinde (eds.), *Relationships within Families: Mutual Influences*, Oxford, Clarendon Press, pp. 83–103.

EISENBERG, J.F. (1962) Studies on the behavior of *Peromyscus maniculatus gambelii* and *Peromyscus californicus parasiticus. Behaviour*, 19, 177–207.

EISENBERG, J.F. (1963) The intraspecific social behavior of some Cricetine rodents of the genus *Peromyscus. American Midland Naturalist*, 69, 240–246.

EKMAN, P. (ed.) (1982) *Emotion in the Human Face*. Second Ed. Cambridge, Cambridge University Press.

ELWOOD, R.W. (1977) Changes in the responses of male and female gerbils *(Meriones unguiculatus)* towards test pups during pregnancy of the female. *Animal Behaviour,* 25, 46–51.

ELWOOD, R.W. (1983) Paternal care in rodents. In: R.W. Elwood (ed.), *Parental Behaviour in Rodents*, New York, Wiley, pp. 235–257.

ELWOOD, R.W. (1986a) The inhibition of infanticide and the onset of paternal care in male mice. *Journal of Comparative Psychology*, 99, 457–467.

ELWOOD, T.W. (1986b) What makes male mice paternal? *Behavioral and Neural Biology*, 46, 54–63.

ELWOOD, T.W. and Ostermeyer, M.C. (1984) Does copulation inhibit infanticide in male rodents? *Animal Behaviour*, 32, 293–305.

EMBER, M.E. and Ember, C. (1979) Male-female bonding: A cross-species study of mammals and birds. *Behavior Science Research*, 1, 37–56.

EMLEN, S.T. and Oring, L.W. (1977) Ecology, sexual selection, and the evolution of mating systems. *Science*, 197, 215–223.

ENGFER, A. (1988) The interrelatedness of marriage and the mother-child relationship. In: R. A. Hinde and J. Stevenson-Hinde (eds.), *Relationships within Families: Mutual Influences*, Oxford, Clarendon Press, pp. 104–139.

EPPLE, G. (1977) Notes on the establishment and maintenance of the pair-bond in *Saguinus fuscicollis*. In: D.G. Kleiman (ed.), *The Biology and Conservation of the Callitrichidae*, Washington, D.C., Smithsonian Institution Press, pp. 231–238.

EPPLE, G. and Alveario, M.C. (1985) Social facilitation of agonistic responses to strangers in pairs of saddle-back tamarins (*Saguinus fuscicollis*). *American Journal of Primatology*, 9, 207–218.

EVANS, S. (1983) The pair-bond of the common marmoset, *Callithrix jacchus jacchus*; An experimental investigation. *Animal Behaviour*, 31, 651–658.

FERGUSON, B., Fuentes, S.M., Sawrey, D.K. and Dewsbury, D.A. (1986) Male preferences for unmated versus mated females in two species of voles (*Microtus ochrogaster* and *M. montanus*). *Journal of Comparative Psychology*, 100, 243–247.

FOLTZ, D.W. (1981) Genetic evidence for long-term monogamy in a small rodent, *Peromyscus polionotus. American Naturalist*, **117**, 665–675.

FRENCH, J.A. and Inglett, B.J. (1989) Female-female aggression and male indifference in response to unfamiliar intruders in lion tamarins. *Animal Behaviour*, **37**, 487–497.

FRENCH, J.A. and Snowdon, C.T. (1981) Sexual dimorphism in response to unfamiliar intruders in the tamarin *Saguinus oedipus. Animal Behaviour,* **29**, 822–829.

FUENTES, S.M. and Dewsbury, D.A. (1984). Copulatory behavior of voles (*Microtus montanus* and *M. ochrogaster*) in multiple-female test situations. *Journal of Comparative Psychology*, **98**, 45–53.

GETZ, L.L., Carter, C.S. and Gavish, L. (1981) The mating system of the prairie vole, *Microtus ochrogaster*: Field and laboratory evidence for pair-bonding. *Behavioral Ecology and Sociobiology*, **8**, 189–194.

GETZ, L.L. and Hofmann, J.E. (1986) Social organization in free-living prairie voles, *Microtus ochrogaster. Behavioral Ecology and Sociobiology*, **18**, 275–282.

GOLDBERG, W. A. and Easterbrooks, M. A. (1984) The role of marital quality in toddler development. *Developmental Psychology*, **20**, pp. 504–514.

GOWATY, P. A. (1983) Male parental care and apparent monogamy among Eastern bluebirds (*Sialia sialis*). *American Naturalist,* **121**, pp. 149–157.

GOWATY, P.A. (1985) Multiple parentage and apparent monogamy in birds. In: P.A. Gowaty and D.W. Mock (eds..), *Avian Monogamy,* American Ornithological Union, Ornithological Monograph 37, Lawrence, Kansas Allen Press, pp. 11–21.

GOWATY, P.A. and Karlin, A.A. (1984) Multiple paternity and maternity in single broods of apparently monogamous Eastern Bluebirds. *Behavioral Ecology and Sociobiology*, **15**, 91–95.

GOWATY, P.A. and Mock, D.W. (eds.) (1985) *Avian Monogamy*. American Ornithological Union, Ornithological Monograph 37, Lawrence, Kansas, Allen Press.

GUBERNICK, D.J. (1988) Reproduction in the California mouse, *Peromyscus californicus. Journal of Mammalogy,* **69**, 857–860.

GUBERNICK, D.J. (1990) A maternal chemosignal maintains paternal behaviour in the biparental California mouse, *Peromyscus californicus. Animal Behaviour*, **39**, pp. 916–923.

GUBERNICK, D.J. and Alberts, J.R. (1987) The biparental care system of the California mouse, *Peromyscus californicus. Journal of Comparative Psychology*, **101**, 169–177.

GUBERNICK, D.J. and Alberts, J.R. (1989) Postpartum maintenance of paternal behaviour in the biparental California mouse, *Peromyscus californicus. Animal Behaviour*, **37**, 656–664.

GUBERNICK, D.J. Schneider, K.A. and Jeannotte, L.A. (In press). Individual differences in the mechanisms underlying the onset and maintenance of paternal behavior and the inhibition of infanticide in the monogamous biparental California mouse, *Peromyscus californicus. Behavioral Ecology and Sociology*.

GYLLENSTEN, U.B., Jakobsson, S., and Temrin, H. (1990) No evidence for illegitimate young in monogamous and polygynous warblers. *Nature*, **343**, 168–170.

HARLOW, H.F., Harlow, M.K. and Suomi, S.J. (1971) From thought to therapy: Lessons from a primate laboratory*American Scientist*, **59**, 536–549.

HARTUNG, T.G. and Dewsbury, D.A. (1979) Paternal behaviour in six species of muroid rodents. *Behavioral and Neural Biology*, **26**, 466–478.

HOLMES, W.G. (1984) The ecological basis of monogamy in Alaskan hoary marmots. In: J.O. Murie and G.R. Michener (eds..), *The Biology of Ground-Dwelling Squirrels*, Lincoln, University of Nebraska Press, pp. 250–274.

JAKUBOWSKI, M. and Terkel, J. (1982) Infanticide and caretaking in non-lactating *Mus musculus:* influence of genotype, family group and sex. *Animal Behaviour,* 30, 1029–1035.

JANNETT, F.J. (1980) Social dynamics of the montane vole, *Microtus montanus*, as a paradigm. *The Biologist*, 62, 3–19.

KLEIMAN, D.G. (1977) Monogamy in mammals. *Quarterly Review of Biology*, 52, 39–69.

KLEIMAN, D.G. (1981) Correlations among life history characteristics of mammalian species exhibiting two extreme forms of monogamy. In: R.D. Alexander and D.W. Tinkle (eds.), *Natural Selection and Social Behavior,* New York, Chiron, pp. 332–344.

KLEIMAN, D.G. and Malcolm, J.R. (1981) The evolution of male parental investment in mammals. In: D.J. Gubernick and P.H. Klopfer (eds.), *Parental Care in Mammals*, New York, Plenum Press, pp. 347–387.

KNOWLTON, N. (1982) Parental care and sex role reversal. In: King's College Sociobiology Group, Cambridge (eds.),*Current Problems in Sociobiology*, Cambridge, Cambridge University Press, pp. 203–222.

KUMMER, H. (1968) *Social Organization of Hamadryas Baboons*, Chicago, Chicago University Press.

KURLAND, J. A. and Gaulin, S. J. C. (1984) The evolution of male parental investment: Effects of genetic relatedness and feeding ecology on the allocation of reproductive effort. In: D. W. Taub (ed.), *Primate Paternalism,* New York, Van Nostrand Reinhold, pp. 259–308.

LACK, D. (1968) *Ecological Adaptations for Breeding in Birds*, London, Metheun.

LEBOEUF, B. J. (1974) Male-male competition and reproductive success in elephant seals. *American Zoologist*, 14, 163–176.

LEIGHTON, D. R. (1987) Gibbons: Territoriality and monogamy. In: B. B. Smuts, D. L. Cheney, R. M. Seyfarth, R. W. Wrangham and T. T. Struhsaker (eds.), *Primate Societies*, Chicago, University of Chicago Press, pp. 135–145.

MCCABE,T.T. and Blanchard, B.D. (1950) *Three Species of Peromyscus*, Santa Barbara, Rood Associates.

MCCARTHY, M.M. and vom Saal, F.S. (1986) Inhibition of infanticide after mating by wild male house mice. *Physiology & Behavior*, 36, 203–209.

MCGUIRE, B.A. and Novak, M. (1984) A comparison of maternal behaviour in the meadow vole (*Microtus pennsylvanicus*), prairie vole (*M. ochrogaster*) and pine vole (*M. pinetorum*). *Animal Behaviour*, 32, 1132–2241.

M'CLOSKY, R. T. (1972) Temporal changes in populations and species diversity in a California rodent community. *Journal of Mammalogy*, 53, 657–676.

MCKINNEY, F., Cheng, K.M. and Bruggers, D.J. (1984) Sperm competition in apparently monogamous birds. In: R.L. Smith (ed.), *Sperm Competition and the Evolution of Animal Mating Systems*, Orlando, FL, Academic Press, pp. 523–545.

MENDOZA, S. P. and Mason, W. A. (1986) Parental division of labour and differentiation of attachments in a monogamous primate (*Callicebus moloch*). *Animal Behaviour*, 34, 1336–1347.

MERRITT, J.F. (1974) Factors affecting the local distribution of *Peromyscus californicus* in northern California. *Journal of Mammalogy*, 55, 102–114.

MERRITT, J.F. (1978) *Peromyscus californicus. Mammalian Species*, **85**, 1–6.

MESERVE, P.C. (1974) Temporary occupancy of a coastal sage shrub community by a seasonal immigrant, the California mouse (*Peromyscus californicus*). *Journal of Mammalogy*, **55**, 836–840.

MINEKA, S. and Suomi, S. J. (1978) Social separation in monkeys. *Psychological Bulletin*, **85**, 1376–1400.

MOCK, D. W. (1985). An introduction to the neglected mating system. In: P. A. Gowaty and D. W. Mock (eds..), *Avian Monogamy*, American Ornithological Union, Ornithological Monograph 37, Lawrence, Kansas, Allen Press, pp. 1–10.

MOCK, D. W. and Fujioka, M. (1990) Monogamy and long-term pair bonding in vertebrates. *Trends in Evolution and Ecology*, **5**, 39–43.

MOEHLMAN, P. D. (1986) Ecology of cooperation in canids. In: D. I. Rubenstein and R. W. Wrangham (eds.),*Ecological Aspects of Social Evolution: Birds and Mammals*, Princeton, N.J., Princeton University Press, pp. 64–86.

MOSS, C. (1988) *Elephant Memories*. New York, Fawcett Columbine.

MURRAY, B.G. (1984) A demographic theory on the evolution of mating systems as exemplified by birds. In: M.K. Hecht, B. Wallace and G.T. Prance (eds.), *Evolutionary Biology*, Vol. 18, New York, Plenum Press, pp. 71–140.

MURRAY, B.G. (1985) The influence of demography on the evolution of monogamy. In: P. A. Gowaty and D. W. Mock (eds.),*Avian Monogamy*, American Ornithological Union, Ornithological Monograph 37, Lawrence, Kansas, Allen Press, pp. 100–107.

NESSE, R. M. (1990) Evolutionary explanations of emotions. *Human Nature*, **1**, 261–289.

OLIVERAS, D. and Novak, M. (1986) A comparison of paternal behaviour in the meadow vole *Microtus pennsylvanicus*, the pine vole, *M. pinetorum,* and the prairie vole, *M. ochrogaster. Animal Behaviour*, **34**, 519–526.

ORIANS, G.H. (1969) On the evolution of mating systems in birds and mammals. *American Naturalist*, **103**, 589–604.

PECK, J.R. and Feldman, M.W. (1988) Kin selection and the evolution of monogamy. *Science*, **240**, 1672–1674.

PLUTCHIK, R. (1984) Emotions: A general psychoevolutionary theory. In: K. R. Scherer and P. Ekman (eds.),*Approaches to Emotion*, Hillsdale, N.J., Lawrence Erlbaum, pp. 197–219.

RASMUSSEN, D.R. (1981) Pair-bond strength and stability and reproductive success. *Psychological Review*, **88**, 274–290.

RATHBUN, G. B. (1979) The social structure and ecology of elephant-shrews. *Advances in Ethology*, **20**, 1–77.

REITE, M. and Field, T. (eds.) (1985) *The Psychobiology of Attachment and Separation*, New York, Academic Press.

RIBBLE, D. (1990) *Social Organization and Population Dynamics of the California Mouse (Peromyscus californicus)*. Unpublished Phd Thesis, University of California, Berkeley.

RIBBLE, D.O. (1991) The monogamous mating system of *Peromyscus californicus* as revealed by DNA fingerprinting. *Behavioral Ecology and Sociobiology*, **29**, 161–166.

RIBBLE, D.O. (1992) Dispersal in a monogamous rodent, *Peromyscus californicus. Ecology*, **73**, 859–866.

RIBBLE, D.O. and Salvioni, M. (1990) Social organization and nest co-occupancy in *Peromyscus californicus*, a monogamous rodent. *Behavioral Ecology and Sociobiology*, **26**, 9–15.

RICHARDSON, P.R.K. (1987) Aardwolf mating system: Overt cuckoldry in an apparently monogamous mammal. *Suid-Afrikaanse Tydskrif vir Wetenskap,* **83**, 405–410.

ROBINSON, J.G., Wright, P. C., and Kinzey, W. G. (1987) Monogamous Cebids and their relatives: Intergroup calls and spacing. In: B. B. Smuts, D. L. Cheney, R. M. Seyfarth, R. W. Wrangham and T. T. Struhsaker (eds.), *Primate Societies,* Chicago, University of Chicago Press, pp. 44–53.

ROOD, J.P. (1966) Observations on the reproduction of *Peromyscus* in captivity. *American Midland Naturalist,* **76**, 496–503.

ROSENBLATT, S.J. and Siegel, H.I. (1981) Factors governing the onset and maintenance of maternal behavior among nonprimate mammals: The role of hormonal and nonhormonal factors. In: D.J. Gubernick and P.H. Klopfer (eds), *Parental Care in Mammals,* New York, Plenum Press, pp. 13–76.

SCHERER, K. R. and Ekman, P. (eds.) (1984) *Approaches to Emotion,* Hillsdale, N.J., Lawrence Erlbaum.

SCOTT, M.P. and Tan, T.N. (1985) A radiotracer technique for the determination of male mating success in natural populations. *Behavioral Ecology and Sociobiology,* **17**, 29–33.

SELANDER, R.K. (1972) Sexual selection and dimorphisms in birds. In: B. Campbell (ed.), *Sexual Selection and the Descent of Man,* Chicago, Aldine, pp. 180–230.

SHAPIRO, L.E., Austin, D., Ward, S.E. and Dewsbury, D.A. (1986) Familiarity and female mate choice in two species of voles (*Microtus ochrogaster* and *Microtus montanus*). *Animal Behaviour,* **34**, 90–97.

SHAPIRO, L.E. and Dewsbury, D.A. (1986) Male dominance, female choice, and male copulatory behavior in two species of voles (*Microtus ochrogaster* and *Microtus montanus*). *Behavioral Ecology and Sociobiology,* **18**, 267–274.

SHAVER, P., Hazan, C., and Bradshaw, D. (1988) Love as attachment. In: R. J. Sternberg and M. L. Barnes (eds.), *The Psychology of Love,* New Haven, CT, Yale University Press, pp. 68–99.

SLOBODCHIKOFF, C. N. (1984) Resources and the evolution of social behavior. In: P.W. Price, C.N. Slobodchikoff and W.S. Gaud (eds.), *A New Ecology: Novel Approaches to Interactive Systems,* New York, John Wiley and Sons, pp. 227–251.

SMUTS, B.B. (1985) *Sex and Friendship in Baboons.* Hawthorne, New York, Aldine.

SMUTS, B.B. (1988) *Primates, Dolphins and Humans: The Evolution of Intimate Relationships.* Helen Homans Gilbert Prize Lectureship, Harvard University.

SMUTS, B.B. and Gubernick, D.J. (1992) Male-infant relationships in nonhuman primates: Paternity certainty or mating effort? In: B. Hewlett (ed.), *Father-child Relations,* Hawthorne, New York, Aldine, pp 1–30.

SNOWDON, C. T. (1990) Mechanisms maintaining monogamy in monkeys. In: D. A. Dewsbury (ed.), *Contemporary Issues in Comparative Psychology,* Sunderland, MA, Sinauer Associates, pp. 225–251.

SNOWDON, C.T. and Suomi, S.J. (1982) Paternal behavior in primates. In: H.E. Fitzgerald, J.A. Mullins, and P. Gage (eds.), *Child Nurturance,* Volume 3: *Studies of Development in Nonhuman Primates,* New York, Plenum Press, pp. 63–108.

SOROKER V. and Terkel, J. (1988) Changes in incidence of infanticidal and parental responses during the reproductive cycle in male and female wild mice *Mus musculus. Animal Behaviour,* **36**, 1275–1281.

SUTCLIFFE, A.G. and Poole, T.B. (1984) An experimental analysis of social interactions in the common marmoset (*Callithrix jacchus jacchus*). *International Journal of Primatology,* **5**, 591–607.

TARDIF, S.D., Carson, R.L. and Gangaware, B.I.. (1986) Comparison of infant care in family groups of the common marmoset (*Callithrix jacchus*) and the cotton-top tamarin (*Saguinus oedipus*). *American Journal of Primatology*, **11**, 103–110.

TAUB, D.M. (1984) Male caretaking behavior among wild Barbary macaques (Macaca sylvanus). In: D.M. Taub (ed), *Primate Paternalism*, New York, Van Nostrand Reinhold, pp. 20–55.

TERBORGH, J. and Goldizen, A.W. (1985) On the mating system of the cooperatively breeding saddle-backed tamarin (*Saguinus fuscicollis*). *Behavioral Ecology and Sociobiology*, **16**, 293–299.

THORNHILL, R. and Alcock, J. (1983) *The Evolution of Insect Mating Systems*, Cambridge, Harvard University Press.

TRIVERS, R.L. (1971) The evolution of reciprocal altruism. *Quarterly Review of Biology*, **46**, 35–57.

TRIVERS, R.L. (1972) Parental investment and sexual selection. In: B.Campbell (ed.), *Sexual Selection and the Descent of Man*, Chicago, Aldine, pp. 136–179.

VAN SCHAIK, C.P. and Dunbar, R.I.M. (1990) The evolution of monogamy in large primates: A new hypothesis and some crucial tests. *Behaviour*, **115**, 30–62.

VOM SAAL, F.S. (1985) Time-contingent change in infanticide and parental behavior induced by ejaculation in male mice. *Physiology & Behavior*, **34**, 7–15.

VOM SAAL, F.S. and Howard, L.S. (1982) The regulation of infanticide and parental behavior: implications for reproductive success in male mice. *Science*, **215**, 1270–1272.

WEINRICH, J.D. (1977) Human sociobiology: pair-bonding and resource predictability. *Behavioral Ecology and Sociobiology*, **2**, 91–118.

WERREN, J.H., Gross, M. R. and Shine, R. (1980) Paternity and the evolution of male parental care. *Journal of Theoretical Biology*, **82**, 619–631.

WICKLER, W. (1976) The ethological analysis of attachment. *Zeitschrift fur Tierpsychologie*, **42**, 12–28.

WICKLER, W. and Seibt, U. (1981) Monogamy in crustacea and man. *Zeitschrift fur Tierpsychologie*, **57**, 215–234.

WICKLER, W. and Seibt, U. (1983) Monogamy: An ambiguous concept. In: P.P.G. Bateson (ed.), *Mate Choice*, Cambridge, Cambridge University Press, pp. 33–50.

WILLIAMS, G. C. (1966) *Adaptation and Natural Selection*, Princeton, N.J., Princeton University Press.

WILLIAMS, J.R., Carter, C.S., and Insel, T. (1992) The development of partner preferences in female prairie voles (*Microtus ochogaster*) is especially facilitated by mating and the central infusion of oxytocin. *Annals of the New York Academy of Sciences*. **652**, 487–491.

WILLIAMS, J., Catania, K.C. and Carter, C.S. (1992) Mating facilitates partner preference development in female prairie voles (*Microtus ochrogaster*). Hormones and Behavior. **26**, 339–346.

WILSON, E.O. (1975) *Sociobiology: The New Synthesis*, Cambridge, MA, Harvard University Press.

WITT, D.M., Carter, C.S., and Walton, D.M. (1990) Central and peripheral effects of oxytocin administration in prairie voles (*Microtus ochrogaster*). *Pharmacology, Biochemistry and Behavior*, **37**, 63–69.

WITTENBERGER, J.F. (1979) The evolution of mating systems in birds and mammals. In: P. Marler and J.G. Vandenbergh (eds.), *Handbook of Behavioural Neurobiology*, Vol. 3, New York, Plenum Press, pp. 271–350.

WITTENBERGER, J.F. and Tilson, R.L. (1980) The evolution of monogamy: Hypotheses and evidence. *Annual Review of Ecology and Systematics*, **11**, 197–232.

WOLFF, J.O. and Cicirello, D.M. (1989) Field evidence for sexual selection and resource competition infanticide in white-footed mice. *Animal Behaviour*, **38**, 637–642.

WRANGHAM, R. W. (1980) An ecological model of female-bonded primate groups. *Behaviour*, **75**, 262–300.

WRANGHAM, R. W. (1982) Mutualism, kinship and social evolution. In: King's College Sociobiology Group, Cambridge (eds.), *Current Problems in Sociobiology*, Cambridge, Cambridge University Press, pp. 269–289.

WRIGHT, P.C. (1984) Biparental care in *Aotus trivirgatus* and *Callicebus moloch*. In: M.F. Small (ed.), *Female Primates: Studies by Women Primatologists*, New York, Liss, pp. 60–75.

ZAHAVI, A. (1976) The testing of a bond. *Animal Behaviour*, **25**, 246–247.

ZENONE, P. G., Sims, M. E. and Erickson, C. J. (1979) Male ring dove behavior and the defense of genetic paternity. *American Naturalist*, **114**, 615–626.

ZEVELOFF, S.I. and Boyce, M. (1980) Parental investment and mating systems in mammals. *Evolution*, **34**, 973–982.

CHAPTER 19

PROTECTION AND ABUSE OF YOUNG IN MICE: INFLUENCE OF MOTHER-YOUNG INTERACTIONS

BRUCE SVARE and MICHAEL BOECHLER

Department of Psychology, State University of New York at Albany,
1400 Washington Ave., Albany, New York 12222, USA

Field and laboratory-based experimental work examining protection and abuse of young has focused primarily upon the abusive behavior that adults can exhibit toward young and the reasons (both proximate and ultimate) for the emergence of this response strategy. A competing behavior which may have evolved as a counter strategy to infant abuse is that of maternal protection exhibited by pregnant and/or lactating female mice. As it is most often studied in the laboratory, this behavior consists of aggressive attacks exhibited by reproductively active females toward intruder males and females that threaten the nest site and the young. This aspect of maternal behavior is often referred to as maternal defense or maternal aggression and, in this chapter, these terms will be used interchangeably with that of maternal protection.

The subject of the present chapter is an examination of maternal protective behavior in mice as it relates to three fundamental questions of importance to behavioral and biological researchers. First, what are the proximate biological and behavioral factors that are known to modulate this behavior? Second, in terms of optimization and parental investment theory, what are the ultimate causation factors that may help us to understand its widespread occurrence as well as its breakdown under certain situations? Third, what theoretical models may help us in the future to explore the relationship between maternal protection on the one hand and

infant abuse on the other? These questions are best answered by reviewing older data as well as more recent information derived from our laboratory. For more extensive reviews on this subject as well as the topic of rodent maternal behavior in general, the reader is referred to a number of additional sources (Numan, 1988; Svare, 1990; Svare, 1989)

DEFINING MATERNAL PROTECTIVE BEHAVIOR

Maternal protective behavior can be defined from a number of different perspectives. It can be characterized by exploring outcomes (*i.e.*, do pups survive or are they killed?), it can be defined in terms of the receiver of such behavior (*i.e.*, are intruders attacked or left alone?), and it can be explored by studying the maternal female (*i.e.*, is the female victorious or defeated?). In situations where we have explored the proximate mechanisms modulating this behavior in mice, we have defined maternal protective behavior in very pragmatic terms. That is, a female is declared to be maternally aggressive if she exhibits attacks or lunges toward a strange intruder male mouse introduced to her cage. Test periods typically are short (only several minutes in duration) and there is no attempt to examine the outcomes listed above. When examining ultimate mechanisms however, questions of behavioral outcomes become more important and it is necessary to broaden our definition of what we will call maternal protective behavior. In these situations (questions of ultimate causation), we will consider maternal protective behavior to be aggressive behaviors exhibited by the female that decrease the probability that an intruder will harm young while simultaneously increasing the likelihood of injury or death to the mother.

THE INCIDENCE OF MATERNAL AGGRESSION IN MAMMALS

Field and laboratory-based experiments indicate that maternal aggressive behavior is widely represented in mammals. The behavior has been observed in rhesus monkeys (Harlow *et al.*, 1963), langurs (Jay, 1963), gelada baboons (Mori and Dunbar, 1985), cows (Altman, 1963), sheep (Hersher *et al.*, 1963), cats (Schneirla *et al.*, 1963), rabbits (Ross *et al.*, 1963), squirrels (Taylor *et al.*, 1966), rats (Erskine *et al.*, 1978), hamsters (Wise, 1974), and mice (Svare and Gandelman, 1973). In human females, pregnancy, parturition, and lactation are accompanied by emotional changes and psychiatric illness (Flemming, 1990). For example, it is not uncommon to see heightened levels of depression, irritability, mood swings, neuroses, as well

as violence and hostility during the peripartum period in humans. Perhaps these behavioral changes in humans represent the vestiges of maternal aggressive behavior observed in other mammals.

Within species phenotypic variation in maternal aggression may help us to understand the basic mechanisms responsible for the behavior. For example, in outbred mice, there is considerable variation in the exhibition of the behavior with some pregnant and lactating female mice highly aggressive and other females nonaggressive (*e.g.*, Svare and Gandelman, 1973). Also, certain inbred strains of house mice exhibit higher levels of the behavior than do other strains (St. John and Corning, 1973; Broida and Svare, 1982). Research utilizing reciprocal hybrid crosses and cross-fostering indicates that these strain differences are not due to maternal environmental factors but rather they are due to genetic influences (Broida and Svare, 1982). Just how genetic influences ultimately exert their effects on aggression is not understood at the present time. Future work exploring neurochemical comparisons between inbred and selectively bred strains that differ in aggressive behavior may help us to understand the underlying mechanisms that control individual variation in maternal aggressive behavior.

Genetic variation in the incidence of maternal protective behavior can also be used to explore questions surrounding the presumed co-evolution of infanticide and maternal aggression. For example, it is parsimonious to assume that selection worked in such a way to favor the evolution of maternal protective behavior in species where the male tendency to kill young is high. By comparing numerous inbred strains of mice with respect to the incidence of both maternal aggressive behavior exhibited by females and infanticide displayed by males, the question of co-evolution could be addressed.

METHODOLOGIES EMPLOYED TO STUDY MATERNAL AGGRESSIVE BEHAVIOR

The methods employed by behavioral biologists studying maternal protective behavior can vary from researcher to researcher. Although this variation is largely determined by the nature of the questions being posed, great care must be taken when interpreting research results derived from widely divergent methods.

In our laboratory, we have utilized a relatively simple procedure for studying the proximate behavioral and biological mechanisms underlying maternal aggressive behavior exhibited by Rockland-Swiss albino mice.

Our animals are generated as an outbred stock and females are used in experiments beginning at 60 days of age. At this time, they are housed with stud males and are isolated when a copulatory plug is found. At parturition (19 days following impregnation) litters typically are adjusted to 6 pups and the young are weaned at 21 days of age. For the purposes of aggression tests, sexually naive males are placed into the cage of the female for 3 minutes and attacks and lunges exhibited by the female are scored. In cases where females are being tested in the postpartum state, pups are removed shortly before the test to avoid their potentially confounding role in the behavior exhibited by the resident female. Adult male intruders that have been group housed (6/cage) are used as opponents since they elicit high levels of attack and usually do not fight back in response to being attacked.

The tactics employed for assessing questions of ultimate causation utilize far different methods including seminatural environments with nest boxes and entrances and exits that can be manipulated in such a way to allow or prevent access by intruders (See Parmigiani, this volume, Boechler, this volume). Other tactics also include the imposition of environmental constraints upon the female, such as foraging requirements. The simultaneous assessment of maternal protective behavior and infanticide in these situations is a powerful tool for exploring questions of ultimate causation.

It is important to note that no single method for assessing maternal protective behavior is necessarily better than any other. Each has its advantages and its limitations. For example, utilizing sexually naive intruder males of unknown rank provides a standard stimulus for aggression tests when examining questions of proximate causation. However, intruder dominance/subordinance status and previous sexual experience may alter the exhibition of maternal aggression and therefore may represent an important variable for consideration (See Parmigiani this volume). Likewise, isolating females in studies designed to examine proximate questions may duplicate what happens in wild female mice that are known to live in a solitary fashion, but they obviously do not replicate the communal rearing situation that is often seen in other mouse social organizations. The point here is that something is lost and something is gained with each strategy that is employed. Oversimplification of the test situation limits our generalizability to the wild while duplication of seminatural environments limits our ability to control single variables and determine cause and effect relationships. Regardless, we must be content as scientists to confine our conclusions to the strategies employed in our testing situations.

PHYSIOLOGY AND AGGRESSIVE BEHAVIOR DURING PREGNANCY, PARTURITION, AND LACTATION

Female mice exhibit pronounced changes in neuroendocrinology and aggressive behavior as they advance through pregnancy, parturition, and lactation. For the uninitiated reader these changes are briefly reviewed here. (See Figure 1).

In Rockland-Swiss albino mice, the subjects of much of our work, aggression is seldom observed during the virgin state. Aggression is first observed in the form of threat behavior during midpregnancy (Gestation Day 10) with peak levels of threats seen toward the end of the gestation period (Gestation Days 14–18) (Mann and Svare, 1982). Aggressive behavior during pregnancy, referred to as pregnancy-induced aggression, is much less intense than the agonistic behavior typically observed during lactation (Mann and Svare, 1982). As noted above, late pregnant females typically exhibit threat behaviors; these responses consist of rapid lunges exhibited toward the intruder which fall short of actual physical contact. Attacks (intense biting and wrestling) also are occasionally displayed by late pregnant females but the primary mode of aggressive responding is threat behaviors.

Immediately following parturition, aggression is absent during the postpartum estrous period when females normally remate with males (Ghiraldi and Svare, unpublished observations). However, after the parturient female has nursed her young for 2 days, intense aggressive behavior, referred to as postpartum aggression, emerges and gradually peaks between Lactation Days 4 to 10 (Svare and Gandelman, 1973; Svare, et al., 1981). This form of aggressive behavior is extremely intense and consists primarily of rapid biting attacks toward the intruder. With advancing lactation and the growth of young, lactating females exhibit a dramatic reduction in the intensity of aggressive behavior to a point where there is little postpartum aggression observed around the time when young are weaned (Svare et al., 1981).

The striking behavioral changes noted above are related to changes in steroid hormone profiles and some evidence indicates that neuroendocrine factors underlie changes in maternal protective behavior.

Four lines of evidence indicate that the threat behavior observed in mid to late pregnant female mice is associated with elevations in circulating progesterone. First, female mice made pseudopregnant by cervical stimulation experience transient increases in threat behaviors and circulating progesterone (Barkley, et al., 1979; Noirot, et al., 1975). Second, gravid females begin to show characteristic threat behaviors when

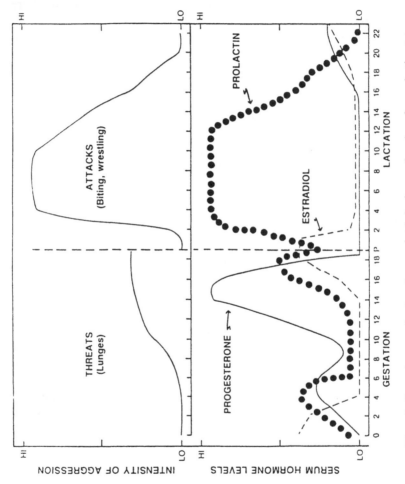

Figure 1 Serum hormone levels and aggression behavior as a function of reproductive state in female mice.

progesterone is beginning to peak in plasma (Mann *et al.*, 1984). Third, though their aggression is not equivalent to that of normal pregnant animals, progesterone-exposed virgin female mice exhibit increases in threat behavior when exposed to chronic injections of progesterone of Silastic implants of the steroid (Mann *et al.*, 1984). Fourth, pregnancy-termination, a surgical procedure which is known to dramatically lower serum progesterone, eliminates the aggressive behavior of gravid mice when it is performed on the 15th day of gestation (Svare, *et al.*, 1986). Silastic implants of P maintain aggression in pregnancy-terminated mice but do not fully restore the behavior to the level of fighting normally observed in pregnant animals (Svare *et al.*, 1986). Thus, while progesterone appears to be playing an important role in pregnancy-induced aggressive behavior, other neuroendocrine factors also may be important. For example, circulating testosterone, a hormone which has repeatedly been implicated in rodent intermale aggressive behavior (Svare, 1983), also increases during pregnancy in the mouse (Barkley, *et al.*, 1979) and therefore may be involved in the aggression exhibited by gravid female mice.

As noted above, aggression is absent during the immediate (24–48 hour) postpartum estrous period when females ordinarily remate with males. Recent research from our laboratory (Ghiraldi and Svare, unpublished observations) suggests that ovarian hormone stimulation just prior to parturition or right at parturition is probably responsible, in part, for the absence of aggression during the early postpartum period. This follows from work showing that ovariectomy just prior to the time of delivery promotes the onset of postpartum aggression at a time earlier (the day of parturition and 24 hours following delivery) than it would normally appear (Ghiraldi and Svare, unpublished observations). Also, estrogen injections reverse the aggression promoting effects of ovariectomy.

Although hormones appear to play an important role in the activation of aggression during pregnancy (progesterone) and the suppression of the behavior during the early postpartum period (estrogen), the neuroendocrine system apparently plays no role in the activation or maintenance of the behavior once the female begins nursing her young. Ovariectomy and adrenalectomy do not alter aggression in animals that have been nursing their young for several days (Svare and Gandelman, 1976b). Also, prolactin, the presumed "maternal" hormone and other pituitary hormones are not involved in postpartum aggressive behavior. For example, aggressive behavior during lactation is not disrupted by drugs that inhibit prolactin release (bromocryptine and ergocornine) (Mann, *et al.*„ 1980) nor is the behavior reduced by hypophysectomy (Svare, *et al.*,

1982). Instead, the activation and maintenance of the behavior is due, in part, to elevated serotonergic function. Serotonin, which has been implicated in many different forms of aggressive behavior (Svare, 1983), exhibits rapid increases in hypothalamic turnover rates during the postpartum period (Kordon, *et al.*, 1973/1974). Treating postpartum mice with drugs that reduce serotonergic function (*e.g.*, methysergide and PCPA) decreases aggressive behavior (Svare and Mann, 1983; Ieni and Thurmond, 1986).

THE IMPORTANCE OF MOTHER-YOUNG INTERACTIONS FOR MATERNAL AGGRESSIVE BEHAVIOR

Intense aggression begins to develop after the postpartum female has received 48 hours of suckling stimulation. It is well-documented that the onset of maternal aggression during the postpartum period is contingent upon two elements. First, estrogen and progesterone induces growth of the overall size and length off the teats during pregnancy (Svare and Gandelman, 1976a; Svare,*et al.*, 1980). The growth of the teats, referred to as the substrate preparation phase (Svare, 1977), is an important process during pregnancy since newborns otherwise would not be able to attach and suckle from the mother if it did not occur. Second, the receipt of suckling stimulation from young is essential for the establishment of fighting behavior during the postpartum period (Svare and Gandelman, 1976a; 1976b). Aggressive behavior does not develop in postparturient female mice if either one of these two elements (growth of the nipples or suckling stimulation from young) is missing (Svare and Gandelman, 1976a; 1976b). Thus, steroid hormone exposure during pregnancy stimulates female agonistic behavior in two different ways: First, they directly promote threat behavior during pregnancy; Second, they indirectly stimulate aggression during the postpartum period by preparing the substrate (growth of the nipples) for attachment and suckling stimulation by young.

The receipt of suckling stimulation during the nursing process is a critical factor in the establishment of aggressive behavior. Not only does this event turn on the synthesis and release of milk (lactation), but it also promotes defensive behavior on the part of the female. As reviewed above, this relationship has been thoroughly established in rodents but information in other mammals is lacking.

Although suckling stimulation is critical for the establishment of aggressive behavior in the postpartum rodent, it apparently is not needed on a continuous basis. For example, if pups are removed for one hour during

early lactation, postpartum aggression in the mouse is not affected (Gandelman, 1972; Svare, 1977; Svare and Gandelman, 1973). However, 5 hours of pup removal all but eliminates aggressive behavior in the postpartum female. Interestingly, if young are replaced for as little as 5 minutes following 5 hr of pup removal, the behavior is restored almost immediately. Research also shows that direct physical contact between the dam and her young is not a prerequisite for the short-term maintenance of the behavior in mice. Placement of the dam's entire litter or a single pup behind a double wire mesh partition in the home cage maintains the behavior at a level indistinguishable from mothers in direct contact with their young (Svare and Gandelman, 1973).

The fact that contact with young is not critical for the short-term maintenance of aggressive behavior in postpartum rodents may have important implications for other aspects of parental behavior. For example, foraging behavior is of obvious importance to both the dam and the young during this period. Time constraints on foraging behavior may be regulated by the young in the sense that long durations of time away from the nest may compromise the aggression defense mechanism. Clearly, however, mother-young reunions of short duration can reactivate the aggression mechanism and exteroceptive cues, to the extent that they are proximate, can maintain the behavior.

MATERNAL AGGRESSION VIEWED AS AN OPTIMIZATION PROBLEM

The foregoing has dealt mainly with understanding the proximate factors involved in postpartum aggressive behavior. Let us now examine maternal aggression in the reproductive female by exploring issues surrounding ultimate causation.

The maternal protective behavior exhibited by the female can be treated as an optimization problem. The intensity of nest defense performed by a parent entails some risk (injury or death) and results in some benefits (increased probability of offspring survival). Therefore, a cost-benefit analysis can be used to predict optimum level of nest defense in a given situation (Montgomerie and Weatherhead, 1988, Sargent and Gross, 1985, Andersson, et al., 1980). The data from Figure 1, which shows the development and decline of aggression over pregnancy and lactation, is somewhat consistent with current parental defense theory. As can be seen in the figure, maternal protective behavior during pregnancy increases as the reproductive value of the young increases, but is still quite low. Thus, the female

should take relatively low risks and, as noted earlier, only threat behavior is exhibited at this time. Risk taking (*e.g.*, aggression) increases after parturition as the reproductive value of the young increases. Thus, increasing benefits to the female are being derived and elevated aggression (and its increasing costs) are evident. However, once the young begin to leave the nest, their value to the female continues to increase but benefits from defense decline because juveniles are more capable of escaping danger and adult male intruders probably are not capable of destroying an entire litter once it has started dispersing from the nest.

Two stages of the reproductive cycle of the female would initially appear to contradict the optimization analysis advanced here. First, in the early postpartum period (postpartum estrous), the female does not protect her young even against a strange male intruder. It must be assumed that enormous benefits can be derived from remating since the female would be able to produce many more young and her fitness would be greatly enhanced. If one assumes that the new male is a dominant male and is therefore more "fit" (*e.g.*, better genes), it could be advantageous for the female to abandon (*i.e.*, not protect) her current offspring and remate. The benefits of this behavior may outweigh the costs to the female if the offspring produced by the second mating are more "fit" than the young produced by the first mating. Second, it would initially appear to be paradoxical that maternal protective behavior declines before young disperse. It is well known that females suckle their young less during this period and that they actively reject them at the nipple. The benefit here also involves the production of new offspring. The female will begin to ovulate sooner if she is suckled less. Thus, she can mate again, produce more young in a shorter period of time, and elevate her lifetime fitness. The benefit of initiating reproduction sooner may outweigh the cost/risk of continued defense.

Some recent studies in our laboratory have attempted to explore some of the many questions concerning ultimate causation and maternal protective behavior. Several of these experiments are briefly reviewed here.

As noted earlier, removal of young during lactation terminates maternal defense. We recently examined a similar question in pregnant females by producing pregnancy terminations on day 16 of gestation (Svare, unpublished observations). From parental defense theory, it of course makes no sense to continue to engage in defense when the reproductive value of your offspring is zero. When compared to the high aggressive behavior of sham-operated pregnant females, pregnancy terminated females failed to engage in any aggressive behavior when challenged with an adult male intruder.

A second recent experiment dealing with parental investment theory concerns what happens to females of different fighting ability and the allocations they make to the two sexes (Svare and Boechler, unpublished observations). Under ideal conditions, parental investment theory predicts that sons should be favored over daughters (Trivers and Willard, 1973). In two different experiments, we recently assessed aggressive behavior during pregnancy and fetal sex ratio. Our findings showed that highly aggressive females produced more males than medium or low aggressive females.

Finally, it is relevant to the topic of the conference and this book that we mention one final experiment recently completed in our laboratory (Svare and Boechler, unpublished observations). This work concerns the relationship between maternal protective behavior and the spontaneous killing of young exhibited by females with large litters. Postpartum female mice kill their own young even when environmental conditions are favorable (*i.e.*, food is freely available). With data on over 100 lactating female mice, our findings show the following: (1) Roughly 35–40% of Rockland-Swiss mice exhibit killing of young; (2) The larger the litter size, the greater the likelihood that females will display infanticide; (3) The killing of young usually occurs early in lactation; (4) Those animals that kill young and those that do not are identical with respect to aggressive behavior, lactation performance, and ability to rear young to weaning. These findings would apparently suggest that the killing of young by the postpartum female is not an aberrant behavior but rather it represents an adaptive strategy that does not compromise the maternal caretaking or defensive behavior of the dam (see also Boechler, this volume).

CONCLUSIONS

The following represent the major conclusions of this chapter: First, maternal protective behavior is regulated by the genetic, hormonal, and neurochemical background of the female; Second, the presence or absence of young during pregnancy and during lactation determines whether or not protective (aggressive) behavior will emerge in the female mouse; Third, protective behavior exhibited by the female is unrelated to the exhibition of spontaneous postpartum infanticide; Fourth, the understanding of mother-young interactions and their influence on protective behavior is probably better studied by employing environmental contingencies that more closely duplicate the natural environment.

REFERENCES

ALTMAN, M. (1963) Naturalistic studies of maternal care in the moose and elk. In: H.L. Rheingold (ed), *Maternal Behavior in Mammals*. New York, Wiley, pp. 233–253.

ANDERSSON, M., Wiklund, C. G., and Rungren, H. (1980) Parental defence of offspring: a model and an example. *Animal Behaviour*, **28**, 536–542.

BARKLEY, M.S., Geschwind, I.I. and Bradford, G.E. (1979) The gestational pattern of estradiol, testosterone, and progesterone secretion in selected strains of mice. *Biology of Reproduction*, **20**, 733–738.

BARKLEY, M.S., Michael, S.D., Geschwind, I.I. and Bradford, G.E. (1977) Plasma testosterone during pregnancy in the mouse. *Endocrinology*, **100**, 1472–1475.

BROIDA, J. and Svare, B. (1982) Postpartum aggression in C57BL/6J and DBA/2J mice: Experiential and environmental influences. *Behavioral and Neural Biology*, **35**, 76–83.

ERSKINE, M., Barfield, R.J., and Goldman, B.D. (1978) Aggression in the lactating rat: Effects of intruder age and test arena. *Behavioral Biology*, **23**, 206–218.

FLEMMING, A. (1990) Hormonal and experiential correlates of maternal responsiveness in human mothers. In: N.A. Krasnegor and R.S. Bridges (eds),*Mammalian Parenting: Biochemical, Neurobiological, and Behavioral Determinants*, New York, Oxford, pp. 184–208.

GANDELMAN, R. (1972) Mice: Postpartum aggression elicited by the presence of an intruder. *Hormones and Behavior*, **3**, 23–28.

HARLOW, H.F., Harlow, M.K. and Hansen E.W. (1963) The maternal affectional system of rhesus monkeys. In: H.L. Rheingold (ed), *Maternal Behavior in Mammals*, New York, Wiley, pp. 254–281.

HERSHER, L., Richmond, J.B. and Moore, A.U. (1963) Maternal behavior in sheep and goats. In: H.L. Rheingold (ed), *Maternal Behavior in Mammals*, New York, Wiley, pp. 203–232.

IENI, J.R. and Thurmond, J.B. (1985) Maternal aggression in mice: Effects of treatments with PCPA, 5-HTP, and 5-HT receptor antagonists. *European Journal of Pharmacology*, **111**, 211–220.

JAY, P. (1963) Mother-infant relations in langurs. In: H.L. Rheingold (ed), *Maternal Behavior in Mammals*, New York, Wiley, pp. 282–304.

KORDON, C., Blake, C.A., Terkel, J., and Sawyer, C.H. (1973–1974) Participation of serotonin-containing neurons in the suckling induced rise in plasma prolactin levels in lactating rats. *Neuroendocrinology*, **13**, 213–223.

MANN, M.A., Michael, S. and Svare, B. (1980) Ergot drugs suppress prolactin and lactation but not aggressive behavior in parturient mice. *Hormones and Behavior*, **14**, 319–328.

MANN, M.A. and Svare, B. (1982) Factors influencing pregnancy-induced aggression in mice. *Behavioral and Neural Biology*, **36**, 242–258.

MANN, M.A., Konen, C. and Svare, B. (1984) The role of progesterone in pregnancy-induced aggression in mice.*Hormones and Behavior*, **18**, 140–160.

MONTGOMERIE, R. D. and Weatherhead, P. J. (1988) Risks and rewards of nest defence by parent birds. *The Quarterly Review of Biology*, **63**, 167–187.

MORI, U. and Dunbar, R.I.M. (1985) Changes in the reproductive condition of female gelada baboons following the takeover of one-male units. *Zeitschrift Tierpsychology*, **67**, 215–224.

NOIROT, E., Goyens, J. and Buhot, M.C. (1975) Aggressive behavior of pregnant mice towards males. *Hormones and Behavior*, **6**, 9–17.

NUMAN, M. (1988) Maternal behavior. In: E. Knobil and J. Neill (ed), *The Physiology of Reproduction*, New York, Raven, pp. 1569–1645.

PARMIGIANI, S. This volume.

ROSS, S., Sawin, P.B., Zarrow, M.X. and Denenberg, V.H. (1963) Maternal behavior in the rabbit. In: H.L. Rheingold (ed), *Maternal Behavior in Mammals*, New York, Wiley, pp. 122–168.

ST. JOHN, R.S. and Corning, P.A. (1973) Maternal aggression in mice. *Behavioral Biology*, **9**, 635–639.

SARGENT, R. C. and Gross, M. R. (1985) Parental investment decision rules and the Concorde fallacy. *Behavioral Ecology and Sociobiology*, **17**, 43–45.

SCHNEIRLA, T.C., Rosenblatt, J.S., and Tobach, E. (1963) Maternal behavior in the cat. In: H.L. Rheingold (ed), *Maternal Behavior in Mammals*, New York, Wiley, pp. 122–168.

SVARE, B. (1977) Maternal aggression in mice: Influence of the young. *Biobehavioral Reviews*, **1**, 151–164.

SVARE, B. (ed.) (1983) *Hormones and Aggressive Behavior*, New York, Plenum.

SVARE, B. (1989) Recent advances in the study of female aggressive behavior. In: P.F. Brain, D. Mainardi, and S. Parmigiani (eds), *House Mouse Aggression: A Model for Understanding the Evolution of Social Behavior*, Harwood, Switzerland, pp. 135–159.

SVARE, B. (1990) Maternal aggression: Hormonal, genetic, and developmental determinants. In: N. Krasnegor and R. Bridges (eds), *Mammalian Parenting: Biochemical, Neurobiological, and Behavioral Determinants*, New York, Oxford, pp. 135–154.

SVARE, B. and Gandelman, R. (1973) Postpartum aggression in mice: Experiential and environmental factors. *Hormones and Behavior*, **4**, 323–334.

SVARE, B. and Gandelman, R. (1976a) Postpartum aggression in mice: The influence of suckling stimulation. *Hormones and Behavior*, **7**, 407–416.

SVARE, B. and Gandelman, R. (1976b) Suckling stimulation induces aggression in virgin female mice. *Nature*, **260**, 606–608.

SVARE, B., Betteridge, C., Katz, D. and Samuels, O. (1981) Some situational and experiential determinants of maternal aggression in mice. *Physiology and Behavior*, **26**, 253–258.

SVARE, B., Mann, M.A., Broida, J. and Michael, S. (1982) Maternal aggression exhibited by hypophysectomized parturient mice. *Hormones and Behavior*, **16**, 455–461.

SVARE, B., Mann, M.A. and Samuels, O. (1980) Mice: Suckling but not lactation is important for maternal aggression. *Behavioral and Neural Biology*, **29**, 453–462.

SVARE, B., Miele, J. and Kinsley, C. (1986) Mice: Progesterone stimulates aggression in pregnancy-terminated females. *Hormones and Behavior*, **20**, 194–200.

TAYLOR, J.C. (1966) Home range and agonistic behavior in the grey squirrel. In: P.A. Jewell and C. Loizos (eds), *Play, Exploration, and Territory in Mammals*, New York, Academic, pp. 229–236.

TRIVERS, R. L. and Willard, D. E. (1973) Natural selection of parental ability to vary the sex ratio of offspring. *Science*, **179**, 90–91.

WISE, D.A. (1974) Aggression in the female golden hamster: Effects of reproductive state and social isolation. *Hormones and Behavior*, **5**, 235–250.

CHAPTER 20

TO PROVISION OR KILL OFFSPRING: A DYNAMIC MODEL OF POSTPARTUM DECISION MAKING

MICHAEL BOECHLER

Department of Psychology, State University of New York at Albany, Albany, N.Y. 12222, USA

INTRODUCTION

Much of the literature on rodent maternal behavior has been devoted to the elucidation of physiological and biobehavioral influences on postpartum female behavior (see Krasnegor and Bridges, 1990; Numan, 1988; Stern, 1990 for several exhaustive reviews). Although this approach has been valuable in determining some of the proximate mechanisms responsible for behavioral change during the peripartum period, little work has been done to formally relate the behavior of the postpartum mouse to its immediate environment, or to its genetic contribution to subsequent generations (*i.e.* its fitness). The first section of this chapter will focus on a dynamic programming model that integrates assumptions about physiology, constraints on the feasible set of behaviors, and lifetime reproductive success of house mice. Several predictions based on the solution of the model are then offered. Finally, a laboratory methodology designed to investigate postpartum behavioral phenomena, and to test model predictions will be discussed.

A DYNAMIC PROGRAMMING MODEL OF MOUSE POSTPARTUM BEHAVIOR

Dynamic programming is an optimization technique which has the following components: a state space, a set of constraints, a strategy set, an optimization criterion, and state dynamics (Mangel and Clark, 1988). An important feature of this technique is that it takes into account the influence of an organism's current state and the current time on its behavior (Houston and McNamara, 1988). The current behavior governs the organism's future state and, consequently, its behavior. This interdependence of state and behavior, mediated by time, allows for the calculation of a feedback control policy, which is a strategy for optimizing some performance criterion.

State Space

The state space is the range of state variables which, in this case, are biological attributes expected to influence the behavioral decision making process. In this example, we are interested in the body masses of a postpartum female mouse and her offspring, as well as the number of offspring. The choice of state variables is based on, 1) the importance of body mass on the bioenergetics of reproduction (Bronson, 1979) and reproductive success (Clutton-Brock, 1988), 2) the reliability of measurement in the laboratory, and 3) a consideration of the tradeoff between the quantity and quality of offspring (Smith and Fretwell, 1974; Parker and Begon, 1986; McGinley and Charnov, 1988; Forbes, 1991). Let $X(t)$ be the mass of a lactating mouse at some time t, during the period of offspring dependence. Note that $X(t)$ does not imply multiplication. Similarly, $Y(t)$ represents the mass of an individual offspring at the beginning of the same time period, while $N(t)$ represents the number of offspring at time t. $X(t)$, $Y(t)$, and $N(t)$ are the state variables assumed to modulate maternal behavior. Any time t is a one hour period; thus a day d extends from $t =1$ to $t =24$, and all offspring are assumed to be weaned after eighteen days of lactation.

Constraint Set

The constraints on the state variables are based on observations of Rockland-Swiss mice. Female body mass has a lethal limit below which the animal perishes due to malnutrition; let this lower limit be denoted by X_0. Likewise, some upper limit on body mass exists; represented by X_C. Thus, the following relationship is true for all time periods of interest: $X_0 \leq X(t) \leq X_C$, where $X_0 = 30$ g and $X_C = 45$ g. Similar constraints exist

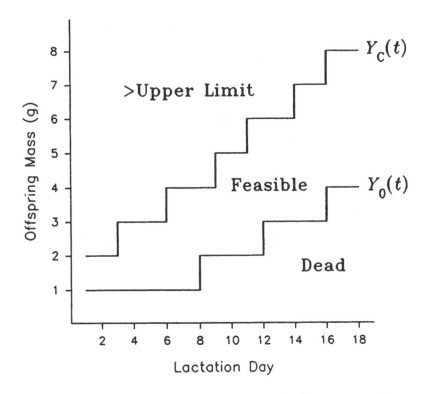

Figure 1 Step plot of the constraints on individual offspring mass throughout lactation. The upper line shows $Y_C(t)$ and the lower depicts $Y_0(t)$.

for the state variable representing individual offspring mass. However, these constraints must also account for offspring growth during the immediate postpartum period. That is, the constraints are time dependent: $Y_0(t) \leq Y(t) \leq Y_C(t)$. Specifically, $Y_0(t) = 1$ and $Y_C(t) = 2$ on $d = 1$. These upper and lower bounds on offspring mass increase throughout lactation to $Y_0(t) = 4$ and $Y_C(t) = 8$ on $d = 18$. Figure 1 shows a step plot of the time dependent constraints on offspring mass. Note that growth occurs in discrete steps rather than continuously. Although this is biologically inaccurate, it is necessary to discretize state variables (*i.e.* they may only take on integer values) in order to construct the computer algorithm used to find the solution to the model (Mangel & Clark, 1988). If $Y(t) < Y_0$, we set $Y(t+1) = 0$ indicating litter lethality. All offspring are assumed to be the

same size at parturition and grow at the same rate during the postpartum period.

For rodents, the upper limit on the number of offspring a female can rear depends on several physiological and ecological variables. For example: social status (Huck *et al.*, 1988), physiological condition (Perrigo, 1987), age and parity (Huck *et al.*, 1988), in addition to food availability (Bronson and Marsteller, 1979, Heasley, 1983, König, 1989) govern her maximal fecundity. However, for simplicity we will assume that the maximal number of offspring a lactating female can bear is sixteen; thus during the postpartum period $0 \leq N(t) \leq 16$.

Strategy Set

Although we acknowledge that the number of plausible behavioral actions for the mouse may be large, only three behaviors are of interest for the present model: 1) foraging, 2) nursing, and 3) cannibalizing one of the offspring.

For any period t the animal must choose a behavioral action A_i, where $i = \{1,2,3\}$. A strategy is a rule for choosing a behavior at time t for $t = 1,2,...T$, from the set of feasible behaviors. The "best" strategy is one that maximizes some performance criterion, which in this case is fitness. A dynamic optimization algorithm can be employed to determine the "best" strategy.

Optimization Criterion

The state variables influence the mother's lifetime fitness, which is the sum of current reproduction plus expected future reproductive success. For simplicity we assume that maternal fitness, via current reproduction, increases with the values of $Y(T)$, the mass of an offspring, and $N(T)$, the number of offspring successfully weaned, at the terminal time period T. Additionally, a concave-convex function describes the relationship between the mother's expected future reproductive success and $X(T)$, her mass at the end of lactation.

Equation 1 relates the mother's fitness at the terminal time period, $\Psi(X, Y, N, T)$, to current reproductive success, ω, and to her expected future reproductive success ω', which increases non-linearly with her mass at the time of weaning;

$$\Psi(X, Y, N, T) = \omega + \{\omega'/[1 + fe^{h(j-X(T))}]\} \tag{1}$$

where:

$$\omega = [\omega'' N(T)]/[1 + ae^{b(c-Y(T))}]$$

That is, current reproductive success, ω, comprises the number of off-spring at weaning weighted: 1) non-linearly by individual offspring mass at weaning, and 2) by ω'' which represents the expected reproductive success of the N offspring of mass Y at the terminal time T. Note that a, b, c, f, h, and j in Equation 1 do not arise from any particular biological considerations, they are simply tuning parameters employed to set the shape of the terminal reward functions. By changing these parameters we change the assumed relationship between the values of the state variables at T and the mother's expected future fitness at T. Figures 2a and 3a graphically depict two terminal reward functions generated by Equation 1. These reward functions were used as starting points for the backward induction algorithm used to solve the model; this will be discussed in detail further below.

State Dynamics

The occurrence of any behavior during period t may result in a change in the value of the state variables, observable at time $t + 1$. Equation 3 describes the transition of the mother's mass $X(t)$ when she leaves the nest to forage. Her mass decreases from t to $t + 1$ due to the metabolic expenditure of running for and handling food. Call this loss in mass α_1; it can be estimated by:

$$\alpha_1 = kM_A + M_B + M_{TR} \tag{2}$$

The first term in Equation 2 represents metabolism due to motor activity; k is a coefficient that modifies metabolism for the posture associated with activity and equals 2.0 for the mouse (Taylor, et al., 1970). The second term is basal metabolism, and the last term is metabolism associated with thermoregulation. All M are expressed in units of ml O_2/g/hr. These terms are calculated as follows (Wunder, 1975);

$$M_A = v_1(8.46X(t)^{-0.40})$$
$$M_B = 3.8X(t)^{-0.25}$$
$$M_{TR} = 1.05X(t)^{-0.50}[(38 - 4X(t)^{0.25}) - T_A]$$

where T_A is air temperature in $°C$, and v_1 is running velocity in km/hr. Equation 2 yields a metabolism estimate that can be converted from moles

of oxygen consumed to mass change per hour period (Eckert & Randall, 1983). X increases due to foraging as a result of the ingestion of food, call this, γ_1. x_1 represents her mass at $t + 1$, given action A_1 (foraging) is undertaken during t.

$$X(t+1|X(t) = x, \ A_1) = x_1 = x - \alpha_1 + \gamma_1 \tag{3}$$

The mass of the litter decreases from t to $t + 1$ by some amount a_1, due to metabolism, when the mother leaves the nest to forage. y_1 is the litter mass at $t + 1$ if the mother performs action A_1 at time t.

$$Y(t+1|Y(t) = y, \ A_1) = y_1 = y - a_1 \tag{4}$$

When the mother nurses the litter, action A_2, $X(t)$ decreases by α_2, the cost of milk production and letdown, which can be estimated allometrically by (Hanwell & Peaker, 1977):

$$\alpha_2 = .0835 X(t)^{0.765}$$

$Y(t)$ increases by some amount g_2 and decreases by a small amount a_2, due to offspring metabolism during a nursing period. Since the offspring are less than 100% efficient at converting milk mass to body mass $g_2 < \alpha_2$. These state transitions are symbolized by:

$$X(t+1|X(t) = x, \ A_2) = x_2 = x - \alpha_2 \tag{5}$$
$$Y(t+1|Y(t), \ A_2) = y_2 = y - a_2 + g_2 \tag{6}$$

Finally, the female may cannibalize one of her young, action A_3. If this behavior occurs, $X(t)$ increases while $Y(t)$ and $N(t)$ decreases from time t to $t + 1$. The amount of maternal mass increase, γ_3, will depend on the size and energetic content of the offspring consumed, both of which are time dependent (Myrcha & Walkowa, 1968):

$$X(t+1|X(t) = x, \ A_3) = x_3 = x + \gamma_3(t)$$

where,

$$\gamma_3(t) = 0.0755 Y(t) + 0.95 kCal/g \quad \text{if } d < 10$$
$$\gamma_3(t) = 1.56 kCal/g \quad\quad\quad\quad\quad \text{if } d \geq 10$$

We will assume that when a mother kills and consumes one of her offspring she leaves the nest to do so, just as she does when she chooses

to forage. Therefore, the mass of an individual pup that is not chosen for cannibalism decreases slightly due to thermoregulatory demands when the mother engages in this behavior; $y_3 = y_1$.

$$N(t + 1|N(t) = n, A_3) = n_3 = n - 1$$

$N(t)$ changes as a result of the mother's behavior only when action A_3 is taken, otherwise;

$$N(t + 1|N(t) = n, A_i) = n_i = n \qquad (7)$$

for $i = 1, 2$; unless $Y(t) = Y_0(t)$, in which case $N(t + 1) = 0$, indicating that the litter has died due to starvation.

Solution

Let us now recapitulate some of the important details covered before deriving the dynamic programming algorithm. Equation 1 provides a set of lifetime fitness values for the mother. They are influenced by: 1) her mass at the end of her first reproductive cycle, 2) the number of offspring reared during her first lactation, and 3) the amount of biomass produced by raising her initial litter. Thus, the terminal reward function is a calculation of fitness that includes recent and expected future reproduction. We are not as interested in the female's fitness *per se* over the course of lactation as we are in predicting her postpartum behavior, which we assume optimizes her fitness.

In our model, lactation comprises eighteen days, and each day consists of 24 time periods, yielding 432 time periods to be modelled. The female must engage in one, and only one, behavioral action chosen from a set of three actions at each time t. The actions may influence the female's future state, and the future states of her offspring; these state transitions were specified above.

Given the number of time periods and possible behaviors, the number of possible behavior sequences equals 3^{432}. The optimal strategy (that which yielded the highest fitness value at the end of the terminal time period) would then have to be chosen from the set of sequences. Computationally, this would be onerous, if not impossible. We can, however, use Bellman's Principle of Optimality (Bellman, 1957) and backward induction to derive a dynamic programming algorithm that will calculate the sequence of behaviors that maximizes the terminal reward.

Let:

$$F(x, y, n, t, T) = \max_A E(\Psi | X(t) = x, \, Y(t) = y, \, N(t) = n, \, A(t), \, t+1, \, T) \tag{8}$$

Equation 8 gives the maximum expectation of the terminal reward at time t, given that the states at t are x, y, and n; and that the optimal action is taken during each period from t to T. To illustrate, we can show how Equation 8 can be used to calculate the optimal behavior, and its corresponding fitness, for the penultimate time period, $T-1$. Assume that the female has two actions to choose from at $T-1$; foraging and nursing, (A_1 and A_2, respectively). Equation 8 then becomes:

$$F(x, y, n, T - 1) = \max \, \{F(x_1, y_1, n_1, T); \, F(x_2, y_2, n_2, T)\} \tag{9}$$

Since Eqns. (3), (4), and (7) specify the state transitions when A_1 is taken; and Eqns. (5), (6), and (7) can be used to calculate the state changes under A_2, Eqn. 9 solves for the behavior that maximizes the value of the terminal reward. Now, knowing the expected value of the terminal reward under optimal behavior at $T-1$, we can similarly evaluate different behaviors at $T-2$. This process is then repeated for $t = T - 3, T - 4, ...1$.

We assume that the number of plausible behaviors changes throughout the day; the dynamic programming algorithm reflects this.

For $1 + 96d \le t < 36 + 96d$; $72 + 96d < t \le T$:

$$\begin{aligned} F(x, y, n, t) = \max \, \{&F(x_1, y_1, n_1, t + 1); \\ &F(x_2, y_2, n_2, t + 1); \\ &F(x_3, y_3, n_3, t + 1)\} \end{aligned}$$

that is, for most of the evening and early morning, the female may forage, lactate, or cannibalize one of her offspring during each time period. During much of the day, however, foraging is not an option:

For $36 + 96d \le t \le 72 + 96d$:

$$\begin{aligned} F(x, y, n, t) = \max \, \{&F(x_2, y_2, n_2, t + 1); \\ &F(x_3, y_3, n_3, t + 1)\} \end{aligned}$$

This model was programmed in FORTRAN and the optimal policies were found for a number of terminal reward (TR) functions by the backward induction method outlined above. Figures 2a and 3a graphically depict the shapes of two of the terminal rewards that were explored, while Figures 2b-g and 3b-g show the results generated under the corresponding reward functions.

Predictions

In both cases, maternal cannibalism is predicted in some portion of the state space, and should decrease in frequency as lactation progresses. Offspring from large litters should be killed more often than young from small litters and small offspring from large litters should be subject to the highest mortality risk due to maternal infanticide, particularly during the latter stages of the postpartum period. Several reports are consistent with these predictions. Gandelman and Simon (1978) and Svare and Boechler (this volume) found that under conditions of *ad lib* food availability, mice with large litters killed more of their young than mice with smaller litters. Furthermore, this killing occurred during the first third of lactation. Gandelman and Simon (1978) also noted that within litters subjected to infanticide, small offspring suffered the greatest mortality risk.

The models differ with respect to the role of maternal mass in the killing of young. When the model was solved with TR1, it was discovered that larger mothers should kill offspring more often than smaller mothers, while the solution of the model under TR2 revealed the opposite prediction, that is, smaller mothers should kill offspring more frequently. Houston and McNamara (1988) have argued that the form of the terminal reward function does not significantly influence the predictions obtained from stochastic dynamic models, particularly as one moves away from the terminal time period. The present (deterministic) model shows that while some relationships between the state variables and optimal behavior remain qualitatively similar, others can change dramatically (*i.e.* the effect of maternal mass on infanticidal behavior). The results of Perrigo (1987) and Boechler (1992) support the predictions obtained from the model solved under TR2; that is, smaller mothers tend to kill more of their offspring than larger mothers.

The expected reproductive success of the mother (ω') and her young (ω'') were also found to influence the predicted incidence of maternal infanticide. Various levels of each of these variables were used to determine the sensitivity of the solution to changes in these parameters. It was discovered that as maternal expected reproductive success decreased so did the proportion of the policy in which maternal infanticide was optimal.

(a)

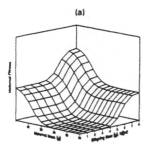

Figure 2 (a) Terminal reward
1. The z-axis shows maternal
fitness at T as a function of
maternal mass (x - axis) and
individual offspring mass (y
- axis) for two offspring. Parameter values for Equation
1 are: f=45, h=0.87, j=37,
ω'=60, a=.001, b=2, c=9,
ω''=25. Figures 2b-g depict
model predictions across lactation under terminal reward
1. In figures 2 and 3, offspring mass is abbreviated by
yhi for large young and ylo
for small young, while the number of offspring is represented
by n.   (b,c) Proportion of
the optimal policy in which
killing and cannibalizing one
of the offspring is optimal for
mothers of mass $\leq$ 37g (b) or
$\geq$ 38g (c).   (d,e) Proportion of the optimal policy in
which foraging is optimal for
mothers of mass $\leq$ 37g (d)
or $\geq$ 38g (e).   (f,g) Proportion of the optimal policy in
which nursing is optimal for
mothers of mass $\leq$ 37g (f) or
$\geq$ 38g (g).

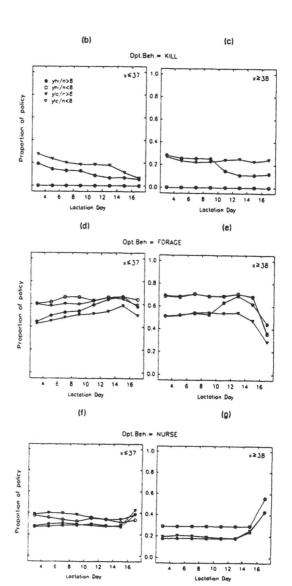

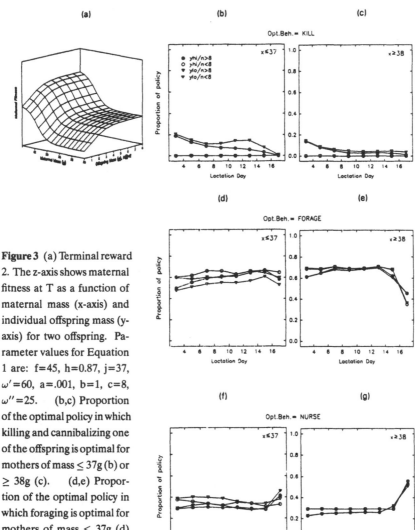

Figure 3 (a) Terminal reward 2. The z-axis shows maternal fitness at T as a function of maternal mass (x-axis) and individual offspring mass (y-axis) for two offspring. Parameter values for Equation 1 are: f=45, h=0.87, j=37, ω'=60, a=.001, b=1, c=8, ω''=25.   (b,c) Proportion of the optimal policy in which killing and cannibalizing one of the offspring is optimal for mothers of mass ≤ 37g (b) or ≥ 38g (c).   (d,e) Proportion of the optimal policy in which foraging is optimal for mothers of mass ≤ 37g (d) or ≥ 38g (e).   (f,g) Proportion of the optimal policy in which nursing is optimal for mothers of mass ≤ 37g (f) or ≤ 38g (g).

Thus, if ω' decreases across the lifespan of a female mouse due to past reproductive effort, she should show a corresponding decrease in the propensity to kill her young. This result is similar to others (eg., Pianka and Parker, 1975, Charlesworth and León, 1976, Clutton-Brock, 1984), and hinges on the same assumption that past and current reproductive effort are negatively correlated with expected future reproduction. In contrast, as the expected reproductive success of the current litter decreases, mothers should be more inclined to engage in cannibalism.

Nursing behavior should be influenced by the state variables across lactation, although not to the extent of infanticidal behavior. Both models predict that small mothers should nurse more than large mothers. That is, the fitness increments that result from promoting offspring growth are larger than the costs of body mass loss to small mothers. This pattern is reversed, however, toward the end of lactation, when large mothers are expected to increase the time spent nursing their young more than small mothers. Both models predict that mothers with small litters should nurse more than mothers with large litters, indicating a predicted tradeoff in the quantity and quality of the young. Offspring size should also affect nursing behavior, although to a lesser extent than maternal size or offspring number. Specifically, small offspring should be nursed more often than large offspring. That is, the increment in fitness to a mother nursing a litter of small offspring is larger than the increment to mother that nurses a litter of large pups; presumably this is due to the constraints on offspring size during the postpartum period.

A careful examination of the results shown in Figures 2b-g and 3b-g reveal that for the majority of lactation, the behavioral tradeoffs should predominantly occur between infanticide and foraging, while nursing should remain fairly stable. Near the end of lactation, however, this pattern should change; the tradeoffs involving nursing and foraging. This result was obtained for virtually all of terminal rewards examined, but not shown here.

In conclusion, the present analysis provides one scenario of the evolution of maternal infanticidal behavior in mice. That is, if the assumptions of the models are met, the selective killing of unweaned young is a behavior that could evolve because it maximizes the expected lifetime reproductive success of the mother. The assumptions required to obtain this result were made from a careful consideration of the biology and natural history of the species involved. It is important to note, however, that an analysis of the generality of these findings to other species was not undertaken. That is, the state dynamics are, in some cases, specific to mice. These dynamics would have to be estimated if we were interested in predicting whether ma-

ternal infanticide could evolve in other species. Wherever possible, care has been taken to specify state dynamics based on allometric relationships. These relationships should hold for a variety of species. Another aspect of the present scenario that limits its generality is the assumption that maternal infanticide is accompanied by cannibalism. Some species (eg., herbivores) may not be equipped morphologically or physiologically to kill or consume their offspring. In such cases, the appropriate behavior of interest may be the abandonment of young. Finally, it is important to remember that a trait that optimizes fitness is not necessarily one that will evolve or has evolved (Gould and Lewontin, 1979).

LABORATORY METHODOLOGY

The predictions made by the models discussed above can be tested relatively easily. We simulate foraging in the laboratory with a caging procedure similar to that described by Perrigo and Bronson (1983). The apparatus consists of a transparent cage large enough to accommodate a running wheel, connected by PVC tubing to two smaller transparent cages, in one of which water and nest material are available. Wheel revolutions, detected by microswitches, are counted by a computer which determines the delivery of standard food pellets. The amount of food consumed, the time course of running activity, and the schedule of food delivery are easily monitored. Subjects live in a closed economy, that is, they have to expend energy, by running, to receive all of the food available to them during lactation.

With this system we can control or observe virtually all of the state variables and parameters discussed in the previous section. For example, we may want to treat the acquisition of food as a random process, in order to more closely approximate the conditions under which wild mice exist. In the lab, this would correspond to exposing subjects to probabilistic food delivery contingent upon wheel running responses. The model presented above could be easily modified by treating maternal mass increase due to foraging (γ_1) as a stochastic variable. We could model a scenario where resources, food pellets in this case, are depleted across the lactation period (patch depletion), γ_1 would then be time dependent. Both the model and the program that controls the delivery of food pellets could easily be modified to accommodate this assumption.

We are working on an extension of these models that includes another behavior commonly exhibited by postpartum female mice; the defense of young against potentially infanticidal male conspecifics. We could assume

that female mice encounter conspecifics who may attempt to kill her off-spring with some probability. In this case, stochastic dynamic programming may be employed to find the strategy that maximizes maternal fitness. This approach may be extended to include payoffs for both the mother and the intruder as a result of behavioral actions. A dynamic game approach may then be applied. Similarly, pairs of lactating females could co-occupy cages and compete for access to a "territory" in which to forage, (i.e. the running wheel). Again, a dynamic game model might be used to predict the strategies of competing females.

ACKNOWLEDGEMENTS

I would like to thank Jonathan Newman for the initial formulation of the models and for programming assistance, Tom Caraco for reviewing the manuscript and providing invaluable advice, and Bruce Svare for making this possible.

REFERENCES

BELLMAN, R. (1957) *Dynamic Programming*. Princeton University Press, Princeton.

BOECHLER, M. (1992) Rodent postpartum behavior: physiology and ecology. Unpublished Ph. D. thesis.

BRONSON, F. H. (1979) The reproductive ecology of the house mouse. *Quarterly Review of Biology*, **54**, 265–299.

BRONSON, F. H. and Marsteller, F. A. (1979) Effect of short-term food deprivation on reproduction in female mice. *Biology of Reproduction*, **33**, 660–667.

CHARLESWORTH, B. and León, J. A. (1976) The relation of reproductive effort to age. *The American Naturalist*, **110**, 449–459.

CLUTTON-BROCK, T. H. (1984) Reproductive effort and terminal investment in iteroparous animals. *The American Naturalist*, **123**, 212–229.

CLUTTON-BROCK, T. H. (1988) Reproductive success. In: Clutton-Brock, T. H. (ed.), *Reproductive Success*, The University of Chicago Press, Chicago, pp. 472–485.

FORBES, L. S. (1991) Optimal size and number of offspring in a variable environment. *Journal of Theoretical Biology*, **150**, 299–304.

GOULD, S. J. and Lewontin, R. C. (1979) The spandrels of San Marco and the Panglossian Paradigm: a critique of the adaptionist programme. *Proceeding of the Royal Society of London B*, **205**, 581–598.

HEASLEY, J. E. (1983) Energy allocation in response to reduced food intake in pregnant and lactating laboratory mice. *Acta Theriologica*, **28**, 55–71.

HOUSTON, A. I. and McNamara, J. M. (1988) A framework for the functional analysis of behaviour. *Behavioral and Brain Sciences*, **11**, 117–163.

HUCK, U. W., Lisk, R. D. and McKay, M. V. (1988) Social dominance and reproductive success in pregnant and lactating golden hamsters (*Mesocricetus auratus*) under seminatural conditions. *Physiology and Behavior*, **44**, 313–319.

HUCK, U. W., Pratt, N. C., Labov, J. B. and Lisk, R. D. (1988) Effects of age and parity on litter size and offspring sex ratio in golden hamsters (*Mesocricetus auratus*). *Journal of Reproduction and Fertility*, **83**, 209–214.

KÖNIG, B. (1989) Kin recognition and maternal care under restricted feeding in house mice (*Mus domesticus*). *Ethology*, **82**, 328–343.

KRASNEGOR, N. and Bridges, R. (1990) *Mammalian Parenting*. Oxford University Press, New York.

MANGEL, M. and Clark, C. W. (1988) *Dynamic Modeling in Behavioral Ecology*. Princeton University Press. Princeton.

MCGINLEY, M. A. and Charnov, E. L. (1988) Multiple resources and the optimal balance between size and number of offspring. *Evolutionary Ecology*, **2**, 77–84.

NUMAN, M. (1988) Maternal behavior. In: Knobil, E. and Neill, J. (eds.), *The Physiology of Reproduction*. Raven Press, New York.

PARKER, G. A. and Begon, M. (1986) Optimal egg size and clutch size: effects of environment and maternal phenotype. *The American Naturalist*, **128**, 573–592.

PERRIGO, G. (1987) Breeding and feeding strategies in deer mice and house mice when females are challenged to work for their food. *Animal Behavior*, **35**, 1298–1316.

PERRIGO, G. and Bronson, F. H. (1983) Foraging effort, food intake, fat deposition, and puberty in female mice. *Biology of Reproduction*, **29**, 455–463.

PIANKA, E. R. and Parker, W. S. (1975) Age-specific reproductive tactics. *The American Naturalist*, **109**, 453–464.

SMITH, C. C. and Fretwell, S. D. (1974) The optimal balance between size and number of offspring. *The American Naturalist*, **108**, 499–506.

STERN, J. (1989) Maternal behavior: sensory, hormonal, and neural determinants. In: Brush, F. B. and Levine, S. (eds.), *Psychoendocrinology*. Academic Press, pp.105–226.

INDEX